T0181067

Planktic Foraminifers in the Modern Ocean

Ralf Schiebel · Christoph Hemleben

Planktic Foraminifers in the Modern Ocean

 Springer

Ralf Schiebel
Climate Geochemistry
Max Planck Institute for Chemistry
Mainz, Rheinland-Pfalz
Germany

Christoph Hemleben
Department of Geoscience
University of Tübingen
Baden-Wuerttemberg
Germany

ISBN 978-3-662-57052-4   ISBN 978-3-662-50297-6   (eBook)
DOI 10.1007/978-3-662-50297-6

Printed on acid-free paper

This Springer imprint is published by Springer Nature
The registered company is Springer-Verlag GmbH Berlin Heidelberg

This volume is dedicated to Allan W.H. Bé

Preface

The purpose of this book is to provide timely and comprehensive information on extant planktic foraminifers. This book is based on 'Modern Planktonic Foraminifera' published by Hemleben et al. (1989). An extensive amount of literature published over the past 26 years adds new information on modern and fossil planktic foraminifers and merits an update of the current knowledge. New chapters review the modern advances on stable isotope geochemistry, element ratios of planktic foraminifer tests, and molecular genetics of planktic foraminifers, the latter being an entirely new field of research developed since the mid-1990s. As a practical guide for students and colleagues, the book provides 35 plates on the classification of the extant morphotypes, most of which include various genotypes. A vast amount of new knowledge on planktic foraminifer ecology, settling dynamics, and carbonate geochemistry is presented over several chapters. Much less new information has been produced on the ultrastructure, ontogeny, and nutrition. In these cases, parts of the book of Hemleben et al. (1989) were rewritten, summarized, or complemented. Finally, we present the current state of the rapidly increasing methodological and technological advances available to our field of research.

This book is meant to provide a tool and new perspective for the application of planktic foraminifers in paleoceanography and climate research, as well as in eco-monitoring, for example, in offshore hydrocarbon prospection and exploitation. This volume presents a review of the recent findings and includes thus far published and unpublished findings of the authors. As much as we have aimed for completeness, we may have missed some papers published over the past decades. Although Internet-based sources have improved awareness, distribution, and accessibility of information, the vast amount of new literature published in the increasingly large number of journals has magnified the challenge of being thoroughly complete.

Reference

Hemleben C, Spindler M, Anderson OR (1989) Modern planktonic Foraminifera. Springer, Berlin

Acknowledgments

There are many colleagues, coworkers, students, friends, and family who are acknowledged for their participation, help, generosity, and motivation over the course of our research and in the preparation of this book. Margret Bayer, Ingrid Breitinger, and Sabine Winter at the University of Tübingen helped to produce many of the new SEM images of tests for the plates on taxonomy and classification. Wolfram Schinko (Meßkirch) and Iris Bambach (MPIC Mainz) helped in selecting and scanning of older SEM images. Geert-Jan Brummer (the Netherlands), Katsunori Kimoto (Japan), Andreas Kiefer (Germany), and Julie Meilland (France) kindly provided additional SEM images. Some of the light micrographs of living planktic foraminifers originate from cooperation between Christoph Hemleben and the professional photographer Manfred Kage (Germany) at Barbados in 1984. Takashi Toyofuku (Japan), Aurore Movellan (France), and Michael Spindler (Germany) provided unpublished data and figures. Hélène Howa (France) is acknowledged for providing images on sampling gear. Gerhard Fischer and Gerhard Schmiedl (both Germany) provided access to some rare data and literature. We are much indebted to John Murray (UK) for inspiring discussions. Richard Olsson and O. Roger Anderson (both USA) are particularly acknowledged for generously proofreading and improving grammar and content of this volume. We are grateful to Vera Schiebel for help with the reference management software and artwork. Vera Hemleben is acknowledged for indispensable advice on the paragraphs on molecular genetics. Both Vera Hemleben and Vera Schiebel have provided continuous support by reading and commenting on various stages of the manuscript. We are particularly indebted to our colleagues and staff at our departments for continuous encouragement and support in the production of this volume. The German Science Foundation provided financial support. Last but not least, we thank Johanna Schwarz from Springer, Heidelberg, for her ceaseless encouragement and editorial support.

Contents

Introduction

Planktic foraminifers are marine protozoans with calcareous shells and chambered tests (Plate 1.1), first appearing in the mid-Jurassic approximately 170 million years ago, and populating the global ocean since the mid-Cretaceous (cf. Frerichs et al. 1972; Caron and Homewood 1983). The scientific and economic value of planktic foraminifers is based on their global marine abundance since the Lower Cretaceous $\sim$110 Million years ago. Owing to the high preservation potential of their calcareous shell, planktic foraminifers provide information on the past environment and climate. Physical conditions and chemical composition of ambient seawater are reconstructed from faunal assemblages, i.e. the presence or absence of foraminifer species, as well as through the chemical composition of their test calcite, including crystallinity of the test wall, and changes in stable isotope and element ratios.

Test: The foraminifer shell is called a test. Shell and test are often used synonymously. Shell may be used for part(s) of the test, and for fragments of the test.

Planktic—planktonic: Planktic and planktonic may be used synonymously. In the strict Greek meaning the word planktic is possibly correct (Burckhardt 1920; Rodhe 1974). In the international literature both planktic and planktonic are used to the same degree, and either term may be applied based on personal preference. In benthic foraminifers, the term benthic has largely been used over the past decades, and benthonic has been out of fashion for some time.

Modern planktic foraminifers evolved from the earliest Tertiary including the first spinose species in Earth history soon after the Cretaceous-Paleogene (K/Pg) boundary (Olsson et al. 1999). Most modern species live in the surface to thermocline layer of the open ocean, and in deep marginal seas as the Mediterranean, Caribbean, South China Sea, and Red Sea. Some species descend to waters as deep as several thousand meters in the tropical to temperate ocean. Planktic foraminifers are largely absent from shallow marginal seas, for example the North Sea where reproduction is impeded. The presence and absence of planktic foraminifer species at the regional scale is related to the quality and quantity of food, physical and chemical properties of ambient seawater, and displays an overall latitudinal pattern at the global scale.

Species abundance varies according to seasons as well as on an interannual scale, and on longer time-scales depending on environmental conditions, and affected by climate change. Symbiont-bearing species depend on light and are restricted to the euphotic zone of the surface ocean. Symbiont-barren species may dwell as

© Springer-Verlag Berlin Heidelberg 2017
R. Schiebel and C. Hemleben, *Planktic Foraminifers in the Modern Ocean*,
DOI 10.1007/978-3-662-50297-6_1

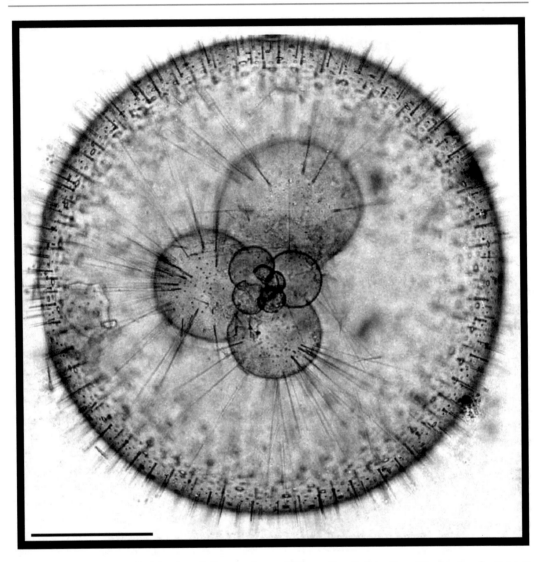

Plate 1.1 The modern planktic foraminifer species *Orbulina universa* seen in transmitted light. The inner trochospiral test of the pre-adult individual is surrounded by the spherical adult test. Spines are protruding from both inner trochospiral and outer spherical test. Pores are visible as tiny dark spots on the inner and outer test. Multiple small *circles* on the outer test wall are the apertures of the adult individual. The opening at the inner trochospiral test is caused by dissolution. Scale bar 200 μm

deep as the abyssal ocean, and have been sampled from below 4000 m water depth. Planktic foraminifers are rather marginal to marine biological research including modern biogeochemistry (Sarmiento and Gruber 2006), although they are major producers of marine calcareous particles (i.e. their tests) deposited on the ocean floor forming the globigerina ooze (e.g., Vincent and Berger 1981). Data compilation of a large variety of marine Plankton Functional Types (see text box below) have shown that planktic foraminifers possibly constitute a minor but ubiquitous component of marine planktic biomass (Buitenhuis et al. 2013). In addition, modeling approaches on the planktic foraminifer population dynamics from the 1990s have contributed to a better understanding of planktic foraminifer ecology and application in paleoceanography (e.g., Signes et al. 1993; Žarić et al. 2006; Fraile et al. 2009; Lombard et al. 2011; Roy et al. 2015).

Plankton functional type (PFT): The expression plankton functional type (PFT) is used in modeling, and includes different conceptual categories of organisms as, for example, organisms of similar ecology, and serving similar roles within an ecosystem (Anderson 2005). The PFTs included in the MAREDAT initiative on the ecology and biomass of marine plankton are picophytoplankton, diazotrophs, coccolithophores, Phaeocystis, diatoms, picoheterotrophs, microzooplankton, planktic foraminifers (which range between micro- and mesozooplankton), mesozooplankton, pteropods, and macrozooplankton (Buitenhuis et al. 2013).

By contributing substantially to the fossil record of marine sediments, planktic foraminifers provide indispensable ecologic information used in paleoecologic, paleoceanographic, and stratigraphic research from the Lower Cretaceous (~110 millions years, Ma). Faunistic and biogeochemical (e.g., stable isotopes) information from the calcareous (calcite, $CaCO_3$) planktic foraminifer tests is used to reconstruct, for example, temperature and salinity of the past surface ocean. Radiocarbon (^{14}C) gives an absolute age of test formation of late Pleistocene and Holocene sediments. Factors determining the modern faunal composition are applied to the interpretation of the fossil assemblages, for example, by multiple regression techniques (i.e. transfer functions), yielding information (proxy data) on ancient environmental parameters. The chemical composition, i.e. stable isotope and element ratios of the calcareous test (calcite, $CaCO_3$) provides an assessment of the chemical and physical state of ambient seawater, and is applied to the reconstruction of temperature, and biological productivity of the past marine environment.

Proxy (*pl.* proxies): A proxy is a measurable feature from which another not directly measurable characteristic can be derived. For example, the test of a planktic foraminifer bears certain stable isotope ratios (e.g., $^{18/16}O$), measurable with a mass spectrometer, from which temperature and other parameters of ambient seawater can be reconstructed by applying empirically derived formulae (see, e.g., Fischer and Wefer 1999).

1.1 A Brief History of Planktic Foraminifer Research

Technological improvement of binocular microscopes allowed the French naturalist Alcide d'Orbigny (1826) to describe the first planktic foraminifer species *Globigerina bulloides* from beach sands of Cuba, but erroneously classifying it with the cephalopods. Alcide d'Orbigny's family lived in the village of Esnandes at the Baie d'Aiguillon north of La Rochelle (France), where Alcide's father Charles Marie d'Orbigny was a renowned 'naturaliste'. Young d'Orbigny was fortunate enough to look at the sediments of the bay, and to find at a rich benthic foraminifer fauna using the first good binocular microscopes available in the 1820s (Vénec-Peyré 2005). D'Orbigny's French contemporary Félix Dujardin (1835), then, correctly described planktic foraminifers as unicellular organisms. Some 30 years later, Owen (1867) suspected the planktic life habit of these organisms. Following the Challenger Expedition from 1872 to 1876, the surface-dwelling habitat of planktic foraminifers was generally recognized thanks to observations provided by John Murray in the Challenger Reports (Brady 1884). Foraminifer biology was described first by Rhumbler (1911). In the first half of the 20th century, foraminifers were widely used for stratigraphic purposes in the search for hydrocarbon reservoirs, and Joseph

Cushman published a plethora of catalogues on foraminifers of all major ocean basins, and from various time-slices (e.g., Cushman 1911; Cushman and Todd 1949).

Distribution and ecology of different living planktic foraminifer species were first studied on plankton samples by Schott (1935). From the 1960s, planktic foraminifers have been used in biostratigraphy to date marine sediments sampled, for example, within the Deep Sea Drilling Programme (DSDP) from 1964 to 1983, followed by the Ocean Drilling Programme (ODP), and the Integrated Ocean Drilling Programme (IODP) from 2003 onward. The taxonomy of modern planktic foraminifers was largely improved by the seminal publication of Frances Parker (1962).

Distribution, ecology, and biology of the live fauna mostly of the western North Atlantic were extensively studied by Bé, Hemleben, Anderson, and co-workers, including graduate students and post-doctoral appointees, between the late 1950s and 1980s. Among these participants were David Caron and Howard Spero who became significant researchers in the field. Other major contributors included Peter Wiebe, Sharon Smith, Susumu Honjo, and Richard Fairbanks at Woods Hole Oceanographic Institution. At about the same time, Esteban Boltovskoy developed new sampling methods, and conducted projects on the production and sedimentation of planktic foraminifers in the South Atlantic. Ecological significance of modern species was applied to paleoecological and paleoceanographic settings to obtain new information on the ancient ocean and Earths' climate. Since the late 1960s, Wolfgang Berger and co-workers supplied ample information in many papers on planktic foraminifer carbonate chemistry and application of proxies to paleoceanography, starting in the eastern north Pacific, and later focusing on the South Atlantic (e.g., Berger 1981; Berger et al. 1989; Kemle-von-Mücke and Hemleben 1999; see also Fischer and Wefer 1999). Population dynamics and carbon turnover of modern planktic foraminifers mostly of the eastern North Atlantic and Indian Ocean including adjacent regions were studied by Christoph Hemleben and

co-workers since the late 1960s (e.g., Hemleben 1969; Hemleben and Spindler 1983; Hemleben et al. 1989; Bijma and Hemleben 1994; Schiebel et al. 1995; Schiebel 2002).

In the early 1970s, a joint group guided by O. Roger Anderson, Allan Bé (both Lamont-Doherty Earth Observatory), Christoph Hemleben, and Michael Spindler (both Tübingen University), came together at the Bermuda Biological Station (BBS) in order to culture planktic foraminifers (e.g., Bé et al. 1977; Hemleben et al. 1989). The BBS is close to blue water locations and thus exceptionally suited to experiment with planktic foraminifers. Living foraminifers were sampled by means of SCUBA collection and net tow sampling, and a sophisticated experimental set up in order to maintain viable planktic foraminifers from early ontogenetic stages to maturity was developed. Almost the entire range of all basic planktic foraminifer behavior was observed and recorded. Analyses of planktic foraminifers from laboratory culture have been substantially advanced by Howard Spero and co-workers at the University of California (e.g., Spero 1986; Spero et al. 2015). Culturing of planktic foraminifers also has been conducted at the Bellairs Research Institute at Barbados (e.g., Caron et al. 1982; Spindler et al. 1984), the Caribbean Marine Research Center on Lee Stocking Island, Bahamas (e.g., Spero and Williams 1988; Spero and Lea 1993), the H. Steinitz Marine Biology Laboratory at Eilat, Gulf of Aquaba (e.g., Erez et al. 1991, and references therein), the Caribbean Marine Biological Institute (CARAMBI) at Curacao (e.g., Bijma et al. 1992), the Isla Magueyes Marine Laboratory at Puerto Rico (e.g., Hönisch et al. 2011; Allen et al. 2011, 2012). However, a second generation of any planktic foraminifer species has never been successfully achieved in laboratory culture, which remains one of the major issues to be solved in the future.

Recent work focuses on planktic foraminifer taxonomy, stratigraphy, evolution, ecology, carbonate chemistry, paleoceanography, population dynamics, and biology. Stratigraphy and paleoceanography were among the original scientific interests in planktic foraminifers, due to their economic and scientific value, respectively.

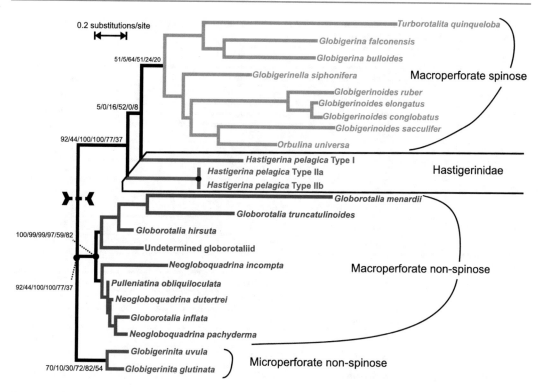

Fig. 1.1 Phylogenetic relationships of the four major groups of modern planktic foraminifers, macroperforate spinose, macroperforate non-spinose, microperforate spinose, and Hastigerinidae, based on a maximum likelihood reconstruction from SSU rDNA. Modified after Aurahs et al. (2009), from Weiner et al. (2012)

Modern techniques of molecular genetics (i.e. DNA sequencing) are currently applied to reveal the taxonomic and phylogenetic relations (Fig. 1.1) of the earlier established morphospecies (Table 1.1) distinguished by their test architecture (e.g., Darling et al. 1997; de Vargas et al. 1999; André et al. 2014). The relation to morphological features of the tests of modern species is reviewed in the fossil species (e.g., Hemleben et al. 1999; Hemleben and Olsson 2006).

Technological development of mass spectrometry analytical systems provides ever more precise measurements of rare elements, stable isotope ratios and 'clumped isotopes'. Based on these advances, new proxies have been developed in paleoceanography (see the review of Katz et al. 2010). Laser ablation-inductively coupled plasma-mass spectrometry (LA-ICP-MS) and secondary ion mass spectrometry (NanoSIMS) allow analyses of single chambers of tests, and hence better interpretation of ontogenetic changes

in planktic foraminifer ecology. Outer and inner shell architecture is analyzed and visualized at high-resolution using X-ray micro-tomography (e.g., Johnstone et al. 2010). Using refined technology, new knowledge has been gained from planktic foraminifer research, and the field has been substantially advanced, but simultaneously a number of intriguing new questions have been raised. Planktic foraminifer assemblages and test properties have become increasingly valuable proxies, and are applied in monitoring climate and environmental change including the position and strength of marine currents and fronts, oxygenation of the water column, and ocean acidification, among others. In 2010, SCOR (Scientific Committee on Oceanic Research) Working Group 138 was formed to synthesize the current knowledge on 'Modern Planktic Foraminifera and Ocean Changes'.

Investigation of modern and geologically ancient planktic foraminifers have diversified

Table 1.1 Modern planktic foraminifer morphospecies sorted by genus, including author and year of first description, and page of detailed description given in Chap. 2

Genus	Species	Author	Year
Beella	*digitata*	(Brady)	1879
Berggrenia	*pumilio*	(Parker)	1962
Bolliella	*adamsi*	Banner and Blow	1959
Candeina	*nitida*	d'Orbigny	1839
Dentigloborotalia	*anfracta*	(Parker)	1967
Gallitellia	*vivans*	(Cushman)	1934
Globigerina	*bulloides*	d'Orbigny	1826
	falconensis	Blow	1959
Globigerinella	*calida*	(Parker)	1962
	siphonifera	(d'Orbigny)	1839
Globigerinita	*glutinata*	(Egger)	1895
	minuta	(Natland)	1938
	uvula	(Ehrenberg)	1861
Globigerinoides	*conglobatus*	(Brady)	1879
	ruber	(d'Orbigny)	1839
	sacculifer	(Brady)	1877
Globoquadrina	*conglomerata*	(Schwager)	1866
Globorotalia	*cavernula*	Bé	1967
	crassaformis	(Galloway and Wissler)	1927
	hirsuta	(d'Orbigny)	1839
	inflata	(d'Orbigny)	1839
	menardii	(d'Orbigny)	1865
	scitula	(Brady)	1882
	theyeri	Fleisher	1974
	truncatulinoides	(d'Orbigny)	1839
	tumida	(Brady)	1877
	ungulata	Bermudez	1960
Globorotaloides	*hexagonus*	(Natland)	1938
Globoturborotalita	*rubescens*	Hofker	1956
	tenella	(Parker)	1958
Hastigerina	*digitata*	(Rhumbler)	1911
	pelagica	(d'Orbigny)	1839
Neogloboquadrina	*dutertrei*	(d'Orbigny)	1839
	incompta	(Cifelli)	1961
	pachyderma	(Ehrenberg)	1861
Orbulina	*universa*	d'Orbigny	1839
Orcadia	*riedeli*	(Rögl and Bolli)	1973
Pulleniatina	*obliquiloculata*	(Parker and Jones)	1865
Sphaeroidinella	*dehiscens*	(Parker and Jones)	1865
Streptochilus	*globigerus*	(Schwager)	1866

(continued)

Table 1.1 (continued)

Genus	Species	Author	Year
Tenuitella	*compressa*	(Fordham)	1986
	fleisheri	Li	1987
	iota	(Parker)	1962
	parkerae	(Brönnimann and Resig)	1972
Turborotalita	*clarkei*	(Rögl and Bolli)	1973
	humilis/cristata	(Brady)/Heron-Allen and Earland 1929	1884
	quinqueloba	(Natland)	1938

substantially since the first discoveries (see, e.g. the reviews and books of Vincent and Berger 1981; Hemleben et al. 1989; Murray 1991; Schiebel and Hemleben 2005; Kucera 2007). An enormous wealth of information is available from textbooks, printed papers, online publications, and various Internet sites (e.g., www.species-identification.org, www.EMIDAS.org, www.eforams.org). Many more researchers and working groups, beyond those referred to above, have added an enormous wealth of knowledge, which is presented in the following topical Chaps. 2–10.

References

Allen KA, Hönisch B, Eggins SM, Rosenthal Y (2012) Environmental controls on B/Ca in calcite tests of the tropical planktic foraminifer species *Globigerinoides ruber* and *Globigerinoides sacculifer*. Earth Planet Sci Lett 351–352:270–280. doi:10.1016/j.epsl.2012.07.004

Allen KA, Hönisch B, Eggins SM, Yu J, Spero HJ, Elderfield H (2011) Controls on boron incorporation in cultured tests of the planktic foraminifer *Orbulina universa*. Earth Planet Sci Lett 309:291–301. doi:10.1016/j.epsl.2011.07.010

Anderson TR (2005) Plankton functional type modelling: running before we can walk? J Plankton Res. doi:10.1093/plankt/fbi076

André A, Quillévéré F, Morard R, Ujiié Y, Escarguel G, de Vargas C, de Garidel-Thoron T, Douady CJ (2014) SSU rDNA divergence in planktonic Foraminifera: molecular taxonomy and biogeographic implications. PLoS ONE. doi:10.1371/journal.pone.0104641

Aurahs R, Göker M, Grimm GW, Hemleben V, Hemleben C, Schiebel R, Kučera M (2009) Using the multiple analysis approach to reconstruct phylogenetic relationships among planktonic Foraminifera from highly divergent and length-polymorphic SSU rDNA sequences. Bioinforma Biol Insights 3:155–177

Bé AWH, Hemleben C, Anderson OR, Spindler M, Hacunda J, Tuntivate-Choy S (1977) Laboratory and field observations of living planktonic Foraminifera. Micropaleontology 23:155–179

Berger WH (1981) Paleoceanography: the deep-sea record. In: Emiliani C (ed) The oceanic lithosphere. The sea. pp 1437–1519

Berger WH, Smetacek V, Wefer G (1989) Productivity of the ocean: present and past. John Wiley & Sons, Chichester

Bijma J, Hemleben C (1994) Population dynamics of the planktic foraminifer *Globigerinoides sacculifer* (Brady) from the central Red Sea. Deep-Sea Res I 41:485–510. doi:10.1016/0967-0637(94)90092-2

Bijma J, Hemleben C, Oberhänsli H, Spindler M (1992) The effects of increased water fertility on tropical spinose planktonic foraminifers in laboratory cultures. J Foraminifer Res 22:242–256

Brady HB (1884) Report on the Foraminifera dredged by the H.M.S. "Challenger" during the years 1873–1876. Report on the Scientific Results of the Voyage of H. M.S. Challenger during the years 1873–1876. Zoology 9:1–814

Buitenhuis ET, Vogt M, Moriarty R, Bednaršek N, Doney SC, Leblanc K, Le Quéré C, Luo YW, O'Brien C, O'Brien T, Peloquin J, Schiebel R, Swan C (2013) MAREDAT: towards a world atlas of marine ecosystem data. Earth Syst Sci Data 5:227–239. doi:10.5194/essd-5-227-2013

Burckhardt G (1920) Zum Worte Plankton. Schweiz Z Hydrol 1:190–192

Caron DA, Bé AWH, Anderson OR (1982) Effects of variations in light intensity on life processes of the planktonic foraminifer *Globigerinoides sacculifer* in laboratory culture. J Mar Biol Assoc U K 62:435–451

Caron M, Homewood P (1983) Evolution of early planktic foraminifers. Mar Micropaleontol 7:453–462. doi:10.1016/0377-8398(83)90010-5

Cushman JA (1911) A monograph of the Foraminifera of the North Pacific Ocean Part II Textulariidae. Smithsonian Institution, United States National Museum

Cushman JA, Todd R (1949) Species of the genus *Chilostomella* and related genera. Contrib Cushman Lab Foraminifer Res 25:84–99

Darling KF, Wade CM, Kroon D, Brown AJL (1997) Planktic foraminiferal molecular evolution and their polyphyletic origins from benthic taxa. Mar Micropaleontol 30:251–266

De Vargas C, Norris R, Zaninetti L, Gibb SW, Pawlowski J (1999) Molecular evidence of cryptic speciation in planktonic foraminifers and their relation to oceanic provinces. Proc Natl Acad Sci 96:2864–2868

D'Orbigny AD (1826) Tableau méthodique de la classe des Céphalopodes. Ann Sci Nat 1:245–314

Dujardin F (1835) Recherches sur les organismes inférieurs. Ann Sci Nat-Zool Biol Anim 2:343–377

Erez J, Almogi-Labin A, Avraham S (1991) On the life history of planktonic Foraminifera: lunar reproduction cycle in *Globigerinoides sacculifer* (Brady). Paleoceanography 6:295–306

Fischer G, Wefer G (1999) Use of proxies in paleoceanography: examples from the South Atlantic. Springer, Berlin, Heidelberg

Fraile I, Schulz M, Mulitza S, Merkel U, Prange M, Paul A (2009) Modeling the seasonal distribution of planktonic Foraminifera during the last glacial maximum. Paleoceanography. doi:10.1029/2008PA001686

Frerichs WE, Heiman ME, Borgman LE, Bé AWH (1972) Latitudal variations in planktonic foraminiferal test porosity: Part 1. Optical studies. J Foraminifer Res 2:6–13

Hemleben C (1969) Ultramicroscopic shell and spine structure of some spinose planktonic Foraminifera. In: Brönniman P, Renz HH (eds) Proceedings of 1st International Conference, Plankt Microfoss. Leiden, pp 254–256

Hemleben C, Olsson RK (2006) Wall textures of Eocene planktonic Foraminifera. In: Pearson PN, Olsson RK, Huber BT, Hemleben C, Berggren WA (eds) Atlas of Eocene planktonic Foraminifera. Cushman Found Spec Publ 41:47–66

Hemleben C, Olsson RK, Berggren WA, Norris RD (1999) Wall texture, classification, and phylogeny. In: Olsson RK, Hemleben C, Berggren WA, Huber BT (eds) Atlas of Paleocene planktonic Foraminifera. Smithsonian Contrib Paleobiology 85:10–19

Hemleben C, Spindler M (1983) Recent advances in research on living planktonic Foraminifera. Utrecht Micropaleontol Bull 30:141–170

Hemleben C, Spindler M, Anderson OR (1989) Modern planktonic Foraminifera. Springer, Berlin

Hönisch B, Allen KA, Russell AD, Eggins SM, Bijma J, Spero HJ, Lea DW, Yu J (2011) Planktic foraminifers as recorders of seawater Ba/Ca. Mar Micropaleontol 79:52–57

Johnstone HJH, Schulz M, Barker S, Elderfield H (2010) Inside story: an X-ray computed tomography method for assessing dissolution in the tests of planktonic Foraminifera. Mar Micropaleontol 77:58–70. doi:10.1016/j.marmicro.2010.07.004

Katz ME, Cramer BS, Franzese A, Hönisch B, Miller KG, Rosenthal Y, Wright JD (2010) Traditional and emerging geochemical proxies in Foraminifera. J Foraminifer Res 40:165–192

Kemle-von-Mücke S, Hemleben C (1999) Planktic Foraminifera. In: Boltovskoy E (ed) South Atlantic zooplankton. Backhuys Publishers, Leiden

Kucera M (2007) Chapter six: planktonic Foraminifera as tracers of past oceanic environments. In: Developments in marine geology. Elsevier, pp 213–262

Lombard F, Labeyrie L, Michel E, Bopp L, Cortijo E, Retailleau S, Howa H, Jorissen F (2011) Modelling planktic foraminifer growth and distribution using an ecophysiological multi-species approach. Biogeosciences 8:853–873. doi:10.5194/bg-8-853-2011

Murray JW (1991) Ecology and distribution of planktonic Foraminifera. In: Lee JJ, Anderson OR (eds) Biology of Foraminifera. Academic Press, London, pp 255–285

Olsson RK, Hemleben C, Berggren WA, Huber BT (1999) Atlas of Paleocene planktonic Foraminifera. Smithsonian Contrib Paleobiology 85, pp 252

Owen SRI (1867) On the surface-fauna of mid-ocean. J Linn Soc Lond Zool 9:147–157

Parker FL (1962) Planktonic foraminiferal species in Pacific sediments. Micropaleontology 8:219–254

Rhumbler L (1911) Die Foraminiferen (Thalamorphoren) der Plankton-Expedition. Die allgemeinen Organisations-Verhältnisse der Foraminiferen, Erster Teil

Rodhe W (1974) Plankton, planktic, planktonic. Limnol Oceanogr 19:360

Roy T, Lombard F, Bopp L, Gehlen M (2015) Projected impacts of climate change and ocean acidification on the global biogeography of planktonic Foraminifera. Biogeosciences 12:2873–2889. doi:10.5194/bg-12-2873-2015

Sarmiento JL, Gruber N (2006) Ocean biogeochemical dynamics. Princeton University Press, Princeton and Oxford

Schiebel R (2002) Planktic foraminiferal sedimentation and the marine calcite budget. Glob Biogeochem Cycles 16(4), 1065. doi:10.1029/2001GB001459

Schiebel R, Hemleben C (2005) Modern planktic Foraminifera. Paläontol Z 79:135–148

Schiebel R, Hiller B, Hemleben C (1995) Impacts of storms on Recent planktic foraminiferal test production and CaCO₃ flux in the North Atlantic at 47°N, 20° W (JGOFS). Mar Micropaleontol 26:115–129

Schott W (1935) Die Foraminiferen des äquatorialen Teil des Atlantischen Ozeans: Deutsche Atlantische Expeditionen Meteor 1925–1927. Wiss Ergeb, 43–134

Signes M, Bijma J, Hemleben C, Ott R (1993) A model for planktic foraminiferal shell growth. Paleobiology 19:71–91

Spero HJ (1986) Symbiosis, chamber formation and stable isotope incorporation in the planktonic foraminifer *Orbulina universa*. PhD Thesis, University of California

Spero HJ, Eggins SM, Russell AD, Vetter L, Kilburn MR, Hönisch B (2015) Timing and mechanism for intratest Mg/Ca variability in a living planktic foraminifer. Earth Planet Sci Lett 409:32–42. doi:10.1016/j.epsl.2014.10.030

Spero HJ, Lea DW (1993) Intraspecific stable isotope variability in the planktic Foraminifera

Globigerinoides sacculifer: results from laboratory experiments. Mar Micropaleontol 22:221–234

Spero HJ, Williams DF (1988) Extracting environmental information from planktonic foraminiferal $\delta^{13}C$ data. Nature 335:717–719

Spindler M, Hemleben C, Salomons JB, Smit LP (1984) Feeding behavior of some planktonic foraminifers in laboratory cultures. J Foraminifer Res 14:237–249

Vénec-Peyré MT (2005) Les Planches inédites de foraminifères d'Alcide d'Orbigny. Publ Scient Mus Paris, MNHN pp 302

Vincent E, Berger WH (1981) Planktonic Foraminifera and their use in paleoceanography. Ocean Lithosphere Sea 7:1025–1119

Weiner A, Aurahs R, Kurasawa A, Kitazato H, Kucera M (2012) Vertical niche partitioning between cryptic sibling species of a cosmopolitan marine planktonic protist. Mol Ecol 21:4063–4073. doi:10.1111/j.1365-294X.2012.05686.x

Žarić S, Schulz M, Mulitza S (2006) Global prediction of planktic foraminiferal fluxes from hydrographic and productivity data. Biogeosciences 3:187–207

Classification and Taxonomy of Extant Planktic Foraminifers

Classification of modern and fossil planktic foraminifers is based on a morphological species concept (i.e. morphotypes) for practical reasons, i.e. a non-destructive enumeration from strew-mounted samples, and economical (i.e. time-saving) analyses. Detailed classification of each test, for example, using scanning electron microscopy or analysis of the molecular genetics in case of live specimens would be too costly. Assuming that most modern planktic foraminifer morphotypes are known to science today, and have been properly described in literature, one would still be left with the problem of intraspecific diversity, as well as aberrant and malformed specimens. Depending on the personal taxonomic understanding and philosophy, one would add those problematic specimens to the most similar known morphotypes, or labelled them 'unidentified species'—'indet. spec.'.

The morphotypical classification of planktic foraminifer species is almost uniformly based on adult and mature specimens. Consequently, the description of tests invariantly relates to adult specimens if not stated differently. The rules of nomenclature are well defined by the International Commission on Zoological Nomenclature (ICZN). In turn, the determination of species and genera is not uniform between the different schools of micropaleontology, and may vary even among colleagues within one working group. The reason for different personal views of the same species is possibly based on experience and different philosophies (see also Scott 2011).

Most extant planktic foraminifer species are ubiquitous, and few are endemic. *Globorotaloides hexagonus*, *Globoquadrina conglomerata*, and *Globigerinella adamsi* are limited to the modern Indian and Pacific Oceans. Morphological ecophenotypes of the same species add to differences in classification, and taxonomic concepts. For example, the pink variety of *Globigerinoides ruber* has been extinct from the Indian and Pacific Oceans from MIS 5.5, ~120 ka (Thompson et al. 1979), and today occurs only in the Atlantic Ocean. The absence and presence of menardiform species in different ocean basins during the late Cenozoic has been affected by ecological conditions, geographic barriers, climate change, and ocean circulation (e.g., Mary and Knappertsbusch 2013; Broecker and Pena 2014).

Accounting for the biogeographic variability of species through time including evolutionary changes, the knowledge of planktic foraminifers (and other organisms) and their taxonomy is inherently limited, and depends on personal experience and perspective. Comprehensive understanding of planktic foraminifer taxonomy

© Springer-Verlag Berlin Heidelberg 2017
R. Schiebel and C. Hemleben, *Planktic Foraminifers in the Modern Ocean*,
DOI 10.1007/978-3-662-50297-6_2

can hence only be achieved through joint effort. The following information may help to complement the existing knowledge of modern planktic foraminifer taxonomy.

Taxonomy—nomenclature—classification: Taxonomy is the rather descriptive science of classifying organisms and naming them. The term 'taxonomy' is derived from the Greek words 'nomia' and 'taxis', meaning 'method' of 'arrangement', respectively. Taxonomy follows strict rules, which vary between zoological and botanical concepts. The taxonomy of planktic foraminifers follows the rules of the 'International Commission on Zoological Nomenclature' (ICZN), which releases every now and then a new edition of the 'International Code of Zoological Nomenclature' (ICZN Code). The newest is the 4th Edition published in 1999, available online via the Natural History Museum, London (http://www.nhm.ac.uk). It is strongly suggested to consult the ICZN Code when working on taxonomy.

2.1 Classification and Taxonomy

The chapter presents the current understanding of the morphological genus and species concept applied in analyses of extant planktic foraminifers. About 50 extant planktic foraminifer morphospecies (Chap. 1, Table 1.1) exist in the modern ocean (Loeblich and Tappan 1988), and about 250 genotypes are now distinguished by molecular methods (de Vargas et al. 2015). Accepting the systematics adopted by Cavalier-Smith (2004; see also Adl et al. 2005), planktic Foraminifera belong to the

Kingdom Protozoa
Subkingdom Biciliata
Infrakingdom Rhizaria
Phylum Sarcomastigophora
Subphylum Sarcodina
Superclass Rhizopodea
Class Granuloreticulosa
Order Foraminiferida
Suborder Globigerinina

Planktic foraminifers are believed to have evolved from benthic species during the Early Jurassic with a simple trochospiral morphology including tiny pores (microperforate, Plate 2.33). As in many other fossil groups planktic foraminifers developed in various steps through the Cretaceous and Tertiary. A first wide radiation occurred during the Mid-Cretaceous culminating in the Upper Cretaceous towards the K/Pg boundary. Until this point in time, approximately 300 species developed within the non-spinose group (cf. Caron 1985). Most species became extinct before the K/Pg boundary event. Only two or three species survived the K/Pg extinction event. These species are regarded ancestors of a new development of planktic foraminifer species during Tertiary and Recent times. *Hedbergella monmouthensis* is regarded as the ancestor of the newly developed spinose group, and *Preamurica taurica* and *Globanomalina archaecompressa* as the first non-spinose species at the base of the Tertiary (Olsson et al. 1999). In Zone P0, the lowermost biostratigraphic zone in the Paleocene (lowermost Paleogene), the first spinose species, *Eoglobigerina eobulloides*, appeared. Since then, this group constitutes the largest group of planktic foraminifer assemblages in the world's ocean. The test morphology, and especially the test surface structure, diversified within the various lineages, and is used in classification, besides features such as 'spinose' *versus* 'non-pinose', and 'normal perforate' *versus* 'microperforate'.

All modern planktic foraminifers (approx. 48 morphospecies) are assumed to belong to the suborder Globigerinina following Loeblich and Tappan (1988), which includes seven superfamilies, out of which only three are extant, (1) Heterohelicoidea Cushman 1927, (2) Globorotaloidea Cushman 1927, and (3) Globigerinoidea Carpenter et al. 1862.

On the taxonomic level of informal morphogroups, it is distinguished between (a) spinose (all Globigerinoidea), (b) non-spinose normal perforate or macroperforate (all Globorotaloidea and their precursors), and (c) non-spinose microperforate (Heterohelicoidea) species. It is assumed that (d) monolamellar Hastigerinidae do not belong to bilamellar Globigerinoidea, which is in contrast to Loeblich and Tappan (1988). It is suggested that Hastigerinidae, and (e) all microperforate species including tenuitellids, and other than Heterohelicoidea are separate taxonomic groups.

According to Loeblich and Tappan (1992), Foraminifera Lee 1990 are placed at the systematic Class level, and planktic foraminifers at the Order level, i.e. Globigerinida Delage and Herouard 1896. In this volume, it is focused on the species level, which is the only systematic level on which biology, ecology, and biogeochemistry can be reasonably discussed in sufficient detail. In addition, the species level is possibly the most conservative systematic level, which allows comparability of results at the long-term, and after revisions at the lower systematic levels.

Test features relevant for the classification of genera are pore size, and position of the primary and secondary apertures. At the species level, relevant features for classification are mode of coiling, shape of test and chambers, position and shape of apertures (e.g., lip and rim), and shell texture (e.g., spines and pustules). On the species level, well-illustrated compilations of Parker (1962), Bé (1967), Saito et al. (1981), Li (1987), Hemleben et al. (1989), and Kemle-von-Mücke and Hemleben (1999) give useful overviews of extant planktic foraminifers. The following plates illustrate tests of modern species to support classification. Images produced with Scanning Electron Microscopy (SEM) and incident microscope show the morphological variability and ontogenetic development of the test.

The following description of species starts with the largest and most abundant modern group of (1) spinose planktic foraminifer species (Sect. 2.2), including globigerinid species, which are trochospiral or quasi-planispiral. *Orbulina universa* is the only species with a spherical test formed by the final chamber in the final ontogenetic stages (adult to terminal). (2) Hastigerinids have a monolamellar test wall, and thus differ from all other extant species. All other extant and extinct genera are bilamellar: (Sect. 2.3). (3) Species of the second group are non-spinose, and include the globorotalids. Globorotalid species are normal perforate (or macroperforate, in contrast to microperforate, see below) and trochospiral (Sect. 2.4). One species, *Pulleniatina obliquiloculata*, forms a streptospiral test in its adult ontogenetic stage. All cone-shaped and disc-shaped species belong to the superfamily Globorotaloidea. The macroperforate *Streptochilus globigerus* is the only biserial modern planktic foraminifer species. (4) The microperforate genera *Globigerinita*, *Tenuitella*, and *Candeina*, and the triserial species *Gallitellia vivans* differ from all other trochospiral modern genera in wall structure (Sect. 2.5).

2.1.1 Molecular Genetics

Combining the morphotypic and the new genotypic concepts provides a great opportunity to refine planktic foraminifer systematics and phylogeny, and to create new tools for ecologic and paleoceanographic application. Wide variability in test morphology of some planktic foraminifer species has been shown to coincide well with different genotypes, a topic, which is under debate (De Vargas et al. 2002; André et al. 2014). In addition, ecophenotypic variation in test morphology allows conclusions on ecology (Spero and Lea 1993).

Genetic information for the self-organization of planktic foraminifers, as in all other live organisms, is stored in the cell in the form of double-stranded deoxyribonucleic acid (DNA).

Fig. 2.2 Conventional SSU rDNA phylogenetic tree showing the relative positions of planktic foraminifer ▶ morphospecies and genotypes. The phylogeny is based on 407 unambiguously aligned nucleotide sites, and is rooted on the benthic foraminifer *Allogromia* sp. Bayesian posterior probabilities (from the last 1000 trees, obtained within MrBayes; see Sect. 10.3.6) and ML bootstraps (expressed as a percentage, 1000 replicates) are shown on the tree (BI posterior probabilities/ML bootstraps). The *scale bar* corresponds to a genetic distance of 2 %. Benthic foraminifer taxa in *grey text*, and planktic foraminifers in *black*. Morphospecies and genotypes sampled in the Arabian Sea are shown on a *grey background*. From Seears et al. (2012)

It contains the genetic instructions used in the physiological functions of the cells, and the formation of the developmental stages of the organisms. DNA is a long polymer, which is formed by four different bases (ACGT, corresponding to adenine, cytosine, guanine, thymine). These bases are bound to sugars (desoxyribose), and to phosphate groups connected by ester bonds, then called desoxynucleotides. The desoxynucleotides A and T, and G and C, respectively, are bound by hydrogen bonds called base pairs (bp), in the double-stranded DNA molecule. The respective sequences of the four desoxynucleotides along the DNA molecule specify the long-term information, by encoding the functional ribonucleic acid (RNA). These are mainly the ribosomal RNAs involved in ribosome structure, and in protein synthesis, the transfer-RNA (tRNA)

transferring the amino acids to the ribosomes, and the messenger-RNA (mRNA) containing the information for the production of polypeptides (proteins). The DNA of a cell is transcribed into the different functional RNA molecules, of which the mRNA specifies the amino acid sequences of the proteins according to the genetic triplet code.

Molecular genetics studies of planktic foraminifers, i.e. determination of the nucleotide sequences, have been performed in various laboratories since the mid 1990s, offering a perfect tool for studies of the evolutionary and phylogenetic relationships among different taxa (Figs. 2.1 and 2.2). Molecular analyses of planktic foraminifers are mainly carried out on the nuclear-encoded and tandemly arranged ribosomal rRNA genes (rDNA). The small and large subunit (SSU and LSU) rDNA, and other regions of the rDNA, are of particular interest in planktic foraminifer

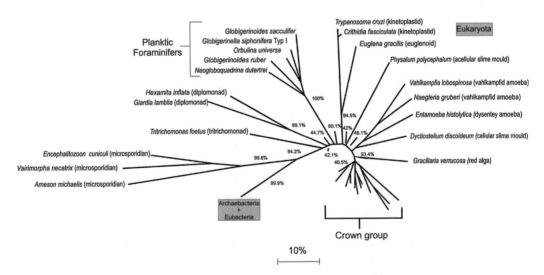

Fig. 2.1 SSU rDNA phylogeny for planktic foraminifers, and representatives of a diverse range of eukaryote, archaebacterial, and eubacterial taxa, reconstructed by NJ ("Neighbor Joining", see Sect. 10.3.6) analyses of 546 unambiguously aligned sites. The "crown group" contains sequences from a wide range of groups from eukaryotic organisms to *Homo sapiens*. Modified after Wade et al. (1996)

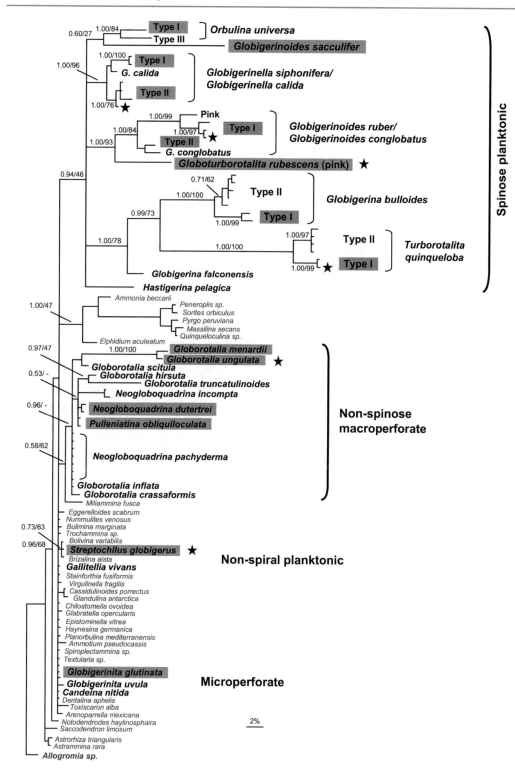

Plate 2.1 *Beella digitata* with rugose surface and highly variable aperture. Digitate chambers occasionally formed ▶ during the adult stage. (*3, 6*) Close-ups of test surface from (*2*), with arrows pointing at spine collars. Bars of overviews 100 μm, bars of close-ups 20 μm. (*1*) *Photo* A. Kiefer and R. Schiebel

molecular studies, corresponding to the 18S and 25/28S rDNA of other eukaryotic organisms (e.g., Pawlowski and Holzmann 2002). The respective methods are briefly described in Chap. 10.

Molecular genetic approaches provide evidence that planktic foraminifers are not a monophyletic group as it was traditionally assumed, but that they have possibly descended from benthic foraminifers more than one time over the past ~ 170 Myrs, from the mid Jurassic onward. Planktic foraminifers would therefore be a polyphyletic group (Darling et al. 1997; Darling and Wade 2008). The genetic and morphological similarity of modern planktic *Streptochilus globigerus* and the benthic *Bolivina variabilis* is a good example of tychopelagic behavior, and later development of a new fully planktic species (Darling et al. 2009). All of the other modern planktic foraminifer morphogroups, spinose, non-spinose macroperforate, and non-spinose microperforate, as well as the monolamellar Hastigerinidae might have adopted their planktic habitat in a similar way as the heterohelicoid species *S. globigerus*.

In general, genetic data confirm the taxa of modern planktic foraminifers to the level of morphospecies (Fig. 2.2). However, the traditionally classified morphotypes often include two or more genotypes, which are referred to as cryptic species. Consequently, new genotypes have been added to the existing portfolio of morphospecies, representing genotype complexes (e.g., Darling et al. 1997; De Vargas and Pawlowski 1998; Kucera and Darling 2002; Darling and Wade 2008; André et al. 2014). This complex constitutes normally very closely related genotypes of one morphospecies (Table 2.2). For example, *Globigerinella siphonifera* forms a complex group of three different types (André et al. 2014), whereas *Globigerinoides sacculifer* displays only one genotype including several different morpho-species (André et al. 2013).

Those differences in the phylogeny of the different species possibly result from independent evolutionary developments, with some species behaving more conservative than others. In addition, various species may be in the (molecularly detectable!) process of speciation, and at present are forming cryptic species.

In the following paragraphs, a detailed description of the test morphology is given, and differences between similar morphospecies are discussed. Ecology and distribution of species are documented. Information on the molecular genetic data is presented if available.

> **Cryptic species:** Cryptic species represent a group of species, which cannot interbreed, and contain individuals that are morphologically identical to each other but can only be differentiated by e.g. molecular genetic methods. Herewith, different genotypes can be distinguished within one morphospecies.

2.2 Bilamellar Spinose Species

Spinose species bear spines, which are plugged into the test wall. Spines may be round in cross-section, triangular, or triradiate at the entire length, or two or all three types may be combined in one spine from base to top. In case of a combination of different cross sections in one spine, round spines develop into more angular spines, i.e. triangular and triradiate from base to top. Spinose species are assumed macroperforate although occasionally producing pores, which embrace the size-limit (1 μm) between microperforate and macroperforate. Hastigerinids are included with the spinose species although they are an exception to the bilamellar spinose species by producing a monolamellar wall, a cytoplasmic bubble capsule, and very large spines.

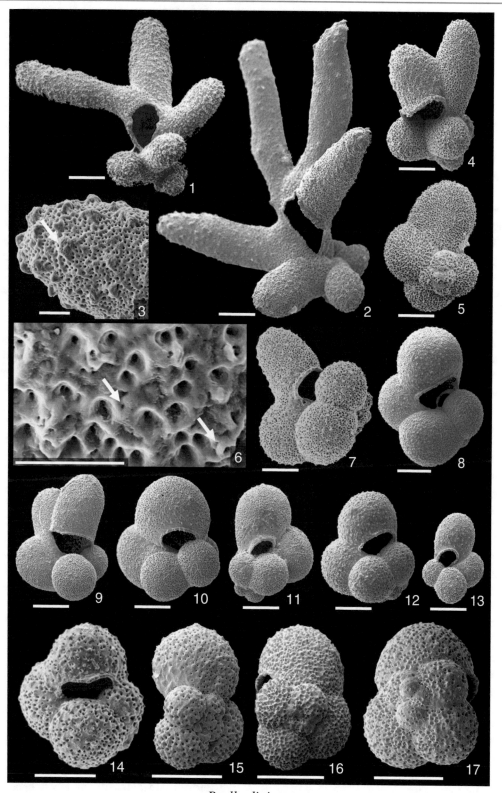

Beella digitata

Plate 2.2 (*1–3*) Adult *Bolliella adamsi* with elongate final chambers. (*4*) Triradiate and flexible (bent) thin round ▶ spines. (*6*) Mature specimen with digitate final chambers. For comparison, (*7*) mature *Beella digitata*, and (*8*) adult *Hastigerinella digitata*. Bars of overviews (*1–3*) 100 µm, (*6–8*) 200 µm. Bar of close-up (*4*) 20 µm, (*5*) 100 µm. (*3,7*) *Photo* A. Kiefer and R. Schiebel

2.2.1   *Beella digitata* (Brady 1879) (Plate 2.1)

Other Generic Assignment (O.G.A.): *Globigerinella* Cushman 1927.

Description: Trochospiral, spinose, normal perforate, rather smooth surface, 4–6 chambers in the last whorl, and a rather large aperture at the base of the final chamber. Adult specimens of *B. digitata* are easy to identify due to their digitate chambers. Pores are rather irregularly distributed and may partially be covered by a smooth calcite veneer.

Molecular genetics: Very few data are available so far. According to André et al. (2014) only one genotype exists, thus no cryptic species of this morphospecies have been described up to date (2015). *Beella digitata* appears to be a well-established sister-group to the plexus of *G. siphonifera*.

Ecology: *Beella digitata* is a rare species of the tropical to subtropical ocean (e.g., Conan and Brummer 2000; Schmuker and Schiebel 2002). Regionally enhanced standing stocks of *B. digitata* may occur in the Mediterranean Sea, and are limited to the subsurface water column (Hemleben et al. 1989). Large numbers of empty tests of non-digitate *B. digitata* occurred at mesobathyal water depths of 300–700 m west of the Canary Islands (29° 59′ 98″N, 21° 59′ 79″W) at a total water depth of 4996 m (Schiebel, unpublished data, 2007). It may hence be assumed that *B. digitata* occupies a rather narrow ecological niche in the modern ocean, and flourishes at places and during times of slightly enhanced nutrient availability following the phytoplankton production. The rather anecdotal evidence of the distribution of *B. digitata* in subsurface waters might identify *B. digitata* as specialized on somehow degraded organic matter as a food source, similar to other subsurface dwelling planktic foraminifer species.

Remarks: According to its round to triangular spines, *B. digitata* would classify with the genera *Globigerinella* or *Orbulina*. Because of its chamber morphology and position of the aperture, Holmes (1984) classifies *G. digitata* with the genus *Beella*. The morphospecies seems to include only one genotype and is referred to as *Globigerinella* (Darling and Wade 2008) or *Beella* (Aurahs et al. 2009a; André 2013). *Globigerinella digitata* is not related to the monolamellar species *Hastigerinella digitata*. The reason why *Globigerinella digitata* is classified with the genus *Beella* is because of the close resemblance of the juvenile tests of both species *B. digitata* and *B. megastoma*. Both *B. digitata* and *B. megastoma* have rather irregular tests, resulting in considerable morphological variability within assemblages (Plate 2.1). Systematic differences between non-digitate tests of two putative types of *Beella* could not be detected among the analyzed individuals from the Atlantic and Indian Ocean. Therefore, *B. digitata* and *B. megastoma* are assumed to be the same species, and given the name of the older synonym *Beella digitata* (Brady 1879). *Beella megastoma* (Earland 1934) was first described as *Globigerina megastoma* from the Drake Passage and Scotia Sea between 3328 and 3959 m water depth by Earland (1934). *Beella megastoma* is rather rare and patchily distributed in subtropical to subpolar waters. Paleoceanographic implications of the presence of *B. megastoma* are discussed by Bauch (1994).

2.2.2   *Bolliella adamsi* Banner and Blow 1959 (Plate 2.2)

O.G.A.: *Hastigerina* Thomson 1876, *Globigerinella* Cushman 1927.

Description: *Globigerinella*-type wall structure. Early chamber arrangement low trochospiral

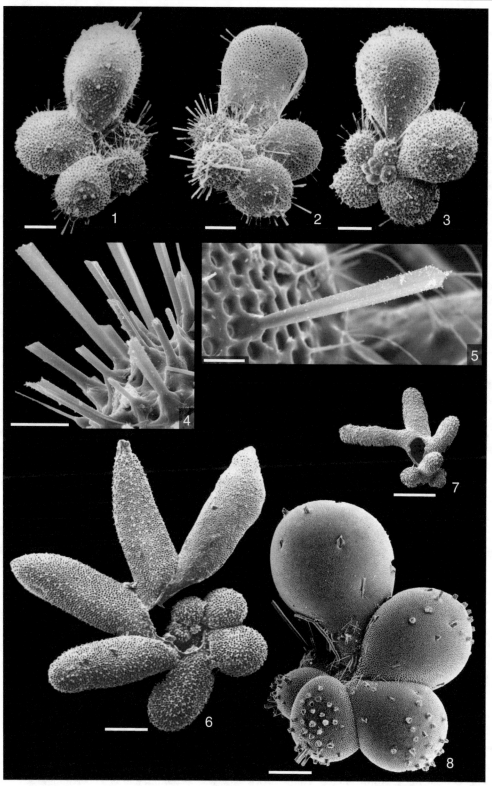

Bolliella adamsi

becoming more becoming more later in ontogeny, with up to 7 chambers in the last whorl. Adult specimens producing one to three elongate club shaped final chambers. The spines may be round and rather thin, or thickened at the base, and develop into triangular to triradiate distal shapes. Pre-adult and early adult specimens are very similar in surface structure and spine morphology to both *G. siphonifera* and *G. calida*.

Molecular genetics: No data available. Morphologically, *Bolliella adamsi* belongs to the *Globigerinella* plexus, and counts as a sister-species of *G. siphonifera, G. radians*, and *G. calida* (e.g., Banner and Blow 1959; Weiner et al. 2012).

Ecology: *Bolliella adamsi* is a rare species, limited to the Indian and Pacific Oceans (Bé 1977, and references therein Hemleben et al. 1989). In the Arabian Sea, *B. adamsi* was sampled from the entire surface water column above 100 m water depth, and was most frequent at the thermocline and immediately below (40–100 m water depth) in oligotrophic and well stratified waters (Schiebel et al. 2004). At a maximum abundance of only 1.8 specimens m^{-3}, i.e. 10.5 % of the entire fauna during the spring intermonsoon, *B. adamsi* is one of the least abundant modern planktic foraminifers. Its particular ecological requirements like food source and reproductive behavior are still unknown. From the types of spines it may be assumed that copepods are part of the diet of *B. adamsi*, similar to *Globigerinella siphonifera* and *Orbulina universa*. Test flux and fossilization potential of *B. adamsi* are low (Dittert et al. 1999; Conan and Brummer 2000).

Further readings: Parker (1962, 1976), Srinivasan and Kennett (1975).

2.2.3   *Globigerina bulloides* d'Orbigny 1826 (Plate 2.3)

Description: *Bulloides*-type wall structure, normal perforate, thin spines supported by spine collars coalescing to form ridges. Rather low

trochospiral, lobulate in outline with globular, slightly embracing chambers increasing rapidly in size, and 4 chambers in the last whorl. Umbilical aperture, which may be slightly out of its centric position, but never facing to either side left or right as in *G. calida*. The aperture is wide open, and the size of the aperture may differ between different genotypes of *G. bulloides* (Darling and Wade 2008), and never as narrow as in *G. glutinata*. In contrast to *G. falconensis*, *G. bulloides* bears no umbilical lip. An apertural rim could result from slightly enhanced thickness of the test wall along the aperture. Pore concentrations range from 70 to 100 pores per 50 μm^2 of test surface area. Pore diameters range from about 0.7–1.2 μm, and hence embrace the pore-size range delimiting 'microperforate' (<1 μm) from 'macroperforate' (>1 μm) species.

The repartition of dextral and sinistral forms of *G. bulloides* seems to be related to temperature (Malmgren and Kennett 1977), and is balanced in contrast to other species, like *G. inflata* or *N. pachyderma*, which are mostly left-coiling. Kummerform final chambers occur rather frequently in *G. bulloides*.

Molecular genetics: In total, 14 genetic types of *G. bulloides* have been distinguished by various authors (e.g., Darling et al. 2000; Darling and Wade 2008; André et al. 2013; Seears et al. 2012; Morard et al. 2013). However, according to André et al. (2014), previously defined genetic types of *G. bulloides*, e.g., Ia-f, IIa-g, and IIIa may have been oversplit. According to two new methods (ABGD: Automatic Barcode Gap Discovery; and GMYC: General Mixed Yule Coalescent; see Sect. 10.3.6, Fig. 2.3) only 7 genotypes of *G. bulloides* are delimitated and qualify a species status (from André et al. 2014, Fig. 2.3). A similar result to that of André et al. (2014) was obtained by Seears et al. (2012, Fig. 2.4). Interestingly, Types IIa and IIb are bipolar distributed. Thus, a gene flow across the tropics must have existed (Darling et al. 2000). The same seems to be true for *T. quinqueloba* and *N. pachyderma*.

Ecology: *Globigerina bulloides* mainly dwells above the thermocline within the upper 60 m of the water column, and is a non-symbiotic species

Fig. 2.3 Ultrametric SSU rDNA tree with GMYC delimitations of *G. bulloides* and the sister-group *G. falconensis* displayed in relation to *T. quinqueloba*. Names of morphospecies are given on the *outer arc*. Plausible biological species are indicated on the *inner arc*. The *scale bar* shows the patristic genetic distance. Distances of the *G. bulloides/G. falconensis* group suggest that these morphospecies may have been oversplit in comparison to *T. quinqueloba*. After André et al. (2014)

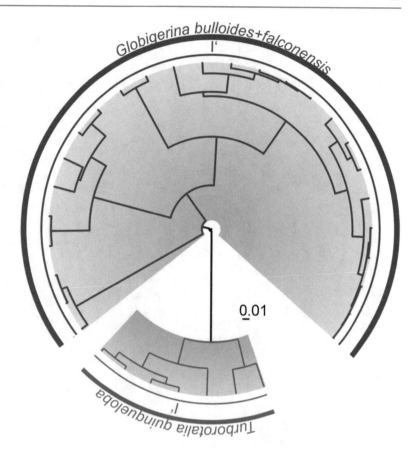

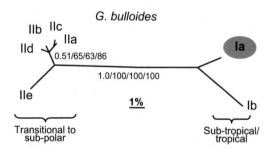

Fig. 2.4 Conventional delimitation of seven genotypes of the morphospecies *G. bulloides* using 669 bp of SSU rDNA. The phylogenetic tree is unrooted, and the genetic distance equals 1 %. Type 1a is found in the Arabian Sea. From Seears et al. (2012)

usually associated with temperate to sub-polar water masses, as well as upwelling. The distribution *G. bulloides* within the surface water column may be modified by hydrologic conditions, and the availability of prey. In addition to ecologic demands, biological prerequisites, i.e. reproduction strategy may determine the depth distribution of *G. bulloides* (Schiebel et al. 1997). *Globigerina bulloides* is equally characteristic of upwelling environments in lower latitudes (e.g., Thiede 1975; Bé and Hutson 1977; Kroon and Ganssen 1988; Naidu and Malmgren 1996a, b; Conan and Brummer 2000; Seears et al. 2012), as of seasonally enhanced primary production at mid and high latitudes (e.g., Bé and Tolderlund 1971; Bé 1977; Ottens 1992; Schiebel and Hemleben 2000; Chapman 2010). *Globigerina bulloides* is an opportunistic species, and often dominates the foraminifer fauna, test flux, and sediment assemblage at the ocean floor (Sautter and Thunell 1989; Sautter and Sancetta 1992), and is therefore an

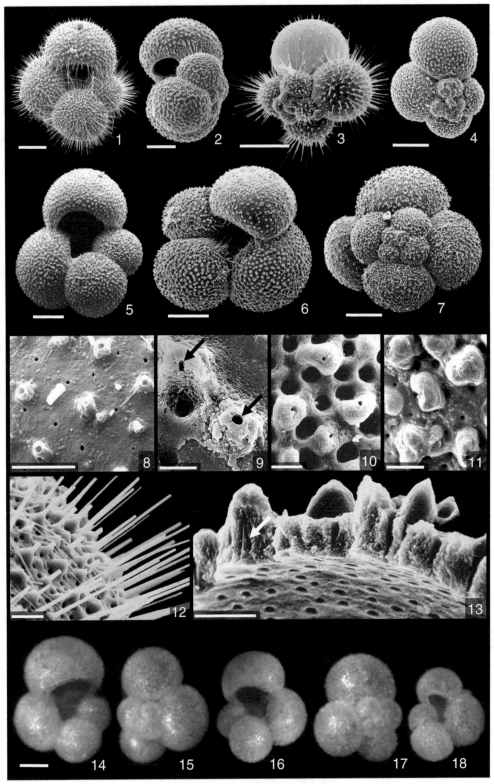

Globigerina bulloides

◀ **Plate 2.3** (*1–7*) Adult *Globigerina bulloides*, with (*8–13*) details of test wall and surface. (*3*) Specimen with newly formed chamber. (*14–18*) *Lower panel* showing incident-light micrographs of specimens from net tows. (*6, 7, 18*) Specimens with final kummerform chamber. (*8*) Test wall with pores and broken spines. (*9*) Spines shed from collars, and starting GAM calcification. (*10*) Pores and spine-collars with spine-holes (*arrows*) and GAM calcite. (*11*) GAM calcite covers spine-collars, spines-holes still open. (*12*) Test surface with spines (mostly broken). (*13*) Pores, spine collars, and spine-remnant (*arrow*) in cross section. Bars of overviews 100 μm, bars of close-ups (*8, 11–13*) 10 μm, (*9, 10*) 2 μm. (*14–18*) CourtesyA. Movellan

important source of geochemical information for paleoceanographic reconstruction (e.g., Fischer and Wefer 1999; Hillaire-Marcel and de Vernal 2007).

In contrast to most spinose species, an important part of the diet of *G. bulloides* consists of algae, as indicated by the olive green to brownish coloration of its cytoplasm in freshly collected specimens, and shown by transmission electron microscopy (TEM). Apart from some biogeographical studies, only a few accounts have been published on its specific ecologic demands (e.g., Lee et al. 1965; Spindler and Hemleben 1980).

Remarks: The environmental and physiological parameters affecting the disequilibrium fractionation of stable isotopes of *G. bulloides* were quantified in a laboratory study (Spero and Lea 1996). Mean natural disequilibria of *G. bulloides* are positive for $\delta^{18}O$ (up to +1 ‰) and negative for $\delta^{13}C$ (−1 to −2 ‰) caused by its ecologic preferences (cool and eutrophic waters), and including ontogenetic and ecologic effects in the <200-μm tests size fraction (Ganssen 1983; Niebler et al. 1999). GAM calcification has been estimated to add up to 10 % to the calcite mass of the test of *G. bulloides* (Schiebel et al. 1997), and to affect its stable isotope and element ratios. *Globigerina bulloides* is a symbiont-barren species, and symbiont induced effects on the stable isotope composition can be excluded. The dissolution susceptibility of *G. bulloides* tests is slightly higher than average among extant species (Dittert et al. 1999). Because of its wide distribution in the global ocean, *G. bulloides* is one of the most analyzed species in paleoceanographic studies.

Further readings: Spero and Lea (1996), Aldridge et al. (2011), Boussetta et al. (2012), André et al. (2014), Darling and Wade (2008), Seears et al. (2012), Morard et al. (2013).

2.2.4   *Globigerina falconensis* Blow 1959 (Plate 2.4)

Description: *Globigerina falconensis* has a *bulloides*-type wall structure, is normal perforate, and bears thin spines. Tests are low-trochospiral, lobulate in outline, with four globular and slightly embracing chambers in the last whorl. Chambers might be slightly ovate in late ontogeny. Subsequent chambers distinctly increase in size. The umbilical aperture may be slightly out of centric position and bears a distinct lip. The apertural area is usually smaller than in *G. bulloides*. Tests of *Globigerina falconensis* resemble those of the genetic sister-taxon *G. bulloides* to a high degree (cf. Aurahs et al. 2009a). The only obvious morphological difference, which provides unequivocal prove for differentiation between the two species, is the apertural lip of adult tests of *G. falconensis* (Malmgren and Kennett 1977).

Molecular genetics: *Globigerina falconensis* appears to be a sister-species of *G. bulloides* and exhibits only one genotype (André et al. 2014).

Ecology: Apart from differences in test features, *G. falconensis* is a symbiont bearing species, and varies from sympatric and symbiont-barren *G. bulloides* by differential stable isotope values. *Globigerina falconesis* is a less opportunistic species than *G. bulloides*, the latter being adapted to an elevated availability of food in surface waters. In the Arabian Sea off Pakistan, *G. falconesis* occurs at maximum abundance in January and February during the NE monsoon. Assuming that *G. falconensis* is adapted to a different kind of nutrition than *G. bulloides*, the abundances of the two species are applied to reconstruct the intensity of NE monsoonal mixing, and SW monsoonal upwelling in the Arabian Sea over the past 24 kyrs (Schulz et al. 2002).

Further readings: Malmgren and Kennett (1977).

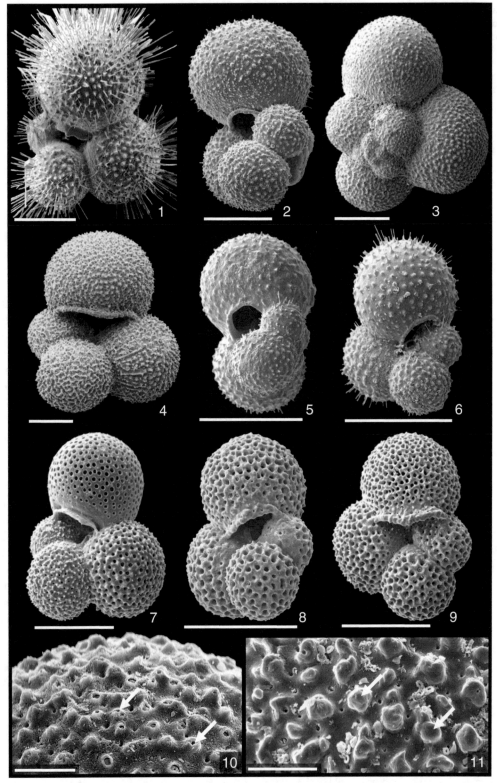

Globigerina falconensis

◀ **Plate 2.4** *Globigerina falconensis* (*1–11*) with normal-sized pores, and (*7–9*) large pores. In comparison to *G. bulloides* (Plate 2.3), the apertural lip and more bulb-like chambers are typical of *G. falconensis*. (*10*) Pores and spine-holes (*arrows*). (*11*) GAM calcification (*arrows*) covering spine-holes. Bars of overviews 100 μm, bars of close-ups 20 μm

2.2.5 *Globigerinella calida* (Parker 1962) (Plate 2.5)

O.G.A.: *Globigerina* d'Orbigny 1826.

Description: *Globigerinella calida* has a low trochospiral to almost planispiral test, with a wall structure similar to the *bulloides*-type. It is normal perforate, has round to triradiate spines. Chambers are globular, and ovate during late ontogeny, i.e. the final and penultimate chamber. In early adult ontogeny, *G. calida* forms 4.5 chambers in the last whorl, and 4.5–6 chambers in the last whorl during the late ontogenetic stages. In the final ontogenetic stage, chambers might detach from the previous whorl. The aperture may be a wide to narrow umbilical to extraumbilical opening. Test and spine architecture of *G. calida* are similar to *G. siphonifera* throughout its ontogeny. In its pre-adult and early adult stages, *G. calida* may be virtually indistinguishable from *G. siphonifera* in incident light microscopy.

Molecular genetics: According to André et al. (2014) the morphospecies *G. calida* belongs to the *Globigerinella siphonifera* plexus, and shows a very low diversity with two or three cryptic species of debated status.

Ecology: *Globigerinella calida* typically occurs at medium frequency in mesotrophic waters of the tropical to temperate ocean (Parker 1962; Bé et al. 1985; Ortiz et al. 1995; Schmuker 2000; Conan and Brummer 2000; Retailleau et al. 2012). In the Arabian Sea, maximum numbers of *G. calida* of up to 3 individuals per m^3 occur throughout the upper 100 m of the water column in the marginal upwelling regions, and during the intermonsoon and NE monsoon season (Conan and Brummer 2000; Schiebel et al. 2004). In neritic waters of the Caribbean Sea off Puerto Rico, *G. calida* was found to be more frequent during more oligotrophic conditions in spring than during the slightly more productive conditions in fall (Schmuker 2000). In the open Caribbean Sea, *G. calida* was found to be less frequent than in neritic waters (Schmuker 2000;

Schmuker and Schiebel 2002). Maximum standing stocks of *G. calida* of up to 150 ind. m^{-3} in the SE Bay of Biscay occurred at a neritic site in an upwelling area over a submarine canyon head in November 2007, and decreased towards the open Bay of Biscay (Retailleau et al. 2011, 2012).

Further readings: Weiner et al. (2014, 2015).

2.2.6 *Globigerinella siphonifera* (d'Orbigny 1839) (Plate 2.6)

O.G.A.: *Hastigerina* Thomson 1876.

Description: *Globigerinella siphonifera* has a low trochospiral to irregular planispiral test, and exhibits wall structure resembling the *bulloides*-type wall but at a higher porosity. The test of *G. siphonifera* is normal perforate, has round to triradiate spines, with ovate chambers during late ontogeny. Early adult individuals have 4.5–5 chambers in the last whorl, and 5–6 in the late ontogenetic stages.

Molecular genetics: According to Weiner et al. (2015) the *G. siphonifera* plexus (morphospecies: *G. siphonifera*, *G. calida*, *G. radians*, *Beella digitata*, and *Bolliella adamsi*) shows a well-defined and clearly separated genetic diversity, which can be characterized by three main lineages, including 12 distinct genetic types for *G. siphonifera/radians/calida* (Fig. 2.5; Table 2.1).

Globigerinella siphonifera includes two morphotypes (Fig. 2.6) and genotypes, Type I (relatively large, evolute, larger pores (av. 4.5 μm), higher spine density than Type II), and Type II (relatively slender, involute, smaller pores (av. 2.2 μm), lower spine density than Type I) reported from the Caribbean Sea (Huber et al. 1997; Darling et al. 1997; Bijma et al. 1998). The aperture ranges from a wide opening in evolute Type I, to a small slit in peripheral (equatorial) to extra-umbilical position in involute Type II. Type I exhibits broad cytoplasmic

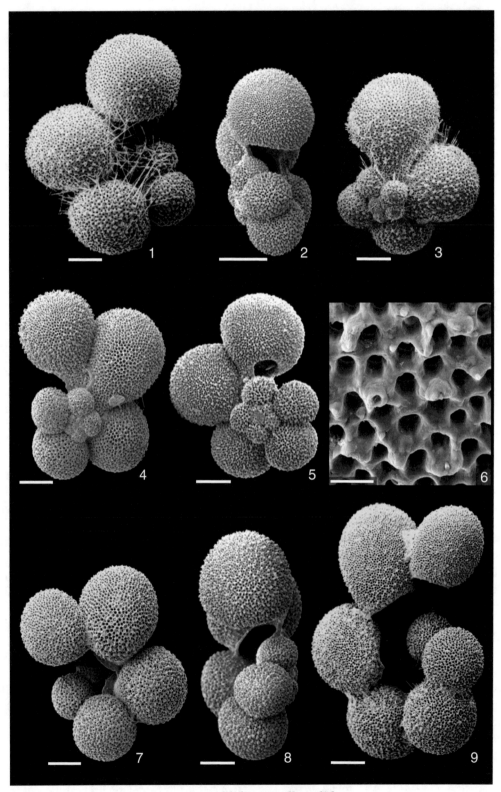

Globigerinella calida

◀ **Plate 2.5** (*1–6*) Adult *Globigerinella calida,* and (*7–9*) mature *G. calida* with final chambers detached from the last whorl. (*6*) Test surface (detail of *5*) with pores, broken spines, and GAM calcification. Bars of overviews 100 μm, bar of close-up 10 μm. (*1,3*) *Photo* A.Kiefer and R.Schiebel

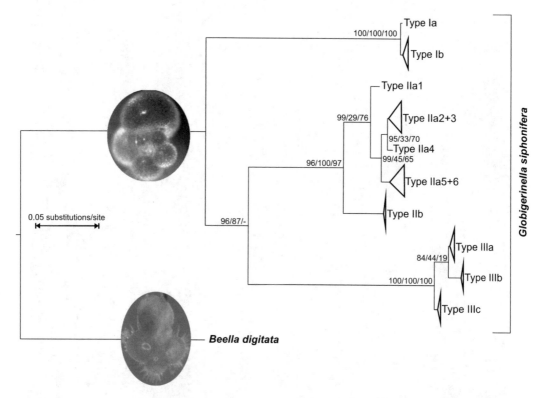

Fig. 2.5 SSU rDNA phylogenetic tree of *G. siphonifera* with *B. digitata* as outgroup. Type Ia and Ib occur within the Indo-Pacific region, but also unevenly distributed in the global Ocean. Types IIa1, IIa2 + 3, IIa4, IIa5 + 6 are cosmopolitans, whereas type IIb occurs in the Atlantic only. Type IIIa had been discovered in the E-Atlantic, type IIIb has been observed in the Caribbean, Mediterranean Sea, Red Sea and Western Indian Ocean, mostly marginal Seas. The distribution pattern may suggest that the regional ecology plays an important role in diversification. Light microscopic images of *G. siphonifera* and *B. digitata* illustrate the gross morphology. Both individuals measure about 250 μm across. From Weiner et al. (2014; see also Weiner et al. 2015)

Table 2.1 Correspondence between genetic diversity and morphological variability within the *Globigerinella siphonifera/G. calida* plexus, including classifications following classical taxonomy (e.g., Parker 1962), and revised taxonomy based on the morphometric measurements from Weiner et al. (2015)

Genetic type	Revised taxonomy	Classical taxonomy
Ia	*G. radians*	*G. calida* or *G. siphonifera*
Ib	*G. siphonifera*	*G. siphonifera*
IIa1	*G. siphonifera*	*G. siphonifera*
IIa2	*G. siphonifera*	*G. siphonifera*
IIa3	*G. siphonifera*	*G. siphonifera*
IIa5	*G. siphonifera*	*G. siphonifera*
IIa6	?	*G. siphonifera*
IIb	*G. siphonifera*	*G. siphonifera*
IIIa	?	*G. calida*
IIIb	*G. calida*	*G. calida*
IIIc	*G. calida*	*G. calida*

Question marks stand for genetic types whose morphology could not be confirmed by quantitative analysis, because no suitable images were available. From Weiner et al. (2015)

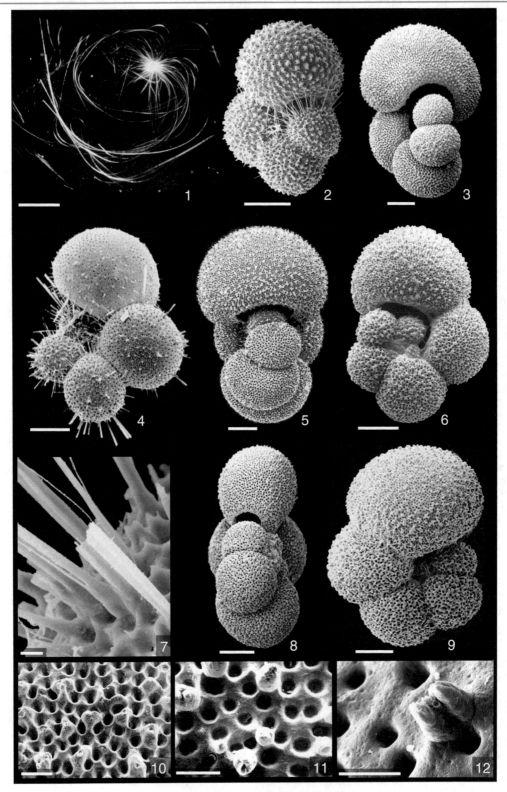

Globigerinella siphonifera

Plate 2.6 (*1*) Live *Globigerinella siphonifera* (Kage Microphotography©, with permission). (*2, 4*) Adult tests of *G. siphonifera* with low porosity, and (*3, 5, 6, 8, 9*) high porosity. Close-ups of (*7*) triradiate and round spines, (*10–12*) spine-collars, pustules, and pores. Bars of overviews (*1*) 1 mm, (*2–9*) 100 μm. Bars of close-ups 20 μm

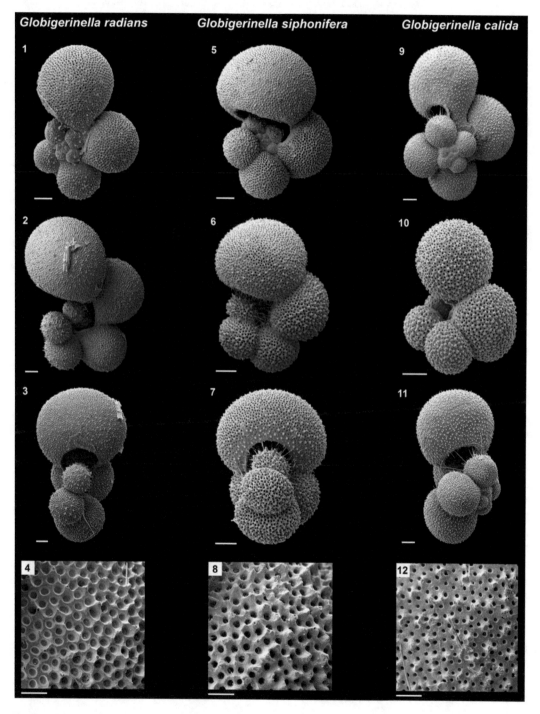

Fig. 2.6 SEM images of the spiral, umbilical, and lateral view, and close-ups of the pores of two individuals of three *Globigerinella* morphotypes, *G. radians*, *G. siphonifera*, and *G. calida*. Bars of overviews 60 μm, close-ups 20 μm. *Globigerinella radians* specimens are from the Mozambique Channel, *G. siphonifera* specimens from the Mozambique Channel and the Arabian Sea, and *G. calida* specimens are from the Mozambique Channel. From Weiner et al. (2015)

Plate 2.7 (*1–9*) Adult and mature *Globigerinoides conglobatus,* with (*7*) crystal growth on edges, (*8*) partly encrusted, ▶ and (*9*) heavily encrusted tests. (*3, 5, 6*) Spiral view showing secondary apertures (*arrows*). Details of (*10*) spine-remnants in aperture. (*11*) Close-up of pores and spine-holes from specimen (*2*). (*12*) Cross section of outer test wall with pores, spine-mold (*arrow*), calcite layers, and inner test wall with pores. Bars of overviews 200 μm, bars of close-ups 20 μm

flanges between the spines, and Type II produces normal rhizopodia along the spines (Bijma et al. 1998). *Globigerinella siphonifera* is the only planktic foraminifer reported up to now to possess two different types of symbionts both being chrysophytes (Faber et al. 1989).

Ecology: *Globigerinella siphonifera* is most frequent in the tropical to subtropical ocean, and less frequent in temperate waters (e.g., Bé 1977). *Globigerinella siphonifera* is rather variable in test morphology, and includes several ecophenotypes (Parker 1962). In the Caribbean, *G. siphonifera* includes two morphotypes. The larger Type I dwells deeper in the surface water column than the more slender Type II (Bijma et al. 1998). Salinity and temperature tolerance of *G. siphonifera* were experimentally determined as to 27–45 PSU and 10–30 °C, respectively (Bijma et al. 1990b). The autecology of *G. siphonifera* is affected by the type of hosted chrysophyte symbiont (Faber et al. 1989). Four genotypes of *G. siphonifera* are assigned to waters of varying trophic conditions (De Vargas et al. 2002).

Remarks: *Globigerinella siphonifera* (d'Orbigny 1839) is the senior synonym of *Globigerinella aequilateralis* (Brady 1897). The name *G. aequilateralis* is still in use possibly due to its descriptive meaning describing the planispiral test architecture of the adult specimens of this species. Weiner et al. (2015) propose to revive the use of the species *G. radians* (Egger 1893), despite the fact that the original material has been lost in Munich (Germany), in World War II. Thus, this species is based only on the figures published by Egger (1893). The morphological variability of the *Globigerinella* plexus according to Weiner et al. (2015) is demonstrated in Fig. 2.6.

Further readings: Darling and Wade (2008), Seears et al. (2012), André et al. (2014).

2.2.7 *Globigerinoides conglobatus* (Brady 1879) (Plate 2.7)

Description: *Globigerinoides conglobatus* has a *ruber*-type wall structure, large pores, and round to triangular spines. The test is low to medium trochospiral and slightly lobulate. One whorl includes four spherical (pre-adult specimens) to compressed (adult specimens) chambers, which may overlap considerably. The aperture forms a rather narrow slit centered over three chambers. The spiral side exhibits two secondary apertures. Most specimens from sediments bear a rather thick calcite crust. Pre-adult tests could be confused with *G. sacculifer*, but in direct comparison have less globular chambers and less incised sutures. Adult specimens have a unique compressed (flat, pillow-like) final chamber (except kummerforms), which distinguishes *G. conglobatus* from any other species.

Molecular genetics: *Globigerinoides conglobatus* appears to include 1 genotype, which is closely related to *G. ruber* (André et al. 2014).

Ecology: *Globigerinoides conglobatus* dwells in the deeper photic zone, and is associated with dinoflagellate symbionts (*Gymnodinium beii*, Spero 1987) similar to those occurring in *G. ruber, G. sacculifer,* and *O. universa. Globigerinoides conglobatus* occurs at low to medium abundances in tropical and subtropical waters, and may be transported by currents to the lower mid-latitude ocean (Kemle-von-Mücke and Hemleben 1999; Schmuker and Schiebel 2002; Schiebel et al. 2004). *Globigerinoides conglobatus* is the only *Globigerinoides* species assumed to form gametogenetic calcite at subsurface waters (Hemleben et al. 1989).

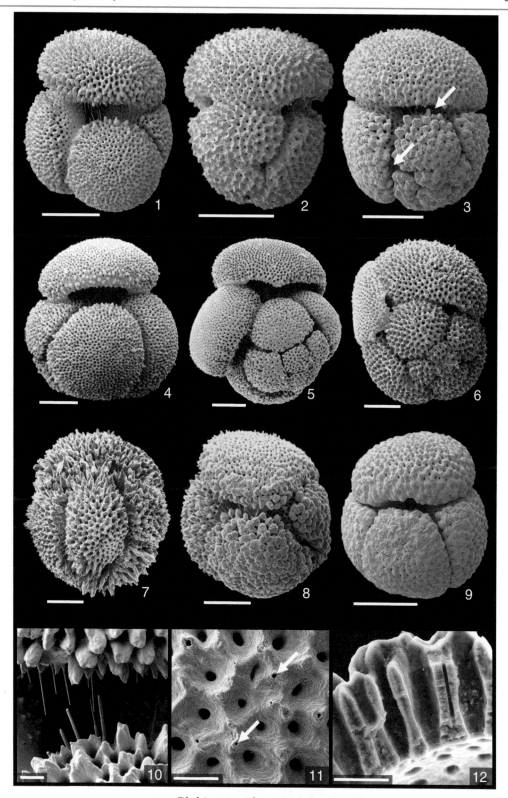

Globigerinoides conglobatus

Plate 2.8 (*1*) Live *Globigerinoides ruber* producing spines and pseudopodia. (*2–4*) Adult, (*5, 6*) juvenile, and (*7*) ▶
neanic *G. ruber*. (*8, 9*) Tests with newly built and thin-shelled final chambers. (*1–11*) Normal morphotype (sensu
stricto, s.s.), and (*12–16*) morphotype variants referred to as 'sensu lato' (s.l.), including (*12, 13*) elongate type ('*G.
elongatus*'), (*14, 15*) pyramidical type ('*G. pyramidalis*'), and (*16*) type with flat kummerform final chamber ('*G.
platys*'). (*17*) Cross-section of test wall showing calcite layers, plate growth, pores withprimary organic membrane
(POM), and spines lodged in the test wall above POM. (*18*) Outer test wall with pores and broken spines, one spine
being repaired (*arrow*). Bars of overviews 100 µm, (*5, 6*) 20 µm, close-ups 10 µm

2.2.8 *Globigerinoides ruber* (d'Orbigny 1839) (Plate 2.8)

Description: *Globigerinoides ruber* has a *ruber*-type wall structure. The test is medium to high trochospiral, with 3 globular chambers per whorl in the adult stage. Ultimate, penultimate, and antepenultimate chambers adjoin the umbilical primary aperture. Two smaller secondary apertures are formed on the spiral side. *Globigerinoides ruber* has normal-sized pores.

Globigerinoides ruber exhibits two phenotypes, a white (*G. ruber*w), and a pink variety (*G. ruber*p) stained by so far unclassified pigments. On average, *G. ruber*p grows about 50 µm larger than the white variety. The white variety is extant in all modern ocean basins. The pink variety became extinct in the Red Sea, and the Indian and Pacific Oceans, in the late Pleistocene (MIS 5.5, around 125 ka), and persisted in the Atlantic Ocean and Mediterranean Sea (Thompson et al. 1979).

The morphospecies of *G. ruber* originally described by d'Orbigny (1839) was referred to as *G. ruber* sensu stricto (s.s.) by Wang (2000). The morphotype *G. ruber* s.s. is symmetrical with spherical chambers formed by the adult specimen, and a high arched primary aperture (Plate 2.8). A second morphotype, *G. elongatus* (d'Orbigny 1826) is referred to as *G. ruber* sensu lato (s.l.) by Wang (2000), and includes compact tests with non-spherical, slightly compressed chambers formed by the adult specimen, and which results in a relatively small primary aperture (Plate 2.8).

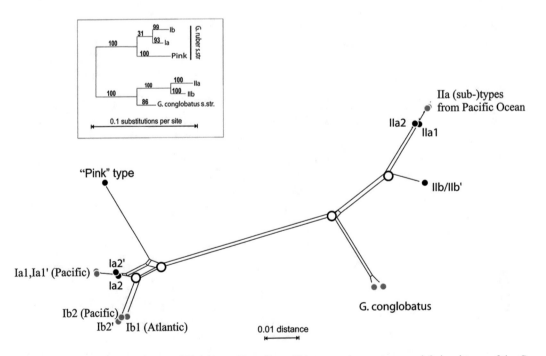

Fig. 2.7 Graphical distances (unrooted Neighbour-Net splitgraph) between six genotypes and their subtypes of the *G. ruber* plexus based on SSU rDNA data (from Aurahs et al. 2009b, 2011)

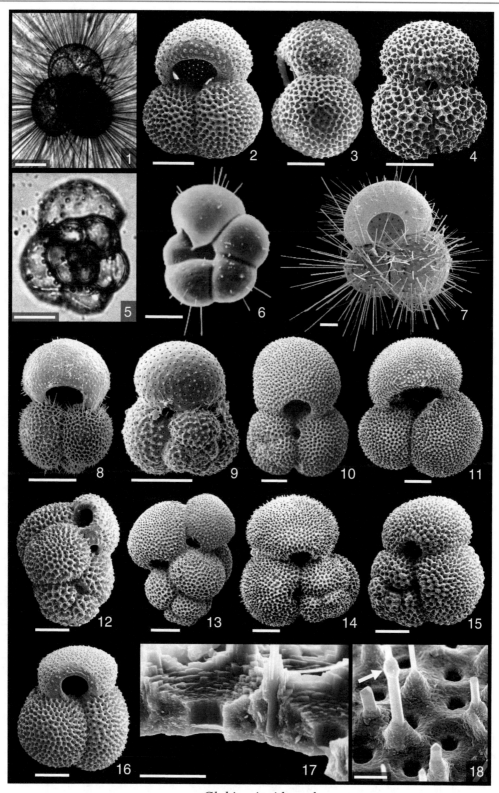

Globigerinoides ruber

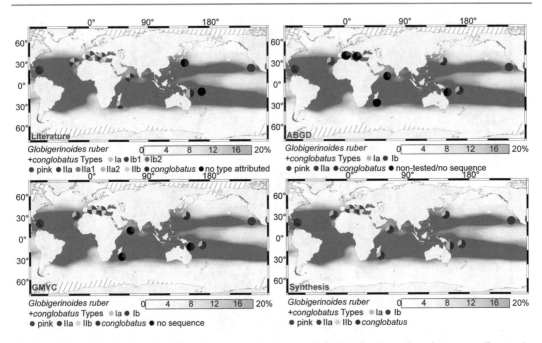

Fig. 2.8 Geographic distribution of genetic types and subtypes of *Globigerinoides ruber* plexus according to the data-synthesis of André et al. (2014)

A third elongate type with a high-trochospiral test named *G. pyramidalis* by van den Broeck (1876, see also Saito et al. 1981) is included in *G. ruber* s.l. (Wang 2000) in the Mediterranean (Numberger et al. 2009) (Plate 2.8).

Molecular genetics: Three morphotypes and 4 cryptic species (genotypes) of *G. ruber*$_w$ are described so far, whereas *G. ruber*$_p$ appears to be formed by only 1 genotype (Darling and Wade 2008; Numberger et al. 2009; André et al. 2014). Following Darling and Wade (2008), 4 genotypes of *G. ruber* may have sympatric or allopatric distribution patterns. According to other analyses (Fig. 2.7; Aurahs et al. 2009b, 2011) the *G. ruber* plexus comprises six genetically defined types in literature, including *G. conglobatus* (Ia, Ib, IIa, IIb, pink, and *conglobatus*), as well as several subtypes (Fig. 2.7). Most of these types are restricted to certain ocean basins (Fig. 2.8). For example, the pink type occurs only in the Atlantic and Mediterranean Sea; Type IIb in the Mediterranean Sea, including Subtypes IIa1 and IIa2; Type Ib Indo-Pacific and Carribean, whereas Ia, Ib, and IIa occur ocean-wide. New

sequences and data from literature were revisited and analyzed using ABGD and GMYC (Fig. 2.9). They offer a synthesis of the *G. ruber* plexus, which confirms the six genotypes of Aurahs et al. (2011). Type pink is characterized by its reddish staining, *G. conglobatus* is clearly distinguishable as a separate morphospecies, and type IIa is known in literature as *G. elongatus* (Aurahs et al. 2011).

Ecology: *Globigerinoides ruber* is the most frequent species in tropical to subtropical waters of the global ocean (e.g., Bé 1977). *Globigerinoides ruber* bears dinoflagellate symbionts similar to those occurring in other *Globigerinoides* species and *O. universa* (Hemleben et al. 1989). Accepting slightly higher ratios of phytoplankton prey than the other modern globigerinoid species (Anderson 1983) (Sect. 4.2.5), *G. ruber* seems to be very adaptable to varying ecological conditions among the modern *Globigerinoides* species. *Globigerinoides ruber*$_w$ may be abundant from upwelling (eutrophic) regions to subtropical (oligotrophic) gyres (Kemle-von-Mücke and Hemleben 1999; Schiebel et al. 2004).

Fig. 2.9 Ultrametric tree based on SSU rDNA of spinose species including *G. ruber* and related species, with significant GMYC delimitations. *Colored branches* correspond to GMYC clusters. The *outer circle* corresponds to the names of the morphospecies. Plausible biological species are given on the inner arc. From André et al. (2014)

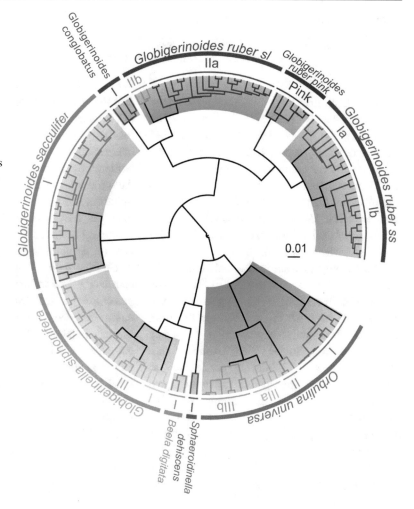

Both pink and white varieties appear to range among the shallowest dwelling planktic foraminifers out of all modern species (e.g., Bé 1977). However, the depth distribution of *G. ruber* varies according to regional ecological conditions. Whereas usually shallow dwelling, *G. ruber* may occur at nutricline depths in less turbid oligotrophic waters (Schiebel et al. 2004). Wang (2000) describes a depth-divide at about 30 m water depth in the South China Sea, with *G. ruber* s.s. dwelling in the upper mixed layer, and *G. ruber* s.l. dwelling below *G. ruber* s.s. in the deeper mixed layer.

Globigerinoides ruber has been found to be the most tolerant species to low Sea Surface Salinity (SSS), caused by continental fresh water runoff into the ocean (Deuser et al. 1988; Guptha et al.

1997; Ufkes et al. 1998; Schmuker 2000; Schmuker and Schiebel 2002; Rohling et al. 2004). The overall wide temperature (14–31 °C) and salinity (22–49 PSU) limits within which *G. ruber* accepts food and reproduces in laboratory cultures (Hemleben et al. 1989; Bijma et al. 1990b) may illustrate the eurythermal and euryhaline nature of planktic foraminifers, although largely variable at the species level (e.g., Bé and Tolderlund 1971; Lombard et al. 2009). Similarly wide temperature and salinity ranges as in *G. ruber* are reported for all of the species analyzed from culture experiments by Bijma et al. (1990b), i.e. *G. sacculifer*, *G. conglobatus*, *G. siphonifera*, *O. universa*, *N. dutertrei*, and *G. menardii*.

The distribution of the white forma extends further into temperate latitudes of the modern

Plate 2.9 *Globigerinoides sacculifer* (*1–3*) forma *trilobus*, (*4*) with kummerform chamber, (*5*) with newly built ▶ kummerform sac chamber, and (*6*) with fistulose chamber. (*7*) *G. sacculifer* forma *quadrilobatus*, (*8*) with sac chamber and final kummerform chamber, and (*9*) fistulose chamber. (*10–11*) *G. sacculifer* forma *sacculifer*, i.e. with final sac chamber. (*12*) *Sphaeroidinella dehiscens* for comparison. *Bars* 100 μm

Atlantic Ocean than that of the pink form (Hemleben et al. 1989; Hilbrecht 1996). *Globigerinoides ruber*$_p$ may be considered a 'summer species' whereas *G. ruber*$_w$ occurs year-round in the tropical to subtropical ocean (e.g., Kemle-von-Mücke and Hemleben 1999). Modern *G. ruber*$_p$ are much less abundant than *G. ruber*$_w$ (Kemle-von-Mücke and Hemleben 1999, South Atlantic; Schmuker and Schiebel 2002, Caribbean; Rigual-Hernández et al. 2012, Mediterranean).

The quantity of tests in the underlying sediment (e.g., van Leeuwen 1989; Kemle-von Mücke and Oberhänsli 1999) suggests that production of tests of *G. ruber*$_w$ results from enhanced (fortnightly) reproduction frequency (cf. Berger and Soutar 1967; Almogi-Labin 1984; Bijma et al. 1990a) in comparison to other shallow-dwelling species (monthly reproduction), as well as the wide acceptance of different food sources. *Globigerinoides ruber* may survive changing ecological conditions for a considerable time when carried towards higher latitudes by currents, and frequently occurs in sediments beyond latitudes of their ecological limits (Mojtahid et al. 2013).

Further readings: Christiansen (1965), Berger (1969), Orr (1969), Glaçon and Sigal (1969), Vergnaud Grazzini et al. (1974), Hecht (1974), Kennett (1976), Brummer et al. (1987), Brummer and Kroon (1988), Gastrich and Bartha (1988), Robbins and Healy-Williams (1991), Oberhänsli et al. (1992), Kemle-von Mücke (1994), Kroon and Darling (1995), Ortiz et al. (1995), Wang et al. (1995), Mulitza et al. (2004), Steinke et al. (2005).

2.2.9   *Globigerinoides sacculifer* (Brady 1877) (Plates 2.9 and 2.10)

O.G.A.: *Trilobatus* Spezzaferri et al. 2015

Description: Low trochospiral test with >3–4 spherical chambers in the last whorl. Final chamber may be elongate and sac-like (*sacculifer*), lobulate (*trilobus*), or rather small (kummerform). Umbilicus narrow, primary aperture interiomarginal, umbilical, forming a distinct arch bordered by a rim. Secondary apertures occur on the spiral side. Sutures slightly curved and incised. Spines are round to slightly triangular. All morphotypes have a *sacculifer*-type wall structure, exhibiting a regular honeycomb-like surface pattern, i.e. regular sub-hexagonal pore pits.

Globigerinoides sacculifer includes four common morphotypes: *Globigerinoides quadrilobatus* (d'Orbigny 1846), *Globigerinoides trilobus* (Reuss 1850), *Globigerinoides sacculifer* (Brady 1877), and *Globigerinoides immaturus* (Leroy 1939), which are produced by only 1 genotype (André et al. 2013). Modern *G. trilobus* and *G. sacculifer* are ubiquitous in the global ocean whereas *G. quadrilobatus* and *G. immaturus* appear to be limited to the Indian and Pacific Oceans (e.g., Hecht 1974; André et al. 2013). A fifth morphotype, *Globigerinoides fistulosus* (Schubert 1910) is rather rare. Fistulose and sac-like final chambers are also formed by the other morphotypes (Plate 2.9). A clear distinction between morphotypes may hence be impossible. Although the morphotype *G. sacculifer* with a sac-like final chamber was described as late as 1877 by Brady, and hence later than *G. quadrilobatus* (1846) and *G. trilobus* (1850), *sacculifer* is kept as the valid species name because it best includes the entire range of morphotypes including mixed types with features of more than one of the five above given morphotype end-members (Plate 2.9).

Forma *G. trilobus* (Reuss 1850): Test low trochospiral, with just over three globular chambers in the last whorl, umbilical aperture forming a narrow arch over antepenultimate chamber, and two to three secondary apertures (one per chamber) on the spiral test side.

Forma *G. immaturus* Leroy 1939: Test low trochospiral, with three and a half globular

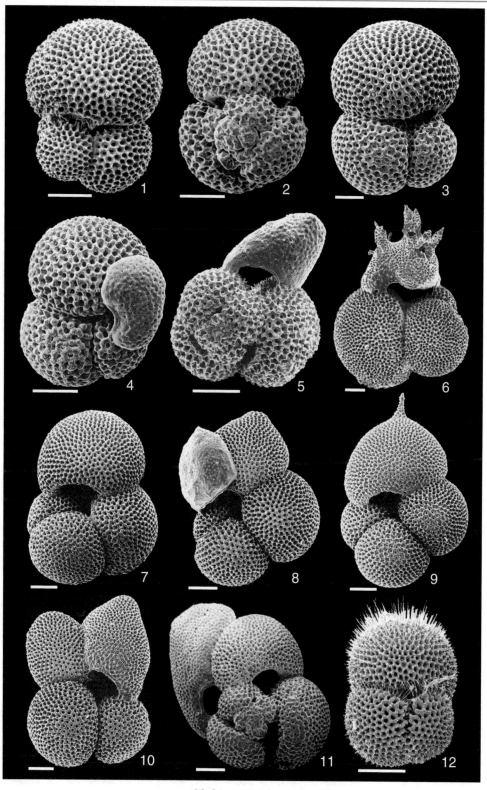

Globigerinoides sacculifer

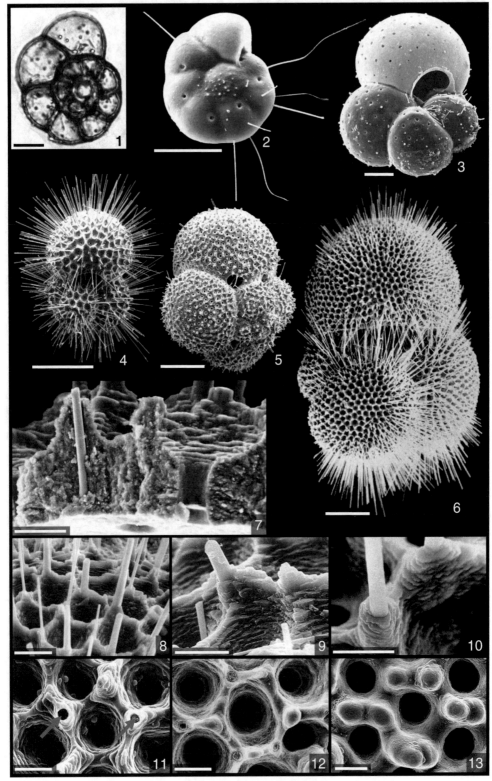

Globigerinoides sacculifer

◄ **Plate 2.10** (*1*) Light micrograph of neanic *Globigerinoides sacculifer*. (*2–6*) Ontogenetic development of *G. sacculifer* from (*2*) juvenile, to (*3*) neanic, and (*4–6*) adult stage. (*7*) Cross-section of test wall with round spine lodged in spine hole, and pore with remains of pore plate. (*8–10*) Round and triangular spines with spines collars. (*11*) Broken spines and spine holes. (*7, 9–11*) Terraced plate-like crystals covering outer test wall. (*11*) Spine holes (*arrows*), (*12*) partly, and (*13*) entirely covered by GAM calcification. Bars (*1–3*) 20 µm, (*4–6*) 100 µm, (*7–13*) 10 µm

chambers in the last whorl, umbilical aperture forming a narrow arch over penultimate and antepenultimate chamber, two secondary apertures on the spiral side of test.

Forma *G. quadrilobatus* (d'Orbigny 1846): Four chambers in the last whorl, rather large aperture centered over the antepenultimate chamber.

Forma *G. sacculifer* (Brady 1877): The final chambers may be different in morphology, exhibiting up to three elongated, sac-like chambers, and often adding a kummerform chamber prior to gametogenesis. Large supplementary apertures on spiral test side. Forma *G. suleki* Bermudez (1961) is considered a variant of *G. sacculifer*.

Forma *G. fistulosa* (Schubert 1910): Final chamber shows a tendency towards forma *G. sacculifer* but exhibits one or more finger-like extensions on the final chamber, which are massive and shows pores. This variety is rare in the water column and sediments. Several specimens growing the finger-like projections had been kept in culture.

Molecular genetics: *Globigerinoides sacculifer* represents only one genotype (i.e. no cryptic species), despite its highly variable adult test morphology (see above). André et al. (2013) found strong "reduced genetic variation within the plexus and no correlation between genetic and morphological divergence, suggesting taxonomical overinterpretation".

Ecology: *Globigerinoides sacculifer* is an abundant tropical to subtropical surface dweller (e.g., Bé 1977; Schmuker and Schiebel 2002). It is one of the most investigated planktic foraminifer species in laboratory culture, and a large amount of experimental ecological data are available for this species (e.g., Hemleben et al. 1977; Spero and Lea 1993). *Globigerinoides*

sacculifer bears dinoflagellate symbionts, feeds mostly on calanoid copepods, and reproduces on a synodic lunar cycle (Hemleben et al. 1989; Bijma et al. 1990a; Erez et al. 1991). *Globigerinoides sacculifer* is a euryhaline species tolerating salinities between 24 and 47 PSU, and temperatures ranging from 14 to 32 °C (Bijma et al. 1990b). *Globigerinoides sacculifer* is one of the most frequent species in oligotrophic surface waters (e.g., Naidu and Malmgren 1996a; Conan and Brummer 2000; Schiebel et al. 2004). Mass flux events of *G. sacculifer* tests may be triggered by favorable ecological condition and cyclic reproduction (Schiebel 2002).

Remarks: The ontogenetic development of *G. sacculifer* might serve as an example for the complex succession of trophic changes, symbiont activity, and test formation (Brummer et al. 1987). *Sphaeroidinella dehiscens* (Parker and Jones 1865) with a honeycomb-like surface texture may resemble *G. sacculifer* (Plate 2.9). However, the ontogeny (including proloculus) and chamber size, as well as the depth habitat (below thermocline in *S. dehiscens*), are different between the two species.

Further readings: Hecht (1974), Scott (1974), Anderson and Bé (1978), Bé (1980), Caron et al. (1982), Duplessy et al. (1981), Bé et al. (1983), Erez (1983), Caron and Bé (1984), Bouvier-Soumagnac and Duplessy (1985), Caron et al. (1987); Brummer et al. (1987), Hemleben et al. (1987), Martinez et al. (1998), Eggins et al. (2003), Mulitza et al. (2004), Williams et al. (2006), Lin and Hsieh (2007), Yamasaki et al. (2008), Lombard et al. (2009, 2011), Dueñas-Bohórquez et al. (2011), Coadic et al. (2013), Schmidt et al. (2013), André et al. (2013), Spezzaferri et al. (2017).

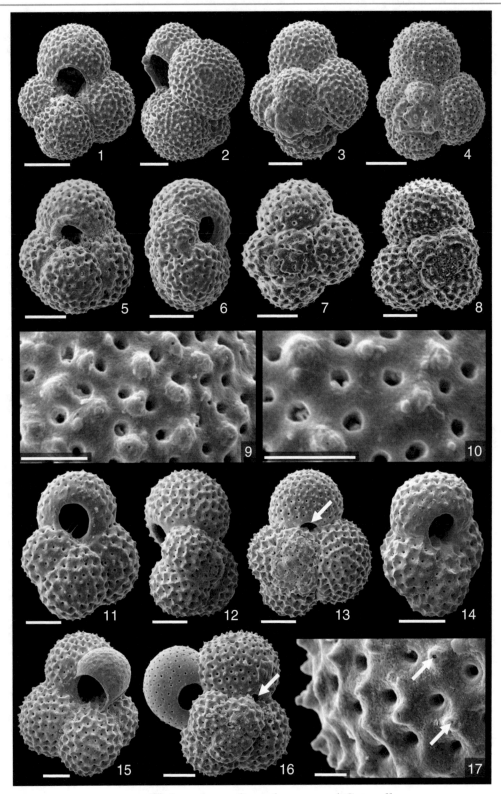

Globoturborotalita rubescens and *G. tenella*

◀ **Plate 2.11** *(1–8) Globoturborotalita rubescens* with *(9, 10)* normal-sized pores and spine-collars showing signs of early GAM calcification. *(11–16) Globoturborotalita tenella* with *(17)* test surface showing signs of corrosion, normal-sized pores, spine-holes *(arrows)*, and spine-collars showing signs of very early GAM calcification. Note the secondary aperture *(arrow* in 13 and 16) on spiral side of *G. tenella*. *(2, 13, 15, 16)* Specimens with kummerform chambers. Bars of overviews 50 μm, bars of close-ups 10 μm

2.2.10 *Globoturborotalita rubescens* Hofker 1956 (Plate 2.11)

Description: *Globoturborotalita rubescens* is a rather small species not much exceeding 250 μm. The normal perforate trochospiral test, with a *ruber*-type wall structure, and 4 globular chambers in the last whorl, has a rather large umbilical aperture, which is arched with a thick rim over the penultimate and antepenultimate chambers. The entire test is reddish pigmented.

Molecular genetics: No data available.

Ecology: *Globoturborotalita rubescens* is ubiquitous in tropical to temperate surface waters (Parker 1962). Usually occurring at moderate standing stocks, *G. rubescens* may be more frequent on a regional scale (Hemleben et al. 1989).

Remarks: Modern *G. rubescens* are rather easy to distinguish from other species by its reddish pigment distributed throughout the test. In contrast, in the pink variety of *G. ruber* usually only the inner whorl is colored reddish. A complete whorl in *G. rubescens* always bears four chambers, and shows no secondary apertures on the spiral side, while *G. ruber*p has clearly visible secondary apertures, and three chambers per whorl. A single secondary aperture on the spiral side of the test of *G. tenella* may be the only feature to distinguish *G. tenella* from *G. rubescens*. Non-pigmented tests of *G. rubescens* are frequently found in bottom sediments underlying temperate waters (e.g., Parker 1962; Hemleben et al. 1989). In the case where tests of *G. rubescens* are not stained red, *G. rubescens* may be distinguished from *G. tenella* only by its secondary aperture.

Further readings: Hofker (1976), Vincent (1976), Schmuker and Schiebel (2002), Seears et al. (2012).

2.2.11 *Globoturborotalita tenella* (Parker 1958) (Plate 2.11)

Description: This species is similar to *G. rubescens* in size, *ruber*-type wall structure, and 4 chambers in the last whorl. The primary aperture has an umbilical position and is often rather high ('loop-shaped'). A small secondary aperture of the final chamber is formed on the spiral test side.

Molecular genetics: No data available.

Remarks: Tests of *G. tenella* are colorless and lack the reddish pigmentation present in most *G. rubescens*. Pre-adult stages of *G. tenella* are difficult to distinguish from *G. rubescens* and *G. ruber* (Hemleben et al. 1989).

Ecology: *Globoturborotalita tenella* occurs in low standing stocks in tropical and sub-tropical and even temperate waters, and is usually sympatric with *G. rubescens* and *G. ruber* (cf. Parker 1962; Schmuker and Schiebel 2002; Yamasaki et al. 2008).

Further readings: Kennett and Srinivasan (1983), Chaisson and Pearson (1997), Chaisson and d'Hondt (2000).

2.2.12 *Orbulina universa* d'Orbigny 1839 (Plate 2.12)

Description: *Orbulina universa* is the only modern species with a spherical test formed at the terminal ontogenetic stage. Large and small openings ('pores') are evenly distributed over the test wall. The large openings act as apertures, and allow exchange of food and other particles including symbionts and cytoplasm. The small openings bear a membrane, and serve the same function as real pores (Spero 1988, and

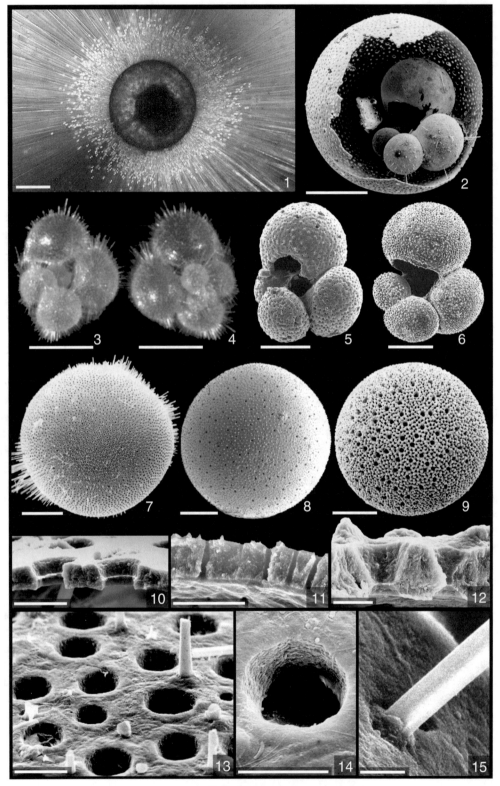

Orbulina universa

◀ **Plate 2.12** (*1*) Live *Orbulina universa* with corona of symbionts (Kage Microphotography©, with permission). (*2*) Pre-adult trochospiral test inside of broken adult spherical test. (*3, 4*) Light micrographs, and (*5, 6*) SEM images of pre-adult tests. (*7, 10*) Adult thin-shelled test with newly formed spherical chamber with large pores. (*8, 11*) Thin-shelled test with small pores. (*9, 12*) Thick-shelled test with funnel-shaped pores. (*7–9*) Apertures (*large openings*) and pores (*small openings*). (*13*) Spines with round, triangular, and triradiate bases. (*14*) Pore and multiple layers of calcite. (*15*) Triradiate spine lodged in test wall. Bars of overviews (*1, 2, 7, 8, 9, 15*) 200 μm, (*3–6*) 100 μm. Bars of close-ups 10 μm. (*3,4*) Courtesy A. Movellan

references therein). During ontogeny, *O. universa* changes its gross architecture from pre-adult trochospiral to adult spherical tests. Pre-adult tests have normal pores and a very thin, smooth, and fragile test wall. Adult spherical tests form tests walls of varying thickness and porosity (Plate 2.12).

Molecular genetics: According to André et al. (2014), the three genotypes I, II, and III are recognized to date. The Type III may be split into two the Subtypes IIIa and IIIb, which are very closely related (Fig. 2.9). Between Types I, II, and III, no overlap in distances to each other within inter and intra species level is observed (De Vargas et al. 1999; André et al. 2014). The three types are regionally separated by their dominance as Caribbean species (Type I), Sargasso species (Type II), and Mediterranean species (Type III). All three types are probably related to certain water bodies and trophic conditions. The Mediterranean genotypes are mostly correlated with nutrient rich waters of the western Mediterranean Sea. In the eastern Atlantic and in the Indian Ocean, genotypes are related to frontal zones, and regions of enhanced productivty. The Sargasso and Caribbean species both occur under more oligotrophic conditions typical of stratified water masses of the subtropical gyres (Morard et al. 2009). However, all three types may occur together at various regions of the transitional to tropical ocean, independent of temperature at water depth.

The three genotypes of *O. universa* differ in the size of pores and apertures (De Vargas et al. 1999). *Orbulina universa* Type I (Caribbean) has large pores and a thick test wall (Fig. 2.10), Type III (Mediterranean) has small pores and a thin test wall, and Type II (Sargasso) has even smaller pores than Type III (De Vargas et al. 1999; Morard et al. 2009). Test porosity also correlates with ecological conditions including sea surface temperature (Bé et al. 1973). Size normalized shell weight, i.e. wall thickness, is difficult to assess in relation to genotypes, because *O. universa* continuously adds calcite to the same sphere (Spero 1988). Ecological and biological signals could hence interfere to some degree with morphometric specifications of the different genotypes of *O. universa*.

Ecology: *Orbulina universa* tolerates wide ranges of ambient water salinity and temperature, and is abundant from tropical to temperate waters (e.g., Hemleben et al. 1989 and references therein; Bijma et al. 1990b; De Vargas et al. 1999; Chapman 2010). *Orbulina universa* might even occur at high latitudes when being transported poleward by currents. Test size of *O. universa* seems to be related to temperature as well as food, and hence the trophic state of surface waters at a regional scale (Bé et al. 1973; Spero and DeNiro 1987). *Orbulina universa* is mostly carnivorous, particularly during its spherical adult ontogenetic stage. Pre-adult stages may prefer herbivorous diet (Anderson et al. 1979). '*Biorbulina*' tests are formed when individuals are 'overfed' in laboratory culture. *Orbulina universa* has been widely employed in different kinds of laboratory experiments, to analyze the effect of hydrologic parameters and symbionts on shell calcification, isotope ratios, and Me/Ca ratios (e.g., Spero et al. 1997; Bemis et al. 1998). Stable isotopic ($\delta^{13}C$ and $\delta^{18}O$) differentition of two morphotypes (thin-shelled and thick-shelled) and genotypes of *O. universa* from the Cariaco Basin are caused by different environmental conditions (Marshall et al. 2015). Reproduction of *O. universa* seems to follow the synodic lunar cycle.

Remarks: Whereas spherical adult tests of *O. universa* are easy to identify, pre-adult tests of *O.*

Plate 2.13 (*1–11*) *Orcadia riedeli.* (*10, 11*) Distal parts of chambers with pores, and round and triangular spine bases. ▶ Note concentration of spines at distal parts of chambers. For comparison: (*12*) *Hastigerina digitata* with (*13*) details of surface of 4th last chamber showing triradiate spines. (*14, 15*) *Turborotalita quinqueloba* with even distribution of spines (and spine bases) on the entire test wall. Bars of overviews 50 μm, (*12*) 200 μm. Bars of close-ups (*10, 11*) 10 μm, (*13*) 50 μm. (2,5,6) Courtesy J. Meilland

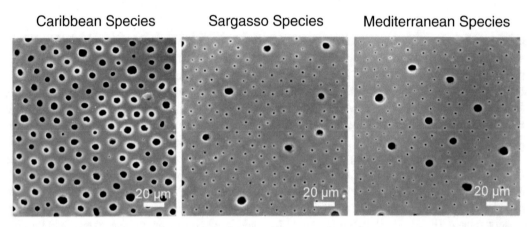

Fig. 2.10 Porosity of *O. universa* Type I (Caribbean species), Type II (Sargasso species), and Type III (Mediterranean species), after Morard et al. (2009)

universa are similar to the tests of other species. During its ontogeny, *O. universa* may attain four to five different stages and test morphologies (cf. Spero 1988).

1. The *Turborotalita*-like juvenile stage of up to six chambers plus proloculus, i.e. the first whorl of the test.
2. The *Globigerina*-like neanic stage. Pre-adult tests of *O. universa* resemble those of *G. bulloides*, but the test of *O. universa* is less rugose and more transparent. Secondary apertures on the spiral side of neanic *O. universa* may be small and difficult to discern under the incident light microscope, though.
3. The *Globigerinoides*-like adult stage. Occasionally small secondary apertures on the spiral side are difficult to discern under the incident light microscope.
4. The *Orbulina* (spherical) terminal stage, formed by the mature individual. The spherical test might serve as protective envelope for cytoplasm and gametes during reproduction (Caron et al. 1987; Spero 1988).
5. The *Biorbulina*-like ecophenotype of the final ontogenetic stage. The more food offered to *O. universa* in laboratory experiments, the larger the spherical test grows. An excess store of energy through high food availability might lead to the formation of a double sphere, forming so-called *Biorbulina bilobata* d'Orbigny 1846 (Hemleben et al. 1989). Those *B. bilobata* have been frequently observed in eutrophic regions like the Arabian Sea during the southwest monsoonal upwelling (Spero 1988; cf. Rossignol et al. 2011).

Further readings: Rhumbler (1911), Robbins (1988), Bijma et al. (1992), Lea and Spero (1992), Spero (1992), Lea et al. (1995, 1999), Ortiz et al. (1995), Hilbrecht and Thierstein (1996), Mashiotta et al. (1997), Rink et al. (1998), Bemis et al. (2000), Schiebel et al. (2001), Eggins et al. (2004), Köhler-Rink and Kühl (2005), Asahi and Takahashi (2007), Hamilton et al. (2008), Ripperger et al. (2008), Lombard et al. (2009), Tsuchiya (2009), Chapman (2010), Friedrich et al. (2012), Morard et al. (2013).

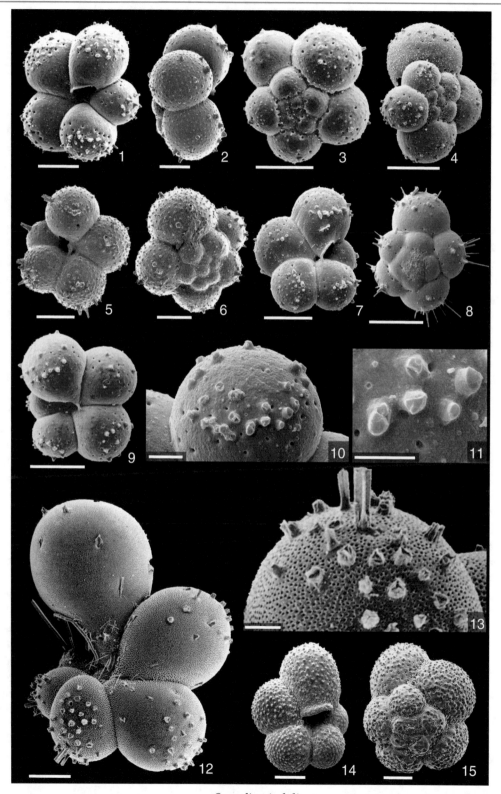

Orcadia riedeli

Plate 2.14   (*1–6*) Adult *Sphaeroidinella dehiscens* with increasing calcite cortex on the outer test from (*1*) to (*6*). (*4*) ▶
Broken test with spines lost at the outside and present at the inside (see Fig. 6.5). (*7*) Calcite addition from left to right
on top of same chamber. (*8*) Remains of spines in outer test wall. (*9*) Pores and smooth calcite layer covering test.
(*10, 13*) Epitactic crystals (*red arrows*) not to be confused with spines (*white arrows*). (*11*) Diagenetic overgrowth, and
(*12*) cross-section of fossil test wall with pores. Bars of overviews 100 μm, bars of close-ups 20 μm

2.2.13   *Orcadia riedeli* (Rögl and Bolli 1973) (Plate 2.13)

O.G.A.: *Hastigerinella* Cushman 1927.

Description: The test of *O. riedeli* is low trochospiral and small sized with an average of 5
thin-walled chambers in the last whorl. Chambers of adult specimens may develop an ampullate shape. The umbilical aperture is high-arched
and bears a small rim. Pores of normal size are
located distally, as well as along sutures on the
spiral side of test. Proximal chamber walls on the
umbilical side of tests are smooth and largely
lack pores. Thin and round spines occur next to
thick and triangular spines at the peripheral
(distal) chamber wall.

Molecular genetics: No data available.

Ecology: *Orcadia riedeli* is a cosmopolitan
though rare species, and dwells in the surface
tropical to polar ocean. *Orcadia riedeli* occurs in
the temperate eastern North Atlantic Ocean (5 m
waters depth), and the Atlantic Sector of the
Southern Ocean (Holmes 1984, and references
therein). Brummer et al. (1988) attribute *O. riedeli* to rather high-productive waters in the high
latitude North Atlantic. In the Indian Ocean
Sector of the Southern Ocean, *O. riedeli* occurs
at up to 7 % of the live planktic foraminifer
assemblage (J. Meilland, University of Angers,
personal communication, 2015).

Further readings: Boltovskoy and Watanabe
(1981), Holmes (1984).

2.2.14   *Sphaeroidinella dehiscens* (Parker and Jones 1865) (Plate 2.14)

O.G.A.: *Globigerinoides* Cushman 1927.

Description: The test of *S. dehiscens* is low
trochospiral and exhibits > 3 to 4 chambers in

the last whorl. The thick *sacculifer*-type wall
bears the same type of round spines as the *Globigerinoides* species. Pre-adult specimens are
similar to *G. sacculifer*. Calcite is added to the
outer test wall during adult ontogeny, and pores
and aperture become increasingly narrow from
the proximal to distal parts of chambers. The
irregular edge of the outer calcite layer along the
sutures and aperture form a slit-like discontinuous depression (looking broken) between chambers of the final whorl, distinguishing *S.
dehiscens* from *G. sacculifer*. When sinking to
the lower mixed layer, *S. dehiscens* loses its
spines starting from the proximal parts of
chambers.

Molecular genetics: Newly attained data of *S.
dehiscens* reveal only one genotype.

Ecology: *Sphaeroidinella dehiscens* is a very
rare tropical to subtropical species. Adult individuals dwell in subsurface waters. *Sphaeroidinella dehiscens* hosts the same dinoflagellate
symbionts as the *Globigerinoides* species. Similarity of *S. dehiscens* to the much more frequent
G. sacculifer, as well as depth habitat of the
former species may add to the fact that *S.
dehiscens* has been reported rare.

Remarks: *Sphaeroidinella dehiscens* might be
confused with encrusted adult *Globigerinoides
sacculifer* of the *trilobus* morphotype. The vertical distribution of *S. dehiscens* in the subsurface
water column (Hemleben et al. 1989) may add to
the confusion by suggesting that a calcite crust or
GAM calcite was formed on top of the outer test
wall of *G. sacculifer*. The juvenile test morphology including the size of proloculus and
chamber arrangement confirm the discrimination
between the two species *S. dehiscens* and *G.
sacculifer* (see Postuma 1971; Hemleben et al.
1989, and references therein).

Further readings: Bé and Hemleben (1970);
Huang (1981); Bolli et al. (1985).

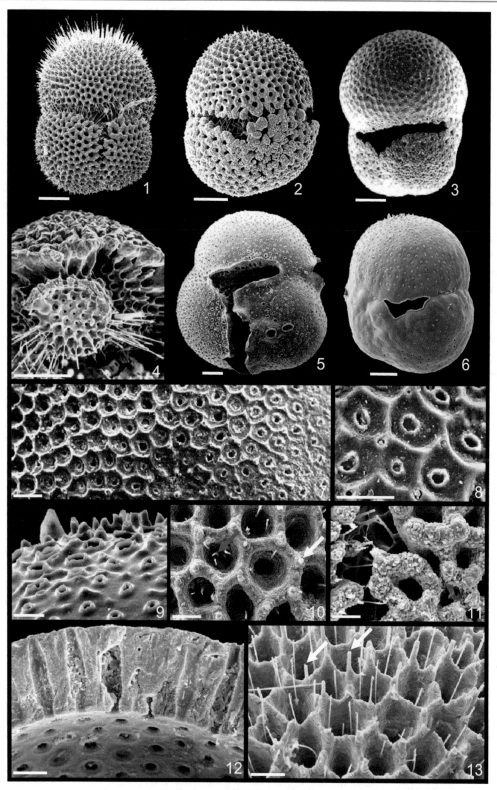

Sphaeroidinella dehiscens

Plate 2.15 (*1–8*) *Turborotalita clarkei* (*1–6*) without and (*7, 8*) with calcite crust. (*9–20*) *Turborotalita humilis.* (*9*) ▶
Light micrograph of live *T. humilis* with symbionts. (*10–19*) Different stages of encrustation, with (*18*) egg-shaped
specimen with outer calcite layer partly taken off. (*20*) *T. humilis* morphotype *T. cristata. Bars* 50 µm

2.2.15 *Turborotalita clarkei* (Rögl and Bolli 1973) (Plate 2.15)

Description: *Turborotalita clarkei* ranges among the smallest modern species (<150 µm). The low trochospiral test shows 4.5 chambers per whorl, and is normal perforate. The final chamber shows the tendency to develop an ampullate shape. The aperture stretches from the umbilicus towards the periphery. The surface is smooth and the spines are rather thin, and placed at distal parts of chambers, similar to *T. humilis*.

Molecular genetics: No data available.

Ecology: *Turborotalita clarkei* is a tropical to temperate species, living at surface to subsurface depths, and below the thermocline. It is presumed to dwell at greater water depth with increasing ontogenetic age. Pre-adult surface-dwelling individuals lack heavy calcite crusts, which are usually formed at depth (Hemleben et al. 1989). When bearing a thick calcite crust, the test of *T. clarkei* is more resistant to dissolution than that of most other species, and is occasionally frequent in the fine fractions of tropical to temperate sediments.

Remarks: The adult *T. clarkei* is difficult to distinguish from the pre-adult stages of *T. quinqueloba*. However, *T. clarkei* tests are smaller than those of *T. quinqueloba*. The terminal ontogenetic stage of *T. quinqueloba* exhibits a pronounced umbilical flange, which may be useful for differentiation from *T. clarkei*.

Further readings: Brummer and Kroon (1988).

2.2.16 *Turborotalita humilis* (Bardy 1884) (Plate 2.15)

Description: *Turborotalita humilis* forms a very low trochospiral small (<250 µm) test with 6–8 chambers in the last whorl. The outline is lobulate and almost circular, and the chambers are globular to ovate. The final chamber often has the tendency to become ampullate and forms a tongue-like flap over the umbilicus. Spines are distributed over the entire test and concentrated distally. *Turborotalita humilis* bears *Globigerina*-type (i.e. round) spines often with conical spine collars. The aperture (interiomarginal to umbilical-extraumbilical) starts at the periphery and opens into a rather deep umbilicus, leaving an open space at the penultimate and antepenultimate chambers (infralaminal apertures), and may bear a small lip. Sutures are radially depressed. Pores arec <1.5 µm in diameter and distally enlarged. When migrating from surface to subsurface waters during ontogeny, *T. humilis* may form a thick calcite crust giving tests an egg-like shape (Plate 2.15-18).

Molecular genetics: No data available.

Ecology: *Turborotalita humilis* is a tropical to subpolar surface dweller (e.g., Holmes 1984). *Turborotalita humilis* bears chrysophytes symbionts similar to *G. siphonifera* and *G. glutinata*. Large numbers of *T. humilis* were sampled from surface waters of the Azores-Front Current-System, i.e. the northern limit of the North Atlantic subtropical gyre in January 1998 (Schiebel et al. 2002). Similar blooms of *T. humilis* were observed during spring 1997, in surface waters off the Canary Islands (H. Meggers, Bremen University, personal communication), and in spring 2006 in the western Mediterranean Sea (Ch. Hemleben, RV Poseidon Cruise 334, March 2006).

Remarks: The ecological range of *T. humilis* is not entirely known. Due to its small size, *T. humilis* might have been missed by plankton net sampling (usually using 100-µm mesh-size), and may be largely remineralized while settling through the water column at low velocity before arriving at the seafloor.

Turborotalita humilis may be a senior synonym of *Globigerina cristata* Heron-Allen and Earland 1929 (Plate 2.15-20), the latter described

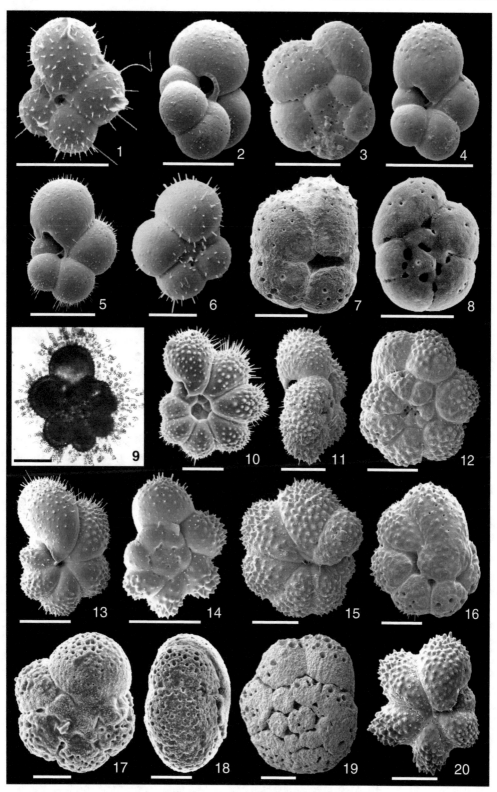

Turborotalita clarkei and *T. humilis*

Plate 2.16 (*1–17*) *Turborotalita quinqueloba* at different stages of encrustation, and formation of calcite cortex. (*6*) ▶
Specimen with apertural rim, (*4, 7, 8*) apertural flaps, and (*12*) bulla-like flap with multiple openings. (*14, 17*) Encrusted
egg-shaped specimens, with (*14*) the outer calcite layers partly taken off. To avoid confusion with *T. clarkei* and *T.
humilis* see Plate 2.15 for comparison. (*18*) Smooth calcite layer partly covering pores and spine-holes. Bars of
overviews 50 μm, bar of close-up 10 μm

as smaller, having five to six more club shaped
chambers in the last whorl, and having more
heavily calcified spine-collars than *T. humilis*.
Molecular genetics should provide proof on the
differentiation between *T. humilis* and *T. cristata*.
Assuming that *T. humilis* and *G. cristata* are
synonyms, the species changes its depth habitat
during its ontogeny from surface waters to deeper and cooler waters. While descending to
depths, it sheds its spines, and grows a thick
calcite crust. A similar change in depth habitat
can be assumed for *S. dehiscens*, *T. clarkei*, and
occasionally *T. quinqueloba*.

2.2.17　*Turborotalita quinqueloba*
　　　　　(Natland 1938)
　　　　　(Plate 2.16)

O.G.A.: *Globigerina* d'Orbigny 1826.

Description: *Turborotalita quinqueloba* is a
small, low trochospiral species with 5 chambers
in the last whorl, and round spines. In the final
ontogenetic stage, an ampullate final chamber
may cover the umbilicus. The aperture has an
umbilical to extraumbilical position, and may
have a rim or apertural flange (Plate 2.16-6 and -
4). During the final ontogenetic stage, specimens
may migrate to deeper waters, and produce a
thick calcite crust, which results in an egg-like
test shape (Plate 2.15-14 and -17), and thus is
difficult to distinguish from *T. clarkei* and *T.
humilis*.

Molecular genetics: Two major genotypes
(Types I and II) and six subtypes (Ia, Ib, IIa, IIb,
IIc, IId; Figs. 2.11 and 2.12) of *T. quinqueloba*
are described from tropical to subtropical waters
(Type I) of the Indian and Pacific Oceans, and
subpolar to polar waters (Type II) in the Atlantic
and North Pacific, as well as subpolar Antarctic
waters (Darling et al. 2000; Darling and Wade
2008; André et al. 2014). Two cryptic species

(Types I and II) have been confirmed by using
ABGD and GMYC methods (Fig. 2.3), indicating that the differentiation of six types would be
an overestimation of the genotypic variability of
T. quinqueloba (cf. André et al. 2014).

Ecology: *Turborotalita quinqueloba* is one of
the most abundant species in the modern ocean.
Standing stocks of *T. quinqueloba* in the Arctic
Ocean reach up to several hundreds of specimens
(>63 μm) per cubic meter at the sea-ice margin
(Carstens et al. 1997), following an overall
enhanced primary production and food availability (Volkmann 2000a). The depth distribution
of *T. quinqueloba* displays the distribution of
water bodies, i.e. colder polar sourced waters
overlying warmer Atlantic sourced waters. In the
relatively warm Atlantic waters of the West
Spitzbergen Current, and in the Barents Sea, *T.
quinqueloba* may comprise up to 85 % of the
shallow-dwelling planktic foraminifer fauna
(Volkmann 2000a, b). Close to the sea-ice margin, highest standing stocks occur at 100–150 m
water depth, and on average at 50–100 m water
depth in open water at some distance from the ice
margin (Carstens et al. 1997; Volkmann 2000b).
Under the sea-ice, maximum standing stocks
were found as deep as 150–200 m water depth
(Carstens et al. 1997). In polar sourced waters of
western Fram Strait and outer Laptev Sea, *T.
quinqueloba* may form only 2–10 % of the
planktic foraminifer fauna dominated by *N.
pachyderma* (Volkmann 2000b). Together with
N. pachyderma, *T. quinqueloba* dominates the
overall small-sized cold-water assemblages
(Carstens et al. 1997; Volkmann 2000b; Schmidt
et al. 2004).

The relative overall abundance of *T. quinqueloba* decreases from high to low latitudes
(e.g., Parker 1962; Vincent and Berger 1981;
Kemle-von-Mücke and Hemleben 1999).
Simultaneous with decreasing relative abundance
of *T. quinqueloba* from high towards low

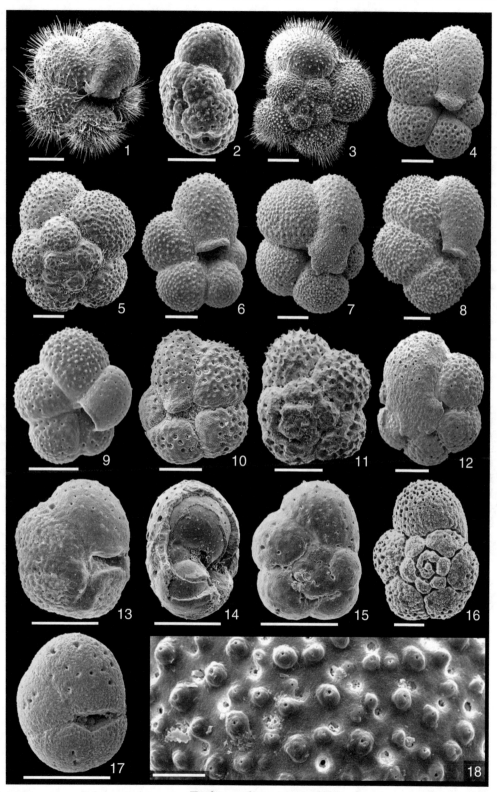

Turborotalita quinqueloba

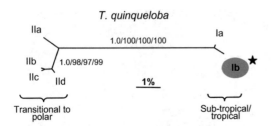

Fig. 2.11 Conventional delimitation of six "genotypes" of the morphospecies *Turborotalita quinqueloba* using 748 bp of the SSU rDNA. The phylogenetic tree is unrooted. The genetic distance equals 1 %. After Seears et al. (2012)

latitudes, the position of pores changes from a more peripheral to a more even distribution over the entire chambers (Hemleben et al. 1989). In low-latitude environments such as the Arabian Sea, *T. quinqueloba* is very rare (Schiebel et al. 2004). *Turborotalita quinqueloba* has been

found to be absent from the Red Sea (cf. Auras-Schudnagies et al. 1989). Tropical to subtropical waters of the Indian and Pacific Oceans, and polar to subpolar waters of the Atlantic and North Pacific, host two different genotypes of *T. quinqueloba*, Type I and Type II, respectively (Fig. 2.12).

A bimodal depth distribution of *T. quinqueloba* associated with surface water turbidity (among other environmental parameter) off the Columbia River (Washington State, USA) plume (Ortiz et al. 1995), as well as the tropical eastern Atlantic Ocean (Oberhänsli et al. 1992) may indicate the presence or absence of symbionts in *T. quinqueloba*. In case of absence of symbionts, the cytoplasm of *T. quinqueloba* has often been found to be colorless and transparent. Cyclical abundance of *T. quinqueloba* observed in the North Atlantic suggests a monthly reproductive

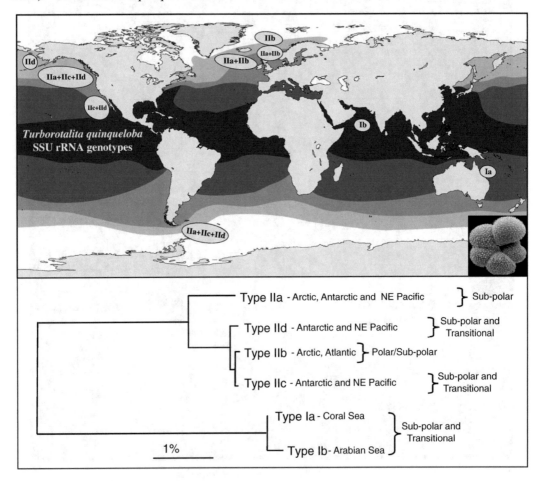

Fig. 2.12 Global distribution of two cryptic species including six subtypes of the morphospecies *T. quinqueloba*, displayed by a SSU rRNA genetic tree (from Darling and Wade 2008)

cycle, possibly triggered by the synodic lunar cycle (Volkmann 2000b).

Further readings: Bauch (1994), Simstich et al. (2003), Asano et al. (1968).

2.3 Monolamellar Spinose Species

2.3.1 *Hastigerina pelagica* (d'Orbigny 1839) (Plate 2.17)

O.G.A.: *Hastigerinella* Cushman 1927.

Description: The genus *Hastigerina* represents one morphospecies and three cryptic species. *Hastigerina pelagica* is unique among the planktic foraminifers. It shows an exceptional coiling sequence, which changes two times over the ontogenetic development, starting with a low trochospiral juvenile form, changing into a streptospiral neanic stage, and ending with a planispiral adult test (Hemleben et al. 1989). The wall structure is monolamellar instead of bilamellar as in all other modern planktic foraminifer genera. Among the spinose species, *Hastigerina* is the only one, which produces spines that are entirely triradial with small barbs on the edge. A unique cytoplasmic bubble capsule surrounds the test suspended from the spines, and extends up to 1.2 mm beyond the edge of the test of live specimens (e.g., Hull et al. 2011). The bubble capsule is an adaptive feature that possibly enables the foraminifer to digest enclosed prey more effectively (Hemleben et al. 1989, and references therein).

Molecular genetics: Genotypes I and II (including Subtypes IIa and IIb) of the cosmopolitan species *H. pelagica* comprises three cryptic species (Chap. 1, Fig. 1.1), which are separated by depth (Weiner et al. 2012). This kind of niche partitioning of a clearly defined morphospecies with two genotypic populations is reported for the first time among planktic foraminifers. Both subtypes of Type II (i.e. IIa and IIb) occur almost globally. These findings support the idea of depth-stratified populations,

and show that speciation does not only occur at the two-dimentional scale in surface waters. A third type (Type I) occurs only in surface waters of the W-Pacific and E-Mediterranean Sea.

Ecology: *Hastigerina pelagica* is exclusively carnivorous, and has never been observed to be associated with symbionts (Hemleben et al. 1989, and references therein). *Hastigerina pelagica* contains the highest size-normalized biomass of all modern planktic foraminifer species (Movellan 2013).

While regionally co-occurring (sympatric), different genotypes of *H. pelagica* are consistently occurring at different water depths (depth-parapatric, Weiner et al. 2012). *Hastigerina pelagica* Type I dwells above 100 m water depth, Type IIa below 100 m, and Type IIb is present from the sea surface to 700 m water depth. The depth-specific distribution pattern is similar to the distribution first described for *G. siphonifera* (Huber et al. 1997), and discussed as diversification and speciation in vertically structured populations ('vertical niche partitioning') by Weiner et al. (2012). *Hastigerina pelagica* is the only species, which evidentially exhibits a synodic lunar periodic reproductive cycle that persists when cultured in the laboratory (Spindler et al. 1979). While undergoing gametogenesis, *H. pelagica* resorbs septa and walls of the initial whorl (see Chap. 5).

Further readings: Alldredge and Jones (1973), Anderson and Bé (1976, 1978), Spindler et al. (1978), Hemleben et al. (1979), Hemleben and Spindler (1983).

2.3.2 *Hastigerinella digitata* (Rhumbler 1911) (Plate 2.17)

O.G.A.: *Hastigerina* Thomson 1876.

Description: *Hastigerinella digitata* is taxonomically closely related with *H. pelagica*, indicated by similarities in ontogeny, wall structure, spine morphology, cytoplasmic bubble

Hastigerina pelagica and *H. digitata*

◀ **Plate 2.17** (*1*) Adult live *Hastigerina pelagica* with cytoplasmic bubble capsule (incident light micrograph). (*2–3*) Live *H. pelagica* with well preserved test and spines. (*4*) Empty test of *H. pelagica* after reproduction, with partly resorbed septae, test wall, and spines. (*5*) Triradiate spines in front of aperture, and (*6–7*) spine with double-spiked hooks. (*8*) Cross-section of outer test walls with pores (view from outside of test), and remains of the pore-plate (*black arrow*) in upper left pore. Low bumps (*white arrows*) are bacteria. (*9*) *Hastigerinella digitata* with spine-remnants, and (*10*) mature live *H. digitata* with red cytoplasm. Bars of overviews 400 μm, bar close-up (*5*) 100 μm, (*6*) 10 μm, (*7, 8*) 2 μm

capsule, and septa resorption during gametogenesis (Banner and Blow 1960; Banner 1982; Hemleben et al. 1989). In contrast to *H. pelagica*, *H. digitata* forms increasingly elongate to finger-shaped (digitate) chambers during ontogeny, and a streptospiral mode of coiling during the late ontogenetic stages. For the distinction between *Hastigerinella* and *Hastigerina* see also Banner (1965).

Molecular genetics: *Hastigerinella digitata* is a sister-species of *Hastigerina pelagica* in the sense of molecular genetics as well as test morphology.

Ecology: *Hastigerinella digitata* is a rather rare subsurface to mesobathyal species of the tropical to subtropical global ocean. Off Bermuda, it was observed only once (Hemleben et al. 1989). Off Monterrey, California (USA), *H. digitata* was found to dominate the planktic foraminifer fauna in waters overlying the oxygen minimum zone (OMZ) between 280 and 358 m, and exceptionally occurring as deep as 1000–3512 m water depth (Hull et al. 2011, analyzing a video time-series survey). Having found only copepods attached to the spines of the observed specimens, Hull et al. (2011) confirm (from live collections) the carnivorous diet of *H. digitata*. Interannual or seasonal cyclicity in the distribution of *H. digitata* could not be detected from a 12-year-long time-series analyzed by Hull et al. (2011), and the reproductive cyclicity remains unknown so far. The lack in statistically significant signals might be explained by the rather low population densities of only one to two specimens per cubic meter of seawater off Monterrey (Hull et al. 2011).

Further readings: Spindler et al. (1979), Hemleben et al. (1979), Weiner et al. (2012).

2.4 Macroperforate Non-spinose Species

2.4.1 *Berggrenia pumilio* (Parker 1962) (Plate 2.18)

O.G.A.: *Globorotalia* Cushman 1927.

Description: *Berggrenia pumilio* is a small (max. 180 μm in diameter) species producing 4.5–6 chambers in the last whorl. The final chamber may be slightly ampullate. Tests with kummerform final chambers are frequent. The aperture has an extraumbilical-umbilical position. The umbilicus is narrow and deep. Typical narrow grooves mostly on the apertural side of test radiate from umbilicus and aperture towards the test periphery (Plate 2.18-1 and -5). Few pores may be scattered over more distal parts of the final three chambers. The surface of the test wall is smooth, and very small pustules may occur on the spiral side. Normal pustules are missing like in most other globorotaliids.

Molecular genetics: No data available.

Ecology: Rarely reported and possibly overlooked due to its small size and rather inconspicuous test characteristics, little information exists on the distribution and ecology of *B. pumilio*. *Berggrenia pumilio* was first described from surface sediments of the deep South Pacific by Parker (1962; cf. Saito et al. 1981).

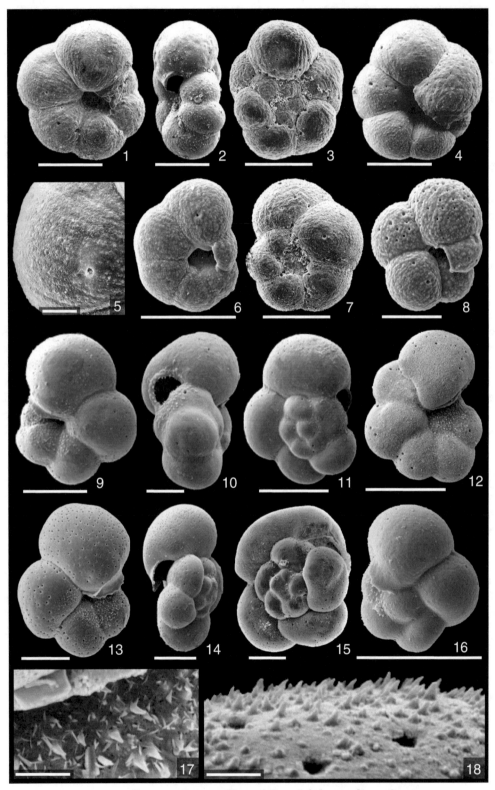

Berggrenia pumilio and *Dentigloborotalia anfracta*

◀ **Plate 2.18** (*1–8*) *Berggrenia pumilio* with (*5*) typical narrow grooves on umbilical test side. (*8*) Test with kummerform chamber. (*9–15*) Adult and (*16*) juvenile *Dentigloborotalia anfracta*. (*17*) Typical shark-teeth shaped pustules of *D. anfracta* off aperture and (*18*) on test surface. Bars of overviews 50 μm, bars close-ups (*5*) 10 μm, (*17, 18*) 3 μm

2.4.2 *Dentigloborotalia anfracta* (Parker 1967) (Plate 2.18)

O.G.A.: *Turborotalita* Blow and Banner 1962, *Tenuitella* Fleisher 1974.

Description: *Dentigloborotalia anfracta* has a low trochospiral test with 4 to 5 chambers in the last whorl on average. The S-shaped sutures give the test an overall lobulate character. The test surface is smooth, and bears typical shark teeth-like pustules in front of the aperture, which are unique to *D. anfracta*, unequivocally identified by SEM imaging (Plate 2.18-17). The aperture is bordered by a thick rim or broad flange. *Dentigloborotalia anfracta* is different from other species and genera by its streptospiral coiled juvenile test. From the juvenile to neanic stage, the test architecture is similar to globorotalid species. In turn, ontogenetic changes in mode of coiling, and the lack of calcite crust and pustules other than those in front of the aperture differentiate *Dentigloborotalia anfracta* from globorotalids.

Molecular genetics: No data available.

Ecology: *Dentigloborotalia anfracta* is a small-sized species of the surface tropical to temperate ocean, and is believed to constitute a major portion of the modern planktic foraminifer assemblage <100 μm. However, due to its small size, *D. anfracta* is rare in test size-fraction >100 μm often analyzed from water and sediment samples, and virtually absent from the >150 μm size fraction (cf. Brummer 1988a; Kemle-von-Mücke and Hemleben 1999).

In the Caribbean Sea, *D. anfracta* has maximum standing stocks of 1.2 individuals per cubic meter (>100 μm) at 60–80 m water depth well above the thermocline (Schmuker and Schiebel 2002). In the Arabian Sea, *D. anfracta* is most abundant at a similar lower mixed layer depth habitat in mesotrophic waters marginal to the upwelling area off Oman (Schiebel et al. 2004). In the upwelling area off Somalia, *D. anfracta* successively increases in numbers towards the final phase of SW monsoonal upwelling, and is present at low standing stocks during the low productive season, i.e. the intermonsoon and NE monsoon (Conan and Brummer 2000).

Further readings: Fleisher (1974), Li (1987).

2.4.3 *Globoquadrina conglomerata* (Schwager 1866) (Plate 2.19)

Description: *Globoquadrina conglomerata* is one of the larger sized species with >3–4 chambers in the last whorl arranged in a medium high trochospire. The umbilical high-arched aperture shows a narrow rim, which is partly enlarged, and forms a so-called tooth. Chambers are almost spherical, and slightly compressed as seen in side view. The surface is strongly cancellate, similar to the spinose species *G. sacculifer*.

Molecular genetics: Very few data are available on the molecular genetics of *G. conglomerata*. Data obtained by André et al. (2014) show that several morphospecies, including *G. conglomerata,* lack cryptic diversity (Fig. 2.13).

Ecology: *Globoquadrina conglomerata* is a rare species in surface waters of the oligotrophic regions of the Indian and Pacific Oceans (e.g., Parker 1962; Bé 1977; Schiebel et al. 2004), and absent from the Atlantic Ocean. Additional information on its distribution might be gathered by future sampling campaigns, since *G. conglomerata* is easy to distinguish from other species, which share the same ecological niche.

Further readings: Parker (1962, 1976).

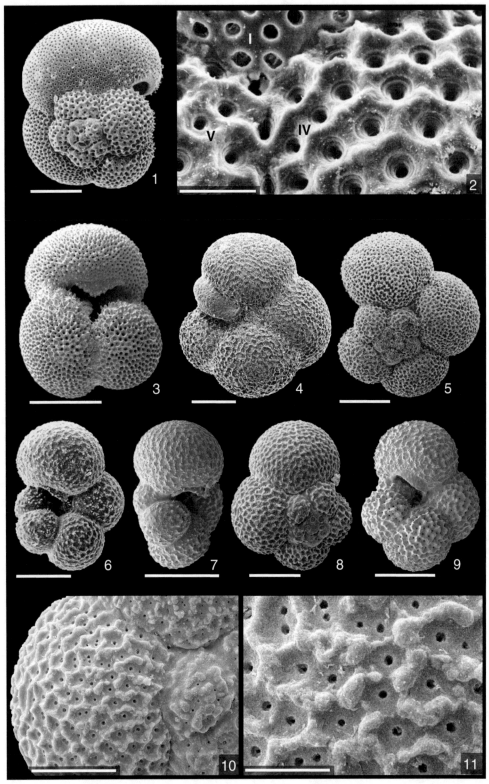

Globoquadrina conglomerata and *Neogloboquadrina incompta*

◀ **Plate 2.19** (*1–3*) *Globoquadrina conglomerata* showing (*2*) ridge-growth between pores from youngest (I) chamber to 4th (IV) and 5th (V) chamber. (*3*) *G. conglomerata* producing apertural teeth. (*4–11*) *Neogloboquadrina incompta* with (*10, 11*) merging pustules forming chains and finally ridges between pore pits. Note contrast to *N. pachyderma* (see Plate 2.21). Bars of overviews (*1, 3*) 200 μm, (*4–9*) 100 μm. Bars of close-ups (*2, 11*) 20 μm, (*10*) 100 μm

Fig. 2.13 Ultrametric tree based on SSU rDNA of the non-spinose species *N. pachyderma, G. inflata, N. incompta, N. dutertrei, G. conglomerata,* and *P. obliquiloculata,* with significant GMYC delimitations. *Colored branches* correspond to GMYC clusters and *outer circles* correspond to the names of the morphospecies, and plausible species are given on the *inner arc. Symbols* associated to specific colors indicate clones sequenced from the same individuals. From André et al. (2014)

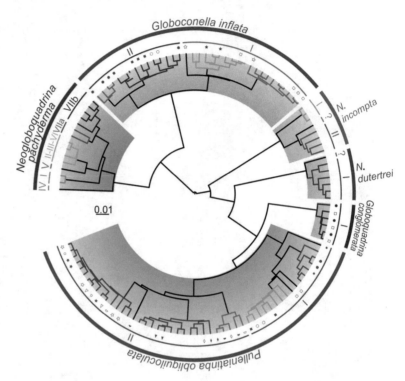

2.4.4 *Neogloboquadrina dutertrei* (d'Orbigny 1839) (Plate 2.20)

O.G.A.: *Globigerina* d'Orbigny 1826, *Globoquadrina* Finlay 1947.

Description: Low trochospiral test with 4.5–6 chambers in the last whorl, and coarse test surface. A lobulate outline results from inflated chambers, and deeply incised sutures. Narrow to rather wide open umbilicus, aperture umbilical to extraumbilical, occasionally bearing a narrow rim. A tooth-plate may occasionally be present. A calcite crust frequently forms while living in the subsurface water column.

Molecular genetics: According to André et al. (2014), only a single genotype of *N. dutertrei* exists (Fig. 2.13). However, intra-individual

variations are assured within the SSU gene repeats (Seears et al. 2012).

Ecology: *Neogloboquadrina dutertrei* is almost exclusively herbivorous, a common diet being unicellular chrysophytes (Anderson et al. 1979). *Neogloboquadrina dutertrei* exhibits a dark greenish cytoplasm at phytoplankton blooms due to ingestion of algae. Those algae may be stored for several days before being digested, or serve as symbionts (Hemleben et al. 1989). The cytoplasm may be rather pale when algae are less abundant. *Neogloboquadrina dutertrei* reproduces at a monthly cycle, which may be linked to the lunar synodic cycle.

Neogloboquadrina dutertrei is frequent in tropical to subtropical waters, and may be present in temperate waters during summer (e.g., Bé 1977; Kemle-von-Mücke and Hemleben 1999;

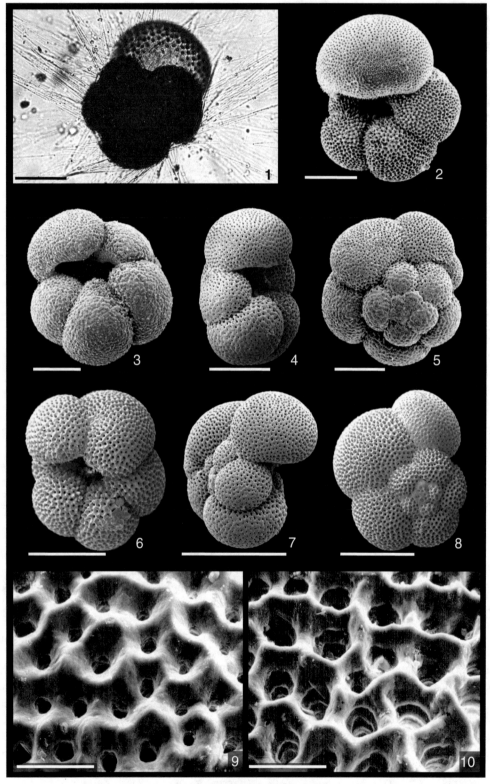

Neogloboquadrina dutertrei

◄ **Plate 2.20** (*1*) Living *Neogloboquadrina dutertrei* with irregular arrangement of pseudopodia caused by disturbance. Adult specimens with (*2, 7*) enlarged final chamber, and (*5, 6, 8*) kummerform final chamber. (*9, 10*) Irregular ridges between pore pits on (*9*) final chamber, and (*10*) 4th last chamber. Bars of overviews 200 μm, bars of close-ups 20 μm

Schiebel and Hemleben 2000). Being related to the initial phase of the SW monsoonal upwelling and enhanced concentration of prey in surface waters of the northwestern Indian Ocean, *N. dutertrei* may adopt an opportunistic behavior (Kroon and Ganssen 1988; Conan and Brummer 2000; Schiebel et al. 2004). In the eastern tropical Atlantic, *N. dutertrei* occurred at maximum standing stocks at the Deep Chlorophyll Maximum, DCM (Ravelo et al. 1990). Along hydrographic fronts of the nutrient rich Congo River fresh water plume, and in the western Caribbean Sea (Amazon/Orinoco River discharge), *N. dutertrei* occurred at increased numbers in surface to thermocline waters, possible displaying an opportunistic behavior to increased food availability at DCM depths (Ufkes et al. 1998; Schmuker and Schiebel 2002).

Neogloboquadrina dutertrei tolerates salinities and temperatures between 25 and 46 PSU, and 13 °C to 33 °C, respectively, under laboratory conditions (Bijma et al. 1990b). Below 15 °C, *N. dutertrei* starts to grow a calcite crust (Hemleben et al. 1989). Consequently, in the natural environment, calcite crusts are frequently found to cover the shell of *N. dutertrei* in the subsurface water column, and thus in sediment assemblages. Thick calcite crusts (forming >60–70 % of the entire test wall) on top of the primary test may cover different chambers to a varying degree (e.g., Steinhardt et al. 2015). Therefore, the resulting Mg/Ca-derived calcification temperatures are lower than ambient sea surface temperatures (cf. Eggins et al. 2003). Mg/Ca and $\delta^{18}O$ derived temperatures would hence indicate a subsurface to deep-water habitat of *N. dutertrei* different from surface to thermocline dwelling depths of live individuals (e.g., Bé et al. 1985; Sautter and Thunell 1991; Schmuker and Schiebel 2002; Schiebel et al. 2004).

Remarks: *Neogloboquadrina dutertrei* and *N. incompta* are genetically (Fig. 2.13) and morphologically closely related (Darling and

Wade 2008; André et al. 2014). However, the two species can be easily distinguished under the light microscope because *N. dutertrei* has a deep umbilicus, teeth-like triangular chamber extensions towards the center of the umbilicus, more inflated chambers and deep incised sutures, and consequently a more inflated and lobulate test than *N. incompta*. Those characteristics of the test of *N. dutertrei* are developed early in ontogeny (Brummer et al. 1987). Whereas the final test diameter of adult individuals of *N. dutertrei* frequently ranges above 600 μm, adult *N. incompta* rarely grow larger than 350 μm. *Neogloboquadrina dutertrei* appears to be the senior synonym of *Globigerina eggeri* Rhumbler 1900.

Further readings: Cifelli (1961), Zobel (1968), Pflaumann (1972), Hecht (1976), Srinivasan and Kennett (1976).

2.4.5 *Neogloboquadrina incompta* (Cifelli 1961) (Plate 2.19)

Description: Low to medium trochospiral test with 4–5 chambers in the last whorl, lobulate outline, extraumbilical aperture with a narrow to broad lip. Typical neogloboquadrinid surface texture producing ridge-growth, i.e. pustules merge and form ridges. *Neogloboquadrina incompta* is the typically right coiling (dextral) relative of the typically left coiling (sinistral) *N. pachyderma* (see discussion on *N. pachyderma*). However, in both species 2–3 % of left and right coiling forms can be observed, respectively (see below).

Molecular genetics: The conventionel tree shows *N. incompta* represented by two genotypes (Fig. 2.13) and their global distribution in relationship to four other non-spinose species (Fig. 2.14).

Ecology: *Neogloboquadrina incompta* is a typical surface dwelling species of the temperate ocean (e.g., Cifelli 1961; Ottens 1991;

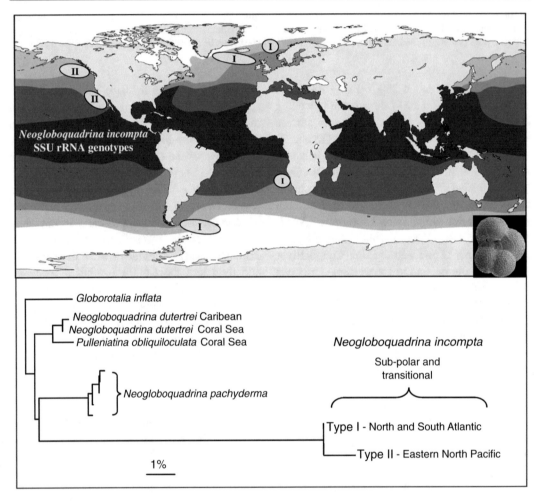

Fig. 2.14 Conventional tree with 2 Types of *N. incompta*, Type I in the Atlantic, and Type II in the Pacific Ocean. The relation *N. pachyderma* as sister-species of *N. incompta* is shown, as well as the relations to *N. dutertrei*, *P. obliquiloculata*, and *G. inflata*. From Darling and Wade (2008)

Kemle-von-Mücke and Hemleben 1999; Kuroyanagi and Kawahata 2004). In the temperate North Atlantic, *N. incompta* is a major faunal component from spring through fall (Schiebel and Hemleben 2000). At times of highest food availability in spring and fall, *N. incompta* may be outnumbered by *G. bulloides* and *T. quinqueloba*, the latter two species being more opportunistic than *N. incompta* (Schiebel et al. 1995, 2001). During low-productive summer conditions, caused by a more stratified surface water column, *N. incompta* was found to dominate the fauna with rather low standing stocks (Schiebel and Hemleben 2000).

Neogloboquadrina incompta appears to be a minor faunal component at low and high latitudes (e.g., Ottens 1992; Schiebel et al. 2002).

Remarks: To facilitate reasonable faunal analyses, all right coiling specimens of the *N. pachyderma/N. incompta* plexus may be classified *N. incompta*, and all left coiling specimens *N. pachyderma*. In case the ratio of the opposite coiled individuals exceeds 3 %, both species *N. pachyderma* and *N. incompta* are possibly present within the same fauna (Darling et al. 2006). In the NE Atlantic, *N. incompta* may be easily distinguished from *N. pachyderma* with 4 chambers in the last whorl, and a rather square outline. Test morphometries of the two species may be more similar in other regions of the ocean (see, e.g., Darling et al. 2006). Even in the North Atlantic, intraspecific variability of test morphologies results in morphotypic end-members of the two species, which may not be distinguished under the binocular microscope.

Further readings: Cifelli (1961, 1971, 1973), Bandy and Theyer (1971), Parker and Berger (1971), Bandy (1972), Vilks (1973), Olsson (1974, 1976); Srinivasan and Kennett (1976), Reynolds and Thunell (1986).

2.4.6 *Neogloboquadrina pachyderma* (Ehrenberg 1861) (Plate 2.21)

O.G.A.: *Globigerina* d'Orbigny 1826, *Globoquadrina* Finlay 1947.

Description: Low trochospiral test with 4–4.5 chambers in the last whorl, a rather squared outline, and straight sutures. The extraumbilical aperture is rather narrow. The surface structure is similar to *N. incompta* when tests are not encrusted. *Neogloboquadrina pachyderma* is typically left coiling (sinistral). However, approximately 2–3 % within a *N. pachyderma* population are right coiling. Test from surface sediments are usually entirely covered by euhedral calcite crystals, forming a typical calcite crust.

Molecular genetics: Up to eight "genotypes" of *N. pachyderma* (Figs. 2.13, 2.14, and 2.15) may be distinguished (Darling et al. 2003, 2004, 2006; Darling and Wade 2008; André et al. 2014). All of the *N. pachyderma* genotypes are typically left coiling, and include <3 % of right coiling morphotypes (Darling et al. 2006). Those <3 % of the morphotypes, which are right coiled, are present in entirely polar samples over glacial-interglacial intervals (e.g., Pflaumann et al. 1996). The typically right coiling *N. incompta* produces rare (<3 %) left coiling specimens. To conclude, coiling 'failure' of <3 % is realized within the same species. Such coiling failure also occurs in other trochospiral species like *G. inflata*. In contrast, a ratio of any coiling direction left or right >3 % may represent different genotypes (cf. Darling et al. 2006). By using ABGD and GMYC, (A. André, oral communication, Angers, 2014) proposes only five and six putative species, respectively. Types II, III, and VI, cluster into a single species (Fig. 2.13). This again demonstrates the danger of oversplitting or undersplitting of genotypes. Some of the cryptic species seem to exhibit a bipolar distribution, although the data are questionable.

Ecology: *Neogloboquadrina pachyderma* dominates polar faunas in the northern and southern hemispheres (e.g., Bé 1977), clearly separated by the tropics (Darling et al. 2004). Among the Antarctic polar to subpolar genotypes, *N. pachyderma* Type IV is interpreted to pursue an overwintering strategy in brine channels within sea ice, tolerating salinities up to 82 PSU (Spindler and Dieckmann 1986; Dieckmann et al. 1991; Darling and Wade 2008). Predominantly large sub-adult individuals of *N. pachyderma* occur in very high standing stocks within the lower layers of the sea ice (Spindler and Dieckmann 1986; Dieckmann et al. 1991). The food source of *N. pachyderma* in sea ice is phytoplankton, consisting almost exclusively of diatoms (Spindler and Dieckmann 1986). Highest standing stocks of any modern species are reported for *N. pachyderma* from melted sea ice samples, amounting to ∼190 individuals per liter (Spindler and Dieckmann 1986), and hence being about 1000 times higher than total planktic foraminifer standing stocks in open waters (Bergami et al. 2009). A similar habitat of *N. pachyderma* within sea ice probably does not exist in the Arctic due to differences in sea ice formation (M. Spindler, Kiel, personal communication). When surface Arctic waters off the ice edge are low in salinity during the short polar summer, *N. pachyderma* moves to subsurface waters (Carstens and Wefer 1992; Carstens et al. 1997; Volkmann 2000a).

In addition to the polar ocean, *N. pachyderma* is frequent in upwelling regions, and marginal seas like the Aegean Sea (e.g., Marchant et al. 1998, 2004; Ivanova et al. 1999; Peeters et al. 1999; Conan and Brummer 2000; Ufkes and Zachariasse 1993; Darling et al. 2006; Darling and Wade 2008; André 2013). The biogeography

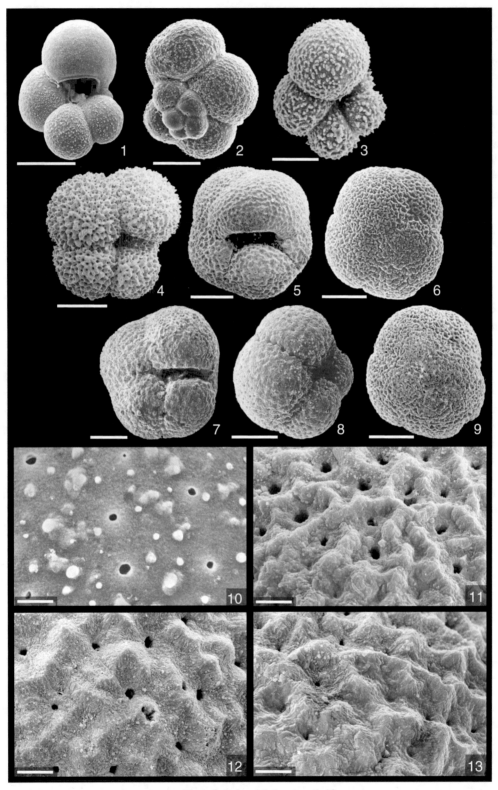

Neogloboquadrina pachyderma

◀ **Plate 2.21** (*1–9*) *Neogloboquadrina pachyderma* with increasing crust formation. (*10*) Test surface of newly formed chamber without calcite crust. (*11*) Calcite ridges are forming, (*12*) calcite crust covering the test surface, and (*13*) euhedral calcite crystals forming the typical thick calcite crust of *N. pachyderma*. Bars of overviews 100 μm, bars of close-ups 10 μm

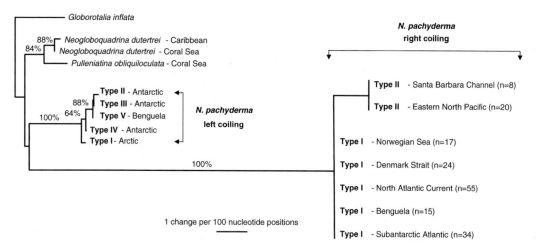

Fig. 2.15 Neighbor-joining SSU rDNA phylogenetic tree (685 nucleotide sites) highlighting the evolutionary relationships among the Neogloboquadrinidae. The phylogeny is rooted on *G. inflata*. The tree shows the highly divergent nature of the classical left and right coiling genotypes of *N. pachyderma*. Bootstrap values are expressed as a percentage and indicate support for branches within the tree. Bootstrap values are only shown for branches that are strongly supported in over 70 % of bootstrap replicates. NGenBank accession numbers are *N. pachyderma* (sin) Types I–V, AY305329, AY305330, AY305331, AF250120, and AY305332 and *N. pachyderma* (dex) Types I–II, AF250117 (Denmark Strait) and AF250118 (Subantarctic) and AY241711. From Darling et al. (2006)

of genotypes in polar-to-subpolar waters and lower latitude upwelling areas is discussed with regard to isolation and exchange of genotypes (Darling et al. 2000; Norris and De Vargas 2000; Darling and Wade 2008; André 2013). Being adapted to rapid consumption of food and reproduction during the short productive season of the polar summer (cf. Jonkers et al. 2010), *N. pachyderma* appears to be competitive also in upwelling areas supported by an opportunistic strategy (cf. Ivanova et al. 1999).

Cyclic abundance of *N. pachyderma* observed in the open northern North Atlantic suggests a monthly reproductive cycle, triggered by the synodic lunar cycle (Volkmann 2000b). Spindler and Dieckmann (1986) suggest overwintering of *N. pachyderma* within the Antarctic sea ice without reproduction. *Neogloboquadrina pachyderma*

may hence be assumed to follow two reproduction strategies depending on environmental conditions.

Remarks: The two *Neogloboquadrina* species *N. pachyderma* and *N. incompta* (see above) have often been confused because of their morphological similarity, and have been referred to as *N. pachyderma* sinistral (left coiling) and *N. pachyderma* dextral (right coiling), respectively. Cifelli (1961) was the first who correctly described the right coiling variety from the North Atlantic, i.e. *N. incompta* as a new species different from *N. pachyderma*. The conclusions on the morphotypic classification of the two *Neogloboquadrina* species *N. pachyderma* and *N. incompta* are confirmed by analyses of the molecular genetics data (Figs. 2.13 and 2.14) of *Neogloboquadrina* (Darling et al. 2000, 2006; Bauch et al. 2003; André et al. 2014).

Plate 2.22 (*1–6*) Adult to pre-adult *Pulleniatina obliquiloculata*. (*3–5*) Specimens with smooth calcite veneer ▶
covering the test. (*7*) Aperture with lip and pustules, and (*9*) pores. Bars of overviews 100 μm, bars of close-ups 20 μm

Further readings: Kohfeld et al. (1996), Simstich et al. (2003), von Langen et al. (2005), Darling et al. (2007).

2.4.7 *Pulleniatina obliquiloculata* (Parker and Jones 1865) (Plate 2.22)

Description: *Pulleniatina obliquiloculata* has a streptospiral coiled test, which is sub-spherical in outline. A long slit-like aperture extends nearly over the entire width of the final chamber. Pre-GAM tests have a smooth surface, except on the area around the aperture, which shows pointed pustules. When migrating to deeper waters, very small crystals form a smooth veneer of calcite covering the whole test.

Molecular genetics: Ujiié et al. (2012) detected three clearly separated cryptic species representing the morphospecies *P. obliquiloculata*, with Type I occurring worldwide, and Types IIa and IIb possibly being restricted to the Pacific region. Using ABGD and GMYC methods (Fig. 2.15), André et al. (2014) show that Type IIa and IIb belong to one putative species with overlapping intra- and inter-type patristic distances.

Ecology: *Pulleniatina obliquiloculata* is a cosmopolitan though rare tropical to subtropical species (e.g., Bé 1977; Li et al. 1997; Kemle-von-Mücke and Hemleben 1999). Maximum standing stocks of *P. obliquiloculata* occur at the upper (juvenile individuals) to lower (adult individuals) surface mixed layer around the thermocline and Deep Chlorophyll Maximum, DCM (e.g., Bé and Tolderlund 1971; Ravelo and Fairbanks 1992; Vénec-Peyré et al. 1995; Watkins et al. 1996). Due to its high preservation potential, the faunal portion of *P. obliquiloculata* in sediment assemblages is much higher than in the live fauna. The diet of *P. obliquiloculata* consists of chrysophytes besides diatoms (Anderson et al. 1979). *Pulleniatina obliquiloculata* is assumed to reproduce at a monthly cycle,

and undergoes gametogenic (GAM) calcification (Hemleben et al. 1989).

Remarks: The outer veneer of *P. obliquiloculata* produces a shell surface similar to *G. inflata*. Pre-adult and early-adult ontogenetic stages suggest a systematic relationship to *Neogloboquadrina*. Both *G. inflata* and *Neogloboquadrina* are genetically closely related to *P. obliquiloculata* (Aurahs et al. 2009a). However, no close morphological relationships exist between these three species.

Further readings: Banner and Blow (1967), Saito et al. (1976).

2.4.8 *Globorotalia cavernula* Bé 1967

Bé (1967) describes *G. carvernula* as new species (Fig. 2.16) sampled from a narrow belt of surface waters between about 46–62°S, south of Australia, and in the Pacific sector of the Southern Ocean. Considerable standing stocks of up to 100 individuals per cubic meter occurred at only three stations between about 80–90°W, east of the Drake Passage, and rarely attained 1 % of the fauna. In the line drawings presented by Bé (1967), *G. carvernula* resembles a high biconvex and left coiled *G. scitula* with a deep umbilicus (cf. Baumfalk et al. 1987). Bé (1967) describes pustules towards the inner whorl of a thin and finely perforate test wall. We have so far not sampled live *G. carvernula* from the water column of any ocean basin. It would be interesting to get more information on *G. carvernula*, in particular on its molecular genetics.

2.4.9 *Globorotalia crassaformis* (Galloway and Wissler 1927) (Plate 2.23)

O.G.A.: *Globorotalia crotonensis* Conato and Follador 1967, *G. crassula* Cushman, Stewart, and Stewart 1930.

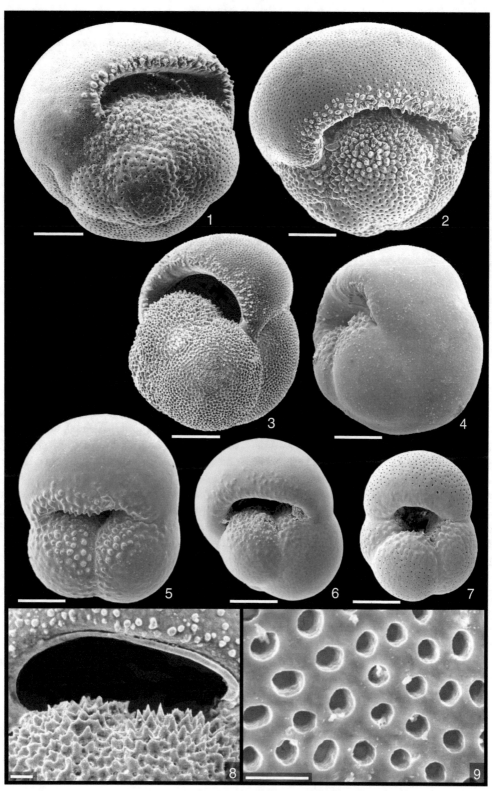

Pulleniatina obliquiloculata

Plate 2.23 (*1–10*) *Globorotalia crassaformis*. (*6, 7*) Tests with final kummerform chambers. (*1, 4, 5, 6, 7*) Pustules, ▶ and calcite crust forming on top of test, mostly on older chambers. (*8*) Gametogenic (GAM) calcification on top of pustules and test wall on *left side* of image. (*9*) Cross-section of test wall showing pores and calcite layers. (*10*) GAM calcification covering wall of fossil test. Bars of overviews 200 μm, bars of close-ups (*8*) 20 μm, (*9, 10*) 10 μm

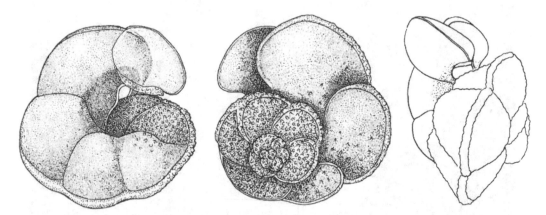

Fig. 2.16 Line drawings of the holotype of *Globorotalia cavernula* Bé 1967 n.sp., from the South Pacific at 55°54′ S, 139°56′W, sampled from the 250–500 m water depth interval. Maximum diameter of the specimen is ∼420 μm. From Bé (1967)

Description: Tests are trochospiral and planoconvex, with 4–4.5 chambers in the last whorl, and may have a squared outline. The extraumbilical aperture is narrow and occasionally bordered by a rim. The periphery is round or slightly angular. The test surface is smooth and peppered with pustules. When pustules merge, they form a thick calcite crust. *Globorotalia crassaformis* may vary considerably in test morphology, i.e. spiral height (cf. Renaud and Schmidt 2003).

Molecular genetics: No data available. Previously established data turned out to be misdetermined and belonging to *Globorotalia inflata*.

Ecology: *Globorotalia crassaformis* is a cosmopolitan species dwelling at subsurface waters around 200–400 m depth in the tropical and subtropical ocean, and ascends to surface waters towards higher latitudes (cf. Parker 1962; Kemle-von-Mücke and Hemleben 1999; Schmuker and Schiebel 2002). The subsurface habitat of *G. crassaformis* has been associated with enhanced biological production in surface waters, and oxygen depleted conditions at habitat

depths of *G. crassaformis* (Kemle-von-Mücke and Hemleben 1999). In the subpolar southern Indian Ocean (off Crozet Islands), the occurrence of *G. crassaformis* is limited to surface waters, and the summer season. The occurrence of exclusively adult specimens leads to the assumption that these individuals were expatriated from lower latitudes by currents. Similar observations exist on *G. crassaformis* from the high latitude South Atlantic.

Remarks: In comparison to *G. truncatulinoides*, the umbilical side of test of *G. crassaformis* is less convex and less pointed. The average number of chambers in the last whorl is 4–4.5 in *G. crassaformis,* and 5–5.5 in *G. truncatulinoides. Globorotalia crassaformis* (Galloway and Wissler 1927) appears to be the senior synonym of *Globorotalia crassula* Cushman and Stewart 1930 (cf. Parker 1962), and *G. scrotonensis* Conato and Follador 1967 (Hemleben et al. 1989, and references therein). The synonym *Globorotalia punctulata* (Deshayes 1832) is doubtful, and appears to be not accepted (Hayward et al. 2014).

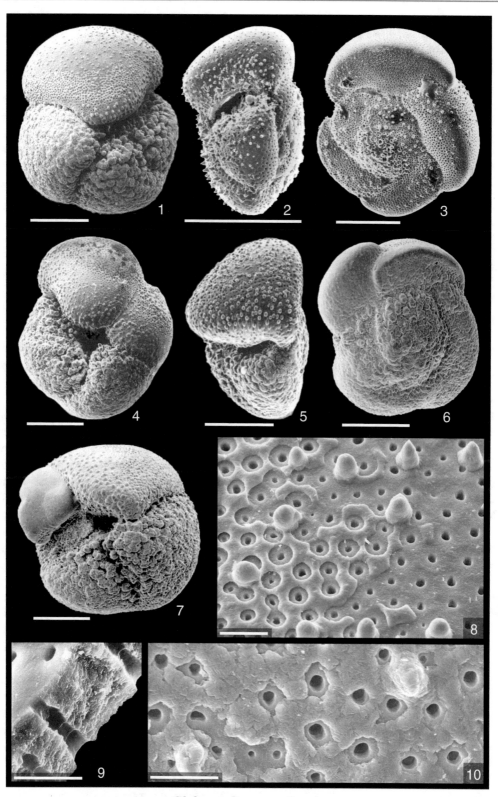

Globorotalia crassaformis

Plate 2.24 (*1*) Living *Globorotalia hirsuta* producing pseudopodia, and attached empty diatom frustrules after ▶ digestion of soft tissue. (*2–11*) Large to small adult specimens. (*3–5*) Test with decreasingly thick calcite crusts from oldest to youngest chamber of the final test whorl. (*12, 13*) Layered pustules overlap pores during lateral growth. (*12*) Fossil and (*13*) living specimen. Bars of overviews 100 μm, bars of close-ups 20 μm

Further readings: Parker and Berger (1971); Lidz (1972); Arnold (1983).

2.4.10 *Globorotalia hirsuta* (d'Orbigny 1839) (Plate 2.24)

O.G.A.: *Hirsutella* Bandy 1972.

Description: Low trochospiral biconvex test with 4–4.5 compressed tetrahedral chambers in the last whorl. Curved to slightly S-shaped sutures give the test an overall lobulate character. The smooth test surface is scattered with pustules. The aperture is extraumbilical-peripheral. The periphery of test is bordered by a keel. A thick calcite crust may occasionally cover the test.

Molecular genetics: According to the few data available, *G. hirsuta* exhibits only one genotype (André et al. 2014).

Ecology: *Globorotalia hirsuta* is a temperate to subtropical species (Tolderlund and Bé 1971; Deuser et al. 1981; Hemleben et al. 1989; Kemle-von-Mücke and Hemleben 1999; Chapman 2010; Harbers et al. 2010; Cléroux et al. 2013). Interpreted as a cosmopolitan species, *G. hirsuta* dwells predominantly in the Atlantic Ocean. Its occurrence in the Indian and Pacific Oceans appears to be limited to small populations in the temperate northern and southern hemisphere (cf. Parker 1962; Bé 1977; Tsuchihashi and Oda 2001; Belyaeva and Burmistrova 2003).

Highest standing stocks of *G. hirsuta* in surface waters in spring, and low standing stocks at subsurface waters in summer possibly display an annual (or biannual) reproduction cycle similar to *G. truncatulinoides* (Hemleben et al. 1985; Schiebel and Hemleben 2000; Schiebel et al. 2002). Ascending to subsurface waters after reproduction in surface waters, the average dwelling depth of *G. hirsuta* ranges at 200–300 m water depth in the Caribbean Sea (Schmuker and Schiebel 2002).

The main food source of *G. hirsuta* consists of diatoms, whose frustrules where consistently observed in food vacuoles of *G. hirsuta* (Hemleben et al. 1985). *Globorotalia hirsuta* probably does not produce any GAM calcite. Secondary calcite crusts may be produced during sedimentation in the deep water column (Hemleben et al. 1985).

Further readings: Glaçon et al. (1973); Boltovskoy (1974).

2.4.11 *Globorotalia inflata* (d'Orbigny 1839) (Plate 2.25)

O.G.A.: *Globigerina* d'Orbigny 1826, *Truncorotalia* Cushman and Bermúdez 1949, *Globoconella* Bandy 1975.

Description: The trochospiral test exhibits >3–4 chambers in the last whorl. The spiral side is rather flat, and the umbilical (apertural) side is high convex. The subspherical tetrahedral chambers are scattered by pustules. Pointed pustules occur in front of the aperture. The aperture is bordered by a narrow rim and forms a low arch extending from the periphery towards the umbilicus. During adult ontogeny, pustules grow larger and finally coalesce to form a calcite crust, which is covered by a fine veneer of very small calcite crystals (Hemleben et al. 1985).

Molecular genetics: Morard et al. (2011, 2013) distinguish two allopatric genotypes of *G. inflata* (Types I and II), which are also recognized as morphotypes: Type I occurs equatorward of the subpolar front. Type II occurs in subpolar to polar waters (Fig. 2.17). Morphotype I has a large aperture in relation to the size of the final chamber, and morphotype II has a relatively small aperture, and a low aperture-to-terminal chamber ratio (Morard et al. 2011).

Ecology: *Globorotalia inflata* is most abundant in the subtropical to the subpolar ocean. Due to its

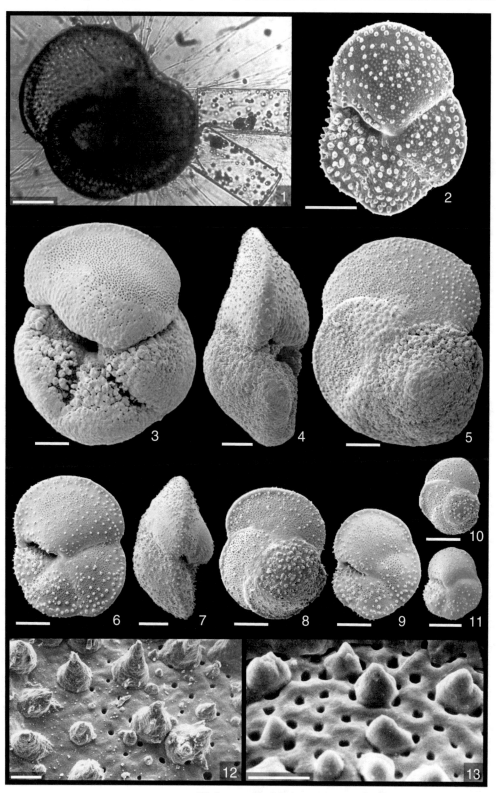

Globorotalia hirsuta

Plate 2.25 (*1*) Living *Globorotalia inflata* producing pseudopodia. (*2–6*) Adult specimens, (*2*) with pristine test ▶ surface, (*3*) with complete calcite veneer, and (*5*) with early calcite crust. (*7, 8*) Early adult, and (*9, 10*) neanic specimens. (*11*) Early pustules, and (*12*) pustules after addition of calcite and lateral growth. (*13*) Calcite crust on top of thickened test wall, (*14*) merging calcite crust, and (*15*) calcite veneer covering the test. (*16*) Cross-section of test wall (*lower part with wide pores*) covered by thick calcite crust (*upper part with narrow pores*). Bars of overviews 100 μm, bars of close-ups (*11–15*) 20 μm, (*16*) 10 μm

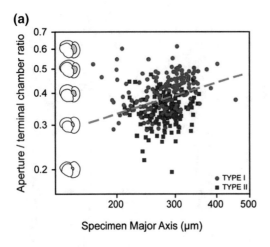

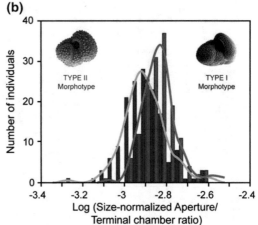

Fig. 2.17 Morphological differences between *G. inflata* genotypes I and II. **a** Log-Log biplot of the ratio of specimen's major axis vs. aperture/terminal chamber length for 306 specimens collected in the South Atlantic. All specimens collected north of the Subpolar Front are considered Type I, all others are considered Type II. The discriminant boundary, which maximizes the separation between the two genotypes is given by the dashed gray line. **b** Histograms and Gaussian kernel densities of the log-ratio between the aperture/terminal chamber length ratio and the specimen's major axis. From Morard et al. (2011)

regionally high standing stocks, and its high fossilization potential, *G. inflata* is of considerable interest as a proxy in paleoceanography (e.g., Dittert et al. 1999; Niebler et al. 1999; Lončarić et al. 2006). *Globorotalia inflata* has often been found to occur in the vicinity of hydrologic fronts and eddies, and has hence been interpreted to display an opportunistic behavior to limited, i.e. mesotrophic conditions in the surface to subsurface water column (cf. Lončarić et al. 2007; Storz et al. 2009; Chapman 2010; Retailleau et al. 2011).

During enhanced phytoplankton production in the spring, the cytoplasm has often been found to be greenish due to consumed chrysophytes, or orange in case of diatom prey (Hemleben et al. 1989). In addition to its abundance in pelagic waters, *G. inflata* may dominate the planktic foraminifer fauna at surface (0–40 m) or subsurface (40–100 m) water depths in neritic waters of enhanced food availability, caused by weak topographically driven upwelling over a submarine canyon head in the SE Bay of Biscay (Retailleau et al. 2012). *Globorotalia inflata* undergoes gametogenic calcification, and is interpreted to have a monthly reproductive cycle.

Remarks: Fossil tests of *G. inflata* from sediment samples often have a shiny appearance caused by the smooth finely crystalline calcite veneer easy to identify under the incident light microscope, and distinguishing *G. inflata* from other globorotalids like *G. crassaformis*. The shiny appearance is similar to *P. obliquiloculata*. *Globorotalia inflata* genotypes are closely related to *P. obliquiloculata*, and the Neogloboquadrinids *N. dutertrei* and *N. pachyderma* (Darling and Wade 2008).

Further readings: Ganssen (1983).

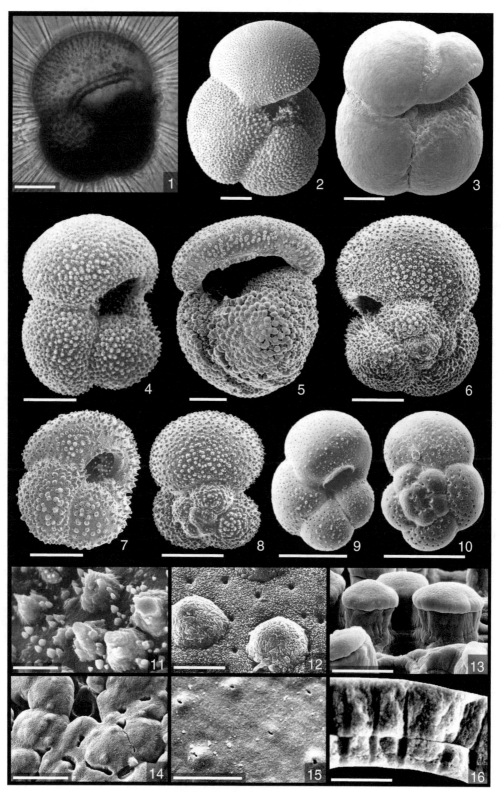

Globorotalia inflata

Plate 2.26 (*1–15*) *Globorotalia menardii*. (*3*) Image combined from multiple SEM micrographs. (*4–6*) Tests with ▶ fimbriated keel, and (*7*) increasingly thick calcite crust from younger to older chambers. (*8, 9*) Pustules and grooves on test surface. (*10*) Sutures and pores on spiral test side. (*11*) Euhedral calcite crystals on test surface. (*12*) Umbilicus and aperture with lip and pustules. (*13*) Pustules on keel. (*14*) Cross-section of test wall with multiple calcite layers, Primary Organic Membrane (POM). (*15*) Cross-section of keel with calcite layers. Bars of overviews 200 μm, bars of close-ups 20 μm

2.4.12 *Globorotalia menardii* (Parker, Jones and Brady 1865) (Plate 2.26)

O.G.A.: *Menardella* Bandy 1972.

Description: *Globorotalia menardii* has a very low trochospiral, large discoidal test with 5–5.5 chambers in the last whorl. The narrow umbilical-extraumbilical aperture is bordered by a prominent lip. Sutures are curved on the spiral side and rather straight on the apertural side. A thick prominent keel is well developed, and occasionally fimbriated. Pustules increase in numbers and size from the younger to older chambers, and are particularly dense in front of the aperture (Hemleben et al. 1977).

Molecular genetics: *Globorotalia menardii* belongs to the group of non-spinose morphospecies (*Globorotalia hirsuta*, *G. tumida*, *G. ungulata*, *G. menardii*, and *Globoquadrina conglomerata*), which are represented by only one genotype each, thus no cryptic species are distinguished (André et al. 2014).

Ecology: *Globorotalia menardii* is a cosmopolitan species most frequent in tropical to subtropical waters of low to medium productivity and food availability (e.g., Kroon and Ganssen 1989; Ufkes et al. 1998; Conan and Brummer 2000; Schmuker and Schiebel 2002; Schiebel et al. 2004). Dwelling predominantly in the surface ocean, maximum standing stocks of *G. menardii* occur at pycnocline/nutricline/DCM depths (e.g., Fairbanks and Wiebe 1980; Ravelo et al. 1990). Due to its large size and occasionally thick calcite crust, *G. menardii* is a major component of surface and Pleistocene sediments, and has been extensively analyzed for its test surface texture, morphometry, and paleoceanographic

significance (e.g., Hemleben et al. 1977; Schweitzer and Lohmann 1991; Mekik et al. 2002; Mekik and François 2006; Knappertsbusch 2007; Mary and Knappertsbusch 2013).

Globorotalia menardii prefers a phytoplankton diet consisting of diatoms and chrysophytes, and occasionally an omnivorous diet is consumed (Anderson et al. 1979). Some of the ingested phytoplankton might be used as symbionts (Hemleben et al. 1989, and references therein). Reproduction of *G. menardii* follows the synodic lunar cycle.

Remark: For a taxonomic discussion of *G. menardii* (Neotype Parker, Jones and Brady 1865, earlier d'Orbigny 1826) versus *Globorotalia cultrata* (d'Orbigny 1839) see Parker (1962).

Further readings: Hemleben et al. (1985), Watkins et al. (1996; 1998), Tedesco and Thunell (2003), Tedesco et al. (2007), Mohtadi et al. (2009), Regenberg et al. (2010), Weijnert et al. (2010, 2013), Schmidt et al. (2013), Broecker and Pena (2014).

2.4.13 *Globorotalia scitula* (Brady 1882) (Plate 2.27)

Description: *Globorotalia scitula* has a low trochospiral test of medium size and 4–5 slightly inflated chambers in the last whorl. Curved and S-shaped sutures on the spiral and umbilical side, respectively, give the test an overall lobulate character. The test surface is smooth and the test wall is rather thin (cf. Parker 1962). Few pustules occur on the apertural side and even less on the spiral side. The slit-like aperture reaches from the periphery towards the umbilicus and is bordered by a narrow lip. Pores are concentrated on the

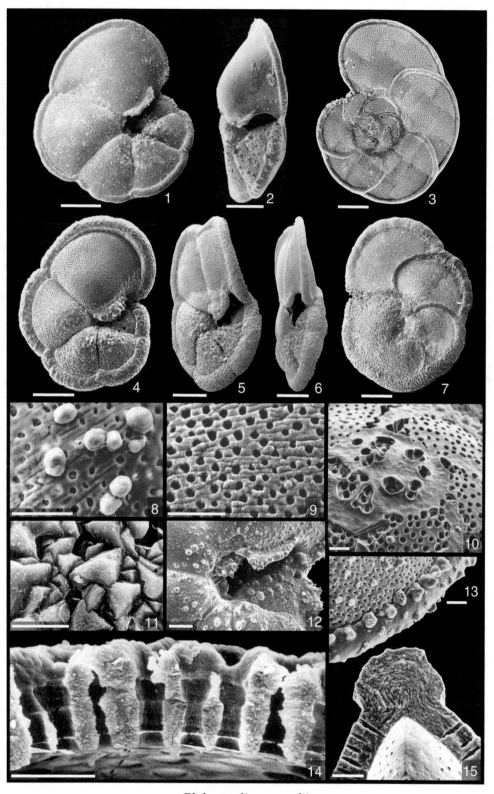

Globorotalia menardii

Plate 2.27 (*1–8*) Adult *Globorotalia scitula* of increasing size from (*1, 2*) to (*3, 4, 5*). (*6*) Pores of final chamber, (*7*) ▶ pores of second last chamber, and (*8*) pustules off the aperture. (*9–12*) Adult *Globorotalia theyeri* with (*10*) low convex and (*11*) high convex umbilical side. Bars of overviews 100 μm, bars of close-ups 10 μm

apertural side, and much fewer pores occur on the spiral side of test. In contrast to *G. hirsuta*, *G. scitula* does not produce a keel.

Molecular genetics: No data available. Previous putative sequences turned out to be misidentified and belong to *G. hirsuta* (André et al. 2014).

Ecology: *Globorotalia scitula* is a cosmopolitan species most frequent at mid-latitude temperate regions during spring and fall, i.e. during times of increased primary productivity (Schiebel and Hemleben 2000; Schiebel et al. 2002; Chapman 2010). From mid latitudes towards low and high latitudes, *G. scitula* decreases in abundance (Schiebel et al. 2002). In pelagic waters, *G. scitula* dwells at subsurface waters below the thermocline to 200–300 m depth (Ottens 1992; Schmuker and Schiebel 2002; Retailleau et al. 2011). Itou et al. (2001) propose a *G. scitula*-to-*N. dutertrei* ratio as a proxy of the mixed layer depth at the Kuroshio-Oyashio confluence off NE Japan. *Globorotalia scitula* may be present in high standing stocks in neritic waters following time-intervals of enhanced primary productivity (Retailleau et al. 2012). In neritic waters <200 m waters depth, *G. scitula* may be most frequent in the surface water column (Retailleau et al. 2011). In turbid shelf waters off Congo, *G. scitula* was the most frequent among the few live planktic foraminifer individuals, which might indicate its opportunistic/robust nature (cf. Ufkes et al. 1998).

Globorotalia scitula co-occurs with *G. hexagonus* in the upper Oxygen Minimum Zone (100–200 m depth) of the central Arabian Sea, which may possibly indicate a preference in diet than low-oxygen conditions (cf. Baumfalk et al. 1987). In the western Arabian Sea off Somalia, *G. scitula* increases in number during the late phase of SW monsoonal upwelling (Conan and Brummer 2000).

Remark: *Globorotalia bermudezi* Rögl and Bolli 1973 is a junior synonym of *G. scitula* (cf. Saito et al. 1981).

Further readings: Baumfalk et al. (1987), Steinhardt et al. (2015).

2.4.14 *Globorotalia theyeri* Fleisher 1974 (Plate 2.27)

Description: *Globorotalia theyeri* has a low trochospiral rather large and thin-shelled test with 4.5–5 chambers in the last whorl. The equatorial periphery exhibits a lobulate outline, and a discontinuous peripheral keel. Sutures on the umbilical side of the test are straight or slightly curved, and more curved on the spiral test side. The aperture is an umbilical-extraumbilical slit bordered by a narrow lip. The test surface is smooth and uniformly penetrated by pores.

Molecular genetics: No data available.

Ecology: *Globorotalia theyeri* is a rare surface dweller in the tropical to subtropical Indian and Pacific Oceans (e.g., Bé 1977; Fairbanks et al. 1982). In the central Arabian Sea, *G. theyeri* is most frequent in oligotrophic above-thermocline waters at 40–60 m depth (Schiebel et al. 2004). *Globorotalia theyeri* occurs in low numbers in the upwelling region in the western Arabian Sea off Somalia (Conan and Brummer 2000, sediment trap samples, 1265 and 1617 m water depth). A change in depths habitat from thermocline to surface waters may be caused by a shoaling thermocline during intensified upwelling in the Panama Basin (Thunell and Reynolds 1984, sediment trap samples, 890, 2590, and 3560 m water depth).

Remarks: *Globorotalia theryeri* can be confused with *G. scitula* but differs by the possession of a keel, and uniformly distributed pores on the spiral and umbilical side of test.

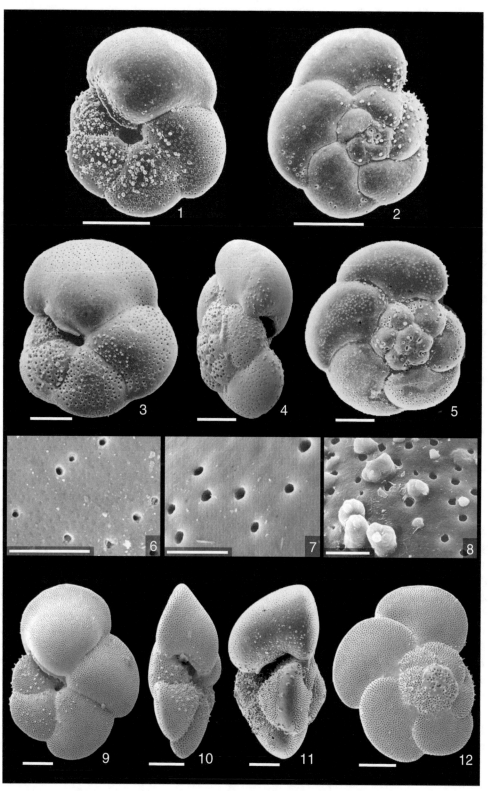

Globorotalia scitula and *G. theyeri*

Plate 2.28 (*1*) Living *Globorotalia truncatulinoides* producing pseudopodia. (*2–6*) Adult specimens with ▶ high-conical umbilical side. (*7–10*) Neanic to pre-adult specimens with low-conical umbilical side. (*11–13*) Adult tests with kummerform final chambers, wide umbilicus, and apertures of earlier chambers with apertural lips. *Bars* 100 μm

2.4.15 *Globorotalia truncatulinoides* (d'Orbigny 1839) (Plates 2.28 and 2.29)

O.G.A.: *Truncorotalia* Cushman and Bermúdez 1949.

Description: The trochospiral test of *Globorotalia truncatulinoides* is medium to high conical on the umbilical side, and flat on the spiral side. *Globorotalia truncatulinoides* has 4.5–5.5 tetrahedral chambers in the last whorl. The narrow aperture spans from the rather deep umbilicus towards the periphery, and is bordered by a lip. Multiple umbilical apertures may occasionally exist. The periphery is keeled from the neanic stage onward. Pores are distributed over the entire surface. Pustules develop from older to younger chambers. A thick calcite crust may be developed by adult specimens when dwelling in subsurface waters below 8–10 °C (see Chap. 6).

Molecular genetics: De Vargas et al. (2001) suggest four cryptic species, two of them right coiling and two of them left coiling, respectively. Ujiié et al. (2010) describe five types distributed in well-defined regions of the global ocean. Quillévéré et al. (2013) confirm the morphospecies *G. truncatulinoides* and its five genotypes from morphometric evidence. André et al. (2014) analyzed all existing genetic data by using ABGD, and show that putative delimitations between Types I and II, and Types III and IV, are somewhat doubtful. When using GMYC, Types I and II can be clearly separated, but the separation between Types III and IV is doubtful (André et al. 2014). Both methods ABGD and GMYC corroborate Type V.

Ecology: *Globorotalia truncatulinoides* probably populates the deepest habitat of all extant species, having been sampled alive from the water

column below 2000 m water depth (Schiebel and Hemleben 2005). *Globorotalia truncatulinoides* was found to reproduce once per year, i.e. in late winter in surface waters at different regions at the poleward margin of the subtropical gyres (Bé and Hutson 1977: Indian Ocean; Weyl 1978: North Atlantic; Hemleben et al. 1985: Bermuda; Kemle-von-Mücke and Hemleben 1999: South Atlantic; Schiebel et al. 2002: Azores), and this is probably true for the Pacific Ocean (cf. Bé 1977).

It is speculated that *G. truncatulinoides* generally reproduces in surface waters to provide sufficient food for offspring (compared to deeper waters), and to avoid competition (or predation?). The offspring descends in the water column, and disperses over the vast expands of the deep ocean, where they spend most of the year growing a mature test and producing a calcite crust. In the subtropical ocean towards higher latitudes, *G. truncatulinoides* dwells at decreasing water depths, possibly driven by the availability of food. A similar shoaling of habitat towards the poles has been observed in other subsurface dwelling globorotalid species like *G. crassaformis* (Hemleben et al. 1985).

Globorotalia truncatulinoides may enter marginal basins like the Mediterranean Sea and the Caribbean Sea through shallow and narrow passages, and occurs at some distance from the passage at subsurface depth (e.g., Schmuker and Schiebel 2002). *Globorotalia truncatulinoides* is absent from the modern Red Sea and Arabian Sea (e.g., Ivanova et al. 2003). In the Arabian Sea, *G. truncatulinoides* was present in low standing stocks during the past glacials, possibly imported by currents from the southern Indian Ocean (e.g., Auras-Schudnagies et al. 1989; Ivanova et al. 2003).

Right and left coiling morphotypes of *G. truncatulinoides* were distinguished from

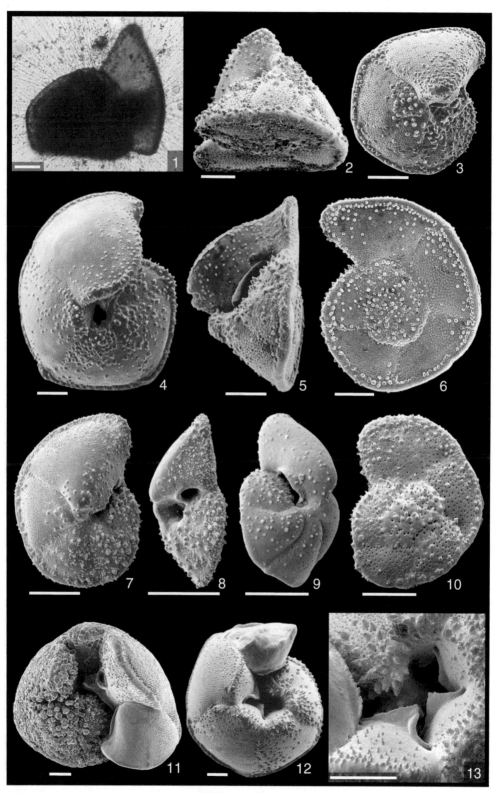

Globorotalia truncatulinoides

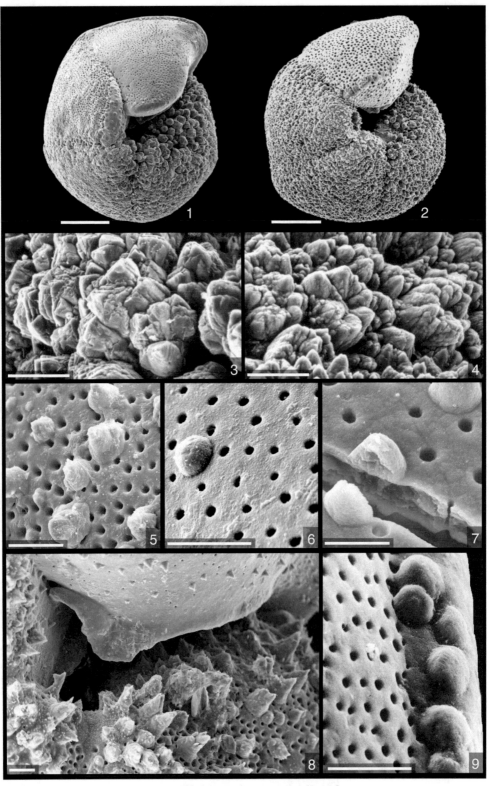

Globorotalia truncatulinoides

◀ **Plate 2.29** (*1–4*) *Globorotalia truncatulinoides* with calcite crusts and euhedral crystals covering the test wall except of final chamber. (*1*, *3*) and (*2*, *4*) from same specimen, respectively. (*5–7*) Test surface with pores and pustules. (*8*) Pores and pustules of variable shape off aperture and on top of apertural flap. (*8*) Pustules on keel. Bars of overviews 100 μm, bars of close-ups 20 μm

surface sediments, and interpreted for their bio-geographic and ecological significance (Ericson et al. 1954). Healy-Williams (1983), and Healy-Williams et al. (1985) were the first to suggest two sub-populations of *G. truncatulinoides*. Stratification and productivity in surface waters are interpreted to affect the distribution of different genotypes of the deep dwelling *G. truncatulinoides* (De Vargas et al. 2001; Darling and Wade 2008). Morphometric variability of the tests of different genotypes in the South Atlantic over the past 140 kyrs is discussed in connotation with environment and glacial-interglacial climate change (Renaud and Schmidt 2003).

The deep habitat of *G. truncatulinoides* makes the species an ideal proxy of surface ocean stratification. In comparison to the stable isotope and element ratio (notably Metal/Ca ratio) of various surface dwelling species, the chemical composition of *G. truncatulinoides* tests provides a measure of environmental conditions above and below the seasonal thermocline (Hemleben et al. 1985; Mulitza et al. 1997; Cléroux et al. 2007, 2008). Care needs to be taken because of long-distance transport of *G. truncatulinoides* within currents, and hence uncertain locations and water depths of calcite precipitation (Deuser et al. 1981; Lohmann and Malmgren 1983; Cléroux et al. 2009). The effect of encrustation and dissolution of *G. truncatulinoides* tests on the isotope signal is discussed by Lohmann (1995).

Remarks: With its distinct conical test morphometry, *G. truncatulinoides* is different from all other extant planktic foraminifers and easy to identify. *Globorotalia crassaformis* may resemble low-conical *G. truncatulinoides* in test outline, but it lacks a keeled test periphery, and has fewer chambers in the last whorl than *G. truncatulinoides*.

Globorotalia truncatulinoides is one of the species most often investigated from sediment samples.

Further readings: Healy-Williams and Williams (1981), Spencer-Cervato and Thierstein (1997), Sexton and Norris (2008), Spear et al. (2011).

2.4.16 *Globorotalia tumida* (Brady 1877) (Plate 2.30)

Description: Trochospiral test with medium to high-convex umbilical side, and low-convex spiral side, and 4.5–6 chambers in the last whorl. The final two chambers often somewhat elongated and twisted. The narrow umbilical-extraumbilical aperture is bordered by a prominent lip. A typical keel is well developed. A thick calcite crust may form from the older to younger chambers on spiral and umbilical side, as in *Globorotalia menardii*. Tests of *G. tumida* are more elongate and higher bi-convex than tests of *G. menardii*. The genetic relation of the two species has not yet been analyzed.

Molecular genetics: Only one genotype has been identified for this species, thus no cryptic species do exist (André et al. 2014).

Ecology: *Globorotalia tumida* is a rare species of the tropical to subtropical ocean (Conan and Brummer 2000). *Globorotalia tumida* occurs in low productive subsurface waters (40–80 m depth) in the northern Arabian Sea (Schiebel et al. 2004), and at low-latitudes around the seasonal thermocline/DCM in the Atlantic Ocean (Ravelo et al. 1990; Ravelo and Fairbanks 1992). Due to its rather low dissolution susceptibility, the faunal portion of *G. tumida* tests in sediment samples is higher than in the water column (cf. Dittert et al. 1999).

Further readings: Malmgren et al. (1983).

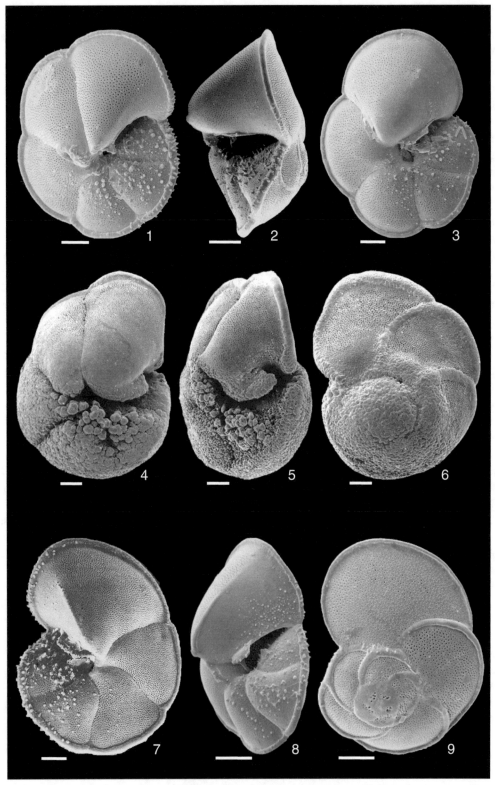

Globorotalia tumida and *G. ungulata*

◄ **Plate 2.30** *Globorotalia tumida* (*1–3*) without and (*4–6*) with calcite crusts mostly on top of older chambers. (*7–9*) *Globorotalia ungulata* producing a ridge on the shoulder of last chamber (*7, 8*). *Bars* 100 μm

2.4.17 *Globorotalia ungulata Bermudez 1960* (Plate 2.30)

Description: *Globorotalia ungulata* has a medium to high-convex trochospiral test. The final chambers are slightly elongate, resulting in an overall elongated adult test. *Globorotalia ungulata* has a prominent keel, and a sharp ridge ("keel") on the shoulder of the final chamber. The rather narrow umbilicus and a slit-like umbilical-extraumbilical aperture are typical of the globorotaliid species.

Molecular genetics: Only one genotype has been identified for this species (André et al. 2014).

Ecology: Rare species with poorly known ecological affinities, probably similar in its distribution to *Globorotalia menardii*. It occurs somewhat sporadically in tropical to subtropical waters, and may attain as many as two individuals per cubic meter in the oligotrophic Arabian Sea.

Remarks: Tests of *G. ungulata* may resemble those of *G. menardii* and *G. tumida,* but the keel on the umbilical shoulder of the final chamber of *G. ungulata* is absent in *G. menardii* and *G. tumida.* In addition, the test wall of *G. ungulata* is thinner and smoother, and the apertural side is more convex and angular than in *G. tumida.*

Further readings: Bermudez (1960), Seears et al. (2012).

2.4.18 *Globorotaloides hexagonus (Natland 1938)* (Plate 2.31)

O.G.A.: *Globoquadrina* Finlay 1947.

Description: Very low trochospiral test with 4.5–6 chambers in the last whorl. The aperture at the base of the final chamber is high arched and bordered by a rim or lip. *Globorotaloides hexagonus* has a coarsely pitted perfect honeycomb surface texture, which is unique among modern planktic foraminifers (cf. Parker 1962). GAM calcification may occur (see Chap. 5).

Molecular genetics: No data available.

Ecology: *Globorotaloides hexagonus* is a rare species restricted to the Indian and Pacific Oceans (Bé and Tolderlund 1971; Thunell and Reynolds 1984). *Globorotaloides hexagonus* tolerates a wide range of environmental conditions, and occurs from the tropical to temperate ocean including a broad depth habitat. In the central equatorial Arabian Sea, and the central equatorial Pacific Ocean, *G. hexagonus* occurs in maximum standing stocks of 2–4 individuals per cubic meter in the sub-thermocline layer of the water column, at 100–200 m (Fairbanks et al. 1982; Zhang 1985; Watkins et al. 1996; Schiebel et al. 2004). Subthermocline waters of 100–200 m depth at both sampling locations in the In the Arabian Sea and Pacific Ocean correspond to the upper limit of prominent Oxygen Minimum Zones (cf. Warren 1994). It is hence interpreted that *G. hexagonus* is adapted to low-oxygen conditions by its wide and numerous pores, which facilitate respiration even at oxygen limitation. In addition, it is assumed that *G. hexagonus* exploits food sources particular to its depth habitat and environmental conditions, i.e. oxygen limitation. The cytoplasm of *G. hexagonus* sampled from the Arabian Sea was colored either dark green or dark orange, which might be indicative of its so far unknown diet.

Further readings: Kennett and Srinivasan (1983), Chaisson and d'Hondt (2000).

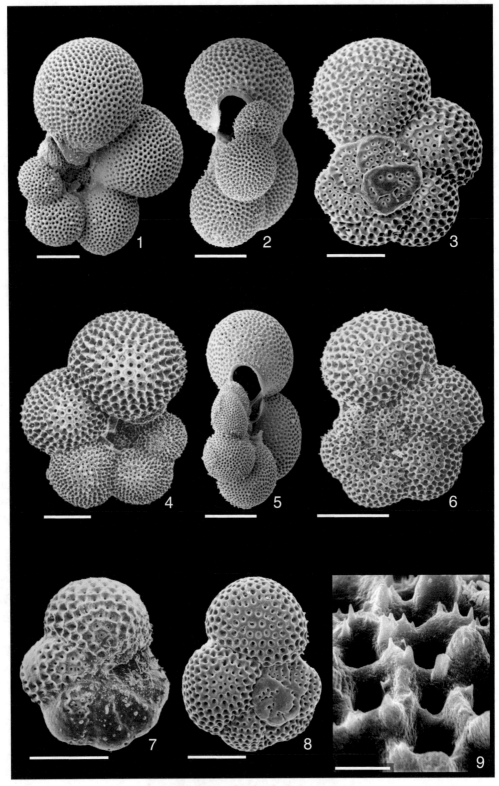

Globorotaloides hexagonus

◀ **Plate 2.31** (*1–9*) Adult *Globorotaloides hexagonus* producing large pore pits and regular hexagonal test surface structure. (*7*) Specimen showing change in test surface structure from neanic to adult stage. (*9*) Oblique view of test surface showing pore pits separated by ridges with small pointed epitactic spikes. Bars of overviews 100 µm, bar of close-up 10 µm. (*4*) *Photo* A. Kiefer and R. Schiebel

2.4.19 *Streptochilus globigerus* (Schwager 1866)

Description: *Streptochilus globigerus* is the only modern planktic foraminifer species with biserial test architecture through its entire ontogenetic development. The test of *S. globigerus* is macroperforate throughout. The test wall is smooth to slightly rugose, and lacks spines and pustules. The aperture is umbilical, and bears a distinct lip on the extraumbilical side, as well as a toothplate (Hemleben et al. 1989). The species resembles the benthic foraminifer *Bolivina variabilis* genetically and morphologically (Darling et al. 2009). However, tests of *S. globigerus* bear a geochemical signature (Mg/Ca ratio) that clearly indicates a planktic habitat (Darling et al. 2009).

Molecular genetics: Darling et al. (2009) show that the modern biserial planktic species *Streptochilus globigerus* "belongs to the same biological species as the benthic *Bolivina variabilis*".

Ecology: Modern *S. globigerus* occur from tropical to temperate regions (cf. de Klasz et al. 1989; Schmuker and Schiebel 2002). Since live *S. globigerus* are rare in plankton samples, and since the species cannot be distinguished from benthic *B. variabilis*, the actual distribution of the planktic tychotype is possibly much larger than so far reported. Mg/Ca derived calcification temperatures of 26–29 °C on average correspond to 30 m and 75 m water depth in the northwestern Indian Ocean, and indicate ontogenetic migration from the shallow to deep surface mixed layer of the ocean (Darling et al. 2009).

Remarks: *Streptochilus globigerus* is possibly the oldest extant planktic foraminifer species, descending from the Aptian/Albian (Lower Cretaceous, ~ 125–100 Ma) ancestors, and is reported from the upper Paleogene and Neogene (Brönnimann and Resig 1971; Resig and Kroopnick 1983;

Poore and Gosnell 1985; Smart and Thomas 2007). The rarely developed tychopelagic lifestyle of *S. globigerus* is discussed for its phylogenetic significance, and the repopulation of empty ecological niches after the Cretaceous-Tertiary mass extinction (Darling et al. 2009).

2.5 Microperforate Species

Microperforate planktic foraminifer species have pores <1 µm in diameter, in contrast to macroperforate (normal perforate) species with pores >1 µm. Microperforate planktic foraminifer species bear no spines.

2.5.1 *Gallitellia vivans* (Cushman 1934) (Plate 2.32)

Description: *Gallitellia vivans* is the only triserial extant planktic foraminifer species. Mature individuals may abandon the triserial test symmetry, and produce a multiserial chamber arrangement. Chambers are globular, with deeply depressed sutures. The test wall is smooth. The aperture is a symmetric arc, occasionally with a tooth-like flap, and has an umbilical position.

Molecular genetics: No data available.

Ecology: The species has been discussed as a surface (from plankton net samples) to subsurface (from $\delta^{18}O$ data) dwelling species in the global subtropical to temperate ocean (Kroon and Nederbragt 1990). In the Tsushima Strait, between Japan and Korea, *G. vivans* occurred at exceptionally high numbers in November 2006 (Kimoto et al. 2009). Similar $\delta^{13}C$ and $\delta^{18}O$ data of tests of *G. vivans* and *G. bulloides* may indicate similar ecological demands (Kimoto et al. 2009).

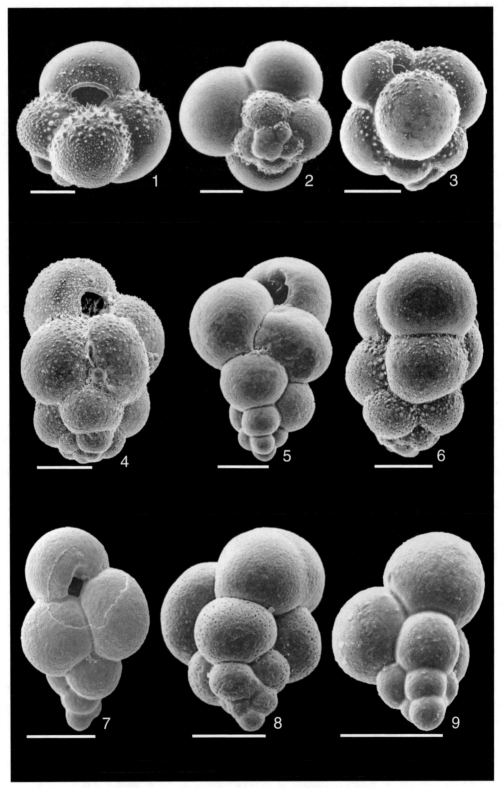

Gallietella vivans and *Globigerinita uvula*

◄ **Plate 2.32** *(1–4)* Adult trochospiral pine-cone shaped *Globigerinita uvula* with pointed pustules off aperture. *(5–9)* Adult triserial *Gallitellia vivans* with smooth-walled apertural face. *Bars 50 μm*

Remarks: Because of its small test size, most of the populations of *G. vivans* are not sampled with the typically employed plankton-nets of >100-μm mesh-size. Therefore, the regional distribution and ecological demands of *G. vivans* are still insufficiently known.

Further readings: Loeblich and Tappan (1986); Kroon and Nederbragt (1988).

2.5.2 *Globigerinita glutinata* (Egger 1893) (Plate 2.33)

O.G.A.: *Globigerina* d'Orbigny 1926.

Description: *Globigerinita glutinata* has a medium trochospiral subglobular test with 4 globular chambers in the last whorl. The final chamber may be slightly compressed. The normal umbilical aperture of adult individuals is frequently covered by a bulla, leaving one or more apertural openings. The entire test surface is covered by numerous small and pointed pustules. Pores are usually smaller than 1 μm in diameter. Tests of juvenile *G. glutinata* are planispiral with an equatorial aperture provided with a large flange. During the neanic stage the aperture migrates to an umbilical position, and the umbilicus closes.

Molecular genetics: Genetic distances within *G. glutinata* suggest that several cryptic species are included in this morphospecies (Ujiié and Lipps 2009; André 2013). André et al. (2014) established four cryptic species (Types I, II, III, and IV) of *G. glutinata*, which are delimitated by regional occurrences.

Ecology: *Globigerinita glutinata* is possibly the most ubiquitous planktic foraminifer in the modern ocean. *Globigerinita glutinata* is most abundant in subtropical to temperate waters, and decreases in frequency towards high latitudes (cf. Ottens 1992; Schiebel and Hemleben 2000; Volkmann 2000a; Schmuker and Schiebel 2002).

In the NE Atlantic, *G. glutinata* is present in surface waters throughout the year and constitutes up to 20 % of the live fauna during spring, following enhanced phytoplankton production in surface waters (Schiebel et al. 1995; Schiebel and Hemleben 2000; Chapman 2010). A second seasonal maximum of *G. glutinata* in the NE Atlantic occurs in fall, when wind-driven nutrient entrainment into surface waters triggers phytoplankton production at the nutricline (Schiebel et al. 2001). Similarly, the occurrence of *G. glutinata* in the Gulf of Aden is related to nutrient entrainment into surface waters during the NE monsoon (Ivanova et al. 2003). With up to 35 individuals per cubic meter, *G. glutinata* is the 2nd most frequent species in the Caribbean Sea after *G. ruber*, being related to eddy-driven nutrient entrainment into surface waters, and phytoplankton production (Schmuker and Schiebel 2002). Calcification depths of *G. glutinata* tests range from surface waters down to thermocline depths (Lončarić et al. 2006; Friedrich et al. 2012). The ratio of bulla vs. non-bulla bearing individuals is possibly not related to ecological conditions.

The cytoplasm of *G. glutinata* is often colored dark red. According to TEM analyses, the diet of *G. glutinata* consists mainly of diatoms, but also of chrysophytes (Anderson et al. 1979; Spindler et al. 1984). Chrysophytes may have facultatively been harbored as symbionts when present in abundance, and were subsequently digested (Hemleben et al. 1989). Gametogenesis and gametogenic calcification have repeatedly been observed (Hemleben et al. 1989).

Remarks: Small *G. glutinata* tests might be confused with *G. minuta*, the latter bearing secondary apertures on the spiral side. *Globigerinita glutinata* may also be misidentified as *Globigerina bulloides*. However, *G. glutinata* exhibits a smooth surface with rather flat but pointed pustules that are scattered over the entire surface, whereas

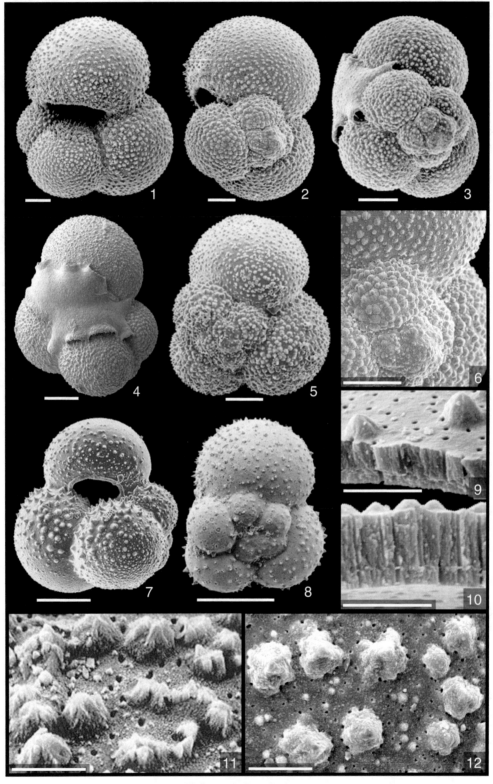

Globigerinita glutinata

◀ **Plate 2.33** (*1–12*) Adult *Globigerinita glutinata*, with (*3, 4*) bulla with multiple openings covering the aperture. (*6*) Smooth pustules on central part of spiral test side. (*7, 8*) Young adult specimens with pointed pustules. (*9, 10*) Cross-sections of bilamellar test wall with POM, and internal and external calcite layers. (*11, 12*) Conical pustules showing lateral growth and approaching pores. Bars of overviews 50 μm, bars of close-ups (*6*) 20 μm, (*9–12*) 10 μm. (*7*) *Photo* A. Kiefer and R. Schiebel

G. bulloides has a rather rough surface, no pustules but spines, and no milky but glassy appearance.

Further readings: Parker (1962), Kahn and Williams (1981), Spindler et al. (1984), Li (1987), Brummer (1988b), Ortiz et al. (1995).

2.5.3 *Globigerinita minuta* (Natland 1938) (Plate 2.34)

O.G.A.: *Globigerinoides* Cushman 1927.

Description: Small-sized and medium high trochspiral test with 4 subspherical chambers in the last whorl. Secondary apertures on the spiral test side. The test of *G. minuta* may be heavily pustulate over the entire surface (Brummer 1988b). A bulla might cover the primary aperture of mature individuals of *G. minuta*, similar to *G. glutinata*.

Molecular genetics: No data available.

Ecology: *Globigerinita minuta* has been recognized as a ubiquitous species in tropical and subtropical assemblages (Schmuker and Schiebel 2002; Schiebel et al. 2004), though possibly is frequently overlooked and misclassified as pre-adult *G. glutinata*. *Globigerinita minuta* occurs at highest numbers of 13 individual per cubic meter (∼5 % of the fauna >100 μm) in the upper 60 m of the mesotrophic water column marginal to the upwelling center off Oman (Schiebel et al. 2004). From subtropical towards higher latitudes, *G. minuta* becomes less frequent. In the temperate NE Atlantic, up to one individual >100 μm per 10 m^3 occurred during late spring (Schiebel and Hemleben 2000).

Remarks: *Globigerinita minuta* is the small-sized relative of *G. glutinata*. In contrast to both *G. glutinata* and *G. uvula*, *G. minuta* bears secondary apertures on the spiral test side (cf. Saito et al. 1981).

Further readings: Brunner and Culver (1992).

2.5.4 *Globigerinita uvula* (Ehrenberg 1861) (Plate 2.32)

Description: High trochospiral pine cone shaped test with 4 chambers in the last whorl, and an umbilical aperture. Adult tests are easy to distinguish from all other extant species. *Globigerinita uvula* is possibly the only microperforate species, which resorbs its septa prior to reproduction (Hemleben et al. 1989).

Molecular genetics: Two genotypes, Type I and II, from subtropical and subpolar waters, respectively, have been identified for this species so far (André 2013; André et al. 2014).

Ecology: *Globigerinita uvula* is a frequent faunal component of the temperate to polar ocean, and decreases in abundance towards lower latitudes (Schiebel and Hemleben 2000; Schiebel et al. 2002; Bergami et al. 2009). *Globigerinita uvula* attains up to 5 % of the live assemblage (>100 μm) during spring and early summer in the temperate NE Atlantic (Schiebel et al. 1995; Schiebel and Hemleben 2000), and about 0.5–1.2 % in marginal Arctic Seas (Volkmann 2000a). The distribution of *G. uvula* in the South Atlantic and southern Indian Ocean resembles that of northern hemisphere waters (Kemle-von-Mücke and Hemleben 1999). In marginal basins like the southern Bay of Biscay, *G. uvula* was most abundant within the surface water column during spring (Retailleau et al. 2011). First data on the molecular genetics of *G. uvula* suggest a subtropical (*G. uvula* Type I) and subpolar (*G. uvula* Type II) genotype (André 2013). In general, *G. uvula* displays an opportunistic behavior to seasonally (i.e. spring) enhanced phytoplankton production in the surface ocean.

Remarks: Pre-adult tests of *G. uvula* are similar to *G. minuta*, the latter often producing a

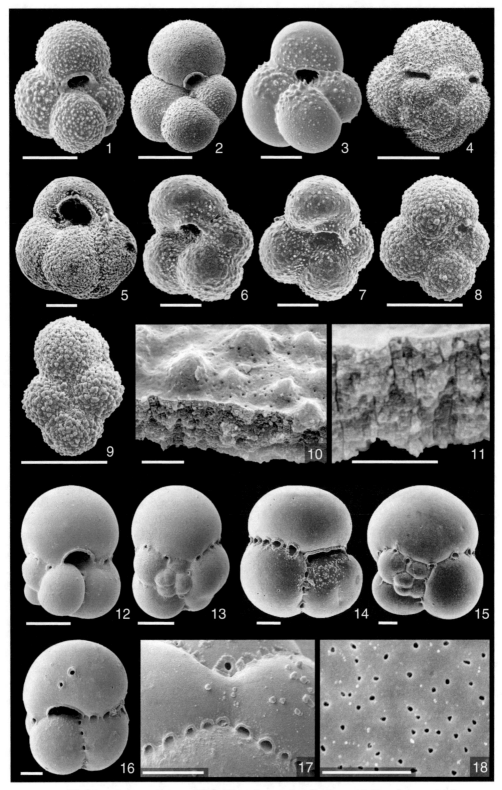

Globigerinita minuta, Globuligerina oxfordiana, and *Candeina nitida*

◀ **Plate 2.34** *(1–4) Globigerinita minuta* with *(4)* secondary apertures on the spiral test side. *(5–11) Globuligerina oxfordiana* from the Jurassic showing test features similar to modern *G. minuta*. *(10, 11)* Cross sections of test wall of *G. oxfordiana* with pores and pustules. *(12–16) Candeina nitida* with *(17)* sutural openings and *(18)* pores. Bars of overviews and *(17)* 50 µm, bars of close-ups *(10, 11)* 5 µm, *(18)* 10 µm

more pointed juvenile spire than the former, and secondary apertures on the spiral test side (Brummer 1988b). Tests of *G. minuta* may show a ('smooth') satin-like reflection under the incident light microscope, whereas *G. uvula* may exhibit a ('hard') glass-like reflection according to dissolution intensity. *Globigerinita bradyi* Wiesner 1931 is a junior synonym of *G. uvula* (Parker 1962; Saito et al. 1981).

Further readings: Parker (1962), Li (1987), Saito et al. (1981), Brummer (1988b).

2.5.5   *Candeina nitida* (d'Orbigny 1839) (Plate 2.34)

Description: Test with 3 chambers per whorl, and a high spiral winding. The primary aperture is situated over the umbilicus. Numerous accessory apertures along the sutures characterize the test of adult specimens of *C. nitida*. Numerous accessory apertures along the sutures gradually replace the primary aperture during the adult ontogenetic development. Being the largest microperforate species, *C. nitida* possesses one of the largest proloculi known in modern planktic foraminifer taxa (Sverdlove and Bé 1985). A very smooth surface characterizes this species at an early adult stage. Pustules are formed only in pre-adult stages, and are lacking on the adult test.

Molecular genetics: According to André et al. (2014) no cryptic species have been found so far.

Ecology: *Candeina nitida* is a rare faunal element in the tropical to subtropical ocean (e.g., Bradshaw 1959; Parker 1962; Bé 1977; Watkins et al. 1996). In the Caribbean Sea, *C. nitida* occurred with a maximum abundance of one individual per cubic meter (Schmuker and Schiebel 2002).

Remarks: *Candeina* is a single species genus, *C. nitida*, and not comparable with other planktic foraminifer species. Producing small pores <1 µm, *C. nitida* is grouped with the microperforate species. Pre-adult test of *C. nitida* are similar to those of the genus *Globigerinita*.

Further readings: Brummer (1988b).

2.5.6   *Tenuitella compressa* (Fordham 1986) (Plate 2.35)

Description: Low trochospiral test compressed on both sides with 4.5–6 chambers in the last whorl. The chambers may develop an ovoid character, and an ampullate final chamber is frequent. The aperture stretches from the umbilicus towards the rounded periphery and shows a medium broad lip. Test surface smooth pustules are scattered over the entire test surface.

Molecular genetics: No data available.

Ecology: *Tenuitella compressa* is assumed cosmopolitan though rare tropical to temperate species occurring in surface waters (cf. Brummer 1988b; Schmuker and Schiebel 2002).

2.5.7   *Tenuitella fleisheri* Li 1987 (Plate 2.35)

Description: Small test with lobulate outline and round periphery composed of 5–6 globular to ovate chambers in the last whorl. The extraumbilical aperture is low-arched. Additional openings may exist at the umbilical ends of sutures. Pustules are more prominent at the proximal than distal parts of the test wall.

Molecular genetics: No data available.

Ecology: *Tenuitella fleisheri* is a small sized species, which has rarely been sampled with plankton nets, which are most often equipped with 100-µm nets, or even larger mesh-sizes. *Tenuitella fleisheri* has been reported from the subtropical to temperate global ocean, but its distribution may well reach beyond the mid latitudes.

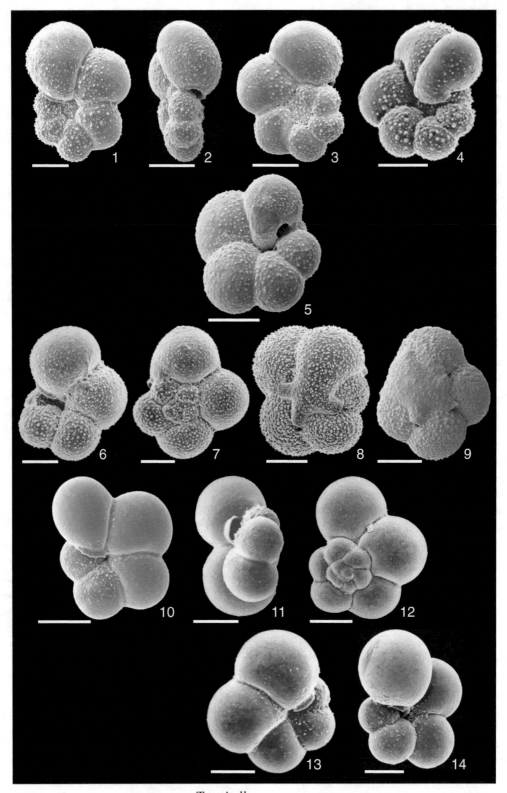

Tenuitella

◀ **Plate 2.35** (*1–4*) *Tenuitella compressa* in (*1*) umbilical, (*2*) lateral, and (*3*) spiral view. (*4*) Final chamber of *T. compressa* with umbilical flap. (*5*) *T. fleisheri* in umbilical view. (*6–9*) *T. iota*, (*6*) in umbilical, (*7*) in spiral view, and (*8, 9*) with bulla. (*10–14*) *T. parkerae*, both specimens in umbilical view. Bars 50 μm. (*7,9,13*) *Photos* A. Kiefer and R. Schiebel

Remarks: The three *Tenuitella* species *T. fleisheri*, *T. iota*, and *T. parkerea* are almost indistinguishable in their pre-adult ontogenetic stages, and classification is only possible in mature tests under the incident light microscope. In contrast to all other microperforate taxa, *T. fleisheri* does not develop an apertural flange during the juvenile stage, and may (rarely) develop a bulla in its terminal ontogenetic stage (Li 1987). *Tenuitella fleisheri* differs from *D. anfracta* by being microperforate, and producing small pustules covering the umbilical to apertural areas of the test wall (Li 1987).

towards the late phase of the SW monsoon-driven upwelling off Somalia (Conan and Brummer 2000).

Remarks: Earlier ontogenetic stages without a bulla are not easy to distinguish under the binocular microscope from the other small-sized tenuitellids *T. fleisheri* and *T. parkerae*. *Globigerina atlantisae* Cifelli and Smith 1970 is a junior synonym of *T. iota* (cf. Cifelli and Smith 1970). Li (1987) attributes *T. iota* to a new genus called *Tenuitellita* when producing a bulla.

Further readings: Parker (1962).

2.5.8 *Tenuitella iota* (Parker 1962) (Plate 2.35)

O.G.A.: *Globigerinita* Brönnimann 1951, *Turborotalita* Blow and Banner 1962.

Description: Low trochospiral test with 4–5 chambers in the last whorl. A bulla with multiple apertures is frequently produced in the final ontogenetic stage. Small pustules are scattered over the entire test. Juvenile stages of this microperforate species show the same planispirally coiled test with an equatorial aperture as in *Globigerinita*.

Molecular genetics: No data available.

Ecology: *Tenuitella iota* frequently occurs at low standing stocks in the surface subtropical to tropical oceans, and sporadically occurs in temperate waters. In the Caribbean, maximum standing stocks of *T. iota* attain one individual per cubic meter at mixed layer depths (40–60 m) well above the thermocline (Schmuker and Schiebel 2002). Similar to *D. anfracta* and *T. quinqueloba*, *T. iota* occurs in highest numbers

2.5.9 *Tenuitella parkerae* (Brönnimann and Resig 1971) (Plate 2.35)

Description: The test of *T. parkerae* is small-sized, and consists of 12–13 chambers in total and 4–5.5 chambers in the last whorl (cf. Brummer 1988b). The aperture is bordered by a large lip (Saito et al. 1981). The final chamber may be slightly ampullate. The test surface is smooth, and some pustules may be present in front of the aperture.

Molecular genetics: No data available.

Ecology: *Tenuitella parkerae* is a cosmopolitan though very rare species in the tropical to temperate ocean (Saito et al. 1981).

Remarks: Due to its inconspicuous test morphology, *T. parkerae* may often be overlooked in samples from the water column and surface sediment. Tests of *T. parkerae* could be confused with *T. iota* without bulla, and any small sized *Tenuitella* and *Globigerinita* species.

Further readings: Brönnimann and Resig (1971), Li (1987).

Appendix

Table 2.2 Morphospecies, genotypes, and subtypes of extant planktic foraminifers

Species/genotype	Geographic distribution	Faunal province	Remarks	Ref. no.
Beella digitata (1 genotype)				
Type I	S Pacific	Subtropical		17
Berggrenia pumilio			No data available	
Bolliella adamsi			No data available	
Candeina nitida (1 genotype)				
Type I	W Atlantic	Tropical		17
Dentigloborotalia anfracta			No data available	
Gallitellia vivans			No data available	
Globigerina bulloides (3 genotypes incl. 16 subtypes)				
Type Ia	Coral Sea, Arabian Sea, Central Pacific, cosmopolitan	Subtropical/tropical		2, 4, 14, 19, 25
Type Ib	Mediterranean Sea, W Indian Ocean, Red Sea,	Subtropical		8, 19, 11
	S North Atlantic Current, Canary Current, Canary Islands	Transitional/subtropical		3, 19
Type Ic	NW Pacific	Transitional		17, 21
Type Id	NW Pacific	Transitional		17, 21
Type Ie	NW Pacific			17,21
Type If				21
Type IIa	North and South Atlantic, NE Pacific	Subpolar		3, 9,19
	Nordic Sea, S Indian Ocean			23
	S North Atlantic Current, Canary Current	Transitional		3, 19
	Santa Barbara Channel	Transitional		14
Type IIb	Drake Passage, Atlantic, Nordic Sea	Subpolar/transitional		3, 9,19
	S Indian Ocean			23
	S North Atlantic Current, Canary Current, Azores Current?	Transitional/subtropical?		3, 19
Type IIc	Drake Passage, Antarctic, S Indian Ocean	Subpolar		9, 14, 19, 23
Type IId	Southern California Bight, NE Pacific	Transitional		3, 4, 14, 19
	Santa Barbara Channel			
Type IIe	NE Pacific, Sea of Okhotsk	Subpolar		19, 21, 18

(continued)

Table 2.2 (continued)

Species/genotype	Geographic distribution	Faunal province	Remarks	Ref. no.
Type IIf	NW Pacific	Subpolar		
Type IIg	S Indian Ocean			23
Type IIIa	E Atlantic			
Globigerina falconensis (1 genotype)				
	Off Canary Islands	Subtropical/transitional		3
Globigerinella calida (2–3 genotypes)				
Type IV of *G. siphonifera*				29
Type IIIb	Red Sea, W Indian Ocean, Atlantic		Ocean margins in the Atlantic	32, 34
Type IIIc	Red Sea, W Indian Ocean, Atlantic		Ocean margins in the Atlantic	32, 34
Globigerinella siphonifera (4 genotypes)				
Type Ia (1)	Caribbean Sea, Atlantic, Arabian Sea	Tropical/subtropical	Only >17.5 °C SST, cosmopolitan	1, 2, 19, 29, 32
	Azores Current, off Canary Islands	Subtropical	Probably confined to tropics/subtropics	3
Type Ia (2)	Indo-Pacific	Tropical/subtropical	Oligotrophic waters	
Type Ib	Red Sea, W Indian Ocean		patchy cosmopolitan	32
Type IIa complex (incl. 6 subtypes)	Caribbean Sea, Arabian Sea, cosmopolitan	Tropical to transitional	Tropical waters at 25 °C	2, 19, 25, 32
	Southern California Bight, Santa Barbara Channel		Waters as cold as 12 °C, mesotrophic waters, DCM	4, 5
	Off Canary Islands, Azores Current, cosmopolitan	Subtropical/transitional		3, 19, 29
Type IIa = II	Coral Sea	Tropical	Slight modification of Type IIa (1)	2, 19
Type Iib = III	E North Pacific, Southern California Bight	Tropical to transitional	Probably occurring together with Type IIa	4, 5, 19, 32
	Santa Barbara Channel, Red Sea, W Indian Ocean	Transitional		13, 19
	North Atlantic Current, Canary, Azores Current	Subtropical/transitional		3, 19, 29
Type IV	North Atlantic Drift	Subtropical/transitional		29
Globigerinita glutinata (4 genotypes)				
Type I	North Atlantic, NW Pacific	Subtropical/transitional		26,17
Type II	NW and SW Pacific	Subtropical		26,17
Type III	North Atlantic, NW Pacific	Subtropical		17
Type IV	Arabian Sea	Subtropical		17
Globigerinita minuta		No data available		
Globigerinita uvula (2 genotypes)				
Type I	Subarctic Atlantic	Polar to transitional		3, 8, 17, 19
Type II	Not specified	Not specified		17
Globigerinoides conglobatus (1 genotype)				
			Closely related to *G. ruber*$_W$	17
Globigerinoides elongatus				
			1 genotype	33
Globigerinoides ruber (5 genotypes)				

(continued)

Table 2.2 (continued)

Species/genotype	Geographic distribution	Faunal province	Remarks	Ref. no.
Pink	Caribbean Sea off Curaçao, Atlantic	Tropical/subtropical		2, 19
	Off Canary Islands, Mediterranean Sea	subtropical	Atlantic only	3, 19
Type Ia	Coral Sea, Pacific, Indian Ocean	Tropical	SST 19–23 °C	2, 12, 13, 25
Type Ia1	Atlantic, off Canary Islands	Subtropical		12, 13
	Caribbean Sea off Puerto Rico, Mediterranean Sea	tropical		7
Type Ib	Atlantic, Indian Ocean	Tropical/subtropical		12, 13, 25
Type Ib1	Caribbean Sea, Arabian Sea			12, 13, 19, 25
Type Ib2	Arabian Sea, Pacific off Japan	Transitional	Probably also present in colder waters	4, 5, 13, 19, 25
Type Ic	Central Pacific	Subtropical/tropical		20
Type IIa	NE Pacific, Southern California Bight	Subtropical	Probably *G. elongatus* plexus	19, 25
	Santa Barbara Channel, Arabian Sea, Atlantic			12, 13, 19
Type IIa1	Pacific, E Mediterranean Sea	Transitional	Probably *G. elongatus* plexus	12, 13
Type IIa2	Pacific, W Mediterranean Sea	Subtropical	Probably *G. elongatus* plexus	12, 13
Type IIb	E Atlantic		*Globigerinoides* sp. (new name aquired)	12, 13
Globigerinoides sacculifer (1 genotype)				
Type I		Subtropical/tropical	Incl. 5 different morpho-types: *G. trilobus*, *G. sacculifer*, *G. immaturus*, *G. quadrilobatus*, *G. fistulosa*	11
Globoquadrina conglomerata (1 genotype)				
Type I	Indian Ocean	Tropical/subtropical		17
Globorotalia cavernula			No data available	
Globorotalia crassaformis			No data available	
Globorotalia hirsuta				
1 Type	Atlantic			8
Globorotalia inflata (2 genotypes)				
Type I	Cosmopolitan	Transitional/subtropical		15, 22
Type II	Antarctic	Subpolar		15, 22
Globorotalia menardii				
Type I				8
Globorotalia scitula (1 genotype)				
Type I				8
Globorotalia theyeri			No data available	
Globorotalia truncatulinoides (4 genotypes)				
Type 1 (I)	Brazil Current, S Indian Ocean, S Pacific	Tropical to transitional	DCM, deep water column, vertical and spatial niche separation	10, 19, 23, 24, 27

(continued)

Table 2.2 (continued)

Species/genotype	Geographic distribution	Faunal province	Remarks	Ref. no.
Type 2 (II)	Brazil Current, Canary Current	Tropical/subtropical	North Equatorial Current	10, 19, 24
	Sargasso Sea, Mediterranean, South Atlantic			27
Type 3 (III)	South Atlantic, Subantartic Convergence	Transitional/subpolar	Strong vertical mixing, high productivity	10, 19, 23, 24
	S Indian Ocean, Subtropical Frontal Zone	Subtropica to subpolar	Subtropical Frontal Zone	27
Type 4 (IV)	Falkland Current, Polar Frotal Zone	Subpolar/polar	Cold, dense, nutrient rich productive	10, 19, 23, 24, 27
Type V	NW Pacific	Subtropical	Off Japan	24
Globorotalia tumida (1 genotype)				
Type I	Cosmopolitan	Tropical/subtropical		17
Globorotalia ungulata (1 genotype)				
Type I				25
Globorotaloides hexagonus			No data available	
Globoturborotalita rubescens			No data available	
Globoturborotalita tenella			No data available	
Hastigerina pelagica (3 genotypes)				
Type I	Mediterranean Sea, W Pacific			17, 30, 33
Type IIa	Cosmopolitan		300–400 m depth	30, 33
Type IIb	Cosmopolitan		Upper 100 m	30, 33
Hastigerinella digitata			No data available	30
Neogloboquadrina dutertrei (3 genotypes)				
Type Ia	Caribbean Sea off Cuaraçao	Tropical	Inconclusive after André et al. (2014)	2, 14, 15
	Azores Current	subtropical		3, 14
Type Ib	Coral Sea	tropical		14,15
Type Ic	Santa Barbara Channel	transitional		14,15
Neogloboquadrina incompta (2 genotypes)				
Type I	Atlantic	Subpolar/transitional		16, 19
Type II	E North Pacific	Subpolar/transitional		16, 19
Neogloboquadrina pachyderma (4 genotypes)				
Type I sin	Arctic/North Atlantic	Polar		16
Type I dex	Drake Passage, North Atlantic, Nordic Sea, Benguela Current	Subpolar/transitional		9, 15, 16
Type II sin	Antarctic			15, 16
Type II dex	Santa Barbara Channel, E North Pacific	Transitional		15, 16
Type III sin	Antarctic, Drake Passage	Subpolar		9, 16
Type IV sin	Antarctic, Drake Passage	Polar	In sea ice	9, 15, 16
Type V sin	Benguela Current	Transitional		15, 16
Type VI	Benguela Current			18
Type VII	North Pacific			18
Orbulina universa (3 genotypes)				

(continued)

Table 2.2 (continued)

Species/genotype	Geographic distribution	Faunal province	Remarks	Ref. no.
Type I	Caribbean Sea off Curaçao, Coral Sea	Tropical	Throughout the Atlantic	2, 19
	Caribbean Sea off Puerto Rico, Atlantic, S Indian Ocean, S Pacific	Subtropical/tropical	Oligotrophic waters	6, 19, 23, 31
Type II	Sargasso Sea, Atlantic, S Pacific	Subtropical/tropical	Spezialized adaptions?	6, 19, 31
Type III	Southern California Bight, Mediterranean Sea, Santa Barbara Channel, S Indian Ocean	Transitional/subtropical	Eutrophic waters throughout the Atlantic	4, 6, 19, 23
	Mediterranean Sea, Atlantic, Red Sea, S Indian Ocean	Subtropical/tropical	Marginal transitional zones	6, 19, 23, 31
Orcadia riedeli			No data available	
Pulleniatina obliquiloculata (3 genotypes)				
Type I	Coral Sea, cosmopolitan	Transitional to tropical		17, 28
Type IIa	equatorial Pacific	Subtropical/tropical		17, 28
Type IIb	Pacific	Transitional to tropical		17, 28
Sphaeroidinella dehiscens (1 genotype)				
Type I	Indo-Pacific	Tropical		17
Streptochilus globigerus				
			Benthic species identical to *Bolivina variabilis*	36
Tenuitella compressa			No data available	
Tenuitella fleisheri			No data available	
Tenuitella iota			No data available	
Tenuitella parkerae			No data available	
Turborotalita clarkei			No data available	
Turborotalita humilis/cristata			No data available	
Turborotalita quinqueloba (5 genotypes)				
Type Ia	Coral Sea	Subtropical/tropical		3, 9, 14, 19
Type Ib	Arabian Sea, N Indian Ocean	Subtropical/tropical		19, 25
Type IIa	Drake Passage, North Atlantic	Subpolar		3, 9, 14, 19
Type IIb	North Atlantic, Nordic Seas	Polar/subpolar		3, 9, 14, 19
Type IIc	Antarctic, NE Pacific	Subpolar/transitional		9, 14, 19
Type IId	Antarctic, NE Pacific	Subpolar/transitional		14, 19

Geographic distribution, faunal province, and remarks as referred (Reference-Number, Ref. No., 1–36). DCM is Deep chlorophyll maximum, and SST is sea surface temperature

(1) Huber et al. (1997), (2) Darling et al. (1997), (3) Stewart et al. (2001), (4) Darling et al. (1999), (5) Kucera et al. (2001), (6) De Vargas et al. (1999), (7) Pawlowski et al. (1997), (8) De Vargas et al. (1997), (9) Darling et al. (2000), (10) De Vargas et al. (2001), (11) André et al. (2013), (12) Aurahs et al. (2009b), (13) Aurahs et al. (2011), (14) Darling et al. (2003), (15) Darling et al. (2004), (16) Darling et al. (2006), (17) André et al. (2014), (18) Darling et al. (2007), (19) Darling and Wade (2008), (20) Kuronayagi et al. (2008), (21) Kurasawa et al. (accepted), (22) Morard et al. (2011), (23) Morard et al. (2013), (24) Quillévéré et al. (2013), (25) Seears et al. (2012), (26) Ujiié and Lipps (2009), (27) Ujiié et al. (2010), (28) Ujiié et al. (2012), (29) De Vargas et al. (2002), (30) Weiner et al. (2012), (31) Morard et al. (2009), (32) Weiner et al. (2014), (33) Weiner (2014), (34) Weiner et al. (2015), (35) Aurahs et al. (2011), (36) Darling et al. (2009)

References

Adl SM, Simpson AGB, Farmer MA, Andersen RA, Anderson OR, Barta JR, Bowser SS, Brugerolle G, Fensome RA, Fredericq S, James TY, Karpov S, Kugrens P, Krug J, Lane CE, Lewis LA, Lodge J, Lynn DH, Mann DG, Mccourt RM, Mendoza L, Moestrup O, Mozley-Standridge SE, Nerad TA, Shearer CA, Smirnov AV, Spiegel FW, Taylor MFJR (2005) The new higher level classification of eukaryotes with emphasis on the taxonomy of protists. J Eukaryot Microbiol 52:399–451. doi:10.1111/j.1550-7408.2005.00053.x

Aldridge D, Beer CJ, Purdie DA (2011) Calcification in the planktonic 7. *Globigerina bulloides* linked to phosphate concentrations in surface waters of the North Atlantic Ocean. Biogeosciences Discuss 8:6447–6472. doi:10.5194/bgd-8-6447-2011

Alldredge AL, Jones BM (1973) *Hastigerina pelagica*: Foraminiferal habitat for planktonic dinoflagellates. Mar Biol 22:131–135

Almogi-Labin A (1984) Population dynamics of planktic Foraminifera and Pteropoda—Gulf of Aqaba, Red Sea. Proc K Ned Akad Van Wet Ser B Palaeontol Geol Phys Chem 87:481–511

Anderson OR (1983) Radiolaria. Springer, New York

Anderson OR, Bé AWH (1976) A cytochemical fine structure study of phagotrophy in a planktonic foraminifer, *Hastigerina pelagica* (d'Orbigny). Biol Bull 151:437–449

Anderson OR, Bé AWH (1978) Recent advances in foraminiferal fine structure research. In: Hedley RH, Adams CG (eds) Foraminifera 1. Academic Press, London

Anderson OR, Spindler M, Bé AWH, Hemleben C (1979) Trophic activity of planktonic Foraminifera. J Mar Biol Assoc U K 59:791–799. doi:10.1017/S002531540004577X

André A (2013) Taxonomies moléculaire et morphologique chez les foraminifères planctoniques: Élaboration d'un référentiel et cas particuliers de *Globigerinoides sacculifer* et *Neogloboquadrina pachyderma*. PhD Thesis, University of Lyon

André A, Weiner A, Quillévéré F, Aurahs R, Morard R, Douady CJ, de Garidel-Thoron T, Escarguel G, de Vargas C, Kucera M (2013) The cryptic and the apparent reversed: lack of genetic differentiation within the morphologically diverse plexus of the planktonic foraminifer *Globigerinoides sacculifer*. Paleobiology 39:21–39. doi:10.1666/0094-8373-39.1.21

André A, Quillévéré F, Morard R, Ujiié Y, Escarguel G, de Vargas C, de Garidel-Thoron T, Douady CJ (2014) SSU rDNA divergence in planktonic Foraminifera: molecular taxonomy and biogeographic implications. PLoS ONE. doi:10.1371/journal.pone.0104641

Arnold AJ (1983) Phyletic evolution in *Globorotalia crassaformis* (Galloway and Wissler) lineage: a preliminary report. Paleobiology 9:390–397

Asahi H, Takahashi K (2007) A 9-year time-series of planktonic foraminifera fluxes and environmental change in the Bering sea and the central subarctic Pacific Ocean, 1990–1999. Prog Oceanogr 72:343–363. doi:10.1016/j.pocean.2006.03.021

Asano K, Ingle JC, Takayanagi Y (1968) Origin and development of *Globigerina quinqueloba* Natland in the North Pacific 2:213–241

Aurahs R, Göker M, Grimm GW, Hemleben V, Hemleben C, Schiebel R, Kučera M (2009a) Using the multiple analysis approach to reconstruct phylogenetic relationships among planktonic Foraminifera from highly divergent and length-polymorphic SSU rDNA sequences. Bioinforma Biol Insights 3:155–177

Aurahs R, Grimm GW, Hemleben V, Hemleben C, Kucera M (2009b) Geographical distribution of cryptic genetic types in the planktonic foraminifer *Globigerinoides ruber*. Mol Ecol 18:1692–1706. doi:10.1111/j.1365-294X.2009.04136.x

Aurahs R, Treis Y, Darling K, Kucera M (2011) A revised taxonomic and phylogenetic concept for the planktonic foraminifer species *Globigerinoides ruber* based on molecular and morphometric evidence. Mar Micropaleontol 79:1–14. doi:10.1016/j.marmicro.2010.12.001

Auras-Schudnagies A, Kroon D, Ganssen G, Hemleben C, van Hinte JE (1989) Distributional pattern of planktonic foraminifers and pteropods in surface waters and top core sediments of the Red Sea, and adjacent areas controlled by the monsoonal regime and other ecological factors. Deep-Sea Res I 36:1515–1533. doi:10.1016/0198-0149(89)90055-1

Bandy OL (1972) Origin and developement of *Globorotalia (Turborotalia) pachyderma* (Ehrenberg). Micropaleontology 18:294–318

Bandy OL, Theyer F (1971) Growth variation in *Globorotalia pachyderma* (Ehrenberg). Antarct J 6:172–174

Banner FT (1965) On *Hastigerinella digitata* (Rhumbler, 1911). Micropaleontology 114–116

Banner FT (1982) A classification and introduction to the Globigerinacea. In: Banner FT, Lord AR (eds) Aspects of micropalaeontology, vol 5. Springer, London, pp 142–239

Banner FT, Blow WH (1959) The classification and stratigraphic distribution of the Globigerinacea. Palaeontology 1–27

Banner FT, Blow WH (1960) The taxonomy, morphology and affinities of the genera included in the subfamily Hastigerininae. Micropaleontology 6:19–31

Banner FT, Blow WH (1967) The origin, evolution and taxonomy of the foraminiferal genus *Pulleniatina* Cushman, 1927. Micropaleontology 13:133–162. doi:10.2307/1484667

Bauch HA (1994) *Beella megastoma* (Earland) in late Pleistocene Norwegian-Greenland Sea sediments: stratigraphy and meltwater implication. J Foraminifer Res 24:171–177. doi:10.2113/gsjfr.24.3.171

Bauch D, Darling K, Simstich J, Bauch HA, Erlenkeuser H, Kroon D (2003) Palaeoceanographic implications of genetic variation in living North Atlantic *Neogloboquadrina pachyderma*. Nature 424:299–302

Baumfalk YA, Troelstra SR, Ganssen G, van Zanen MJL (1987) Phenotypic variation of *Globorotalia scitula* (Foraminiferida) as a response to Pleistocene climatic fluctuations. Mar Geol 75:231–240

Bé AWH (1967) *Globorotalia carvernula*, a new species of planktonic Foraminifera from the subantarctic Pacific Ocean. Cushman Found Foraminifer Res Contrib 18:128–133

Bé AWH (1977) An ecological, zoogeographic and taxonomic review of recent planktonic Foraminifera. In: Ramsay ATS (ed) Oceanic micropaleontology. Academic Press, London, pp 1–100

Bé AWH (1980) Gametogenic calcification in a spinose planktonic foraminifer, *Globigerinoides sacculifer* (Brady). Mar Micropaleontol 5:283–310. doi:10.1016/0377-8398(80)90014-6

Bé AWH, Hemleben C (1970) Calcification in a living planktonic foraminifer, *Globigerinoides sacculifer* (Brady). Neues Jahrb Geol Paläontol 134:221–234

Bé AWH, Hutson WH (1977) Ecology of planktonic Foraminifera and biogeographic patterns of life and fossil assemblages in the Indian Ocean. Micropaleontology 23:369. doi:10.2307/1485406

Bé AWH, Tolderlund DS (1971) Distribution and ecology of living planktonic Foraminifera in surface waters of the Atlantic and Indian Oceans. In: Funell BM, Riedel WR (eds) The micropalaeontology of oceans. University Press, Cambridge, pp 105–149

Bé AWH, Harrison SM, Lott L (1973) *Orbulina universa* (d'Orbigny) in the Indian Ocean. Micropaleontology 19:150–192

Bé AWH, Anderson OR, Faber WW, Caron DA (1983) Sequence of morphological and cytoplasmic changes during gametogenesis in the planktonic foraminifer *Globigerinoides sacculifer* (Brady). Micropaleontology 29:310. doi:10.2307/1485737

Bé AWH, Bishop JKB, Sverdlove MS, Gardner WD (1985) Standing stock, vertical distribution and flux of planktonic Foraminifera in the Panama Basin. Mar Micropaleontol 9:307–333

Belyaeva N, Burmistrova I (2003) Planktonic Foraminifera in the recent sediments of the sea of Okhotsk. Oceanology 43:206–214

Bemis BE, Spero HJ, Bijma J, Lea DW (1998) Reevaluation of the oxygen isotopic composition of planktonic Foraminifera: experimental results and revised paleotemperature equations. Paleoceanography 13:150–160

Bemis BE, Spero HJ, Lea DW, Bijma J (2000) Temperature influence on the carbon isotopic composition of *Globigerina bulloides* and *Orbulina universa* (planktonic Foraminifera). Mar Micropaleontol 38:213–228

Bergami C, Capotondi L, Langone L, Giglio F, Ravaioli M (2009) Distribution of living planktonic Foraminifera in the Ross Sea and the Pacific sector of the Southern Ocean (Antarctica). Mar Micropaleontol 73:37–48. doi:10.1016/j.marmicro.2009.06.007

Berger WH (1969) Ecologic patterns of living planktonic Foraminifera. Deep-Sea Res 16:1–24. doi:10.1016/0011-7471(69)90047-3

Berger WH, Soutar A (1967) Planktonic Foraminifera: field experiment on production rate. Science 156:1495–1497

Bermudez PJ (1960) Foraminiferos planctonicos del Golfo de Venezuela. Mem Soc Cienc Nat Salle 20:58–76

Bijma J, Erez J, Hemleben C (1990a) Lunar and semi-lunar reproductive cycles in some spinose planktonic foraminifers. J Foraminifer Res 20:117–127

Bijma J, Faber WW, Hemleben C (1990b) Temperature and salinity limits for growth and survival of some planktonic foraminifers in laboratory cultures. J Foraminifer Res 20:95–116. doi:10.2113/gsjfr.20.2.95

Bijma J, Hemleben C, Oberhänsli H, Spindler M (1992) The effects of increased water fertility on tropical spinose planktonic foraminifers in laboratory cultures. J Foraminifer Res 22:242–256

Bijma J, Hemleben C, Huber BT, Erlenkeuser H, Kroon D (1998) Experimental determination of the ontogenetic stable isotope variability in two morphotypes of *Globigerinella siphonifera* (d'Orbigny). Mar Micropaleontol 35:141–160

Bolli HM, Saunders JB, Perch-Nielsen K (1985) Plankton stratigraphy. Cambridge University Press, Cambridge

Boltovskoy E (1974) *Globorotalia hirsuta eastropacia* n. subsp. planktonic subspecies (Foraminiferida) from the tropical Pacific Ocean. Rev Esp Micropaleontol 6:127–133

Boltovskoy E, Watanabe S (1981) Foraminiferos de los sedimentos cuatarios entre Tierra del Fuego e Islas Georgias del Sur. Rev Mus Argent Cienc Nat (Bernardino Rivadavia) Geol 8:95–124

Boussetta S, Kallel N, Bassinot F, Labeyrie L, Duplessy JC, Caillon N, Dewilde F, Rebaubier H (2012) Mg/Ca-paleothermometry in the western Mediterranean Sea on planktonic foraminifer species *Globigerina bulloides*: constraints and implications. Comptes Rendus Geosci 344:267–276. doi:10.1016/j.crte.2012.02.001

Bouvier-Soumagnac Y, Duplessy JC (1985) Carbon and oxygen isotopic composition of planktonic Foraminifera from laboratory culture, plankton tows and Recent sediment; implications for the reconstruction of paleoclimatic conditions and of the global carbon cycle. J Foraminifer Res 15:302–320. doi:10.2113/gsjfr.15.4.302

Bradshaw JS (1959) Ecology of living planktonic Foraminifera in the North and Equatorial Pacific Ocean. Contrib Cushman Found Foraminifer Res 10:25–64

Broecker WS, Pena LD (2014) Delayed Holocene reappearance of *G. menardii*. Paleoceanography 29:291–295. doi:10.1002/2013PA002590

Brönnimann P, Resig J (1971) A Neogene globigerinacean biochronologic time-scale of the southwestern Pacific. Initial Rep Deep Sea Drill Proj 7:1235–1469

Brummer GJA (1988a) Comparative ontogeny and species definition of planktonic foraminifers: a case study of *Dentigloborotalia anfracta* n. gen. In: Brummer GJA, Kroon D (eds) Planktonic foraminifers as tracers of ocean-climate history. Free University Press, Amsterdam, pp 51–75

Brummer GJA (1988b) Comparative ontogeny of modern microperforate planktonic foraminifers. In: Brummer GJA, Kroon D (eds) Planktonic foraminifers as tracers of ocean-climate history. Free University Press, Amsterdam, pp 77–129

Brummer GJA, Kroon D (1988) Planktonic foraminifers as tracers of ocean-climate history: ontogeny, relationships and preservation of modern species and stable isotopes, phenotypes and assemblage distribution in different water masses. Free University Press, Amsterdam

Brummer GJA, Hemleben C, Spindler M (1987) Ontogeny of extant spinose planktonic Foraminifera (Globigerinidae): a concept exemplified by *Globigerinoides sacculifer* (Brady) and *G. ruber* (d'Orbigny). Mar Micropaleontol 12:357–381. doi:10.1016/0377-8398(87)90028-4

Brummer GJA, Troelstra SR, Kroon D, Ganssen GM (1988) Ontogeny, distribution and geologic record of the extant planktonic foraminifer *Orcadia riedeli* (Rögl & Bolli 1973). In: Brummer GJA, Kroon D (eds) Planktonic foraminifers as tracers of ocean-climate history. Free University Press, Amsterdam, pp 149–162

Brunner CA, Culver SJ (1992) Quaternary Foraminifera from the Walls of Wilmington, South Wilmington, and North Heyes Canyons, U.S. East Coast: implications for continental slope and rise evolution. Palaios 7:34–66. doi:10.2307/3514795

Caron M (1985) Cretaceous planktic Foraminifera. In: Bolli HM, Saunders JB, Perch-Nielsen K (eds) Plankton stratigraphy. Cambridge University Press, Cambridge

Caron DA, Bé AWH (1984) Predicted and observed feeding rates of the spinose planktonic foraminifer *Globigerinoides sacculifer*. Bull Mar Sci 35:1–10

Caron DA, Bé AWH, Anderson OR (1982) Effects of variations in light intensity on life processes of the planktonic foraminifer *Globigerinoides sacculifer* in laboratory culture. J Mar Biol Assoc UK 62:435–451

Caron DA, Faber WW, Bé AWH (1987) Effects of temperature and salinity on the growth and survival of the planktonic foraminifer *Globigerinoides sacculifer*. J Mar Biol Assoc UK 67:323–341

Carstens J, Wefer G (1992) Recent distribution of planktonic Foraminifera in the Nansen Basin, Arctic Ocean. Deep-Sea Res I 39:507–524

Carstens J, Hebbeln D, Wefer G (1997) Distribution of planktic Foraminifera at the ice margin in the Arctic (Fram Strait). Mar Micropaleontol 29:257–269

Cavalier-Smith T (2004) Only six kingdoms of life. Proc R Soc B Biol Sci 271:1251–1262. doi:10.1098/rspb.2004.2705

Chaisson WP, d' Hondt S (2000) Neogene planktonic foraminifer biostratigraphy at Site 999, Western Caribbean Sea. In: Leckie RM, Sigurdsson H,

Acton GD, Draper G (eds) Proceedings of the ocean drilling program, scientific results. Ocean Drilling Program, pp 19–56

Chaisson WP, Pearson PN (1997) Planktonic foraminifer biostratigraphy at Site 925: middle Miocene–Pleistocene. In: Shackleton NJ, Curry WB, Richter C, Brawlower TJ (eds) Proceedings of the ocean drilling program, scientific results. Ocean Drilling Program, pp 3–31

Chapman MR (2010) Seasonal production patterns of planktonic Foraminifera in the NE Atlantic Ocean: implications for paleotemperature and hydrographic reconstructions. Paleoceanography. doi:10.1029/2008PA001708

Christiansen BO (1965) A bottom form of the planktonic foraminifer *Globigerinoides rubra* (d'Orbigny, 1839). Publicacione Stazione Zool Napoli 34:197–202

Cifelli R (1961) *Globigerina incompta*, a new species of pelagic Foraminifera from the North Atlantic. Cushman Found Foram Res Contr 12:83–86

Cifelli R (1971) On the temerature relationship of planktonic Foraminifera. J Foraminifer Res 1:170–177

Cifelli R (1973) Observations on *Globigerina pachyderma* (Ehrenberg) and *Globigerina incompta* Cifelli from the North Atlantic. J Foraminifer Res 3:157–166

Cifelli R, Smith RK (1970) Distribution of planktonic Foraminifera in the vicinity of the North Atlantic Current. Smithsonian Contrib Paleobiology 4:1–52

Cléroux C, Cortijo E, Duplessy JC, Zahn R (2007) Deep-dwelling Foraminifera as thermocline temperature recorders. Geochem Geophys Geosystems. doi:10.1029/2006GC001474

Cléroux C, Cortijo E, Anand P, Labeyrie L, Bassinot F, Caillon N, Duplessy J (2008) Mg/Ca and Sr/Ca ratios in planktonic Foraminifera: proxies for upper water column temperature reconstruction. Paleoceanography. doi:10.1029/2007PA001505

Cléroux C, Lynch-Stieglitz J, Schmidt MW, Cortijo E, Duplessy JC (2009) Evidence for calcification depth change of *Globorotalia truncatulinoides* between deglaciation and Holocene in the western Atlantic Ocean. Mar Micropaleontol 73:57–61

Cléroux C, deMenocal P, Arbuszewski J, Linsley B (2013) Reconstructing the upper water column thermal structure in the Atlantic Ocean. Paleoceanography 28:503–516

Coadic R, Bassinot F, Dissard D, Douville E, Greaves M, Michel E (2013) A core-top study of dissolution effect on B/Ca in *Globigerinoides sacculifer* from the tropical Atlantic: potential bias for paleo-reconstruction of seawater carbonate chemistry. Geochem Geophys Geosystems 14:1053–1068. doi:10.1029/2012GC004296

Conan SMH, Brummer GJA (2000) Fluxes of planktic Foraminifera in response to monsoonal upwelling on the Somalia Basin margin. Deep-Sea Res II 47:2207–2227

D'Orbigny AD (1839) Foraminifères. In: de la Sagra R (ed) Histoire Physique, Politique et Naturelle de I'Ile de Cuba. Arthus Bertrand, Paris, pp 1–51

Darling KF, Wade CM (2008) The genetic diversity of planktic Foraminifera and the global distribution of ribosomal RNA genotypes. Mar Micropaleontol 67:216–238

Darling KF, Wade CM, Kroon D, Brown AJL (1997) Planktic foraminiferal molecular evolution and their polyphyletic origins from benthic taxa. Mar Micropaleontol 30:251–266

Darling KF, Wade CM, Kroon D, Brown AJL, Bijma J (1999) The diversity and distribution of modern planktic foraminiferal small subunit ribosomal RNA genotypes and their potential as tracers of present and past ocean circulations. Paleoceanography 14:3–12

Darling KF, Wade CM, Stewart IA, Kroon D, Dingle R, Brown AJL (2000) Molecular evidence for genetic mixing of Arctic and Antarctic subpolar populations of planktonic foraminifers. Nature 405:43–47

Darling KF, Kucera M, Wade CM, von Langen P, Pak D (2003) Seasonal distribution of genetic types of planktonic foraminifer morphospecies in the Santa Barbara Channel and its paleoceanographic implications. Paleoceanography. doi:10.1029/2001PA000723

Darling KF, Kucera M, Pudsey CJ, Wade CM (2004) Molecular evidence links cryptic diversification in polar planktonic protists to Quaternary climate dynamics. Proc Natl Acad Sci USA 101:7657–7662

Darling KF, Kucera M, Kroon D, Wade CM (2006) A resolution for the coiling direction paradox in Neogloboquadrina pachyderma. Paleoceanography. doi:10.1029/2005PA001189

Darling KF, Kucera M, Wade CM (2007) Global molecular phylogeography reveals persistent Arctic circumpolar isolation in a marine planktonic protist. Proc Natl Acad Sci 104:5002–5007

Darling KF, Thomas E, Kasemann SA, Seears HA, Smart CW, Wade CM (2009) Surviving mass extinction by bridging the benthic/planktic divide. Proc Natl Acad Sci 106:12629–12633

De Klasz I, Kroon D, van Hinte JE (1989) Notes on the foraminiferal genera Laterostomella de Klasz and Rerat and Streptochilus Brönnimann and Resig. J Micropalaeontology 8:215–225

De Vargas C, Pawlowski J (1998) Molecular versus taxonomic rates of evolution in planktonic Foraminifera. Mol Phylogenet Evol 9:463–469

De Vargas C, Zaninetti L, Hilbrecht H, Pawlowski J (1997) Phylogeny and rates of molecular evolution of planktonic Foraminifera: SSU rDNA sequences compared to the fossil record. J Mol Evol 45:285–294

De Vargas C, Norris R, Zaninetti L, Gibb SW, Pawlowski J (1999) Molecular evidence of cryptic speciation in planktonic foraminifers and their relation to oceanic provinces. Proc Natl Acad Sci 96:2864–2868

De Vargas C, Renaud S, Hilbrecht H, Pawlowski J (2001) Pleistocene adaptive radiation in Globorotalia truncatulinoides: Genetic, morphologic, and environmental evidence. Paleobiology 27:104–125

De Vargas C, Bonzon M, Rees NW, Pawlowski J, Zaninetti L (2002) A molecular approach to biodiversity and biogeography in the planktonic foraminifer Globigerinella siphonifera (d'Orbigny). Mar Micropaleontol 45:101–116

De Vargas C, Audic S, et al. (2015) Eukaryotic plankton diversity in the sunlit ocean. Science 348. doi:10.1126/science.1261605

Deuser WG, Ross EH, Hemleben C, Spindler M (1981) Seasonal changes in species composition, numbers, mass, size, and isotopic composition of planktonic Foraminifera settling into the deep Sargasso Sea. Palaeogeogr Palaeoclimatol Palaeoecol 33:103–127

Deuser WG, Muller-Karger FE, Hemleben C (1988) Temporal variations of particle fluxes in the deep subtropical and tropical North Atlantic: Eulerian versus Lagrangian effects. J Geophys Res Oceans 1978–2012(93):6857–6862

Dieckmann G, Spindler M, Lange MA, Ackley SF, Eicken H (1991) Antarctic sea ice: a habitat for the foraminifer Neogloboquadrina pachyderma. J Foraminifer Res 21:182–189

Dittert N, Baumann KH, Bickert T, Henrich R, Huber R, Kinkel H, Meggers H (1999) Carbonate dissolution in the deep-sea: methods, quantification and paleoceanographic application. In: Fischer G, Wefer G (eds) Use of proxies in paleoceanography. Springer, Berlin, pp 255–284

Dueñas-Bohórquez A, Da Rocha RE, Kuroyanagi A, De Nooijer LJ, Bijma J, Reichart GJ (2011) Interindividual variability and ontogenetic effects on Mg and Sr incorporation in the planktonic foraminifer Globigerinoides sacculifer. Geochim Cosmochim Acta 75:520–532

Duplessy JC, Bé AWH, Blanc PL (1981) Oxygen and carbon isotopic composition and biogeographic distribution of planktonic Foraminifera in the Indian Ocean. Palaeogeogr Palaeoclimatol Palaeoecol 33:9–46

Earland A (1934) Foraminifera: Falklands sector of the Antarctic (excluding South Georgia). University Press, Cambridge

Eggins S, De Dekker P, Marshall J (2003) Mg/Ca variation in planktonic Foraminifera tests: implications for reconstructing palaeo-seawater temperature and habitat migration. Earth Planet Sci Lett 212:291–306

Eggins S, Sadekov A, De Deckker P (2004) Modulation and daily banding of Mg/Ca in Orbulina universa tests by symbiont photosynthesis and respiration: a complication for seawater thermometry? Earth Planet Sci Lett 225:411–419

Erez J (1983) Calcification rates, photosynthesis and light in planktonic Foraminifera. In: Westbroek P, de Jong EW (eds) Biomineralization and biological metal accumulation. Reidel Publishing Company, Dordrecht, pp 307–312

Erez J, Almogi-Labin A, Avraham S (1991) On the life history of planktonic Foraminifera: lunar reproduction cycle in Globigerinoides sacculifer (Brady). Paleoceanography 6:295–306

Ericson DB, Wollin G, Wollin J (1954) Coiling direction of Globorotalia truncatulinoides in deep-sea cores. Deep Sea Res 1953 2:152–158

Faber WW, Anderson OR, Caron DA (1989) Algal-foraminiferal symbiosis in the planktonic foraminifer *Globigerinella aequilateralis*: II. Effects of two symbiont species on foraminiferal growth and longevity. J Foraminifer Res 19:185–193

Fairbanks RG, Wiebe PH (1980) Foraminifera and chlorophyll maximum: vertical distribution, seasonal succession, and paleoceanographic significance. Science 209:1524–1526

Fairbanks RG, Sverdlove M, Free R, Wiebe PH, Bé AWH (1982) Vertical distribution and isotopic fractionation of living planktonic Foraminifera from the Panama Basin. Nature 298:841–844

Fischer G, Wefer G (1999) Use of proxies in paleoceanography: examples from the South Atlantic. Springer, Berlin

Fleisher RL (1974) Cenozoic planktonic Foraminifera and biostratigraphy, Arabian Sea, Deep Sea Drilling Project, Leg 23B. In: Whitmarsh RB, Weser OE, Ross DA (eds) Initial Report of the Deep Sea Drilling Project, Leg 23. Washington, pp 1001–1071

Friedrich O, Schiebel R, Wilson PA, Weldeab S, Beer CJ, Cooper MJ, Fiebig J (2012) Influence of test size, water depth, and ecology on Mg/Ca, Sr/Ca, δ^{18}O and δ^{13}C in nine modern species of planktic foraminifers. Earth Planet Sci Lett 319–320:133–145. doi:10.1016/j.epsl.2011.12.002

Ganssen G (1983) Dokumentation von küstennahem Auftrieb anhand stabiler Isotope in Rezenten Foraminiferen vor Nordwest-Afrika. Meteor Forschungsergebnisse Reihe C: 1–46

Gastrich MD, Bartha R (1988) Primary productivity in the planktonic foraminifer *Globigerinoides ruber* (d'Orbigny). J Foraminifer Res 18:137–142

Glaçon G, Sigal J (1969) Morphological precision on wall of *Globorotalia truncatulinoides* (d'Orbigny), *Globigerinoides ruber* (d'Orbigny) and *Globigerinoides trilobus* (Reuss). Comptes Rendues Hebd Seances Acad Sci Ser D 269:987

Glaçon G, Vergnaud-Grazzini C, Leclaire L, Sigal J (1973) Presence des foraminifères: *Globorotalia crassula* Cushman et Steward et *Globorotalia hirsuta* (d'Orbigny) en mer Mediterranée. Rev Esp Micropaleontol 5:373–401

Guptha MVS, Curry WB, Ittekkot V, Muralinath AS (1997) Seasonal variation in the flux of planktic Foraminifera; sediment trap results from the Bay of Bengal, northern Indian Ocean. J Foraminifer Res 27:5–19

Hamilton CP, Spero HJ, Bijma J, Lea DW (2008) Geochemical investigation of gametogenic calcite addition in the planktonic Foraminifera *Orbulina universa*. Mar Micropaleontol 68:256–267. doi:10.1016/j.marmicro.2008.04.003

Harbers A, Schönfeld J, Rüggeberg A, Pfannkuche O (2010) Short term dynamics of planktonic foraminiferal sedimentation in the Porcupine Seabight. Micropaleontology 56:259–274

Hayward B, Cedhagen T, Kaminski M, Gross O (2014) *Globigerina punctulata* Deshayes, 1832. In: World Foraminifera Database Accessed World Regist. Mar.

Species. http://www.marinespecies.org/aphia.php?p=taxdetails&id=590647

Healy-Williams N (1983) Fourier shape analysis of *Globorotalia truncatulinoides* from late quaternary sediments in the southern Indian Ocean. Mar Micropaleontol 8:1–15

Healy-Williams N, Williams DF (1981) Fourier analysis of test shape of planktonic Foraminifera. Nature 289:485–487

Healy-Williams N, Ehrlich R, Williams DF (1985) Morphometric and stable isotopic evidence for subpopulations of *Globorotalia truncatulinoides*. J Foraminifer Res 15:242–253

Hecht AD (1974) Intraspecific variation in recent populations of *Globigerinoides ruber* and *Globigerinoides trilobus* and their application to paleoenvironmental analysis. J Paleontol 48:1217–1234

Hecht AD (1976) An ecologic model for test size variation in Recent planktonic Foraminifera: applications to the fossil record. J Foraminifer Res 6:295–311

Hemleben C, Spindler M (1983) Recent advances in research on living planktonic Foraminifera. Utrecht Micropaleontol Bull 30:141–170

Hemleben C, Bé AWH, Anderson OR, Tuntivate S (1977) Test morphology, organic layers and chamber formation of the planktonic foraminifer *Globorotalia menardii* (d'Orbigny). J Foraminifer Res 7:1–25

Hemleben C, Bé AWH, Spindler M, Anderson OR (1979) "Dissolution" effects induced by shell resorption during gametogenesis in *Hastigerina pelagica* (d'Orbigny). J Foraminifer Res 9:118–124

Hemleben C, Spindler M, Breitinger I, Deuser WG (1985) Field and laboratory studies on the ontogeny and ecology of some globorotaliid species from the Sargasso Sea off Bermuda. J Foraminifer Res 15:254–272

Hemleben C, Spindler M, Breitinger I, Ott R (1987) Morphological and physiological responses of *Globigerinoides sacculifer* (Brady) under varying laboratory conditions. Mar Micropaleontol 12:305–324

Hemleben C, Spindler M, Anderson OR (1989) Modern planktonic Foraminifera. Springer, Berlin

Hilbrecht H (1996) Extant planktic Foraminifera and the physical environment in the Atlantic and Indian Oceans: an atlas based on Climap and Levitus (1982). In: Mitteilungen aus dem Geologischen Institut der Eidgen. Technischen Hochschule und der Universität Zürich. Zürich, 93 pp

Hilbrecht H, Thierstein HR (1996) Benthic behavior of planktic Foraminifera. Geology 24:200–202

Hillaire-Marcel C, de Vernal A (2007) Proxies in Late Cenozoic paleoceanography. Elsevier, New York

Hofker J (1976) La familie Turborotaliidae n. fam. Rev Micropaléontol 19:47–53

Holmes NA (1984) An emendation of the genera *Beella* Banner and Blow, 1960, and *Turborotalita* Blow and Banner, 1962, with notes on *Orcadia* Boltovskoy and Watanabe, 1982. J Foraminifer Res 14:101–110

Huang CY (1981) Observations on the interior of some late Neogene planktonic Foraminifera. J Foraminifer Res 11:173–190

Huber BT, Bijma J, Darling K (1997) Cryptic speciation in the living planktonic foraminifer *Globigerinella siphonifera* (d'Orbigny). Paleobiology 23:33–62

Hull PM, Osborn KJ, Norris RD, Robison BH (2011) Seasonality and depth distribution of a mesopelagic foraminifer, *Hastigerinella digitata*, in Monterey Bay, California. Limnol Oceanogr 56:562–576. doi:10.4319/lo.2011.56.2.0562

Itou M, Ono T, Oba T, Noriki S (2001) Isotopic composition and morphology of living *Globorotalia scitula*: a new proxy of sub-intermediate ocean carbonate chemistry? Mar Micropaleontol 42:189–210

Ivanova E, Conan SMH, Peeters FJ, Troelstra SR (1999) Living *Neogloboquadrina pachyderma* sin and its distribution in the sediments from Oman and Somalia upwelling areas. Mar Micropaleontol 36:91–107

Ivanova EM, Schiebel R, Singh AD, Schmiedl G, Niebler HS, Hemleben C (2003) Primary production in the Arabian Sea during the last 135,000 years. Palaeogeogr Palaeoclimatol Palaeoecol 197:61–82

Jonkers L, Brummer GJA, Peeters FJC, van Aken HM, de Jong MF (2010) Seasonal stratification, shell flux, and oxygen isotope dynamics of left-coiling *N. pachyderma* and *T. quinqueloba* in the western subpolar North Atlantic. Paleoceanography. doi:10.1029/2009PA001849

Kahn MI, Williams DF (1981) Oxygen and carbon isotopic composition of living planktonic Foraminifera from the northeast Pacific Ocean. Palaeogeogr Palaeoclimatol Palaeoecol 33:47–69. doi:10.1016/0031-0182(81)90032-8

Kemle-von Mücke S, Oberhänsli H (1999) The distribution of living planktic Foraminifera in relation to southeast Atlantic oceanography. In: Fischer G, Wefer G (eds) Use of proxies in paleoceanography. Springer, Berlin, pp 91–115

Kemle-von-Mücke S (1994) Oberflächenwasserstruktur und -zirkulation des Südostatlantiks im Spätquartär. Berichte aus dem Fachbereich Geowiss Univ Bremen 55, PhD Thesis University of Bremen

Kemle-von-Mücke S, Hemleben C (1999) Planktic Foraminifera. In: Boltovskoy E (ed) South Atlantic zooplankton. Backhuys Publishers, Leiden, pp 43–67

Kennett JP (1976) Phenotypic variation in some Recent and late Cenozoic planktonic Foraminifera. In: Hedley RH, Adams CG (eds) Foraminifera 2. London, pp 111–170

Kennett JP, Srinivasan MS (1983) Neogene planktonic Foraminifera. Hutchinson Ross Publishing Company, Stroudsburg

Kimoto K, Ishimura T, Tsunogai U, Itaki T, Ujiié Y (2009) The living triserial planktic foraminifer *Gallitellia vivans* (Cushman): distribution, stable isotopes, and paleoecological implications. Mar Micropaleontol 71:71–79

Knappertsbusch M (2007) Morphological variability of *Globorotalia menardii* (planktonic Foraminifera) in two DSDP cores from the Caribbean Sea and the Eastern Equatorial Pacific. Carnets Géologie 4:1–34

Kohfeld KE, Fairbanks RG, Smith SL, Walsh ID (1996) *Neogloboquadrina pachyderma* (sinistral coiling) as paleoceanographic tracers in polar oceans: evidence from Northeast Water Polynya plankton tows, sediment traps, and surface sediments. Paleoceanography 11:679–699

Köhler-Rink S, Kühl M (2005) The chemical microenvironment of the symbiotic planktonic foraminifer *Orbulina universa*. Mar Biol Res 1:68–78. doi:10.1080/17451000510019015

Kroon D, Darling K (1995) Size and upwelling control of the stable isotope composition of *Neogloboquadrina dutertrei* (d'Orbigny), *Globigerinoides ruber* (d'Orbigny) and *Globigerina bulloides* (d'Orbigny); examples from the Panama Basin and Arabian Sea. J Foraminifer Res 25:39–52

Kroon D, Ganssen G (1988) Northern Indian Ocean upwelling cells and the stable isotope composition of living planktic Foraminifera. In: Brummer GJA, Kroon D (eds) Planktonic foraminifers as tracers of ocean-climate history. Free University Press, Amsterdam, pp 219–238

Kroon D, Ganssen G (1989) Northern Indian Ocean upwelling cells and the stable isotope composition of living planktonic foraminifers. Deep-Sea Res I 36:1219–1236

Kroon D, Nederbragt AJ (1988) Morphology and (paleo) ecology of living and fossil triserial planktonic Foraminifera. In: Brummer GJA, Kroon D (eds) Planktonic Foraminifera as tracers of ocean-climate history. Free University Press, Amsterdam, pp 181–201

Kroon D, Nederbragt AJ (1990) Ecology and paleoecology of triserial planktonic Foraminifera. Mar Micropaleontol 16:25–38

Kucera M, Darling KF (2002) Cryptic species of planktonic Foraminifera: their effect on palaeoceanographic reconstructions. Philos Trans R Soc Lond Ser Math Phys Eng Sci 360:695–718

Kucera M, Darling KF, Wade CM, von Langen P, Pak D (2001) Seasonal dynamics of cryptic species of planktonic Foraminifera in Santa Barbara Channel during 1999. ICP VII Sapporo Japan Program Abstracts, pp 137–138

Kuroyanagi A, Kawahata H (2004) Vertical distribution of living planktonic Foraminifera in the seas around Japan. Mar Micropaleontol 53:173–196

Kuroyanagi A, Tsuchiya M, Kawahata H, Kitazato H (2008) The occurrence of two genotypes of the planktonic foraminifer *Globigerinoides ruber* (white) and paleo-environmental implications. Mar Micropaleontol 68:236–243. doi:10.1016/j.marmicro.2008.04.004

Lea DW, Spero HJ (1992) Experimental determination of barium uptake in shells of the planktonic Foraminifera *Orbulina universa* at 22 °C. Geochim Cosmochim Acta 56:2673–2680

Lea DW, Martin PA, Chan DA, Spero HJ (1995) Calcium uptake and calcification rate in the planktonic foraminifer *Orbulina universa*. J Foraminifer Res 25:14–23

Lea DW, Mashiotta TA, Spero HJ (1999) Controls on magnesium and strontium uptake in planktonic Foraminifera determined by live culturing. Geochim Cosmochim Acta 63:2369–2379

Lee JJ, Freudenthal HD, Kossoy V, Bé AWH (1965) Cytological observations on two planktonic Foraminifera, *Globigerina bulloides* (d'Orbigny), 1826, and *Globigerinoides ruber* (d'Orbigny, 1839) Cushman, 1927. J Protozool 12:531–542

Li Q (1987) Origin, phylogenetic development and systematic taxonomy of the *Tenuitella* plexus (Globigerinitidae, Globigerinina). J Foraminifer Res 17:298–320

Li B, Jian Z, Wang P (1997) *Pulleniatina obliquiloculata* as a paleoceanographic indicator in the southern Okinawa trough during the last 20,000 years. Mar Micropaleontol 32:59–69

Lidz B (1972) *Globorotalia crassaformis* morphotype variations in Atlantic and Caribbean deep-sea cores. Micropaleontology 18:194–211

Lin HL, Hsieh HY (2007) Seasonal variations of modern planktonic Foraminifera in the South China Sea. Deep-Sea Res II 54:1634–1644

Loeblich AR, Tappan H (1986) Some new revised genera and families of hyaline calcareous Foraminiferida (Protozoa). Trans Am Microsc Soc 105:239–265

Loeblich AR, Tappan H (1988) Foraminiferal genera and their classification. van Nostrand Reinhold, New York

Loeblich AR, Tappan H (1992) Present status of foraminiferal classification. In: Takayanagi Y, Saito T (eds) Studies in benthic Foraminifera. Tokai University Press, Tokyo, pp 93–102

Lohmann GP (1995) A model for variation in the chemistry of planktonic Foraminifera due to secondary calcification and selective dissolution. Paleoceanography 10:445–457

Lohmann GP, Malmgren BA (1983) Equatorward migration of *Globorotalia truncatulinoides* ecophenotypes through the late Pleistocene: gradual evolution or ocean change? Paleobiology 9:414–421

Lombard F, Labeyrie L, Michel E, Spero HJ, Lea DW (2009) Modelling the temperature dependent growth rates of planktic Foraminifera. Mar Micropaleontol 70:1–7. doi:10.1016/j.marmicro.2008.09.004

Lombard F, Labeyrie L, Michel E, Bopp L, Cortijo E, Retailleau S, Howa H, Jorissen F (2011) Modelling planktic foraminifer growth and distribution using an ecophysiological multi-species approach. Biogeosciences 8:853–873. doi:10.5194/bg-8-853-2011

Lončarić N, Peeters FJC, Kroon D, Brummer GJA (2006) Oxygen isotope ecology of Recent planktic Foraminifera at the central Walvis Ridge (SE Atlantic). Paleoceanography. doi:10.1029/2005PA001207

Lončarić N, van Iperen J, Kroon D, Brummer GJA (2007) Seasonal export and sediment preservation of diatomaceous, foraminiferal and organic matter mass fluxes in a trophic gradient across the SE Atlantic. Prog Oceanogr 73:27–59

Malmgren BA, Kennett JP (1977) Biometric differentiation between recent *Globigerina bulloides* and *Globigerina falconensis* in the southern Indian Ocean. J Foraminifer Res 7:130–148

Malmgren BA, Berggren WA, Lohmann GP (1983) Evidence for punctuated gradualism in the Late Neogene *Globorotalia tumida* lineage of planktonic Foraminifera. Paleobiology 9:377–389

Marchant M, Hebbeln D, Wefer G (1998) Seasonal flux patterns of planktic Foraminifera in the Peru-Chile Current. Deep-Sea Res I 45:1161–1185

Marchant M, Hebbeln D, Giglio S, Coloma C, González HE (2004) Seasonal and interannual variability in the flux of planktic Foraminifera in the Humboldt Current System off central Chile (30°S). Deep-Sea Res II 51:2441–2455

Marshall BJ, Thunell RC, Spero HJ, Henehan MJ, Lorenzoni L, Astor Y (2015) Morphometric and stable isotopic differentiation of *Orbulina universa* morphotypes from the Cariaco Basin, Venezuela. Mar Micropaleontol 120:46–64. doi:10.1016/j.marmicro.2015.08.001

Martinez JI, Taylor L, De Deckker P, Barrows T (1998) Planktonic Foraminifera from the eastern Indian Ocean: Distribution and ecology in relation to the Western Pacific Warm Pool (WPWP). Mar Micropaleontol 34:121–151

Mary Y, Knappertsbusch MW (2013) Morphological variability of menardiform globorotalids in the Atlantic Ocean during Mid-Pliocene. Mar Micropaleontol 101:180–193

Mashiotta TA, Lea DW, Spero HJ (1997) Experimental determination of cadmium uptake in shells of the planktonic Foraminifera *Orbulina universa* and *Globigerina bulloides*: implications for surface water paleoreconstructions. Geochim Cosmochim Acta 61:4053–4065. doi:10.1016/S0016-7037(97)00206-8

Mekik F, François R (2006) Tracing deep-sea calcite dissolution: agreement between the *Globorotalia menardii* fragmentation index and elemental ratios (Mg/Ca and Mg/Sr) in planktonic foraminifers. Paleoceanography. doi:10.1029/2006PA001296

Mekik FA, Loubere PW, Archer DE (2002) Organic carbon flux and organic carbon to calcite flux ratio recorded in deep-sea carbonates: demonstration and a new proxy. Glob Biogeochem Cycles 16:25-1–25-15. doi:10.1029/2001GB001634

Mohtadi M, Steinke S, Groeneveld J, Fink HG, Rixen T, Hebbeln D, Donner B, Herunadi B (2009) Low-latitude control on seasonal and interannual changes in planktonic foraminiferal flux and shell geochemistry off south Java: a sediment trap study. Paleoceanography. doi:10.1029/2008PA001636

Mojtahid M, Jorissen FJ, Garcia J, Schiebel R, Michel E, Eynaud F, Gillet H, Cremer M, Diz Ferreiro P, Siccha M, Howa H (2013) High resolution Holocene record in the southeastern Bay of Biscay: global versus regional climate signals. Palaeogeogr Palaeoclimatol Palaeoecol 377:28–44. doi:10.1016/j.palaeo.2013.03.004

Morard R, Quillévéré F, Escarguel G, Ujiie Y, de Garidel-Thoron T, Norris RD, de Vargas C (2009) Morphological recognition of cryptic species in the planktonic foraminifer *Orbulina universa*. Mar Micropaleontol 71:148–165

Morard R, Quillévéré F, Douady CJ, de Vargas C, de Garidel-Thoron T, Escarguel G (2011) Worldwide genotyping in the planktonic foraminifer *Globoconella inflata*: implications for life history and paleoceanography. PLoS ONE. doi:10.1371/journal.pone.0026665

Morard R, Quillévéré F, Escarguel G, de Garidel-Thoron T, de Vargas C, Kucera M (2013) Ecological modeling of the temperature dependence of cryptic species of planktonic Foraminifera in the Southern Hemisphere. Palaeogeogr Palaeoclimatol Palaeoecol 391:13–33. doi:10.1016/j.palaeo.2013.05.011

Movellan A (2013) La biomasse des foraminifères planctoniques actuels et son impact sur la pompe biologique de carbone. PhD Thesis, University of Angers

Mulitza S, Dürkoop A, Hale W, Wefer G, Niebler HS (1997) Planktonic Foraminifera as recorders of past surface-water stratification. Geology 25:335–338

Mulitza S, Donner B, Fischer G, Paul A, Pätzold J, Rühlemann C, Segl M (2004) The South Atlantic oxygen isotope record of planktic Foraminifera. In: Wefer G, Mulitza S, Ratmeyer V (eds) The South Atlantic in the late quaternary: reconstruction of material budgets and current Systems. Springer, Berlin, pp 121–142

Naidu PD, Malmgren BA (1996a) A high-resolution record of late Quaternary upwelling along the Oman Margin, Arabian Sea based on planktonic Foraminifera. Paleoceanography 11:129–140

Naidu PD, Malmgren BA (1996b) Relationship between late Quaternary upwelling history and coiling properties of *Neogloboquadrina pachyderma* and *Globigerina bulloides* in the Arabian Sea. J Foraminifer Res 26:64–70

Niebler HS, Hubberten HW, Gersonde R (1999) Oxygen isotope values of planktic Foraminifera: a tool for the reconstruction of surface water stratification. In: Fischer G, Wefer G (eds) Use of proxies in paleoceanography: examples from the South Atlantic. Springer, Berlin, Heidelberg, pp 165–189

Norris RD, de Vargas C (2000) Evolution all at sea. Nature 405:23–24

Numberger L, Hemleben C, Hoffmann R, Mackensen A, Schulz H, Wunderlich JM, Kucera M (2009) Habitats, abundance patterns and isotopic signals of morphotypes of the planktonic foraminifer *Globigerinoides ruber* (d'Orbigny) in the eastern Mediterranean Sea since the marine isotopic stage 12. Mar Micropaleontol 73:90–104. doi:10.1016/j.marmicro.2009.07.004

Oberhänsli H, Bénier C, Meinecke G, Schmidt H, Schneider R, Wefer G (1992) Planktonic foraminifers as tracers of ocean currents in the eastern South Atlantic. Paleoceanography 7:607–632

Olsson RK (1974) Pleistocene paleoceanography and *Globigerina pachyderma* (Ehrenberg) in Site 36, DSDP, northeastern Pacific. J Foraminifer Res 4:47–60

Olsson RK (1976) Wall structure, topography and crust of *Globigerina pachyderma* (Ehrenberg). In: Takayanagi Y, Saito T (eds) Selected papers in honor of Prof. Kiyoshi Asano: Progress in Micropaleontology, Spec Pub. New York, pp 244–257

Olsson RK, Hemleben C, Berggren WA, Huber BT (1999) Atlas of Paleocene planktonic Foraminifera. Smithsonian Contrib Paleobiology 85, pp 252

Orr WN (1969) Variation and distribution of *Globigerinoides ruber* in the Gulf of Mexico. Micropaleontology 15:373–379

Ortiz JD, Mix AC, Collier RW (1995) Environmental control of living symbiotic and asymbiotic Foraminifera of the California Current. Paleoceanography 10:987–1009

Ottens JJ (1991) Planktic Foraminifera as North-Atlantic water mass indicators. Oceanol Acta 14:123–140

Ottens JJ (1992) Planktic Foraminifera as indicators of ocean environments in the northeast Atlantic. PhD Thesis, Free University, Amsterdam

Parker FL (1962) Planktonic foraminiferal species in Pacific sediments. Micropaleontology 8:219–254

Parker FL (1976) Taxonomic notes on some planktonic Foraminifera. In: Selected Papers in honor of Prof. Kiyoshi Asano. Spec Pub, New York, pp 258–262

Parker FL, Berger WH (1971) Faunal and solution patterns of planktonic Foraminifera in surface sediments of the South Pacific. Deep-Sea Res 18:73–107. doi:10.1016/0011-7471(71)90017-9

Pawlowski J, Holzmann M (2002) Molecular phylogeny of Foraminifera: a review. Eur J Protistol 38:1–10

Pawlowski J, Bolivar I, Fahrni JF, de Vargas C, Gouy M, Zaninetti L (1997) Extreme differences in rates of molecular evolution of Foraminifera revealed by comparison of ribosomal DNA sequences and the fossil record. Mol Biol Evol 14:498–505

Peeters F, Ivanova E, Conan S, Brummer GJA, Ganssen G, Troelstra S, van Hinte J (1999) A size analysis of planktic Foraminifera from the Arabian Sea. Mar Micropaleontol 36:31–63

Pflaumann U (1972) Porositäten von Plankton-Foraminiferen als Klimaanzeiger? Meteor Forsch-Ergeb Reihe C 7:4–14

Pflaumann U, Duprat J, Pujol C, Labeyrie LD (1996) SIMMAX: a modern analog technique to deduce Atlantic sea surface temperatures from planktonic Foraminifera in deep-sea sediments. Paleoceanography 11:15–35

Poore RZ, Gosnell LB (1985) Apertural features and surface texture of upper Paleogene biserial planktonic foraminifers; links between *Chiloguembelina* and *Streptochilus*. J Foraminifer Res 15:1–5

Postuma JA (1971) Manual of planktonic Foraminifera. Elsevier, Amsterdam

Quillévéré F, Morard R, Escarguel G, Douady CJ, Ujiié Y, de Garidel-Thoron T, de Vargas C (2013) Global scale same-specimen morpho-genetic analysis of *Truncorotalia truncatulinoides*: a perspective on the morphological species concept in planktonic Foraminifera. Palaeogeogr Palaeoclimatol Palaeoecol 391:2–12

Ravelo AC, Fairbanks RG (1992) Oxygen isotopic composition of multiple species of planktonic Foraminifera: recorders of the modern photic zone temperature gradient. Paleoceanography 7:815–831

Ravelo AC, Fairbanks RG, Philander SGH (1990) Reconstructing tropical Atlantic hydrography using planktontic Foraminifera and an ocean model. Paleoceanography 5:409–431

Regenberg M, Nielsen SN, Kuhnt W, Holbourn A, Garbe-Schönberg D, Andersen N (2010) Morphological, geochemical, and ecological differences of the extant menardiform planktonic Foraminifera *Globorotalia menardii* and *Globorotalia cultrata*. Mar Micropaleontol 74:96–107

Renaud S, Schmidt DN (2003) Habitat tracking as a response of the planktic foraminifer *Globorotalia truncatulinoides* to environmental fluctuations during the last 140 kyr. Mar Micropaleontol 49:97–122

Resig JM, Kroopnick PM (1983) Isotopic and distributional evidence of a planktonic habitat for the foraminiferal genus *Streptochilus* Brönnimann and Resig, 1971. Mar Micropaleontol 8:235–248

Retailleau S, Schiebel R, Howa H (2011) Population dynamics of living planktic foraminifers in the hemipelagic southeastern Bay of Biscay. Mar Micropaleontol 80:89–100

Retailleau S, Eynaud F, Mary Y, Abdallah V, Schiebel R, Howa H (2012) Canyon heads and river plumes: How might they influence neritic planktonic Foraminifera communities in the SE Bay of Biscay? J Foraminifer Res 42:257–269

Reynolds LA, Thunell RC (1986) Seasonal production and morphologic variation of *Neogloboquadrina pachyderma* (Ehrenberg) in the northeast Pacific. Micropaleontology 32:1–18

Rhumbler L (1911) Die Foraminiferen (Thalamorphoren) der Plankton-Expedition. Erster Teil: Die allgemeinen Organisations-Verhältnisse der Foraminiferen. Ergeb Plankton-Exped Humbold-Stift 1909 3:331

Rigual-Hernández AS, Sierro FJ, Bárcena MA, Flores JA, Heussner S (2012) Seasonal and interannual changes of planktic foraminiferal fluxes in the Gulf of Lions (NW Mediterranean) and their implications for paleoceanographic studies: Two 12-year sediment trap records. Deep-Sea Res I 66:26–40

Rink S, Kühl M, Bijma J, Spero HJ (1998) Microsensor studies of photosynthesis and respiration in the symbiotic foraminifer *Orbulina universa*. Mar Biol 131:583–595

Ripperger S, Schiebel R, Rehkämper M, Halliday AN (2008) Cd/Ca ratios of in situ collected planktonic foraminiferal tests. Paleoceanography. doi:10.1029/2007PA001524

Robbins LL (1988) Environmental significance of morphologic variability in open-ocean versus ocean-margin assemblages of *Orbulina universa*. J Foraminifer Res 18:326–333

Robbins LL, Healy-Williams N (1991) Toward a classification of planktonic Foraminifera based on biochemical, geochemical, and morphological criteria. J Foraminifer Res 21:159–167

Rohling EJ, Sprovieri M, Cane T, Casford JSL, Cooke S, Bouloubassi I, Emeis KC, Schiebel R, Rogerson M, Hayes A, Jorissen FJ, Kroon D (2004) Reconstructing past planktic foraminiferal habitats using stable isotope data: a case history for Mediterranean sapropel S5. Mar Micropaleontol 50:89–123. doi:10.1016/S0377-8398(03)00068-9

Rossignol L, Eynaud F, Bourget J, Zaragosi S, Fontanier C, Ellouz-Zimmermann N, Lanfumey V (2011) High occurrence of *Orbulina suturalis* and "*Praeorbulina*-like specimens" in sediments of the northern Arabian Sea during the Last Glacial Maximum. Mar Micropaleontol 79:100–113

Saito T, Thompson PR, Breger DL (1976) Skeletal ultra-microstructure of some elongate-chambered planktonic Foraminifera and related species. In: Takayanagi Y, Saito T (eds) Selected papers in honor of Prof. Kiyoshi Asano. Progress in Micropaleontology Spec Pub, New York, pp 278–304

Saito T, Thompson PR, Breger D (1981) Systematic index of Recent and Pleistocene planktonic Foraminifera. Tokyo University Press, Tokyo

Sautter LR, Sancetta C (1992) Seasonal associations of phytoplankton and planktic Foraminifera in an upwelling region and their contribution to the seafloor. Mar Micropaleontol 18:263–278

Sautter LR, Thunell RC (1989) Seasonal succession of planktonic Foraminifera; results from a four-year time-series sediment trap experiment in the Northeast Pacific. J Foraminifer Res 19:253–267

Sautter LR, Thunell RC (1991) Seasonal variability in the $\delta^{18}O$ and $\delta^{13}C$ of planktonic Foraminifera from an upwelling environment: sediment trap results from the San Pedro Basin, Southern California Bight. Paleoceanography 6:307–334

Schiebel R (2002) Planktic foraminiferal sedimentation and the marine calcite budget. Glob Biogeochem Cycles. doi:10.1029/2001GB001459

Schiebel R, Hemleben C (2000) Interannual variability of planktic foraminiferal populations and test flux in the eastern North Atlantic Ocean (JGOFS). Deep-Sea Res II 47:1809–1852

Schiebel R, Hemleben C (2005) Modern planktic Foraminifera. Paläontol Z 79:135–148

Schiebel R, Hiller B, Hemleben C (1995) Impacts of storms on Recent planktic foraminiferal test production and CaCO$_3$ flux in the North Atlantic at 47°N, 20°W (JGOFS). Mar Micropaleontol 26:115–129

Schiebel R, Bijma J, Hemleben C (1997) Population dynamics of the planktic foraminifer *Globigerina bulloides* from the eastern North Atlantic. Deep-Sea Res Part Oceanogr Res Pap 44:1701–1713

Schiebel R, Waniek J, Bork M, Hemleben C (2001) Planktic foraminiferal production stimulated by chlorophyll redistribution and entrainment of nutrients. Deep-Sea Res I 48:721–740

Schiebel R, Waniek J, Zeltner A, Alves M (2002) Impact of the Azores Front on the distribution of planktic foraminifers, shelled gastropods, and coccolithophorids. Deep-Sea Res II 49:4035–4050

Schiebel R, Zeltner A, Treppke UF, Waniek JJ, Bollmann J, Rixen T, Hemleben C (2004) Distribution of diatoms, coccolithophores and planktic foraminifers along a trophic gradient during SW monsoon in the Arabian Sea. Mar Micropaleontol 51:345–371

Schmidt DN, Renaud S, Bollmann J, Schiebel R, Thierstein HR (2004) Size distribution of Holocene planktic foraminifer assemblages: Biogeography, ecology and adaptation. Mar Micropaleontol 50:319–338

Schmidt DN, Rayfield EJ, Cocking A, Marone F (2013) Linking evolution and development: synchrotron radiation X-ray tomographic microscopy of planktic foraminifers. Palaeontology 56:741–749

Schmuker B (2000) Recent planktic Foraminifera in the Caribbean Sea: distribution, ecology and taphonomy. PhD Thesis, ETH Zürich

Schmuker B, Schiebel R (2002) Planktic foraminifers and hydrography of the eastern and northern Caribbean Sea. Mar Micropaleontol 46:387–403

Schulz H, von Rad U, Ittekkot V (2002) Planktic Foraminifera, particle flux and oceanic productivity off Pakistan, NE Arabian Sea: modern analogues and application to the palaeoclimatic record. In: Clift PD, Kroon D, Gaedicke C, Craig J (eds) The tectonic and climatic evolution of the Arabian Sea region. Geological Society London Spec Pub, London, pp 499–516

Schweitzer PN, Lohmann GP (1991) Ontogeny and habitat of modern menardiiform planktonic Foraminifera. J Foraminifer Res 21:332–346

Scott GH (1974) Pustulose and honeycomb topography in *Globigerinoides trilobus*. Micropaleontology 20:466–472

Scott GH (2011) Holotypes in the taxonomy of planktonic foraminiferal morphospecies. Mar Micropaleontol 78:96–100. doi:10.1016/j.marmicro.2010.11.001

Seears HA, Darling KF, Wade CM (2012) Ecological partitioning and diversity in tropical planktonic Foraminifera. BMC Evol Biol 12:54. doi:10.1186/1471-2148-12-54

Sexton PF, Norris RD (2008) Dispersal and biogeography of marine plankton: long-distance dispersal of the foraminifer *Truncorotalia truncatulinoides*. Geology 36:899. doi:10.1130/G25232A.1

Simstich J, Sarnthein M, Erlenkeuser H (2003) Paired $\delta^{18}O$ signals of *Neogloboquadrina pachyderma* (s) and *Turborotalita quinqueloba* show thermal

stratification structure in Nordic Seas. Mar Micropaleontol 48:107–125

Smart CW, Thomas E (2007) Emendation of the genus *Streptochilus* Brönnimann and Resig 1971 (Foraminifera) and new species from the lower Miocene of the Atlantic and Indian Oceans. Micropaleontology 53:73–103

Spear JW, Poore RZ, Quinn TM (2011) *Globorotalia truncatulinoides* (dextral) Mg/Ca as a proxy for Gulf of Mexico winter mixed-layer temperature: evidence from a sediment trap in the northern Gulf of Mexico. Mar Micropaleontol 80:53–61

Spencer-Cervato C, Thierstein HR (1997) First appearance of *Globorotalia truncatulinoides*: cladogenesis and immigration. Mar Micropaleontol 30:267–291

Spero HJ (1987) Symbiosis in the planktonic foraminifer, *Orbulina universa*, and the isolation of its symbitic dinoflagellate, *Gymnodinium béii* sp. nov. J Phycol 23:307–317

Spero HJ (1988) Ultrastructural examination of chamber morphogenesis and biomineralization in the planktonic foraminifer *Orbulina universa*. Mar Biol 99:9–20

Spero HJ (1992) Do planktic Foraminifera accurately record shifts in the carbon isotopic composition of seawater CO_2? Mar Micropaleontol 19:275–285

Spero HJ, DeNiro MJ (1987) The influence of symbiont photosynthesis on the $\delta^{18}O$ and $\delta^{13}C$ values of planktonic foraminiferal shell calcite. Symbiosis 4:213–228

Spero HJ, Lea DW (1993) Intraspecific stable isotope variability in the planktic Foraminifera *Globigerinoides sacculifer*: results from laboratory experiments. Mar Micropaleontol 22:221–234

Spero HJ, Lea DW (1996) Experimental determination of stable isotope variability in *Globigerina bulloides*: implications for paleoceanographic reconstructions. Mar Micropaleontol 28:231–246

Spero HJ, Bijma J, Lea DW, Bemis BE (1997) Effect of seawater carbonate concentration on foraminiferal carbon and oxygen isotopes. Nature 390:497–500

Spezzaferri S, Kucera M, Pearson PN, Wade BS, Rappo S, Poole CR, Morard R, Stalder C (2015) Fossil and genetic evidence for the polyphyletic nature of the planktonic Foraminifera "*Globigerinoides*", and Description of the new genus *Trilobatus*. PLoS ONE 10(5): e0128108. doi:10.1371/journal.pone.012810

Spezzaferri S, Olsson RK, Hemleben C (2017) Taxonomy, biostratigraphy, and phylogeny of Oligocene to Lower Miocene Globigerinoides and Trilobatus. Cushman Found Spec Publ, Chapter 9

Spindler M, Dieckmann GS (1986) Distribution and abundance of the planktic foraminifer *Neogloboquadrina pachyderma* in sea ice of the Weddell Sea (Antarctica). Polar Biol 5:185–191

Spindler M, Hemleben C (1980) Symbionts in planktonic Foraminifera (Protozoa). In: Schwemmler W, Schenk HEA (eds) Endocytobiology, endosymbiosis and cell biology. Walter de Gruyter & Co, Berlin, pp 133–140

Spindler M, Anderson OR, Hemleben C, Bé AWH (1978) Light and electron microscopic observations of gametogenesis in *Hastigerina pelagica* (Foraminifera). J Protozool 25:427–433

Spindler M, Hemleben C, Bayer U, Bé AWH, Anderson OR (1979) Lunar periodicity of reproduction in the planktonic foraminifer *Hastigerina pelagica*. Mar Ecol Prog Ser 1:61–64

Spindler M, Hemleben C, Salomons JB, Smit LP (1984) Feeding behavior of some planktonic foraminifers in laboratory cultures. J Foraminifer Res 14:237–249

Srinivasan MS, Kennett JP (1975) The status of *Bolliella*, *Beella*, *Protentella* and related planktonic Foraminifera based on surface ultrastructure. J Foraminifer Res 5:155–165

Srinivasan MS, Kennett JP (1976) Evolution and phenotypic variation in the late Cenozoic *Neogloboquadrina dutertrei* plexus. In: Takayanagi Y, Saito T (eds) Selected papers in honor of Prof. Kiyoshi Asano: Progress in Micropaleontology, Spec Pub. New York, pp 329–354

Steinhardt J, de Nooijer LLJ, Brummer G-J, Reichart G-J (2015) Profiling planktonic foraminiferal crust formation. Geochem Geophys Geosystems. doi:10.1002/2015GC005752

Steinke S, Chiu HY, Yu PS, Shen CC, Löwemark L, Mii HS, Chen MT (2005) Mg/Ca ratios of two *Globigerinoides ruber* (white) morphotypes: implications for reconstructing past tropical/subtropical surface water conditions. Geochem Geophys Geosystems. doi:10.1029/2005GC000926

Stewart IA, Darling KF, Kroon D, Wade CM, Troelstra SR (2001) Genotypic variability in subarctic Atlantic planktic Foraminifera. Mar Micropaleontol 43:143–153. doi:10.1016/S0377-8398(01)00024-X

Storz D, Schulz H, Waniek JJ, Schulz-Bull DE, Kučera M (2009) Seasonal and interannual variability of the planktic foraminiferal flux in the vicinity of the Azores Current. Deep-Sea Res I 56:107–124

Sverdlove MS, Bé AWH (1985) Taxonomic and ecological significance of embryonic and juvenile planktonic Foraminifera. J Foraminifer Res 15:235–241

Tedesco KA, Thunell RC (2003) Seasonal and interannual variations in planktonic foraminiferal flux and assemblage composition in the Cariaco Basin, Venezuela. J Foraminifer Res 33:192–210

Tedesco KA, Thunell RC, Astor Y, Muller-Karger F (2007) The oxygen isotope composition of planktonic Foraminifera from the Cariaco Basin, Venezuela: seasonal and interannual variations. Mar Micropaleontol 62:180–193

Thiede J (1975) Distribution of Foraminifera in surface waters of a coastal upwelling area. Nature 253:712–714

Thompson PR, Bé AWH, Duplessy J-C, Shackleton NJ (1979) Disappearance of pink-pigmented *Globigerinoides ruber* at 120,000 yr BP in the Indian and Pacific Oceans. Nature 280:554–558

Thunell RC, Reynolds LA (1984) Sedimentation of planktonic Foraminifera: seasonal changes in species flux in the Panama Basin. Micropaleontology 30:243–262

Tolderlund DS, Bé AWH (1971) Seasonal distribution of planktonic Foraminifera in the western North Atlantic. Micropaleontology 17:297–329

Tsuchihashi M, Oda M (2001) Seasonal changes of the vertical distribution of living planktonic Foraminifera at the main axis of the Kuroshio off Honshu, Japan. Fossils 70:1–17

Tsuchiya M (2009) Origin of bipolar distribution of planktonic foraminifers inferred from molecular phylogenetic analyses: a review. Fossils 85:14–24

Ufkes E, Zachariasse WJ (1993) Origin of coiling differences in living neogloboquadrinids in the Walvis Bay region, off Namibia, southwest Africa. Micropaleontology 39:283–287

Ufkes E, Jansen JHF, Brummer GJA (1998) Living planktonic Foraminifera in the eastern South Atlantic during spring: Indicators of water masses, upwelling and the Congo (Zaire) River plume. Mar Micropaleontol 33:27–53

Ujiié Y, Lipps JH (2009) Cryptic diversity in planktic Foraminifera in the northwest Pacific Ocean. J Foraminifer Res 39:145–154

Ujiié Y, de Garidel-Thoron T, Watanabe S, Wiebe P, de Vargas C (2010) Coiling dimorphism within a genetic type of the planktonic foraminifer *Globorotalia truncatulinoides*. Mar Micropaleontol 77:145–153

Ujiié Y, Asami T, de Garidel-Thoron T, Liu H, Ishitani Y, de Vargas C (2012) Longitudinal differentiation among pelagic populations in a planktic foraminifer. Ecol Evol 2:1725–1737

Van den Broeck E (1876) Étude sur les foraminifères de la Barbade (Antilles). Ann Soc Belge Microsc 1:55–152

Van Leeuwen RJW (1989) Sea-floor distribution and late quaternary faunal patterns of planktonic and benthic foraminifers in the Angola Basin. Utrecht Micropal Bull 38, pp 288

Vénec-Peyré MT, Caulet JP, Grazzini Vergnaud C (1995) Paleohydrographic changes in the Somali Basin (5°N upwelling and equatorial areas) during the last 160 kyr, based on correspondence analysis of foraminiferal and radiolarian assemblages. Paleoceanography 10:473–491

Vergnaud Grazzini C, Letolle CPR, Glacon G (1974) Note préliminaire à l'étude isotopique (^{18}O/^{16}O) du groupe des *Globigerinoides ruber* (d'Orbigny) au Pleistocene inférieur en mer Ionienne. Coll Int Centre Natl Recherche Sci 219:223–231

Vilks G (1973) A study of *Globorotalia pachyderma* (Ehrenberg in the Canadian Arctic. Halifax, Nova Scotia, pp 216

Vincent E (1976) Planktonic Foraminifera, sediments and oceanography of the late quaternary Southwest Indian Ocean. Mar Biol 9:235

Vincent E, Berger WH (1981) Planktonic Foraminifera and their use in paleoceanography. Ocean Lithosphere Sea 7:1025–1119

Volkmann R (2000a) Planktic foraminifer ecology and stable isotope geochemistry in the Arctic Ocean: implications from water column and sediment surface studies for quantitative reconstructions of oceanic parameters. Reports on Polar Research. PhD Thesis, AWI Bremerhaven

Volkmann R (2000b) Planktic foraminifers in the outer Laptev Sea and the Fram Strait-modern distribution and ecology. J Foraminifer Res 30:157–176

Von Langen PJ, Pak DK, Spero HJ, Lea DW (2005) Effects of temperature on Mg/Ca in neogloboquadrinid shells determined by live culturing. Geochem Geophys Geosystems. doi:10.1029/2005GC000989

Wade CM, Darling KF, Kroon D, Leigh Brown AJ (1996) Early evolutionary origin of the planktic Foraminifera inferred from small subunit rDNA sequence comparisons. J Mol Evol 43:672–677. doi:10.1007/BF02202115

Wang L (2000) Isotopic signals in two morphotypes of *Globigerinoides ruber* (white) from the South China Sea: Implications for monsoon climate change during the last glacial cycle. Palaeogeogr Palaeoclimatol Palaeoecol 161:381–394

Wang L, Sarnthein M, Duplessy JC, Erlenkeuser H, Jung S, Pflaumann U (1995) Paleo sea surface salinities in the low-latitude Atlantic: the $\delta^{18}O$ record of *Globigerinoides ruber* (white). Paleoceanography 10:749–761

Warren BA (1994) Context of the suboxic layer in the Arabian Sea. In: Lal D (ed) Biogeochemistry of the Arabian Sea. Indian Academy of Sciences, Bangalore, pp 301–314

Watkins JM, Mix AC, Wilson J (1996) Living planktic Foraminifera: tracers of circulation and productivity regimes in the central equatorial Pacific. Deep-Sea Res II 43:1257–1282

Watkins JM, Mix AC, Wilson J (1998) Living planktic Foraminifera in the central tropical Pacific Ocean: articulating the equatorial "cold tongue" during La Niña, 1992. Mar Micropaleontol 33:157–174

Weiner AKM (2014) Genetic diversity, biogeography and the morpho-genetic relationship in extant planktonic Foraminifera. PhD Thesis, University of Bremen

Weiner A, Aurahs R, Kurasawa A, Kitazato H, Kucera M (2012) Vertical niche partitioning between cryptic sibling species of a cosmopolitan marine planktonic protist. Mol Ecol 21:4063–4073. doi:10.1111/j.1365-294X.2012.05686.x

Weiner AKM, Weinkauf MFG, Kurasawa A, Darling KF, Kucera M, Grimm GW (2014) Phylogeography of the tropical planktonic Foraminifera lineage *Globigerinella* reveals isolation inconsistent with passive dispersal by ocean currents. PLoS ONE 9:e92148. doi:10.1371/journal.pone.0092148

Weiner AKM, Weinkauf MFG, Kurasawa A, Darling KF, Kucera M (2015) Genetic and morphometric evidence for parallel evolution of the *Globigerinella calida* morphotype. Mar Micropaleontol 114:19–35. doi:10.1016/j.marmicro.2014.10.003

Wejnert KE, Pride CJ, Thunell RC (2010) The oxygen isotope composition of planktonic Foraminifera from the Guaymas Basin, Gulf of California: seasonal, annual, and interspecies variability. Mar Micropaleontol 74:29–37

Wejnert KE, Thunell RC, Astor Y (2013) Comparison of species-specific oxygen isotope paleotemperature equations: sensitivity analysis using planktonic Foraminifera from the Cariaco Basin, Venezuela. Mar Micropaleontol 101:76–88. doi:10.1016/j.marmicro.2013.03.001

Weyl PK (1978) Micropaleontology and ocean surface climate. Science 202:475–481

Williams M, Schmidt DN, Wilkinson IP, Miller CG, Taylor PD (2006) The type material of the Miocene to Recent species *Globigerinoides sacculifer* (Brady) revisited. J Micropalaeontology 25:153–156

Yamasaki M, Matsui M, Shimada C, Chiyonobu S, Sato T (2008) Timing of shell size increase and decrease of the planktic foraminifer *Neogloboquadrina pachyderma* (sinistral) during the Pleistocene, IODP Exp. 303 Site U1304, the North Atlantic Ocean. Open Paleontol J 1:18–23

Zhang J (1985) Living planktonic Foraminifera from the eastern Arabian Sea. Deep-Sea Res I 32:789–798

Zobel B (1968) Phenotypische Varianten von *Globigerina dutertrei* d'Orbigny (Foram.) ihre Bedeutung für die Stratigraphie in Quartären Tiefsee-Sedimenten. Geol Jahrb 85:97–122

Cellular Ultrastructure

Although the shell is one of the most obvious features of the living and dead planktic foraminifers, the cytoplasm is clearly of central biological importance as the site of the dynamic properties of life including transformation of the information stored in the genetic code into substantive morphology, metabolism, sensitivity, and activity, as well as shell deposition and reproduction. Moreover, comparative cytoplasmic fine structural studies of living planktic and benthic foraminifers and their relations to other Sarcodina may provide evidence of phylogenetic affinities, and help to elucidate evolutionary origins in concert with data on shell morphology. To a large extent, the shell and cytoplasm are complementary in form, and dynamically interrelated both in the process of calcite deposition, which is clearly determined by the activity of the cytoplasm, and subsequently in the intimate topological association of the cytoplasm with the surfaces and enclosing spaces of the shell. The close morphological complementarity of shell and cytoplasm is visualized in an overall view of a light microscopic and electron microscopic image of a section through a planktic foraminifer

(Plate 3.1). Much of the inner shell space is filled by the vacuolated cytoplasm, but the final chamber may be incompletely filled with cytoplasm, owing either to poor nutrition or lack of health, thus reducing the total cytoplasmic mass. In normal growth and development of the organism, the final chamber is also devoid of large quantities of cytoplasm immediately after new chamber addition. If the organism is well-nourished, the cytoplasm soon enlarges to partially or completely fill the final chamber. Under optimal feeding conditions, a newly constructed final chamber will be replete with cytoplasm within 24 h. The dynamic flowing of the cytoplasm into and out of the shell does not permit a definitive delineation of zones of cytoplasm within the cell mass, but for purposes of convenient reference, the term internal cytoplasm is used to denote that part of the cell mass enclosed within the shell and external cytoplasm for all remaining cytoplasm outside of the shell. Consequently, three major zones of intergrading cytoplasm can be identified: (1) Compact internal cytoplasm, (2) frothy or reticulate cytoplasm often observed in the final chamber or at the

R. Schiebel and C. Hemleben, *Planktic Foraminifers in the Modern Ocean*,
DOI 10.1007/978-3-662-50297-6_3

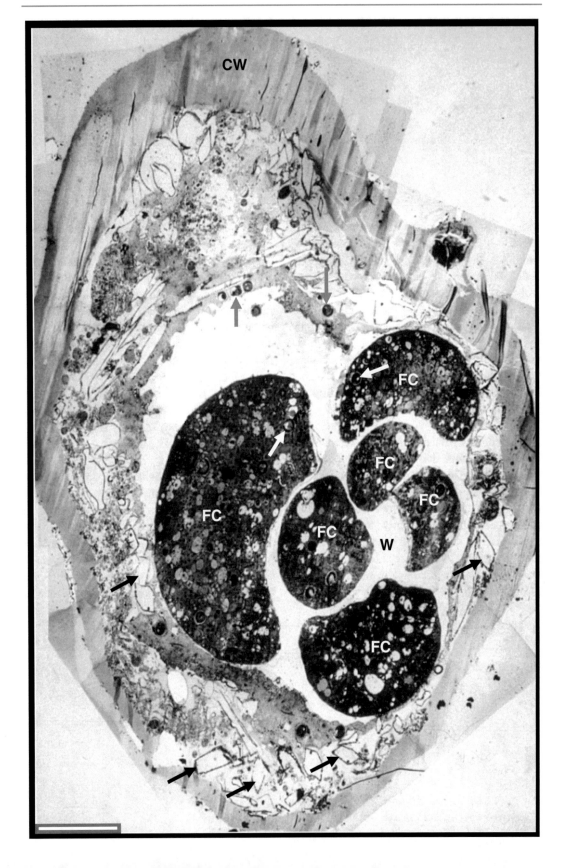

◀ **Plate 3.1** TEM micrograph of oblique thin section through entire specimen of *Globigerinita glutinata* in a 'feeding cyst', fixed immediately after sampling, with the test wall (W) being removed. Numerous empty diatom frustules (*black arrows*) are present in the space between cyst wall (CW) and cytoplasm (FC). Isolated diatom cytoplasts (*grey arrows*) are placed next to empty frustules, and others are enclosed in digestive vacuoles within the foraminifer cytoplasm (*white arrows*). Bar 20 μm. From Spindler et al. (1984)

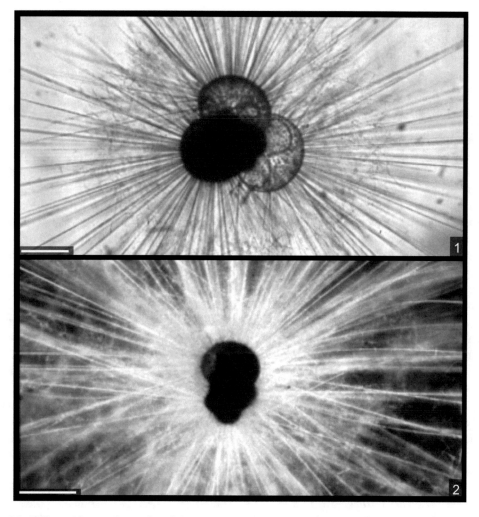

Plate 3.2 Different forms of pseudopodial networks. (1) *Globigerinoides ruber* with finely reticulate network of rhizopodia between calcareous spines. (2) *Globigerinoides* *sacculifer* with long web-like rhizopodia. Bars (1) 100 m, (2) 500 μm. From Hemleben et al. (1989)

transitional zone between internal space and external space near the shell aperture, and (3) alveolate masses of cytoplasm or reticulate to filose strands of rhizopodia engulfing the outer surface of the shell and radiating outward into the surrounding environment as a rhizopodial net (Plate 3.2). The most peripheral structures of the rhizopodial array are typically long thread-like

Plate 3.3 (1) Straight rhizopodia, and (2) reticulate rhizopodial web of *G. truncatulinoides*. (3) Straight rhizopodia in *G.* ▶
inflata attached to the surface of the culture vessel. (4) Club-shaped tips of rhizopodia in *G. truncatulinoides* (from
Hemleben et al. 1985). (5) Rhizopodia on freeze dried *G. truncatulinoides*. (6) Continuous sheath of cytoplasm
(CY) covering the outside of the decalcified test wall (W) of *G. siphonifera*. (7) Longitudinal section through rhizopod of
O. universa containing microtubules (MT) and mitochondria (M). (8) Filamentous strands (F) in rhizopodium of
P. obliquiloculata. Bars (1) 20 μm, (4,5) 50 μm, (2,3) 100 μm, (6-8) 1 μm

rhizopodia (filopodia) that create a halo of sticky
cytoplasmic filaments surrounding the organism
(e.g., Plate 3.3-1 to -4).

3.1 Cytoplasmic Streaming

The fine structural features of cells as observed
with the transmission electron microscope provide a detailed fixed view of the organization of
cellular components (e.g., Plate 3.1). This general organizational scheme of the cytoplasm
rather provides a convenient static perspective
representing a moment in time during a very fluid
state of existence. The cytoplasm is seldom stationary and even in the innermost chambers,
where the cytoplasm is compact and more slug-
gish in streaming, it is rare to find a region where
the cytoplasm is fully quiescent. Typically the
nucleus and surrounding cytoplasm are fairly
stationary and located in one of the inner chambers well protected from the surrounding environment (Plate 3.1). The remainder of the
cytoplasm is mobile and increasingly so as one
progresses from the internal chambers toward the
aperture and surrounding external cytoplasm.
Indeed, the external cytoplasmic strands (net-like
reticulopodia and thread-tike filopodia) exhibit a
most remarkable state of streaming activity. The
fluid behavior of the rhizopodia permit them to
be moved in all directions in space, not only
flowing forward to expand into the surrounding
space but also retracting and spreading laterally
to form compact masses or sheet-like layers of
flowing cytoplasm covering the shell, and in
spinose species, spanning the spaces in a
web-like pattern between the radially disposed

spines. Thus, at any moment some parts of the
rhizopodial mass may be extending to form long
filopodia projecting outward into the surrounding
environment while other parts are coalescing into
a reticulate or sheet-like mass of cytoplasm on
the shell surface. In all aspects, this process is
remarkably protean, yet ordered, clearly providing constant communication between the external
environment and the internal mass of cytoplasm.
At other places, the constant streaming motion of
the rhizopodial network may result in an undulating motion of the network, the total array
expanding and withdrawing within a limited
space among the spines, thus constantly moving
the cytoplasm throughout the space. Likewise,
the ever moving cytoplasmic system ensures that
the planktic foraminifer can, among other
essential life activities, advantageously snare
passing food, efficiently organize its digestive
vacuoles to maximize assimilation of food, and
dispose of waste material at the conclusion of the
digestive process. It is not uncommon to see
particles of waste carried by cytoplasmic
streaming to the tips of rhizopodial strands and
ejected into the surrounding environment. Concurrently, the rhizopodial strands exhibit bidirectional streaming, a form of countercurrent
flow, whereby cytoplasm not only flows outward, but simultaneously flows inward thus
constantly conserving the mass of cytoplasm
surrounding the organism. This bidirectional
flow within a rhizopodial strand, typically exhibits a surface layer moving toward the periphery
and an inner concentrically situated layer moving
centripetally. Intracytoplasmic particles can be
seen flowing within the cytoplasmic stream at
high magnifications with the light microscope

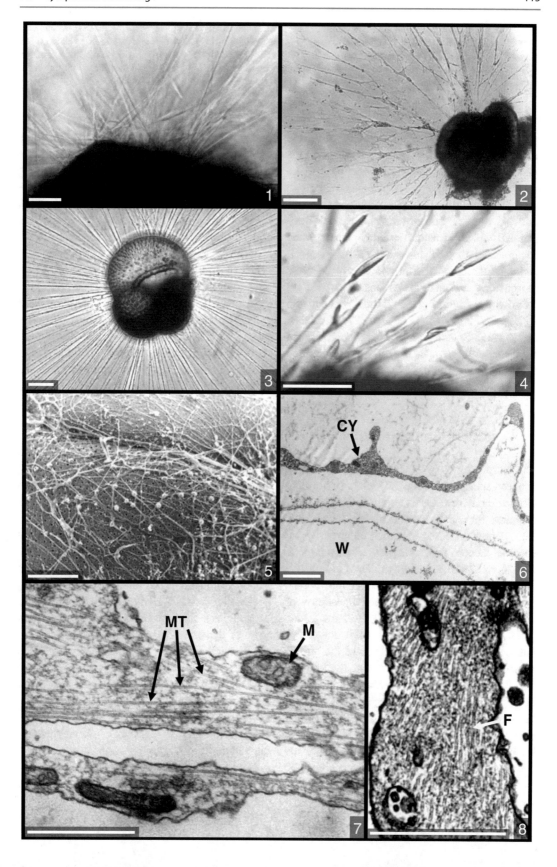

Plate 3.4 (1) Section through lobate nucleus (N), and (2) nuclear envelope (NE) of *H. pelagica*. NU is nucleolar ▶ material. (3) NE of *G. menardii* with pores (PO) visible in tangential view. (4) NE of *O. universa* in cross-section with polyribosomes (P). (5) Dense masses of chromatin (C) close to the NE in *H. pelagica*. (6) Fission of mitochondria (M) in *H. pelagica*, and tubular cristae (arrows) originating from inner mitochondrial membrane. Fibrous strands in the center of the mitochondria are possibly mitochondrial DNA. Bars (1,4,5) 2 μm, (2,3,6) 0.2 μm

(200x or more). At some locations within the rhizopodial network, cytoplasmic flow within the various strands occurs in all directions in the spatial array, permitting cytoplasmic contents to be intermixed and constantly deployed throughout the complex web of living substance. Substantial fine structural data have been accumulated that elucidate the detailed organization of the cytoplasm in relation to the gross morphology and activity as observed by light microscopy (Zucker 1973; Anderson and Bé 1976a, 1976b, 1978; Bé et al. 1977; Hemleben et al. 1977; Spindler et al. 1978; Spindler and Hemleben 1982; Anderson 1984; Anderson and Tuntivate-Choy 1984).

3.2   Peripheral Cytoplasm and Rhizopodial Morphology

The morphology of peripheral cytoplasm differs between specimens (Plate 3.3) while performing different activities, for example, collecting food, chamber formation, and external disturbance. The variety of forms is rather common in all species, except of *H. pelagica*. In *H. pelagica*, the large mass of internal cytoplasm produces a distinctive, alveolate envelope called the "bubble capsule". This array gives the appearance of closely packed soap bubbles (see Plates 2.17-1 and 4.4). The content of the alveoli is not known but they may aid buoyancy. Loss of the bubble capsule in laboratory cultures usually leads to sinking in the water. However, a bubble capsule in *H. pelagica* is not necessary to maintain buoyancy as some specimens have been observed in the open ocean floating without a bubble capsule. The fibrillar bodies within the cytoplasm as also observed in all other planktic foraminifer species (see Sect. 3.5.5 below)

probably enhance buoyancy in the absence of the bubble capsule. The external cytoplasm in other spinose and non-spinose species exhibits varying forms of networks, being somewhat finely reticulate in *Globigerinoides* spp. (Plates 3.3-1 and 4. 3), but more densely web-like in *G. siphonifera* and *G. truncatulinoides* (Plate 3.3-2). In some forms of *G. siphonifera*, the rhizopodial web becomes further organized into arching arrays interlacing the spines yielding a halo-like appearance in trans mitted light. The peripheral cytoplasm of *G. ruber* and *G. sacculifer* frequently exhibits a distinctive zonal distribution when viewed by transmitted light, including (1) a peripheral halo formed by the spine tips and fine strands of rhizopodia, (2) a more proximal fringe of denser rhizopodia, and (3) an inner symbiont-rich cytoplasmic layer immediately surrounding the shell.

In general, in planktic foraminifers, the peripheral rhizopodia are slightly tapered, elongate, thin filopodia projecting almost linearly into the surrounding environment (Plate 3.3-3). The typical form of the rhizopodium in spinose species is a rather wand-like straight, slightly tapered filopodium radiating outward into the surroundings, or as often observed in non-spinose species forming an enlarged lip (Plate 3.3-4) resembling a wheat kernel supported on a rather stocky shaft of cytoplasm (Hemleben et al. 1985). The main mass of cytoplasm that issues from the aperture of the shell in non-spinose species of planktic foraminifers forms a thin network of cytoplasm covering the outer surface of the shell from which the elaborate rhizopodial network is developed (Plate 3.3-5 and -6). During culture in Pyrex laboratory dishes, the peripheral rhizopodia become firmly attached to the bottom of the dish and rather straight as though under tension. It is not known whether this condition is the

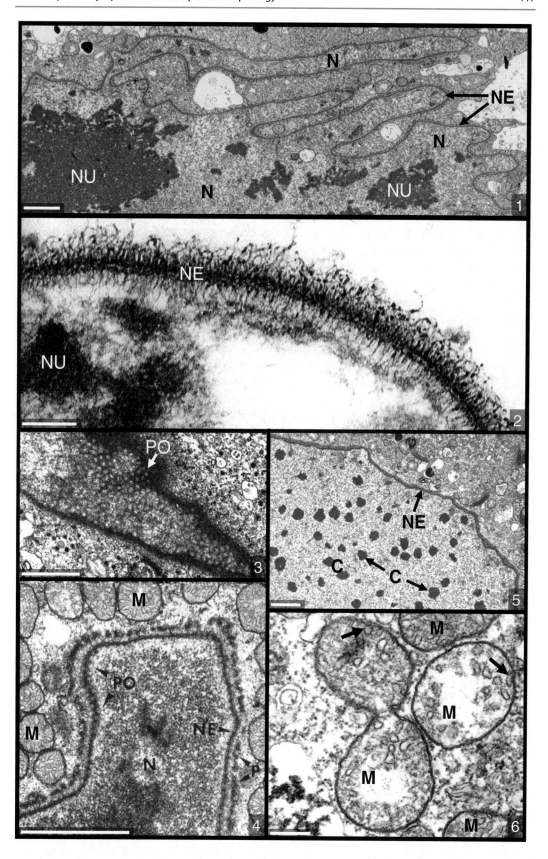

Plate 3.5 (1) Peroxisomes (arrows) in cytoplasm of *G. ruber*. (2) Vesicles (arrows) containing fibrillar material and ▶ adhesive substance secreted from Golgi (G) complex, and (3) schematic drawing of vesicles being secreted from Golgi complex. (4) Golgi complex in *H. pelagica* showing the surface (black arrow) where vesicles fuse to produce the cisternae, and the dispersal surface (red arrow) on the opposite side where secretory bodies are released, as well as endoplasmatic reticulum (E). (5) Cytoplasm of *G. siphonifera* with Golgi bodies. (6) Rhizopodium of *O. universa* with vesicle containing adhesive substance. (7) Two stacks of annulate lamellae (AL) with 7 and 14 membranous layers in *H. pelagica* 30 h before gamete release (from Spindler and Hemleben1982). (8) AL in *G. scitula*. Bars 0.5 µm, (5) 1 µm

result of microtubules forming an intracytoplasmic skeleton, augmented by contractile microfilaments, or if it is due to attachment of the rhizopodia to the glass while they are under tension. If the rhizopodial strands are perturbed, they contract into a flaccid strand indicating contractile processes are involved. Subsequently they elongate apparently by microtubular extension within the cytoplasm (see Bereiter-Hahn et al. 1987). When subjected to turbulence or agitation, the organisms secure themselves firmly to the glass surface by forming peripheral radial arrays of triangular masses of rhizopodia that effectively anchor the organism in all directions around the shell. The external cytoplasm of some non-spinose species with pustules (e.g. *G. menardii*) forms thick strands of rhizopodia that are organized around the surface of the pustule and radiate into the surrounding environment.

In some cases, the surface layer of cytoplasm surrounding the shell of spinose species may be quite uniformly sheath-like making a thin cytoplasmic envelope, while under most other conditions, the layer may be very reticulate forming a web of rhizopodia enshrouding the shell. These variations may reflect different physiological states of the organism. For example, when the organism is undernourished and much of the external cytoplasm is deployed as feeding rhizopodia, the amount of cytoplasm on the shell surface may be quite limited. When, however, the organism is well-nourished and especially if it is adding additional calcite to the surface of the shell, then the surface coat of cytoplasm forms a thin and rather continuous sheath.

3.3 Cytoskeletal Structures

Transmission microscopic examination of ultrathin sections of rhizopodia and filopodia in spinose and non-spinose species frequently reveals the presence of axially arranged microtubules (Plate 3.3-7). These ultramicroscopic tubules (ca. 30 nm diameter) composed of protein form a rigid scaffolding within the cytoplasm and provide structural support for the elongate strands. The microtubules provide intracytoplasmic surfaces guiding the flow of organelles as they are translocated from one part of the cell to another (see, e.g., Travis et al. 1983, for microtubules in benthic foraminifers). Production of microtubules critically depends on the protein 'Type 2' β-tubulin (β2), which occurred before the divergence of benthic foraminifers and radiolarians, i.e. at some 300 Ma before the first planktic foraminifers (Hou et al. 2013).

Microtubules are not regularly observed in ultrathin sections of planktic foraminifers fixed by conventional means as they require special preparations to stabilize them and protect them against calcium ion disruptive effects during initial fixation. Also, fixation with cold medium can result in destabilization of labile microtubules (Hemleben et al. 1989). Hence, the absence of microtubules in conventionally fixed preparations is not sufficient evidence to declare that they do not exist in that part of the cell. Even so, well-preserved elongated rhizopodia frequently exhibit internal parallel arrays of microtubules, which undoubtedly account in part for the remarkable rigidity and strength of these cytoplasmic extensions.

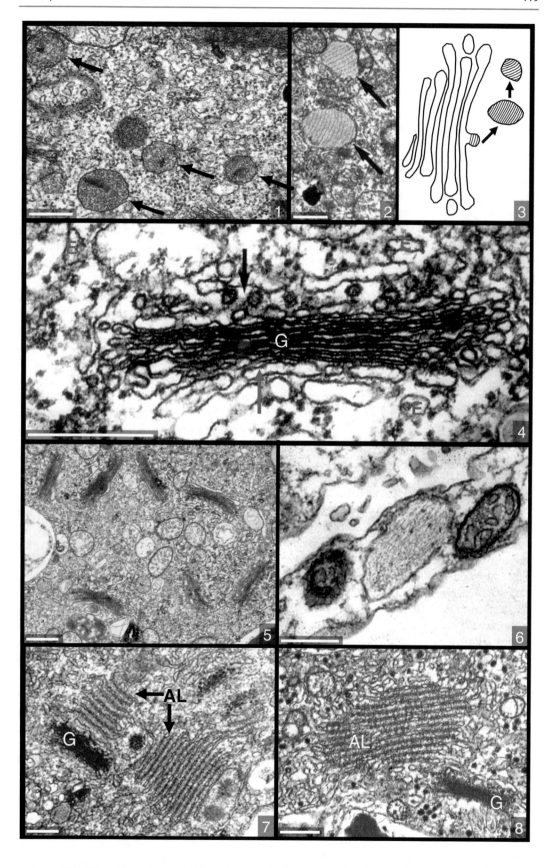

Plate 3.6 Fibrillar bodies in spinose species (1) in longitudinal section in *H. pelagica*, and (2) in *O. universa*. ▶
(3) Fibrillar bodies in cross section in *O. universa*, and (4) in *G. sacculifer* in oblique and cross sections. (5) The hollow
structure of the tubules is visible in larger magnification in *G. ruber*. (6) Fibrillar bodies in non-spinose *N. dutertrei*.
Bars (1-4,6) 3 μm, (5) 0.5 μm

3.4 Filaments

Contractile protein fibrils (microfilaments), ca.
5 nm in diameter, are commonly observed in
rhizopodia, and are particularly apparent as
widely dispersed filamentous strands within
contracted rhizopodia (Plate 3.3-8). In most
ultrathin sections viewed by electron microscopy, the microfilaments present a somewhat
felt-like appearance due to the numerous fine
filaments that have been cut obliquely or in
cross-section. Careful observation of high magnification views, however, will reveal the presence of some segments of the filaments that have
been sectioned more longitudinally.

3.5 Fine Structure of Cytoplasmic Organelles

3.5.1 Nucleus

The nucleus of planktic foraminifers is large
compared to other protozoa (ca. 50 μm or greater
than 200 μm as in *H. pelagica*) and varies in
form from a spheroid to an elongated, lobate
body. In some cases, it is bilobate and extends
within the cytoplasm between two chambers. The
two lobes are connected by a thin strand that
passes through the septal aperture between the
chambers. Otherwise, the planktic foraminifer
nucleus exhibits a typical fine structure of
eukaryotic cells. It is enclosed by a double-
membrane envelope (Plate 3.4-1 and -2) containing numerous pores (Plate 3.4-3 and -4),
whereby continuity is established between the
nucleoplasm and the surrounding cytoplasm.
The nucleoplasm is lightly granular and contains finely dispersed strands of chromatin

(euchromatin) filling the intranuclear space.
More dense masses of chromatin (heterochromatin) often occur near the periphery of the
nucleoplasm in close contact with the inner surface of the nuclear envelope (Plate 3.4-5). At
other places, masses of heterochromatin are also
observed suspended within the central part of the
nucleoplasm. The chromatin composed of DNA,
forms the chromosomal material in the cell
containing the genetic information for control of
cellular activity. Nucleolar material is often
observed aggregated into granular electron-dense
masses either scattered throughout the nucleoplasm as spheroids, organized into a massive
body filling over half of the nucleoplasm, or as a
centrally located nucleolus. The nucleolus contains ribonucleoprotein (RNA and protein) and
among other functions produces ribosomal RNA
to be exported into the cytoplasm where protein-
synthesizing ribosomes are assembled. The outer
membrane of the nuclear envelope is commonly
shrouded by a thin layer of filamentous material
resembling nucleic acid polymers (perhaps mes-
senger RNA) with clusters of polyribosomes
attached to them (Plate 3.4-2 and -4). It is not
known how the presence of this fringe of
polyribosomes correlates with cellular function
(Hemleben et al. 1989). Further research is needed to examine the nuclear envelope during
various phases of cellular metabolism to determine if there are correlated changes. These
observations may be made following feeding,
expansion of the cytoplasm after chamber addition, or during cytoplasmic changes accompanying chamber formation and reproduction.

In all species examined using transmission
electron microscopy, only a single nucleus was
observed, except at the time of reproduction,
when multiple nuclei (see Plate 5.2) fill the
cytoplasm (Hemleben et al. 1989). Although Lee

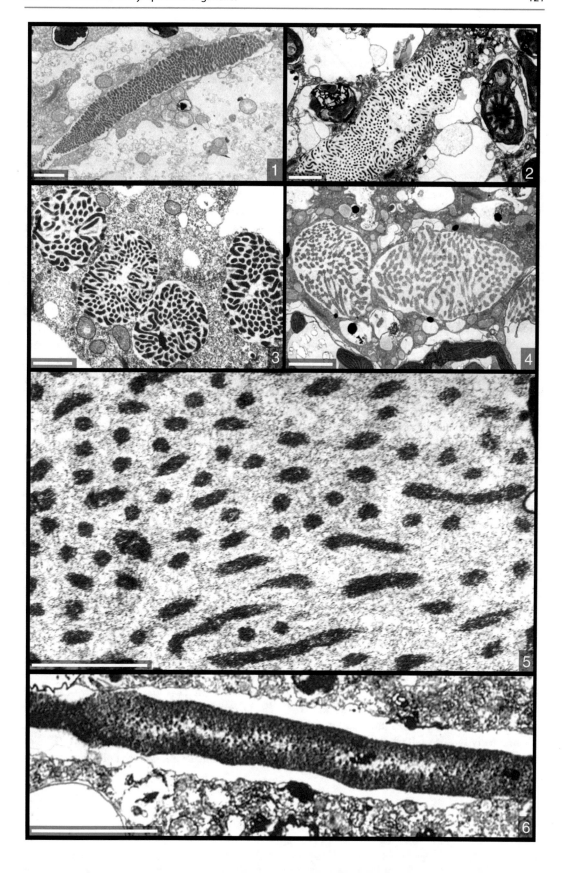

et al. (1965) reported Feulgen-positive bodies in the cytoplasm of *G. ruber* and *G. bulloides* suggesting that there may be more than one nucleus during vegetative growth, no multiple nuclei, except at the time of reproduction, were found in more than 20 specimens of each of these species (Hemleben et al. 1989). It may be possible that Lee et al. (1965) collected their specimens close to reproduction, which in both species is cyclic. Another explanation may be that the stained nuclei are those of recently ingested food organisms (e.g., dinoflagellates).

3.5.2 Mitochondria

The surrounding cytoplasm contains organelles typical of eukaryotic protists. Numerous mitochondria (ca. 1–2 μm diameter) surrounded by a double membrane contain tubular cristae appearing as small finger-like protrusions from the inner membrane (Plate 3.4-4 and -6). The majority of the cristae appear as circular bodies in ultrathin sections. The cross-sections through these sinuous tubular structures produce only an oval to circular profile. The mitochondria in all planktic foraminifers observed thus far exhibit this typical protistan ultrastructure. The distribution of the mitochondria tends to be rather uniform throughout the external and internal cytoplasm. However, during chamber formation, mitochondria congregate near the pores in the shell wall. This may enhance uptake of oxygen diffusing through the pore plate. Mitochondrial division by binary fission is frequently observed in this peripheral location. The dividing mitochondria present a distinctive bilobate or dumb-bell configuration during fission. The function of the mitochondria in part is to produce adenosine triphosphate (ATP), a high-energy storage compound, which is distributed throughout the cell and provides energy to drive cellular metabolic processes and sustain cytoplasmic activity. The mitochondrion is also the site of major oxygen consumption in the process of oxidation of food substances with the concomitant production of ATP. These are not the only functions of the mitochondrion, but represent some of the more typically recognized ones.

3.5.3 Peroxisomes

Cytochemical fine structure evidence clearly confirms the presence of peroxisomes in many species of planktic foraminifers, for example, *H. pelagica* and *G. ruber* (Anderson and Tuntivate-Choy 1984). These organelles (ca. 0.2–0.5 μm diameter) are surrounded by a single membrane, and contain a finely granular matrix exhibiting in some sections a centrally located ensemble of parallel arranged membranous tubules (Plate 3.5-1). Peroxisomes are scattered throughout the cytoplasm of planktic foraminifers, and are often observed in close association with mitochondria. The biochemical activity of peroxisomes as determined generally in other eukaryotic cells is gluconeogenesis (production of storage carbohydrate, frequently from lipid precursors), and conversion of potentially toxic waste products into metabolically useful substances as, for example, the conversion of alcohol to acetaldehyde and lactate to pyruvate. Their enzyme content and physiological functions in planktic foraminifers, however, have not been fully investigated. The cytochemical evidence reported by Anderson and Tuntivate-Choy (1984) indicates the presence of peroxidases, a group of enzymes that convert potentially toxic peroxides into water and regenerate oxidized intermediate compounds (e.g., nicotinamide adenine dinucleotide, NAD) to sustain metabolism.

Peroxisomes with similar internal membranous structures, sometimes forming closely spaced lattices, occur in *G. bulloides* (Fèbvre-Chevalier 1971). Single internal membranous tubules have also been observed in peroxisomes of the benthic sarcomastigophoran *Gromia oviformis* (Hedley and Wakefield 1969), but are not typical in other eukaryotic cells where the internal inclusion is an electron dense particle often exhibiting a very fine crystalline protein lattice structure.

3.5.4 Endoplasmic Reticulum, Golgi Complex, and Vacuolar System

Among the membrane-bound cisternal and vacuolar spaces within the planktic foraminifer cell, a wide variety of forms are observed. The endoplasmic reticulum is quite uniformly organized into networks of flattened or cylindrical tubules of varying cisternal width, but in the order of 30 to 50 nm as observed in other eukaryotic cells (e.g., Grell 1973; Anderson 1987). The vacuolar system, however, presents a wide variety of single membrane bound structures including, (1) electron lucent vacuoles of undetermined content varying in diameter from several hundred nanometers to several microns, (2) vacuoles filled with organic or mineral matter of undetermined function, (3) perialgal vacuoles enclosing the algal symbionts and structurally segregating them from the host cytoplasm (see Chap. 4), (4) food vacuoles enclosing recently ingested prey, (5) primary lysosomes (Golgi-derived vesicles) containing digestive enzymes produced by budding off of the Golgi peripheral saccules, and (6) digestive vacuoles containing food particles that are being digested by enzymes contributed by the primary lysosomes. A transmission electron microscopic view of lysosomal vesicles is presented in Plate 3.5-2, and a schematic drawing of the Golgi-lysosomal vacuolar system is presented in Plate 3.5-3. Smaller membrane-bound bodies, ca. 30-50 nm in diameter, are usually classified as vesicles. Some of these are primary lysosomes, but others of unknown composition are observed throughout the cytoplasm. The microanatomy of these structures is presented in greater detail.

A fine network of the endoplasmic reticulum is commonly observed scattered throughout the cytoplasm of the cell (Plate 3.5-4). Both rough endoplasmic reticulum (RER) containing ribosomes on the cytoplasmic surface, and smooth endoplasmic reticulum lacking ribosomes, occur in most species observed thus far. The rough endoplasmic reticulum is the site of protein synthesis in the cell, while the smooth endoplasmic reticulum is involved in lipid synthesis among other functions. Free ribosomes scattered in the cytoplasm are sometimes observed. The proteins released by ribosomes and endoplasmic reticulum are glycosylated and sulphated by Golgi bodies. Resulting glycosaminoglycans (polysaccharides) are major component of organic linings and organic cements, and assumed of structural importance in the foraminifer shell (Langer 1992).

The Golgi bodies (secretory organelles) exhibit a typical eukaryotic profile in ultrathin sections (Plate 3.5-4 and -5), consisting of a stack of membranous cisternae producing inflated saccules (swollen enlargements of the cisternae at the perimeter of the Golgi complex) containing secreted substances. Depending on the physiological status of the cell, these secretions may be digestive enzymes or other macromolecules synthesized by the cell and concentrated within the Golgi complex, packaged into vesicles and secreted into the cytoplasm. In the case of digestive enzymes, the peripheral saccules of the Golgi complex become concentrated with the hydrolytic enzymes, and then are budded off as small vesicles called primary lysosomes. Cytochemical finestructure evidence confirms that the Golgi complex in *H. pelagica*, for example, is the site of lysosomal enzyme production as evidenced by the dense reaction product for the lysosomal acid phosphatase (digestive vacuole marker enzyme) deposited in the cisternae and peripheral saccules (Anderson and Bé 1976b).

During feeding, one of the most prominent features of internal and external cytoplasm is the abundant appearance of digestive vacuoles containing a variety of prey matter in various stages of decomposition (see Plate 4.1). These organelles varying in size from less than one micron to several microns in diameter are quickly identified by the presence of a single peripheral membrane surrounding a vacuolar space containing clearly degraded or recently engulfed food material. When the vacuoles contain only food material and have not yet formed digestive stages, they are known as food vacuoles or phagosomes. It is difficult to distinguish between phagosomes and early lysosomes without special cytochemical stains, as it is not possible to detect the presence of the digestive enzymes in standard fine structure preparations. The phagosome (food vacuole) is converted to a digestive vacuole by fusion of its membrane with a Golgi-derived lysosomal vesicle, which empties its digestive enzyme contents into the phagosome thus initialing enzymatic degradation of the food. The pH of the phagosome is initially alkaline, but becomes acidic usually before or concurrent with fusion of the lysosomal vesicle (Anderson 1987). Thus, the digestive vacuole is fundamentally a small stomach within the cell containing an acid environment and digestive enzymes to render the large food molecules into smaller, more readily assimilated compounds. After digestion is complete, the vacuoles contain non-digestible debris and are known as residual bodies. These waste vacuoles or residual vacuoles as they are sometimes called, are typically transported by cytoplasmic streaming to the periphery of the cell, where their contents are expelled into the surrounding environment. Hence, these waste vacuoles are among the most commonly observed larger cytoplasmic particles streaming toward the periphery of the cell when viewed by light microscopy.

Mitochondria and other membranous organelles are also observed in the streaming cytoplasm both in the region near the shell aperture and within the rhizopodial strands surrounding the shell or covering the spines. Upon death of the cell, as occurs due to poor health, or when there is some residual cytoplasm remaining in the shell after gametogenesis, the masses of cytoplasm become moribund and usually decay rather quickly within a few hours. This is caused partially by action of residual digestive enzymes within the cytoplasm that are released when the digestive vacuoles decay, and also due to invasion of the cell by bacteria, microflagellates, and other lytic scavengers. Under some conditions,

however, the dead cytoplasm may linger for considerably longer periods of time. It is interesting to note that tests of *G. hirsuta*, *G. truncatulinoides*, and *G. inflata*, containing dead cytoplasm, have been collected from the seabed at a depth of 4000 m (Hemleben and Auras 1986). These either were living there or sank to the sea-floor while still filled with dead cytoplasm.

When planktic foraminifers capture prey, vesicles containing adhesive substance (Plate 3.5-6) are commonly observed in the peripheral cytoplasm. These vesicles (ca. 1.0 μm diameter) contain a clearly identifiable fine flocculent substance sometimes presenting a fan-shaped configuration in ultrathin sections. The vesicles are distributed throughout the rhizopodial strands, and when the rhizopodial plasma membrane is in contact with a prey surface, the vesicles fuse with the membrane and release their contents by exocytosis onto the prey surface. The chemical composition of the adhesive substance and its mechanism of forming the masses of flocculent matter surrounding the prey are not known.

During nuclear division accompanying preparation for reproduction, numerous multi-lamellar membranous bodies, called annulate lamellae, are observed in the cytoplasm close to the Golgi complex of some species of planktic foraminifers (Plate 3.5-7 and -8). As evidenced by Spindler and Hemleben (1982), these annulate lamellae are the origin of the double-membranes used to form the nuclear envelope during rapid division and dispersal of reproductive nuclei into the cytoplasm (see Chap. 5.1).

3.5.5 Fibrillar Bodies

Large vacuoles filled with a fibrillar to fluffy appearing substance, often of species-specific morphology (Plate 3.6), are observed in the cytoplasm of spinose and non-spinose species (e.g., Hansen 1975; Anderson and Bé 1976a; Leutenegger 1977). These "fibrillar bodies" or vesicular reticulum (Hansen 1975) originate in the internal cytoplasm, and occur first as small vacuoles with a densely packed mass of intertwined fibrillar material and tubules of varying electron opacity. Subsequent stages enlarge and the fibrillar contents become expanded, and appear more fluffy forming a finely filamentous mass, at places loosely laced with more substantial tubules of electron opaque material (Plate 3.6-5). The degree of expansion and the general electron opacity of the fibrillar system may be species specific. For example, *H. pelagica* possesses a compact array (Plate 3.6-1), whereas *O. universa* and *G. sacculifer* exhibit more expanded and fluffy fibrillar material in mature stages (Plate 3.6-2, -3, and -4). In non-spinose species, generally, the fibrillar bodies exhibit a more electron-dense composition with the tubules closely spaced within the oblong vacuoles (Plate 3.5-6). The function of these fluffy fibrillar bodies is unknown, although, Hansen (1975) and Anderson and Bé (1976a; 1976b) suggested that among other functions they may aid flotation.

Spero (1986, 1988) has hypothesized that the fibrillar bodies may be a site of calcium storage prior to shell calcification, based on transmission electron microscopic observations that some fibrillar bodies were released from cytoplasmic vacuoles in the region of calcification during spherical chamber deposition in *O. universa*. Therefore, he concludes that these bodies are calcification devices.

The large volume of the final stages of the vacuoles and the apparent low density of the organic matter filling the vacuolar space suggest that the fibrillar bodies may enhance buoyancy. With the possible exception of *H. pelagica*, none of the planktic foraminifers possess a clearly detectable external flotation device. The substantial weight of the calcite shell and spines when present clearly produce negative buoyancy and require some mechanism to sustain flotation and permit the foraminifer to adjust its position in the water column. Moreover, spines are presumed to provide anchorage for the rhizopodia

that radiate outward from the shell to enhance food capture, and of themselves do not aid buoyancy (Hemleben et al. 1989, and references therein). At present, the fibrillar bodies appear to be the most likely cytoplasmic structures mediating buoyancy. They are present in all species of planktic foraminifers examined by transmission electron microscopy. They appear to be unique to the planktic foraminifers as they have never been observed in benthic foraminifers. In addition to the fibrillar bodies, it is likely that buoyancy may be enhanced by the presence of lipid droplets dispersed throughout the cytoplasm.

3.5.6 Lipids and Various Cytoplasmic Inclusions

Lipid droplets (food reserve bodies), vacuoles of varying size and translucency, pigment granules, and vesicles of unknown chemical composition are commonly observed distributed throughout the cytoplasm of many planktic foraminifers. Lipid stores are frequently observed more densely packed in the cytoplasm of the innermost chambers, and become increasingly less abundant in the peripheral cytoplasm. Likewise, within a given chamber, lipid droplets are more likely to be concentrated near the central cytoplasm and less abundantly at the periphery near the wall. During the initial stages of gametogenesis, the larger lipid bodies are fragmented into smaller droplets and dispersed throughout the cytoplasm. Moreover, in *H. pelagica*, a distinctive red pigment appears prior to gametogenesis and spreads throughout the internal cytoplasm concurrent with lipid dispersal (see Chap. 5.1.2).

3.6 Summary and Concluding Remarks

The living planktic foraminifers exhibit characteristic cytoplasmic features and streaming activity within and outside the calcareous test. In both regions, the cytoplasm is in constant movement, but more so in the external than in the internal cytoplasm. The internal cytoplasm can be differentiated into three types of pseudopodia: (1) Rhizopods (anastomosing and branching forms), (2) filopodia (long, thin and straight), and (3) reticulopodia (net-like). The filopodia are supported mostly by internal filaments and sparsely grouped microtubules. The cell organelles (nucleus, mitochondria, peroxisomes, Golgi complex, endoplasmic reticulum, annulate lamellae, vacuolar system, lipids, and other inclusions) are typical of those observed in other eukaryotic cells. A fibrillar system, floating device or possibly calcification organelle, seems to be unique among all known protozoa. Still, many open questions on the ultrastructure of the planktic foraminifers remain to be answered. For example, high-resolution TEM analyses and molecular genetic data should clarify the reproduction mode of planktic foraminifers, i.e. sexual versus asexual reproduction.

References

Anderson OR (1984) Cellular specialization and reproduction in planktonic Foraminifera and Radiolaria. In: Plankton Marine (ed) Steidinger KA, Walker LM. Life, Cycle strategies, pp 35–66

Anderson OR (1987) Comparative Protozoology: Ecology, Physiology. Life History, Springer, Heidelberg

Anderson OR, Bé AWH (1976a) The ultrastructure of a planktonic foraminifer, *Globigerinoides sacculifer* (Brady), and its symbiotic dinoflagellates. J Foraminifer Res 6:1–21. doi:10.2113/gsjfr.6.1.1

Anderson OR, Bé AWH (1976b) A cytochemical fine structure study of phagotrophy in a planktonic foraminifer, *Hastigerina pelagica* (d'Orbigny). Biol Bull 151:437–449

Anderson OR, Bé AWH (1978) Recent advances in foraminiferal fine structure research. In: Hedley RH, Adams CG (eds) Foraminifera 1. Academic Press, London, New York, San Francisco

Anderson OR, Tuntivate-Choy S (1984) Cytochemical evidence for peroxisomes in planktonic Foraminifera. J Foraminifer Res 14:203–205. doi:10.2113/gsjfr.14.3.203

Bé AWH, Hemleben C, Anderson OR, Spindler M, Hacunda J, Tuntivate-Choy S (1977) Laboratory and field observations of living planktonic Foraminifera. Micropaleontology 23:155–179

Bereiter-Hahn J, Anderson OR, Reif WE (1987) Cytomechanics - The mechanical basis of cell form and structure, cytomechanics. Springer, Heidelberg

Fèbvre-Chevalier C (1971) Constitution ultrastructurale de *Globigerina bulloides* d'Orbigny, 1826 (Rhizopoda-Foraminifera). Protistologica 7:311–324

Grell KG (1973) Protozoology. Springer, Berlin

Hansen HJ (1975) On feeding and supposed buoyancy mechanism in four Recent globigerinid Foraminifera from the Gulf of Elat, Israel. Rev Esp Micropaleontol 7:325–339

Hedley RH, Wakefield JSJ (1969) Fine structure of *Gromia oviformis* (Rhizopoda: Protozoa). Bull Br Mus Nat Hist 18:71–89

Hemleben C, Auras A (1986) Zooplankton mit Kalkskelett, sedimentierende Partikel und Sediment. 32 86:34–39

Hemleben C, Bé AWH, Anderson OR, Tuntivate S (1977) Test morphology, organic layers and chamber formation of the planktonic foraminifer *Globorotalia menardii* (d'Orbigny). J Foraminifer Res 7:1–25

Hemleben C, Spindler M, Anderson OR (1989) Modern planktonic Foraminifera. Springer, Berlin

Hemleben C, Spindler M, Breitinger I, Deuser WG (1985) Field and laboratory studies on the ontogeny and ecology of some globorotaliid species from the Sargasso Sea off Bermuda. J Foraminifer Res 15:254–272

Hou Y, Sierra R, Bassen D, Banavali NK, Habura A, Pawlowski J, Bowser SS (2013) Molecular Evidence for β-tubulin neofunctionalization in Retaria (Foraminifera and Radiolarians). Mol Biol Evol 30:2487–2493. doi:10.1093/molbev/mst150

Langer MR (1992) Biosynthesis of glycosaminoglycans in Foraminifera: A review. Mar Micropaleontol 19:245–255

Lee JJ, Freudenthal HD, Kossoy V, Bé AWH (1965) Cytological observations on two planktonic Foraminifera, *Globigerina bulloides* (d'Orbigny), 1826, and *Globigerinoides ruber* (d'Orbigny, 1839) Cushman, 1927. J Protozool 12:531–542

Leutenegger S (1977) Ultrastructure de foraminifères perdorés et imperforés ainsi que de leurs symbiotes. Cah Micropaléontol 3:1–52

Spero HJ (1988) Ultrastructural examination of chamber morphogenesis and biomineralization in the planktonic foraminifer *Orbulina universa*. Mar Biol 99:9–20

Spero HJ (1986) Symbiosis, chamber formation and stable isotope incorporation in the planktonic foraminifer *Orbulina universa*. PhD Thesis, University of California

Spindler M, Anderson OR, Hemleben C, Bé AWH (1978) Light and electron microscopic observations of gametogenesis in *Hastigerina pelagica* (Foraminifera). J Protozool 25:427–433

Spindler M, Hemleben C (1982) Formation and possible function of annulate lamellae in a planktic foraminifer. J Ultrastruct Res 81:341–350

Spindler M, Hemleben C, Salomons JB, Smit LP (1984) Feeding behavior of some planktonic foraminifers in laboratory cultures. J Foraminifer Res 14:237–249

Travis JL, Kenealy J, Allen RD (1983) Studies on the motility of the Foraminifera. II. The dynamic microtubular cytoskeleton of the reticulopodial network of *Allogromia laticollaris*. J Cell Biol 97:1668–1676

Zucker WH (1973) Fine structure of planktonic Foraminifera and their endosymbiontic algae. PhD Thesis, New York

Nutrition, Symbionts, and Predators

4

The temporal and spatial distribution of diet is presumably a major cause for the regional distribution of planktic foraminifer species, by providing the basis for growth and affecting their fecundity. Species capable of subsisting on a broad range of prey and efficiently assimilating prey biomass more likely survive environmental change, and more readily invade and adapt to new environments than less tolerant species (Hemleben et al. 1989). Planktic foraminifers are basically omnivorous. Spinose species prefer a wide variety of animal prey, including larger metazoans such as copepods, pteropods, and ostracods (Rhumbler 1911; Caron and Bé 1984; Spindler et al. 1984). In addition, cannibalism has been reported (Hemleben et al. 1989, and references therein). Non-spinose species are largely herbivorous, and accept animal prey. In addition to prey organisms such as diatoms, dinoflagellates, thecate algae, and eukaryotic algae, muscle tissue and other animal tissue have been found as contents of food vacuoles in non-spinose species (Anderson et al. 1979). Subsurface dwelling species like *Globorotalia scitula* may feed on settling organic matter, and may be characteristic of vertical flux of organic matter within tropical to temperate waters (Itou et al. 2001). Little is known about the possible role of bacteria in the diet of planktic foraminifers. The position of planktic foraminifers in the marine food web differs from that of other protozoans, and ranges above the base of heterotrophic consumers (Hemleben et al. 1989).

Predators specialized on planktic foraminifers are not known, but tests have been found in pteropods, salps, shrimps, and other metazooplankton (e.g., Berger 1971).

4.1 Capture and Digestion of Prey

When capturing prey (Plate 4.1-1 and -2), the foraminifer rhizopodia engulf the major appendages and broad surfaces (Plate 4.1-7) of the prey (Spindler et al. 1984). Masses of adhesive substance originating from the Golgi apparatus surround the prey, and apparently enhance attachment and aid in subduing the struggling prey (Anderson and Bé 1976a). Subsequently, the carapace of the prey is ruptured, and rhizopodial streaming carries lipids, muscle tissue (Plate 4.1-3 and -8), and other soft tissues toward the aperture of the foraminifer (Hemleben and Spindler 1983). Digestive vacuoles formed in the external and internal cytoplasm (Plate 4.1) contain prey tissue in various stages of digestion. Lysosomal (digestive) enzymes may be secreted by primary lysosomes (digestive vesicles), and are concentrated within the larger and smaller digestive vacuoles distributed throughout the cytoplasm (Anderson and Bé 1976a). Large quantities of lysosomal enzymes are also observed in extracellular spaces surrounding the prey, and may be secreted as a means of predigesting some of the prey tissue before it is enclosed within digestive vacuoles.

© Springer-Verlag Berlin Heidelberg 2017
R. Schiebel and C. Hemleben, *Planktic Foraminifers in the Modern Ocean*,
DOI 10.1007/978-3-662-50297-6_4

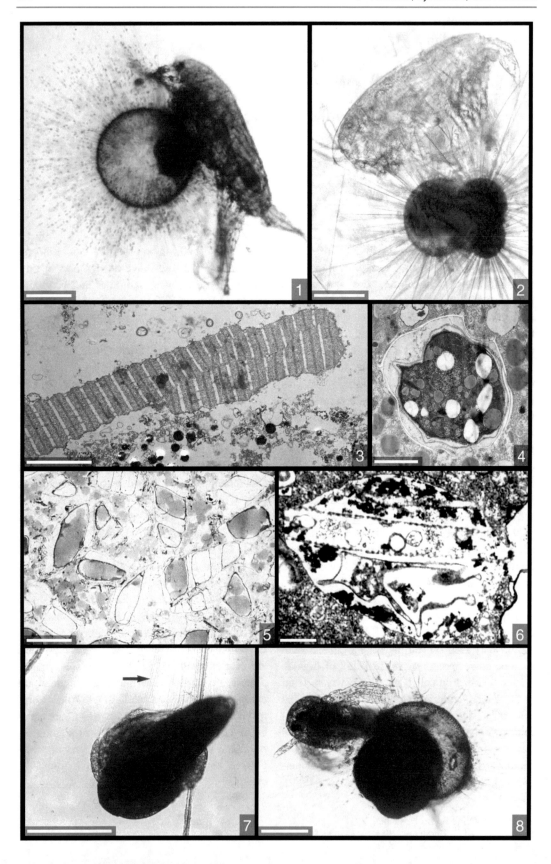

◀ **Plate 4.1** (*1*) Adult *O. universa* with fresh copepod prey captured in the natural environment. (*2*) Ingested tissues of a copepod and empty carapace to be discarded by *G. sacculifer*. (*3*) Copepod muscle tissue in *G. siphonifera*. (*4*) Largely undigested thecale dinoflagellate in food vacuole of *G. truncatulinoides*. (*5*) Empty diatom frustrules in *G. inflata*. (*6*) Digestive vacuole of *N. dutertrei* with an empty diatom frustule and unidentified material. (*7*) *Artemia salina* snared by *H. pelagica*, with bundles of rhizopodia (*arrow*) carrying the prey toward the foraminifer test (from Spindler et al. 1984). (*8*) *Artemia salina* snared by *N. dutertrei* and transported toward the aperture. Bars (*1, 2, 7, 8*) 250 μm, (*3–5*) 3 μm, (*6*) 0.5 μm

4.1.1 Natural Prey

Planktic foraminifers similar to other protozoa rapidly form digestive vacuoles and quickly consume their prey. Larger prey organisms may be visible in the corona of rhizopodia and spines. Empty carapaces of copepods may remain for several hours within the rhizopodial net of some foraminifer species, for example, *H. pelagica* and *G. sacculifer* for some hours (Spindler et al. 1984). Light microscopic examination of specimens immediately after sampling, and rapid fixation for transmission electron microscopy facilitates identification of large and small prey (Plate 4.1) within the peripheral and internal cytoplasm (Anderson and Bé 1976a; Bé et al. 1977; Anderson et al. 1979; Spindler et al. 1984).

In general, spinose planktic foraminifers prefer zooplankton protein (Table 4.1) over phytoplankton protein (Anderson 1983). In turn, non-spinose species are more adapted to herbivorous than carnivorous diet, as deduced from field and laboratory observation (Anderson et al. 1979; Hemleben and Auras 1984; Hemleben et al. 1985; Hemleben and Spindler 1983). Metazoan tissue in the digestive vacuoles of non-spinose species may be obtained from inactive or dead animals, since the ability of non-spinose species to catch motile prey has been observed (in laboratory culture) to be rather limited.

Prey of spinose species collected from the Sargasso Sea near Bermuda, and open ocean locations off the West Indies included copepods, hyperiid amphipods, and tunicates (Anderson et al. 1979). Spinose planktic foraminifer species examined for prey contained both animal and algal prey, with the exception of presumably exclusively carnivorous *H. pelagica* (Hemleben

Table 4.1 Relative share of food organisms in the diet of various planktic foraminifers. Data on *G. sacculifer* after Caron and Bé (1984).

	G. sacculifer		*G. siphonifera*	*O. universa*	*G. ruber*	*H. pelagica*
Number of observations	1124	812	198	456	207	625
Specimens with prey (%)	17.0	29.6	26.8	36.0	24.6	43.2
Copepods	44.0	45.4	47.2	41.5	39.2	*
Ciliates	26.7	27.5	30.1	33.6	19.6	
Tunicates	6.3	2.1	–	1.8	2.0	
Pteropods	1.0	2.5	1.9	–	2.0	
Chaetognaths	0.5	1.3	1.9	–	–	
Radiolarians	6.8	2.5	5.7	1.2	3.9	
Crustacean & Polychaete larvae	2.1	4.2	5.7	4.9	7.8	
Ostracods	0.5	0.8	–	–	2.0	
Siphonophores	1.0	–	–	–	–	
Various eggs	–	–	2.1	5.5	9.8	
Undeterminable	11.0	11.7	7.5	11.6	13.7	

*In *H. pelagica*, remains of copepods comprised > 90 % of food items. Larvae, non-tintinnid ciliates, and tunicate remains were rare

et al. 1989). Laboratory studies and examination of specimens collected near Barbados have shown that *G. sacculifer* is omnivorous, and consumes a substantial amount of tintinnids and diatoms (Spindler et al. 1984). Juvenile and neanic (around 80 μm test size) spinose planktic foraminifers collected from the natural environment, and those observed in laboratory cultures, mostly consume phytoplankton prey. Individuals collected with a 10-μm net may be reared to maturity when only fed algae (Hemleben et al. 1989).

> **Feeding in culture:** In some cases, pre-adult foraminifer individuals accept the algal prey when simply added to the culture vessel. Foraminifer individuals, which do not accept or reach the offered prey may be gently collected with a wide pipette and released just beneath the water surface in the culture vessel. When settling through the water column, the rhizopodia will capture prey. The feeding procedure may be repeated twice a day to ensure adequate nutrition, depending on the objective of the experiment. Natural prey from filtered (with a 10-μm filter to remove large particles) seawater provides a sufficient source of food to pre-adult planktic foraminifers. After 10–20 min in the feeding vessel, the individuals may be removed with a pipette and returned to the maintenance culture vessel.

Non-spinose species are omnivorous with a preference for herbivorous food (Spindler et al. 1984; Hemleben and Auras 1984; Hemleben et al. 1985). Diatoms are major part of the diet of many non-spinose species including *G. scitula*, *G. truncatulinoides*, *G. hirsuta*, *G. inflata*, *N. dutertrei*, *P. obliquiloculata*, and *G. glutinata* (Table 4.2). Tintinnid loricas were observed in the digestive vacuoles of *G. menardii*, indicating that the species preys also upon protozoa in addition to algae and larger zooplankton prey (Hemleben et al. 1977). Ingestion of protzoan prey has been assumed abundant but indiscernible in both spinose and non-spinose species, since the cytoplasm of the protozoa may be quickly consumed in the digestive vacuoles of planktic foraminifers, and would appear as merely non-identifiable animal biomass (Hemleben et al. 1989). Ciliary membranes, mucocysts, and ejectible organelles were observed (TEM imagery) among digested cytoplasmic components, and indicate the presence of ciliate prey (Caron and Bé 1984). Algal prey is identifiable by plastids, which are rather resistant to digestion until the late stages of digestion (Hemleben et al. 1989).

The average ratio of planktic foraminifers containing prey is rather variable between species, and highest in *H. pelagica* and *O. universa* (Table 4.1). Copepods are the major group of identifiable prey organisms observed in spinose species (Caron and Bé 1984; Hemleben et al.

Table 4.2 Prey of non-spinose planktic foraminifers as evidenced from contents of food vacuoles by transmission electron microscopy (including data from Anderson et al. 1979; from Hemleben et al. 1989).

	Algal prey	Animal prey
G. glutinata	D	
N. dutertrei	D, E, T	A
N. pachyderma	D, E	
P. obliquiloculata	D, Dn	A, M
G. inflata	D, Dn	A, M
G. truncatulinoides	D, E, T	A, M
G. hirsuta	D, Dn, E, T	A, M
G. menardii	D, E, T	A, M

A is unidentifiable animal tissue, D are diatoms, Dn are Dinoflagellates, E are eukaryotic algae, M is muscle tissue, T are thecate algae

1989). Non-spinose species consistently contained phytoplankton prey, i.e. mainly diatoms in their digestive vacuoles (Table 4.2), visualized by TEM images (Anderson et al. 1979; Spindler et al. 1984). Food remains may be stored in feeding cysts as, for example, in *G. glutinata*. A thick organic wall encloses numerous residues of digested diatoms, i.e. empty frustrules within the cavity of the cyst, and within digestive vacuoles in the cytoplasm of the foraminifer (Spindler et al. 1984).

4.1.2 Laboratory Studies on Trophic Activity

Culture experiments to examine the feeding behavior were done with spinose planktic foraminifers collected off Barbados, and non-spinose species collected near Bermuda (Spindler et al. 1984). Five spinose species were cultured using a modular system (see Chap. 10, Methods) of constant temperature baths (26.5 °C or 29.5 °C equivalent to open ocean conditions) with fluorescent illumination simulating a water depth of 10–30 m. Five non-spinose species were cultured at 15 to 20 °C equivalent to the temperatures in their natural habitat (Hemleben et al. 1989). Not all of the foraminifer species were present throughout the period of the experiment as seasonal abundances varied (Tables 4.3 and 4.4).

The food organisms offered to the planktic foraminfers were copepods from the suborders Calanoida (*Calocalanus pavo* (Dana), *Euchaeta marina* (Prestandrea), *Clausocalunus* sp., *Undinula vulgaris* (Dana), and *Acartia spinata*

Esterly, Cyclopoida (*Oncaea venusta* Philippi, *Oncaea mediterranea* Claus, *Farranula gracilis* (Dana)), and *Corycaeus speciosus* Dana, and Harpacticoida (*Macrosetella gracilis* (Dana), and *Miracia efferata* Dana).

Globigerinoides sacculifer was observed to capture and digest three of the four offered calanoid copepods. *Euchaeta marina* offered twice to *G. sacculifer* was the only calanoid copepod species, which escaped the predator after some minutes, and could not be devoured. In general, the acceptance rate of cyclopoid copepods was much lower than that of calanoids (Table 4.3). Out of 75 feeding trials only 18 of the trials with *F. gracilis* were successful (24 %). *Oncaea venusta*, *O. mediterranea*, and *C. speciosus* were offered 38 times to *G. sacculifer*, but were never accepted. Likewise, harpacticoid copepods were never accepted. A variety of other zooplankton were readily accepted and digested including pteropods, tunicates (*Oikopleura*), polychaete larvae, ostracods, heteropods, gastropod larvae, unidentified eggs, tintinnids, radiolarians, and acantharians.

Globigerinoides ruber was least adapted among the spinose species to feeding on copepods in laboratory culture. The acceptance rate of calanoid copepods was 20 %, and lowest among the examined species examined. *G. ruber* seems not to be as robust as *G. sacculifer* in laboratory culture, and tends to shorten or lose its spines. Hence, its low food acceptance rate in laboratory culture may not be representative of its behavior in the open ocean.

Globigerinella siphonifera rejected half of the offered copepod individuals (Table 4.3).

Table 4.3 Acceptance rates of copepod suborders by planktic foraminifers.

	Calanoida		Cyclopoida		Harpacticoida	
	ac:re	% ac	ac:re	% ac	ac:re	% ac
G. ruber	5:19	20	0:7	0	0:6	0
G. siphonifera	25:24	51	3:28	10	0:16	0
H. pelagica	6:3	67	3:11	21	0:10	0
G. sacculifer	66:16	80	18:57	24	0:28	0
O. universa	19:22	61	4:6	33	1:3	25

Adopted from Spindler et al. (1984). ac signifies accepted, re signifies rejected

Table 4.4 Digestion time (DT [hours]) of different food organisms.

Foraminifer species	Food organisms					
	Food remains (%)	*A. salina* Nauplius	Calanoid Copepod	Cyclopoid Copepod	Harpacticoid Copepod	DT total
O. universa	36	2:26	7:07	16:32	11:05	28:53
G. sacculifer	30	3:34	7:10	9:39	(+)	26:00
G. ruber	25	3:55	7:45	(−)	(−)	31:12
G. siphonifera	27	3:47	7:57	9:27	(−)	30:44
H. pelagica	53	3:34	8:54	25:49	(−)	24:20

Digestion times for *Artemia salina* nauplii only account for the time from catching to total digestion. Times for other copepods include the ejection time of the empty carapace (adopted from Spindler et al. 1984). Plus (+) indicates digestion observed without exact times, minuses (−) indicate that food was not accepted

Cyclopoid copepods were digested in only 10 % of the trials, and harpacticoids were always refused. *Hastigerina pelagica* digested both calanoid and cyclopoid copepods, whereas harpacticoids were also refused (data on *H. pelagica* are based on a rather limited set of 24 observations). The average digestion time (Table 4.4) for cyclopoid copepods is much longer than for calanoids, even though some of the cyclopoids were smaller (Spindler et al. 1984). The longer average digestion time for cyclopoids may be due to a different structure of the carapace compared to that of calanoids (Hemleben et al. 1989). Average feeding intervals of about 26–31 h indicate total food requirements (DT total) at the species level (Table 4.4). The actual feeding intervals in the natural environment are assumed shorter than in the laboratory experiments. A mixed diet of smaller prey including juvenile stages of copepods typical of the natural environment probably requires less time for digestion than less diverse laboratory food.

Orbulina universa appears to be best adapted to copepod predation among the five spinose species studied (Table 4.3), capturing and digesting at least five species from the three suborders calanoid, cyclopoid, and harpacticoid copepods (Spindler et al. 1984). Among other zooplankton-prey, *O. universa* consumed tunicates (*Oikopleura*), copepod (*Artemia*) nauplii, and acantharians (Anderson et al. 1979). Algal prey contained in the digestive vacuoles of *O. universa* included the colorless dinoflagellate *Cryptothecodinium cohnii* (Seligo), the chrysomonad flagellate *Dunaliella* sp., and the diatom species *Thalassiosira pseudonana* (Hasle and Heimdale). The diatom *Skeletonema costatum* (Greville) was offered to *O. universa* but not devoured. However, the diet of *O. universa* changes over the course of ontogeny from more herbivorous in pre-adult, to more carnivorous in adult individuals. *O. universa* is clearly omnivorous considering the range of prey consumed and digested.

4.1.3 Laboratory Experiments on Omnivorous Feeding

The adaptability of planktic foraminifers to a wide variety of environmental conditions, and ability to survive changes in food-availability may depend in part on the extent of their omnivorous behavior, i.e. the dependence on zooplankton versus phytoplankton carbon (e.g., Anderson et al. 1979). In a laboratory experiment, dinoflagellate prey (*Amphidinium carteri*) and crustacean prey (*Artemia* nauplii) was offered at discretion to the omnivorous and symbiont-bearing species *G. siphonifera*, *G. sacculifer*, and *G. ruber* collected near Barbados (Anderson 1983). The prey consumed over a two-hour feeding interval, and subsequently digested over the ensuing 24-hour period was determined from the algal and animal protein biomass consumed by each foraminifer species (Table 4.5).

Table 4.5 Comparative data on zooplankton and phytoplankton predation of three planktic foraminifer species (after Anderson 1983)

Species	Prey protein consumed[a] (µg)		Z/P ratio[b]
	Zooplankton	Phytoplankton	
G. siphonifera	5.56	0.02	278
G. sacculifer	4.82	0.026	158
G. ruber	3.74	0.032	117

[a]Based on a standard aliquot of *Amphidinium carteri* as a phytoplankton prey and 1-day old *Artemia* nauplii as zooplankton prey offered in laboratory cultures
[b]Ratio of zooplankton protein to phytoplankton protein consumed during a period of 2 h exposure to prey

Globigerinoides ruber seems to be less dependent on zooplankton consumption than *G. siphonifera* and *G. sacculifer* (cf. Spindler et al. 1984). The relatively enhanced consumption of algal protein by *G. ruber* may indicate its capability of obtaining more energy input from primary producers than the other two species, and is thus more competitive in regions of low primary productivity. Consequently, the abundance of *G. ruber* may in part be attributed to its capacity to efficiently feed on primary producers, and thus establish an advantage in competing for energy resources in regions of limited productivity by preying at the base of the trophic pyramid.

The efficiency of spinose planktic foraminifers in capturing and digesting zooplankton prey is possibly supported by their spines, as demonstrated in laboratory studies, and might account in part for their success in inhabiting a wide range of marine environments. Those environments accommodate diverse groups of phytoplankton and zooplankton serving as food sources of the planktic foraminifers.

The rhizopodial net of non-spinose species including *G. truncatulinoides*, *G. hirsuta*, *G. inflata*, *P. obliquiloculata*, and *G. glutinata* is possibly not suited to capture living prey like copepods (laboratory observations). Small pieces of prey produced by chopping the copepods into small servings are accepted by non-spinose species when being moved near the rhizopodia. However, non-spinose species might prey on some small zooplankton in their natural environment. Muscle tissue and other metazoan remains were identified in the digestive vacuoles of non-spinose species collected near Bermuda (Anderson et al. 1979). The behavior of non-spinose species in laboratory culture is likely biased by the fact that they often adhere to the bottom of the culture vessel, and are rarely floating as in the open ocean. Therefore, their rhizopodial net may not be freely extended, and is often spread out on the glass surface of the culture vessel. Nonetheless, there is evidence that generally omnivorous non-spinose species prefer herbivorous over carnivorous prey (Spindler et al. 1984). A mixed diet of diatoms (*Nitzschia* sp.) and *Artemia* nauplii (chopped) seems to support growth rates and extended survival of cultured non-spinose species (Anderson et al. 1979).

4.1.4 Cannibalism

Cannibalism is a particular case of a carnivorous diet, and may occur when two or more specimens of non-spinose species come into contact. The rhizopodia of involved individuals become closely intertwined, and the larger individual usually invades and consumes the cytoplasm of the smaller one. In laboratory cultures, juvenile specimens are often cannibalized if placed in the same dish with an adult organism regardless of the species of the two organisms. Cannibalism of adult individuals feeding on juveniles has been observed in laboratory cultures of the non-spinose species *G. hirsuta*, *G. inflata*, and *G. truncatulinoides*, and has been suspected in the spinose species *G. ruber*, *G. sacculifer*, and *O. universa*.

4.1.5 Effect of Food Availability on Test Development

The amount of food available during different ontogenetic stages was shown in culture experiments to affect the test development in *O. universa*. Unfed juvenile specimens construct very small final spherical chambers, while well-fed individuals develop larger spherical chambers, independent of the temperature at which they are cultured. Excess feeding of adult individuals with a spherical chamber can induce the construction of a second spherical chamber. Either a complete sphere may be attached to an earlier smaller sphere, or an incomplete second sphere may be added, intersecting the first sphere (Plate 5.3-9 and -10, see Sect. 5.1.3). These forms are termed 'Biorbulina' morphotypes (e.g., Hemleben and Spindler 1983). Although those culture experiments provide basic information on the food requirements in planktic Foraminifera, they may not be fully representative of optimal nutritional conditions in the natural environment.

4.1.6 Feeding Frequency

Based on observations of natural prey density and an empirical quantitative model, *G. sacculifer* is assumed to capture on average one copepod every 3.3 days (Caron and Bé 1984), which is significantly longer than the 26 h deduced by Spindler et al. (1984). However, about 56 % of the natural prey of *G. sacculifer* were found to be organisms other than copepods, such as chaetognaths, acantharians, and ciliates, and with digestion times shorter than those for copepods (Caron and Bé 1984). If fed only copepods in laboratory culture, one specimen per day again seems reasonable to satiate *G. sacculifer*.

Planktic foraminifers are capable of capturing and digesting considerable quantities of prey often exceeding their own size by several times. Foraminifers of about 300 μm in test diameter may feed on copepods, which are two to three times their size. Several prey organisms can be digested simultaneously if the planktic foraminifer is in need of food. *G. sacculifer* was observed in laboratory culture to digest four *Artemia* nauplii at the same pace required for digestion of one nauplius. In turn, spinose species can survive several weeks without food supply in laboratory culture. *H. pelagica* was observed to survive for an average of 16.4 days without food supply (Anderson et al. 1979).

4.1.7 Trophic Activity and Longevity

Survival rates and vitality of *G. truncatulinoides* and *H. pelagica* are affected by the quality of algal food, as assessed in laboratory cultures (Anderson et al. 1979). *Globorotalia truncatulinoides* shows clear preference for *E. huxleyi* over diatom and dinoflagellate prey (Table 4.6). *Emiliania huxleyi* is abundant also in the natural environment off Bermuda (Hulburt et al. 1960), where the *G. truncatulinoides* individuals for the laboratory cultures were obtained. Vitality of *G. truncatulinoides* assessed by the number of chambers filled with cytoplasm (Table 4.7).

Daily feeding of *G. truncatulinoides* with food as large as *Artemia* nauplii results in a mass of moribund cytoplasm around the aperture of the foraminifer test, and causes premature death.

Table 4.6 Average survival times (AS Time) of four groups of *G. truncatulinoides* of 20 specimens each, fed with different food items. Data from Anderson et al. (1979)

Food Item	Species	AS Time (days)
Diatom	*Thalassiosira pseudonana*	11.4
Dinoflagellate	*Gymdoninium* sp.	11.4
Coccolithophore	*Emiliania huxleyi*	21.6
Unfed control group	none	16.9

Table 4.7 Median survival times and vitality scores for *G. truncatulinoides* as a function of feeding interval.

Feeding interval	Mean survival (days)	Mean vitality score (days)
Three days	34.6	4.0
Twelve days	22.3	3.2
Starved	16.3	1.6

The vitality score is based on the number of chambers filled with cytoplasm. A score of 4 was given in case all chambers were filled, 3 if the final chamber was empty. When two, three, or four chambers were empty, the score of 2, 1, and 0 was applied, respectively. The overall score assigned was the most frequent score during the life span of each specimen. Ten specimens were analyzed per group. After Anderson et al. (1979); from Hemleben et al. (1989)

In turn, three-day feeding intervals yield maximum average longevity of 34.6 days (Table 4.7). When *G. truncatulinoides* are allowed to feed at discretion from naturally grown diatoms (*Nitzschia* spp.), the foraminifers establish a regular temporal pattern of ingestion, and longevity is up to 16.4 days. Having occupied a site on the bottom of the dish, they gather diatoms by rhizopodial streaming, and digest the prey in a feeding cyst. When the prey is consumed, food remain are discarded as a ring of debris around the specimen, and the foraminifer moves to a new location by rhizopodial extension and contraction, and the feeding cycle is repeated. Finally, a series of waste disposals indicate the former feeding sites of the foraminifer on the bottom of the culture vessel.

When *G. truncatulinoides* is allowed to feed at discretion, food consumption leads to an increase in cytoplasmic mass, and a new chamber may be added every 24–48 h. In turn, if the foraminifer is over-fed, excess cytoplasm may not be included in a single new chamber, and fragments of the excess cytoplasm are discarded. Those fragments of cytoplasm may exist for several days as amoeboid-like bodies exhibiting regular cycles of expansion and contraction, and their rhizopodial cytoplasm appears to stream in a normal way. However, those cytoplasm fragments have never been observed to develop into a mature test-bearing individual in laboratory culture, and may not contain a nucleus.

Vitality of *H. pelagica* is indicated by the presence of rhizopodia and a well-developed bubble capsule, i.e. the abundance of cytoplasm in floating specimens (Table 4.8). When fed at a daily schedule, gametogenesis (see Chap. 5) of *H. pelagica* is more regular and more prompt than at longer feeding-intervals, resulting in shorter average survival times than at a six-day feeding schedule (Table 4.8). Six-day feeding intervals appear to merely keep *H. pelagica* at a basic subsistence level, and longer feeding intervals even cause significantly reduced average survival times (Table 4.8). Assuming that daily feeding is more likely than a six-day feeding regime in the natural environment (Spindler et al. 1984), it may be concluded that gametogenesis of *H. pelagica* follows a rather prompt periodic timing within the lunar synodic cycle under normal open ocean conditions.

Table 4.8 Median survival time and vitality score (mean days floating) for *H. pelagica* as a function of feeding interval. 30 specimens analyzed per group.

Feeding interval	Mean survival (days)	Mean floating time (days)
Daily	23.2	17.7
Six days	26.8	21.7
Twelve days	17.4	12.9
Starved	16.4	13.7

After Anderson et al. (1979); from Hemleben et al. (1989)

Table 4.9 Influence of light intensity on chamber formation.

Light regime	N	Survival time	Frequency of gametogenesis	Rates of chamber formation			
				Fed daily	Ev. 3. day	Ev. 7. day	Unfed
H	189	15.6	90	0.30	0.21	0.008	0.01
L	189	9.1	92	0.37	0.28	0.21	0.07
D	189	4.2	85	0.34	0.25	0.18	0.08

High light intensity (H), low light intensity (L), and darkness (D). Feeding intervals and survival time in days, occurrence of gametogenesis (%), ev. means every. Shell growth given as rates of chamber formation per day, calculated from the total number of chambers formed during the survival time of *G. sacculifer*. Adopted from Caron et al. (1982)

Longevity as well as test size of planktic foraminifer individuals is affected by the availability of food. Daily feeding of *G. sacculifer* with *Artemia* nauplii results in more rapid growth rates and earlier gametogenesis (i.e. decreased longevity) than feeding at longer time-intervals, at varying temperature, salinity, and illumination (Table 4.9) under laboratory conditions (Bé et al. 1981; Caron et al. 1982; Caron and Bé 1984; Hemleben et al. 1987). Hence, it may be assumed that food deprivation in the natural environment may result in slower growth rates and prolonged life spans of *G. sacculifer*, and survival until more favorable ecological conditions arise for reproduction and enhanced survival rates of the offspring.

4.2 Biomass

Quantity and distribution of planktic foraminifer biomass are indispensible measures for the assessment of their effect on the modern and past marine carbon turnover and biological carbon pump. The ratio of organic carbon and inorganic $CaCO_3$ bound carbon of planktic foraminifers provides a proxy for the reconstruction of the ancient biological carbon pump in addition to the $\delta^{13}C$ proxy of their tests (e.g., Sigman and Haug 2003). Most of the planktic foraminifer biomass is included in the cytoplasm, and amounts to ~ 2.8 times the carbon mass included in the test calcite. Put the other way round, the calcite-carbon mass of the test of living specimens is approximately 36 % of the protein-C biomass (Schiebel and Movellan 2012). In addition, biomass contained in

organic tissues within the planktic foraminifer test wall amounts to ~ 0.025 % by weight (King 1977), although considerably varying between species, and is negligible in comparison to carbon-biomass of cytoplasm. About 10 % of cytoplasm-bound carbon of planktic foraminifers is at depths below the surface ocean (i.e. export production, see, e.g., Koeve 2002) and results basically from the downward flux of cytoplasm filled tests, and to some degree from deep-dwelling species (see Chap. 7).

The size-normalized protein-biomass of different species, and the planktic foraminifer assemblage biomass from different latitudes and different months and seasons (Fig. 4.1) are affected by trophic conditions, i.e. chlorophyll *a* concentration and availability of prey (Meilland 2015). Lower assemblage biomass in the western North Pacific off northern Honshu (Japan) than in the eastern North Atlantic results from a small-size fauna dominated by *N. incompta* (Movellan 2013). The same is true also for high latitudes and High-Nutrient Low-Chlorophyll (HNLC) regions (see Glossary), as shown by first data from the southern Indian Ocean (Meilland 2015). Higher assemblage biomass in surface waters, and decreasing biomass towards the deeper water column results from decreasing standing stocks (in addition to flux-related effects). Differences in individual planktic foraminifer biomass are assumed to result from the quantity and quality of prey. However, effects exerted on the planktic foraminifer biomass production by ecological conditions are far from understood, and need to be determined for the ontogenetic development of

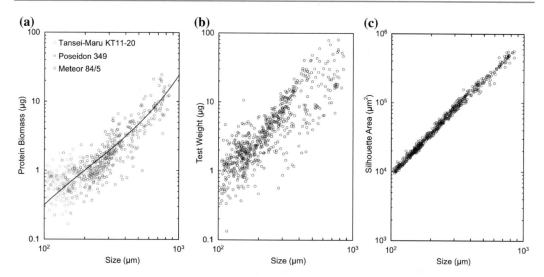

Fig. 4.1 The increase in protein biomass and test mass with size (minimum diameter) in individuals from different ocean basins follows the same trends. (**a**) In general, variability of biomass is higher in small-sized than large-sized individuals (n = 561, r^2 = 0.745 (exponential fit), $p < 0.00001$, standard deviation of the residuals = 1.612). (**b**) Variation of test weight with size relates to species with different test architectures (n = 646). Different size-to-weight ratios of different species result in a low r^2 = 0.571 (linear fit), $p < 0.00001$, standard deviation of the residuals = 6.623. (**c**) Relation of size and silhouette area, the latter of which has been shown to constrain size-and-weight changes to a high degree (Beer et al. 2010), with n = 660, r^2 = 0.974, $p < 0.00001$. From Schiebel and Movellan (2012)

species, before being used in modeling, and applied as a proxy of the biological carbon pump in paleoceanography (Meilland 2015).

In contrast to assemblage biomass, variation of the protein and carbon biomass of individuals planktic foraminifers is rather limited (Fig. 4.2), as shown by data of 21 different morphospecies, and a total of 2570 samples from different ocean basins (Schiebel and Movellan 2012; Meilland et al. 2016). Exceptions are species with different test architecture like adult spherical *O. universa*, as well as very large specimens of various species (Fig. 4.2). Although the ontogenetic development of biomass in *H. pelagica* is similar to other species (Fig. 4.2a), the biomass-to-weight ratio ranges above the ratio in other species due to the thin and light test of *H. pelagica* (Movellan 2013).

The ontogenetic development of biomass follows a logarithmic increase with test volume typical of allometric relationship in planktic foraminifers (Fig. 4.3). Remarkably, small individuals (>100–125 μm in test size) of various species produce the same or even larger quantity of biomass as individuals of the next larger test-size interval, >125–150 μm (Movellan 2013). The phenomenon might be explained due to changes in the metabolism during ontogenetic development, and transition from the juvenile to neanic, or neanic to adult stage depending on species (see Chap. 6). The phenomenon may also be explained by methodical/statistical effects, and a higher frequency (and thus more data) of small than large individuals (cf. Peeters et al. 1999; Schiebel and Hemleben 2000). The size-related development of biomass also differs between the trochospiral and spherical tests produced by *O. universa* (Fig. 4.4). Test mass in spherical *O. universa* seems to be larger than in trochospiral individuals (Fig. 4.4a), whereas the protein-biomass is larger in the latter (Fig. 4.4b, c). Although none of those relations is statistically significant, the trend in biomass-distribution in *O. universa* is assumed to be real, and the magnitude follows the average overall ontogenetic development (Fig. 4.3). Data from additional species and from are wider variety of regions and environmental settings, are needed to

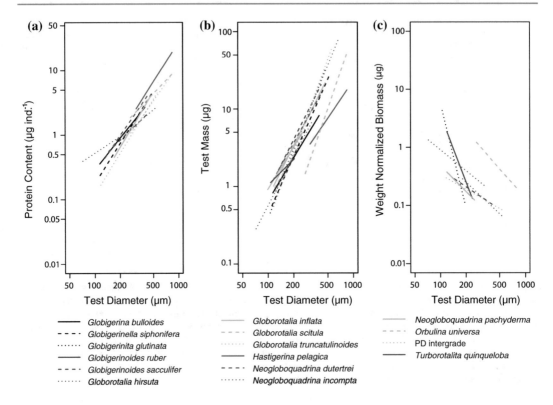

Fig. 4.2 Protein-biomass of 16 planktic foraminifer species from different hydrologic and trophic settings in the Atlantic and Pacific Oceans, sampled between the sea surface and 2500 m water depth. (**a**) Size-to-protein, (**b**) size-to-test mass, and (**c**) size-to-test weight relations show that the biomass distribution in most species is largely similar. Exceptions are very small (see *N. incompta*) and very large specimens (see *O. universa* and *H. pelagica*), and species with different test architecture. Data from additional five species (*G. calida, G. falconensis, G. rubescens, G. tenella, G. uvula*) are not shown here because too few data points are available. Regression-lines are given on a double-logarithmic scale. The entire legend is valid for all three panels of graphs **a–c**. From Movellan (2013)

better understand the production of biomass during the ontogenetic and metabolic development of planktic foraminifers.

4.3 Symbiosis

Associations of algae and spinose planktic foraminifers are visible with the naked eye, and were first reported as early as the late 19th century (Murray 1897). Species such as *G. sacculifer* appear colored distinctly yellowish-brown due to the abundance of dinoflagellates within the rhizopodial system and in the internal cytoplasm. In the absence of algae, the cytoplasm is colorless or only faintly colored amber, reddish, or greenish, depending on the type of food consumed by the foraminifer. The significance of algae as symbionts of planktic foraminifer hosts was recognized in the mid 20th century (Hemleben et al. 1989, and references therein). The widespread occurrence of algal associations especially with the spinose planktic foraminifers suggests that those relations are of profound significance in the physiology and possibly phylogeny of species. Photosymbiosis in planktic foraminifers was possibly developed in the late Cretaceous as indicated by the isotopic composition of fossil tests (e.g., Houston and Huber 1998; Houston et al. 1999).

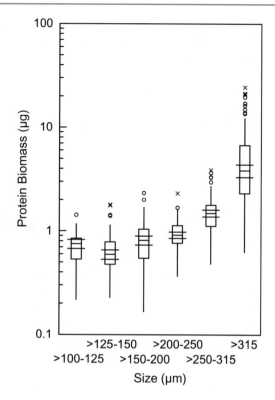

Fig. 4.3 The development of biomass with increasing size is largely logarithmic. However, the smallest analyzed test-size fraction >100–125 μm generally bears slightly more biomass than the next larger test size-fraction >125–150 μm. Average protein-biomass versus minimum diameter displayed as median values, notches, and the upper and lower quartiles for the respective size bins. The arithmetic mean of protein-biomass of the two smallest size bins is similar at 0.7 μg C per specimen. Circles and crosses indicate outside and far outside values, respectively. From Schiebel and Movellan (2012)

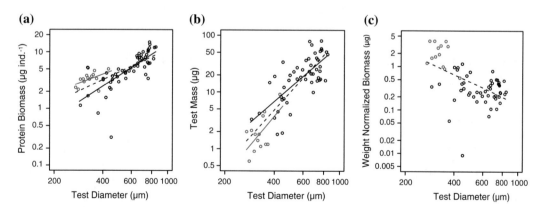

Fig. 4.4 (**a**) The size-to-protein development in *O. universa* changes between the production of the pre-adult trochospiral test (*red circles*), and the adult spherical test (*black circles*), and is related to (**b**) size-to-test mass changes between the two stages in test formation. (**c**) The test weight-normalized biomass in *O. universa* is larger in the trochospiral than the spherical stage. All of those differences are statistically not significant. The dashed line gives the regression for all data. From Movellan (2013)

Most spinose planktic foraminifers are associated with algal symbionts. The cellular structure of algal associates indicates only one type of dinoflagellate symbiont, and at least two kinds of chrysophycophytes, which differ in the organization of plastids at the periphery of the cell (Hemleben et al. 1989). Some non-spinose species appear to facultatively harbor symbionts, which are capable of photosynthesizing within the perialgal vacuoles, and may eventually be digested by the foraminifer. In comparison to planktic foraminifers, benthic foraminifers harbor a wide variety of algal symbionts including dinoflagellates, diatoms, and red algae, most of which are non-motile endobionts within the cytoplasm and perialgal vacuoles.

Data from culture experiments suggest that photo-receptive algae are intimately involved in the daily rhythm of the foraminifer including cytoplasmic activity, and the diel pattern of algal symbiont distribution in the external cytoplasm (cf. Bé et al. 1977). Symbiont activity in some planktic foraminifer species affects oxygen levels and pH, and hence potentially calcification (Jørgensen et al. 1985; Rink et al. 1998; Köhler-Rink and Kühl 2000, 2001, 2005; Hönisch et al. 2003; Lombard et al. 2009).

Whereas symbiosis is relatively well known in planktic foraminifers, relations and processes of commensalism and parasitism are less well understood. The presence of algae within or near the rhizopodial array of planktic foraminifers does not provide sufficient evidence for active symbiotic association. Algae may simply hover near the planktic foraminifer to obtain metabolic products as commensals, others ingest foraminifer cytoplasm, and some mutually benefit from the planktic foraminifer as symbionts.

4.3.1 Host-to-Symbiont Associations

Dinoflagellates and chrysophytes are the predominant type of symbiont associated with spinose planktic foraminifers (Table 4.10, Fig. 4.5) (e.g., Spero and Parker 1985; Spero 1987). The factors determining the association of particular algae with particular hosts are not known. Pigment analyses using liquid chromatography of extracts of symbionts abundant in *G. ruber* confirm the presence of dinoflagellates (Knight and Mantoura 1985). Pigment analyses of symbionts in *G. sacculifer*, *G. ruber*, and *G. siphonifera* reveal the presence of chlorophyll *a*, chlorophyll *c*, and carotenoids, i.e. peridinin, and absence of chlorophyll *b* (Bijma 1986; Huber et al. 1997; Bijma et al. 1998). While *G. ruber* may bear more symbiont-contained chlorophyll than *G. sacculifer* of equivalent size, *G. sacculifer* may absorb more light (i.e. lower light transmittance) than *G. ruber* (Hemleben et al. 1989). The phenomenon might be caused by various reasons including differences in pigment composition, and chemical differences associated with the photosynthetic pigments in the plastids of the dinoflagellate symbionts.

Among spinose species, *G. bulloides* and *H. pelagica* are barren of symbionts. Algal cells observed in the cytoplasm of presumably symbiont-barren spinose and non-spinose species have not shown any symbiotic connection to the host (Fig. 4.5).

Table 4.10 Kind of symbionts identified in planktic foraminifers according to Gastrich and Bartha (1988), Faber et al. (1988), and Hemleben et al. (1989)

Dinoflagellates	Chrysophycophytes	Facultative Chrysophytes
G. conglobatus	*G. siphonifera*	*G. inflata*
G. ruber	*G. humilis*	*G. menardii*
G. sacculifer		*N. dutertrei*
O. universa		*P. obliquiloculata*
		G. glutinata

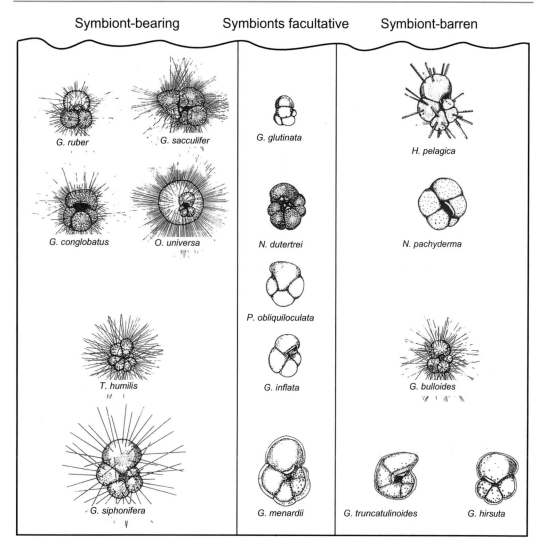

Symbiont-bearing   **Symbionts facultative**   **Symbiont-barren**

Fig. 4.5 Relationship between planktic foraminifers and symbiotic algae. Some species are obligatorily symbiont-bearing, others are consistently symbiont-barren, and some specimens of a few species are found to either possess or lack symbionts. The latter species are listed as facultative symbiont-bearing. Modified from Hemleben et al. (1989)

The presence of algal symbionts in most spinose planktic foraminifers, and their absence from most non-spinose species, raises questions about the physiological adaptations of species that favor a symbiotic mode of life, and about sarcodine evolution in general. Whereas planktic foraminifers harbor only few types of dinoflagellate and chrysophycophyte symbionts, other sarcodines such as benthic foraminifers and radiolarians harbor a much wider range of symbiont species including diatoms. In turn, most larger (benthic) foraminifers host only one species of symbiont at a time (Lee 1980, 2006), and most radiolarian species associate with only one species of algal symbiont (Anderson 1983). The variation in types of symbionts across host species may result in part from the physiological capacity of the symbiont to initially invade a host and then prevent digestion within the host cytoplasm. Successful host-symbiont associations require particular

Plate 4.2 (*1*) Coccoid dinoflagellate symbiont in *O. universa* surrounded by a perialgal membrane (*arrow*) within the ▶
internal cytoplasm. (*2*) Dividing symbiotic dinoflagellate. (*3*) Dinoflagellate symbiont within internal cytoplasm of *G.
sacculifer*. (*4*) Type I and (*5*) Type II symbionts in *G. siphonifera* (from Faber et al. 1988). (*6*) Section of final chamber
of *N. dutertrei* with symbionts accumulated beneath the test wall. Ch are chromosomes, ER is an endoplasmatic
reticulum, GC is a Golgi complex, M is a mitochondrion, N is the nucleus, P are plastids, PV is a perialgal vacuole, Py
is a pyrenoid, S is a starch sheath around a pyrenoid, TV is a thecal membranous vesicle, V is a vacuole with waste
products, VM is a perialgal vacuolar membrane. Bars (*1–5*) 1 μm, (*6*) 5 μm

nutrient concentrations, and light conditions
within the host, which coincide with the needs of
the symbiont. Variations of symbiont species
across geographical ranges and water depths
possibly result from a combination of factors
including their capacity to adapt to particular
ecological conditions, and may affect the distribution of planktic foraminifers at the regional to
global scale, and on short (seasonal) to long (geological) time-scales. Additional information on
the geographical and synecological relation of
symbionts and hosts may be obtained from
comparative analyses of different genotypes of the
same planktic foraminifer morphotype (e.g.,
Darling and Wade 2008).

4.3.2 Acquisition of Symbionts During Ontogeny

Symbionts are possibly not transferred from
parent to offspring during sexual reproduction
(see Chap. 5 on Reproduction). Gametes are
undoubtedly too small to hold dinoflagellate
symbionts. Being not transferred from parent to
offspring during sexual reproduction, symbionts
are digested by the host or expelled from the
foraminifer shortly before the gametes are
released (Bé et al. 1983, for *G. sacculifer*). In
turn, no aposymbiotic individuals of any species
known to harbor symbionts have been found in
the natural environment. Given the lack of algal
symbionts in gametes of planktic foraminifers,
symbionts are presumably acquired after fertilization (e.g., Hemleben et al. 1989; Shaked and
De Vargas 2006). Juvenile planktic foraminifers
apparently acquire algal symbionts from ambient
seawater when they reach the two-chambered to
three-chambered stage, i.e. within the first few

days of ontogeny (Brummer et al. 1986). Even
adult individuals rendered aposymbiotic in the
laboratory are able to acquire symbionts, and
re-establish symbiosis with symbionts offered
from donor individuals of the same species (Bé
et al. 1982).

Juveniles with two to three chambers may
already host three to five symbionts. During
maturation the number of symbionts increases by
cell division (Plate 4.2-2) concomitant with
increasing size of the host. An average number
of ∼3200 symbionts was estimated in a mature
spherical *O. universa* of 350–720 μm test
diameter (Spero and Parker 1985). In one large
specimen with a spherical test of 892 μm about
23,000 algal symbionts were found.

Since all symbiont-bearing planktic foramin-
fer species maintain perpetual species-specific
association with only one kind of symbiont
(Table 4.10), they must rely on encounters with
algae of the appropriate species in sufficient
density to ensure uptake and establishment of the
symbiosis. This mechanism very likely occurs at
the deep chlorophyll maximum (DCM) associated with the pycnocline where both sufficient
food and appropriate symbiotic algae can be
obtained by planktic foraminifers (Fairbanks and
Wiebe 1980). The probability of an encounter
between the host and the symbiotic algae
depends on the distribution of both partners,
which is subject to daily, seasonal (or other)
patterns of variation in temperature, salinity, or
trophic conditions. The distribution and abundance in time and space of potentially acceptable
symbiotic algae has not been quantified. This is
due in part to the incomplete knowledge of the
taxonomic position of some of the symbionts.
Since the engulfed symbionts within the foraminifer cytoplasm are coccoid, and typically lack

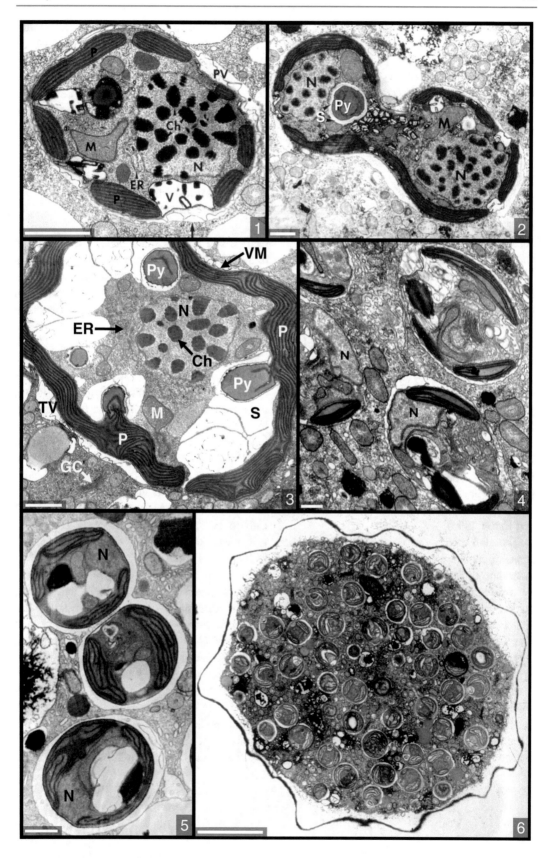

structures needed for identification at the species level such as thecae, frustules, and flagella, they need to be isolated, cultured, and analyzed for their molecular genetics to permit definitive identification (Spero 1987; Gast and Caron 1996; Gast et al. 2000).

4.3.3 Structural Host-to-Symbiont Associations

Algal symbionts associated with planktic fora-minifers are predominantly dinoflagellates in a coccoid state, i.e. non-flagellated and athecate, and about 5–10 μm in size (Plate 4.2-1 and -2). The fine structure of dinoflagellate symbionts is identical in the planktic foraminifer species *G. conglobatus*, *G. ruber*, *G. sacculifer*, and *O. universa*. Spero (1987) isolated and cultured symbionts of *O. universa*, and characterized the flagellated gymnoid form with epicone and hypocone of equal size, and assigned them to the new species *Gymnodinium beii*. Analyses of the small subunit ribosomal DNA (srDNA) have confirmed the classification as *G. beii* (Gast and Caron 1996). According to molecular analyses, the two types of symbionts (Type I and Type II) hosted by *G. siphonifera* (Faber et al. 1988) are a prymnesiophyte, and presumably a chrysophy-cophyte or chrysophyte (Gast et al. 2000; Gast and Caron 1996).

The non-motile endosymbionts of planktic foraminifers are enclosed within perialgal vac-uoles surrounded by a thin layer of cytoplasm in the rhizopodial system, thus permitting visual-ization with the light microscope (e.g., Spindler and Hemleben 1980; Hemleben et al. 1985; Spero and Parker 1985). From a biological per-spective, the thin cytoplasmic sheath facilitates light penetration to the symbiont for photosyn-thesis, while also separating the host cytoplasm from direct contact with the symbiont. When the symbionts are withdrawn by rhizopodial streaming into denser regions of cytoplasm near the test or into the internal cytoplasm, the

perialgal vacuolar membrane (Plate 4.2-3) fur-ther serves as a structural barrier separating the symbiont from the host. Apparently, this barrier regulates physiological processes between host and symbiont, and permits exchange of essential chemical products (cf. Jørgensen et al. 1985; Hemleben et al. 1989; Uhle et al. 1997, 1999).

4.3.4 Physiological Interactions Between Symbiont and Host

Carbon and nitrogen compounds are transferred to the planktic foraminifer host by the symbionts (Uhle et al. 1997). Osmiophilic dense deposits immediately adjacent to perialgal vacuoles within the planktic foraminifer cytoplasm are interpreted as photosynthates released by the symbiont to the host (Anderson and Bé 1976b). Carbon uptake by *G. sacculifer* during photo-synthsis of its symbionts was quantified in ra-diocarbon experiments (Bijma 1986). Radiocarbon is incorporated into the cytoplasm by *G. ruber* when exposed to light, while incorporation in the dark is negligible (Gastrich and Bartha 1988). The photosynthetic activity of the symbiotic algae in *O. universa* and *G. sac-culifer* affect the $\delta^{18}O$ than the $\delta^{13}C$ ratio of the test calcite, and change with test size (Spero and DeNiro 1987; Spero and Lea 1993). High $\delta^{13}C$ values were obtained at intense illumination, and lower values under low-light conditions or in darkness (Spero and Lea 1993).

Close interaction between algae and host in planktic foraminifers is indicated by the diel cycle of symbiont distribution within the host cytoplasm (Anderson and Bé 1976b; Bé et al. 1977; Hemleben and Spindler 1983). Symbionts are withdrawn into the test of the host by rhi-zopodial streaming before sunset. The perialgal vacuoles are carried by cytoplasmic flow along the rhizopodial strands centripetally toward the host's central cell mass, and many are withdrawn into the internal cytoplasm through the aperture.

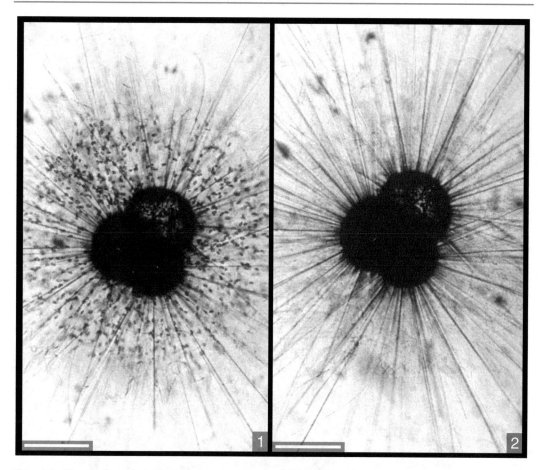

Plate 4.3 Circadian symbiont distribution controlled by the host *G. ruber*. (*1*) Symbionts carried out by rhizopodial streaming along the spines during light period, and (*2*) withdrawn into test during darkness. Both pictures taken from the same specimen. Bars 200 μm. From Bé et al. (1977)

Reversely, the host disperses the symbionts towards the outside into the peripheral cytoplasm at dawn (Plate 4.3). Since the symbionts lack flagella and are enclosed within the host's vacuoles, they are entirely controlled by the cytoplasmic activity of the host (Hemleben et al. 1989). In turn, the diel cycle is triggered by the light sensitivity of the symbionts (Caron et al. 1982; Bé et al. 1982). When a photosynthetic inhibitor such as 3-(3,4 dichlorophenyl)-l, l-dimethylurea (DCMU) is applied to inhibit light reception, the symbionts are continuously withdrawn as though in darkness, even though a dark-light cycle is maintained. If the phase of the diel dark-light cycle is altered by one half cycle, i.e. illumination at night and darkness during the day, the cycle is usually changed to the new schedule within 48–72 h. The altered cycle may be restored to the former ('normal') phase by returning the individual to the regular day-and-night schedule (Hemleben et al. 1989). If aposymbiotic DCMU-treated specimens are subsequently reinfected with new symbionts from a donor *G. sacculifer*, the results are similar to control-group of untreated specimens (Fig. 4.6).

In prolonged darkness, symbionts are gradually digested over a period of several days even if the host is fully nourished by externally supplied prey, and the host commences early gametogenesis (Hemleben et al. 1989). Finally, DCMU-induced inhibition of symbiont activity or continuous darkness in *G. sacculifer* results in significantly decreased survival times.

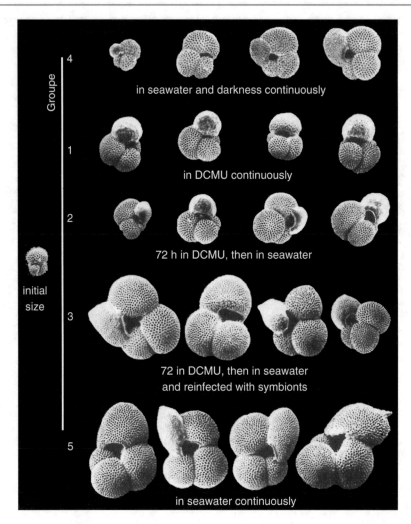

Fig. 4.6 Final test size of *G. sacculifer* treated with DCMU (Groups 1, 2, and 3), kept in the dark (Group 4), and of a control Group 5 grown in untreated seawater at normal diel illumination. The left-hand test represents the initial average size of 230 μm of the 202 individuals of *G. sacculifer* used in the experiments. Four individuals of each group were randomly selected to represent variations in test morphologies and test sizes produced under the five different experimental conditions. Modified from Bé et al. (1982)

Photosynthetic activity of the symbionts (Fig. 4.7) is also essential for calcification and test formation, as indicated by suppressed calcification and chamber growth during prolonged darkness. A similar effect is induced if the symbionts are treated with DCMU, as shown by laboratory cultures of *G. sacculifer* (Bé et al. 1982). However, high DCMU concentration may also directly affect the role of light in calcification rather than through symbionts and photosynthesis (Erez 1983). Reduced final test sizes of light-deprived specimens compared to naturally grown specimens were observed in *G. sacculifer* and *O. universa* (Bé et al. 1982; Spero 1986; cf. also Hemleben et al. 1987). Consequently, water depth and turbidity, and resulting illumination and symbiont activity may also affect the size of chambers and tests of symbiont-bearing species grown in the natural habitat (cf. Spero 1986). Reduced test sizes observed in specimens treated with DCMU and kept in darkness may represent terminal deposition of residual calcium within the calcium pool accumulated during photosynthesis. The calcium pool would normally be

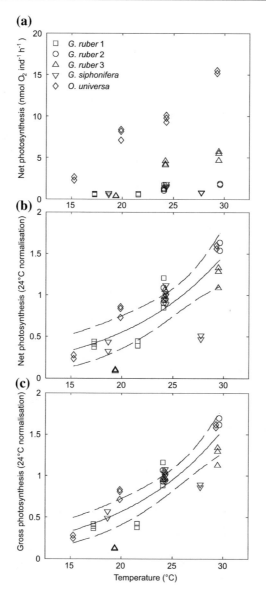

Fig. 4.7 Photosynthesis rate of *G. ruber*, *G. siphonifera*, and *O. universa*. **a** Net photosynthesis rate (nmol O_2 ind.$^{-1}$ h^{-1}) in relation to temperature. **b** Net photosynthesis rate in relation to temperature of the different specimens normalized by the mean observed value at 24 °C. **c** Gross photosynthesis rate in relation to temperature of the different specimens normalized by the mean observed value at 24 °C. *Solid lines*: least-squares regression for data fitted with Arrhenius relationships. *Dashed lines* give 95 % confidence intervals for the regressions. From Lombard et al. (2009)

replenished during illumination after each chamber addition, and is suppressed in the DCMU-treated and non-illuminated specimens (Hemleben et al. 1989).

The physiology of *G. sacculifer* including symbiont photosynthesis and respiration (Fig. 4.8) was first assessed by Jørgensen et al. (1985) using microelectrodes to probe the O_2 concentrations, and *p*H at varying positions peripheral to the test of the host with a resolution of 50–100 μm (see Chap. 10, Methods). Photosynthetic rates were mapped by moving the microelectrodes

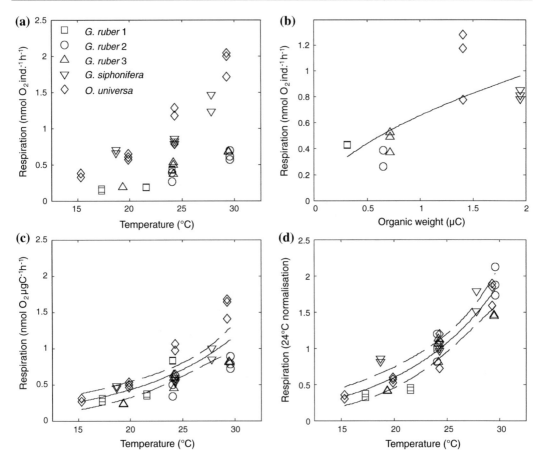

Fig. 4.8 Respiration rate of *G. ruber* (of different test diameter: 1 = 189 μm, 2 = 241 μm, and 3 = 249 μm), *G. siphonifera* (347 μm test diameter), and *O. universa* (521 μm test diameter). **a** Respiration rate (nmol O_2 ind.$^{-1}$ h^{-1}) in relation to temperature. **b** Respiration rate at 24 °C in relation to the organic weight (μg C) calculated from test size and a conversion factor from Michaels et al. (1995). *Solid line* gives least-squares regression for data fitted with a power model with a 0.57 ± 0.18 exponent. **c** Respiration rate in relation to temperature of the different specimens calculated for a 1 μg C individual using the precedent relationship (**b**). **d** Respiration rate in relation to temperature of the different specimens normalized by the mean observed value at 24 °C. *Solid line*: least-squares regression for data fitted with Arrhenius relationship. *Dashed lines* give 95 % confidence intervals for the regressions. From Lombard et al. (2009)

carefully around within the halo of symbionts in the rhizopodia surrounding the test. Under illumination, O_2 concentrations increased to 2.5 times air saturation, while *p*H increased to 8.62, and hence well above the *p*H of 8.23 of ambient water. In darkness, and at a temperature of 24 to 25 °C, planktic foraminifer respiration lowered the O_2 concentration at the test surface to 50 % of air saturation, while the *p*H lowered from the ambient value to 8.15. The compensation light intensity of the algal-to-host system was determined at 26–30 μE m^{-2} s^{-1}, and light saturation intensity was 160–170 μE m^{-2} s^{-1}. Gross photosynthesis at light saturation was 18.1 nmol O_2 h^{-1} per foraminifer individual, and respiration rates ranged between 2.7 and 3.3 nmol O_2 h^{-1} per individual under dark and light saturation, respectively. Such high photosynthetic activity of the symbionts could in theory supply all of the organic carbon required for growth and metabolism of the host, but limited concentrations of dissolved nitrogen and phosphorus necessitates capture of prey to supply those essential elements (Jørgensen et al. 1985).

Respiration rates of the symbiont-bearing planktic foraminifers are possibly related to solar irradiation and photosynthetic activity of their symbionts, as well as temperature (Rink et al. 1998; Lombard et al. 2009) (Figs. 4.7 and 4.8). Respiration rates of the symbiont-bearing species *G. ruber, G. siphonifera, O. universa,* and *G. sacculifer* vary by several orders of magnitude (~ 0.15 to 6 nmol O_2 h^{-1} per individual) possibly resulting from differences in ontogenetic stage and size of the host, number of symbionts hosted, state of 'health' of the host after sampling, differences in experimental set-up (e.g., temperature and light intensity), as well as differences between the foraminifer species (Spero and Parker 1985; Jørgensen et al. 1985; Gastrich and Bartha 1988; Rink et al. 1998; Lombard et al. 2009, and references therein). However, more data would be needed from the same species, and from other species including symbiont-barren species, different size-classes, and different environmental conditions to better constrain planktic foraminifer respiration rates.

4.3.5 Dinoflagellate Symbiont Fine Structure

Dinoflagellate symbionts (Plate 4.2-3) exhibit characteristic cytoplasmic features, which distinguish them from the planktic foraminifer host. The nucleus of dinoflagellates contains dense whorls of chromosomes (DNA) typical of mesokaryotic algae. The surrounding cytoplasm contains mitochondria with tubular cristae, endoplasmic reticulum, food vacuoles, and other vacuoles of varying size. At the periphery of the symbiont, lobes of plastids (light-trapping organelles) with internal lamellae containing thylakoids are visible in TEM images. Pyrenoids surrounded by a starch sheath projecting from the plastid surface, are commonly observed in dinoflagellate symbionts. Pyrenoids are sites of carbohydrate deposition during photosynthesis. Additional starch grains may be scattered throughout the cytoplasm. Crystalline waste products contained in vacuoles where identified as guanine and calcium oxalate (Hemleben et al. 1989). The peripheral membranes surrounding the symbionts are complex owing in part to the several layers of membranes associated with the dinoflagellate cell surface, and the additional layer of cytoplasm formed by the host's rhizopodial sheath enclosing the symbiont. In coccoid dinoflagellate symbionts the organic plates, which normally form the dinoflagellate thecae, are absent and only vesicles with electron translucent cisternae may be observed at the periphery of the cell (Plate 4.2-3).

4.3.6 Chrysophycophyte Symbiont Fine Structure

Chrysophycophytes are eukaryotic algae with a characteristic fine structure. Those small yellow-green algae of about 2–3 μm size are associated with *G. siphonifera* (Plate 4.2-4 and -5), *G. glutinata,* and *Turborotalita humilis* (Hemleben et al. 1989). In general, chrysophycophytes exhibit a prominent eukaryotic nucleus (Plate 4.2-4 and -5) containing finely dispersed strands of euchromatin surrounded by a double-membrane nuclear envelope. The peripherally arranged plastids are enclosed within the cisterna of the nuclear envelope, and contain internal laminae each composed of a stack of three thylakoids. Mitochondria, endoplasmic reticulum, and small vesicles occur within the central mass of cytoplasm.

Two types of the small yellow-green symbionts were observed in *G. siphonifera* collected near Barbados and Jamaica. Each of the two types (Plate 4.2-4 and -5) possibly exerts a different effect on the host. The mean final test size of *G. siphonifera* was consistently larger when Type I algae were present compared to those with Type II algae both grown under the same experimental conditions (Faber et al. 1988, 1989). Starved individuals containing Type I symbionts formed very few chambers and died without

Plate 4.4 (*1*) Commensal *Pyrocystis robusta* dinoflagellates within the bubble capsule of *H. pelagica*, with (*2*) ▶
rhizopodia attached to the *P. robusta* at the *lower left side*. (*3*) Commensal *P. noctiluca* within the bubble capsule of *H.
pelagica*. (*4*) Colorless dinoflagellate attached to spine. (*5–8*) Parasitic sporozoans infesting and digesting *H. pelagica*
with (*5*) bubble capsule, and sporozoans close to the test. (*6*) Sporozoans withdraw from the test 26 h later. (*7*) Again
10 h later, all sporozoans have evacuated the test. Bars (*1, 3, 5–7*) 400 µm, (*2*) 100 µm, (*4, 8*) 20 µm

undergoing gametogenesis, while foraminifers
with Type II algae produced more chambers and
most of them produced gametes. The role of those
algae in other foraminifer species like *G. glutinata*, *G. hirsuta*, *G. menardii*, *N. dutertrei*
(Plate 4.2-6), *P. obliquiloculata*, *G. inflata*, and
Candeina nitida is uncertain. Although algae
were observed in stages of division within vacuoles of those planktic foraminifer species, symbiotic relations could not yet be proven
(Hemleben et al. 1989, and references therein).

4.4   Commensalism

Dinophytes are abundant commensals within the
cytoplasm of planktic foraminifer hosts. Commensals like the cocale dinophytes *Pyrocystis
noctiluca* and *P. robusta* were found immersed
within the bubble capsule of *H. pelagica*
(Plate 4.4-1 and -2), or loosely associated with
the rhizopodial net of *G. sacculifer* (Plate 4.4-3
and -4), *G. ruber*, *O. universa*, and occasionally
G. truncatulinoides. *Pyrocystis* commensals are
possibly present in most *H. pelagica*. More than
200 specimens of *P. robusta* were found hosted
by an individual *H. pelagica* (Spindler and
Hemleben 1980). Those large autotrophic dinophytes (150–400 µm) within the external cytoplasm of the foraminifer host apparently do not
provide any organic products from photosynthesis to the host, as indicated by ^{14}C experiments (Hemleben et al. 1989). The filamentous
diazotroph (nitrogen fixing) blue-green algae
(cyanobacteria) *Trichodesmium*, which typically
occurs in oligotrophic waters, was found harbored between the spines of *G. siphonifera*
Type I (but not *G. siphonifera* Type II, see
Chap. 2), and assumed in extracellular commensal association with the foraminifer host
(Huber et al. 1997; Bijma et al. 1998).

Feeding experiments indicate that *Pyrocystis*
dinophytes neither harm nor support the host
when subjected to starvation (Anderson et al.
1979). The large numbers of commensals
observed in the bubble capsule of *H. pelagica*
suggest that a substantial amount of metabolic
by-products and undigested prey particles may
be present in the cytoplasm of the host. *Pyrocystis* increases in numbers when the host
(*H. pelagica*) is well fed, and commensals may
simply utilize the waste products of their hosts as
food source. Further potential commensals in the
bubble capsule of *H. pelagica* include
P. fusiformis, *Dissodinium lunula*, and *D. ele-
gans* (Bé et al. 1977; Elbrächter et al. 1987).

The mechanism of protection of dinophytes
against digestion by the planktic foraminifer is
unknown. The robust theca or any other mechanism may help to protect the dinophytes against
digestion by the host. However, empty thecae of
Pyrocystis found within *G. sacculifer* indicate
that the protection may not always work (Hemleben et al. 1989).

4.5   Parasitism

Small free-swimming dinoflagellates (Plate 4.4-4)
of the orders gymnodiniales and peridiniales are
assumed parasites of spinose planktic foraminifers. Those dinoflagellates have frequently
been observed hovering in and around the rhizopodial array of *G. siphonifera*, *G. ruber*,
G. sacculifer, *H. pelagica*, and *O. universa*,
and feeding on the foraminifer cytoplasm
(Spindler and Hemleben 1980). *H. pelagica*
heavily infested with sporozoans were found to be
in poor health. The foraminifer may eventually be
digested by the sporozoans leaving an empty test
(Plate 4.4-5 to -8).

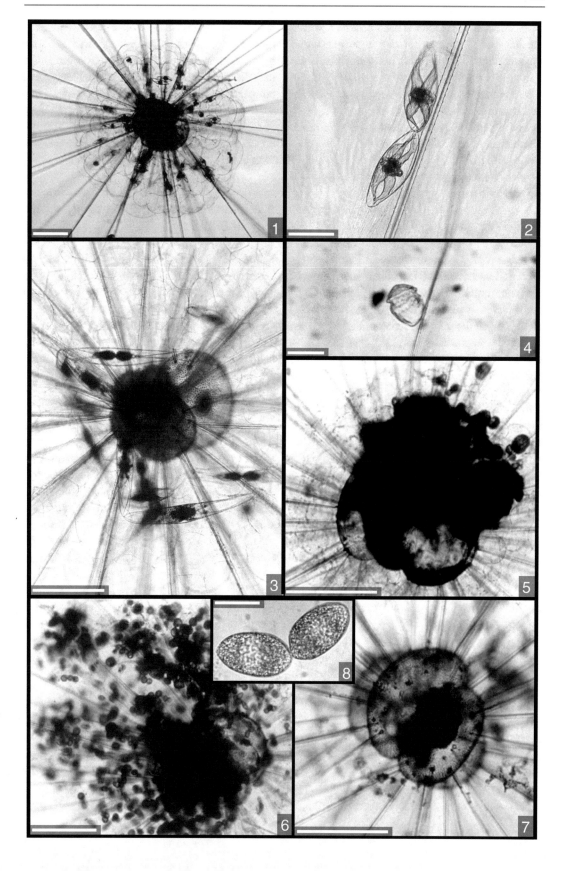

Bacteria have frequently been observed within vacuoles of *G. ruber* both at the periphery and the more internal cytoplasm, but parasitism is not proven. Those bacteria could be prey of the foraminifer. In turn, it has been observed that bacteria rapidly invade and consume the cytoplasm of unhealthy and deceased foraminifers (Hemleben et al. 1989).

4.6 Predation

One of the most frequently asked questions concerns the predators of planktic foraminifers. Whereas a large variety of predators of benthic foraminifers have been identified, the nature of planktic foraminifer predators is largely enigmatic. Even if planktic foraminifers are eventually found in the guts of predators like tunicates, pteropods, euphausids, sergestid prawns, polychaetes, and holothurians (Bradbury et al. 1971; Brand and Lipps 1982) predation might have been active or accidental. Selective predators of living planktic foraminifers have not yet been reported. Random ingestion of planktic foraminifers is assumed for large non-selective filter feeders like salps, and other large predators like shrimps and crabs, as well as suspension feeding invertebrates from subtropical to polar environments (Berger 1971; Brand and Lipps 1982). Raptorial predation of planktic foraminifers was suggested but has never been observed (Culver and Lipps 2003). Somehow quantitative observations are based on evidence, i.e. the contents of fecal pellets of salps, which were mainly composed of juvenile planktic foraminifer tests (Wiebe et al. 1976; Bé 1977).

Observations from laboratory cultures, including attacks from ciliates, may not be transferable to healthy and well-fed individuals in natural environments (Hemleben et al. 1989; Culver and Lipps 2003, and references therein). Round borings of 10–20 µm in diameter in planktic foraminifer tests sampled from the water column are rare, and may indicate predation by nematodes or gastropods similar to observations made on benthic foraminifers (Sliter 1971). Other types of presumably predator-inflicted

bioerosion have not been observed in planktic foraminifer tests sampled from the water column by the authors. Since predation and cannibalism may affect planktic foraminifer carbon budgets of both calcite bound carbon and cytoplasm-carbon at the regional to global scale, quantitative observations would be important for a better understanding of their role within the pelagic food chain, carbon turnover, and taphonomy.

4.7 Summary and Concluding Remarks

Planktic foraminifers are basically omnivorous, and consume a wide variety of phytoplankton and zooplankton prey, but during the earliest ontogenetic stages, they are most likely herbivorous. A clear preference for animal prey exists among the spinose species, as deduced from culture experiments and observation of natural prey in individuals collected from the natural environment. Non-spinose species are largely herbivorous. Planktic foraminifers may hence be positioned at the base of heterotrophic consumers within the marine food web. However, spinose species prey to some extent on larger metazoans such as copepods, and may therefore be placed at a trophic level different from other protozoans. Food availability has been found to affect test size and survival time of specimens. The individual size-normalized protein-biomass of different species, and the planktic foraminifer assemblage biomass are mostly affected by trophic conditions and availability of food. Future experiments should investigate the food source of juvenile individuals by means of culture experiments and electron microscopy analyses. Predators selectively feeding on planktic foraminifers have not yet been observed.

Most spinose planktic foraminifers are associated with dinoflagellate or chrysophycophyte algal symbionts. Some non-spinose species appear to facultatively harbor symbionts, which are capable of photosynthesising within the perialgal vacuoles, or are digested. Only one type of dinoflagellate, and at least two kinds of chrysophycophyte symbionts have been identified in planktic foraminifers so far. Experimental data suggest that symbiontic algae are involved in the daily rhythm of the foraminifer, including cytoplasmic activity and diel pattern of symbiont distribution in the external cytoplasm. Exchange products between symbiont and host include oxygen, carbon, and nitrogen compounds, supporting the foraminifer's metabolism, and affecting the stable isotope ratios of the foraminifer's organic products (e.g., fatty acids) and test calcite. Experimental evidence points toward a significant role of the algal symbionts also in the calcification processes and chamber formation. Additional research is needed on the complex physiological interdependence and possible effects of symbiont activity on growth, calcification, and the test morphology of the foraminifer.

References

Anderson OR (1983) Radiolaria. Springer, New York

Anderson OR, Bé AWH (1976a) A cytochemical fine structure study of phagotrophy in a planktonic foraminifer, *Hastigerina pelagica* (d'Orbigny). Biol Bull 151:437–449

Anderson OR, Bé AWH (1976b) The ultrastructure of a planktonic foraminifer, *Globigerinoides sacculifer* (Brady), and its symbiotic dinoflagellates. J Foraminifer Res 6:1–21. doi:10.2113/gsjfr.6.1.1

Anderson OR, Spindler M, Bé AWH, Hemleben C (1979) Trophic activity of planktonic Foraminifera. J Mar Biol Assoc U K 59:791–799. doi:10.1017/S002531540004577X

Bé AWH (1977) An ecological, zoogeographic and taxonomic review of Recent planktonic Foraminifera. In: Ramsay ATS (ed) Oceanic micropaleontology. Academic Press, London, pp 1–100

Bé AWH, Anderson OR, Faber WW, Caron DA (1983) Sequence of morphological and cytoplasmic changes during gametogenesis in the planktonic foraminifer *Globigerinoides sacculifer* (Brady). Micropaleontology 29:310. doi:10.2307/1485737

Bé AWH, Caron DA, Anderson OR (1981) Effects of feeding frequency on life processes of the planktonic foraminifer *Globigerinoides sacculifer* in laboratory culture. J Mar Biol Assoc U K 61:257–277

Bé AWH, Hemleben C, Anderson OR, Spindler M, Hacunda J, Tuntivate-Choy S (1977) Laboratory and field observations of living planktonic Foraminifera. Micropaleontology 23:155–179

Bé AWH, Spero HJ, Anderson OR (1982) Effects of symbiont elimination and reinfection on the life processes of the planktonic foraminifer

Globigerinoides sacculifer. Mar Biol 70:73–86. doi:10.1007/BF00397298

Beer CJ, Schiebel R, Wilson PA (2010) Technical Note: On methodologies for determining the size-normalised weight of planktic Foraminifera. Biogeosciences 7:2193–2198. doi:10.5194/bg-7-2193-2010

Berger WH (1971) Sedimentation of planktonic Foraminifera. Mar Geol 11:325–358. doi:10.1016/0025-3227(71)90035-1

Bijma J (1986) Observations on the life history and carbon cycling of planktonic Foraminifera. Gulf of Eilat/Aquaba. Master Thesis, University of Groningen

Bijma J, Hemleben C, Huber BT, Erlenkeuser H, Kroon D (1998) Experimental determination of the ontogenetic stable isotope variability in two morphotypes of *Globigerinella siphonifera* (d'Orbigny). Mar Micropaleontol 35:141–160

Bradbury MG, Abbott DP, Bovbjerg RV, Mariscal RN, Fielding WC, Barber RT, Pearse VB, Proctor SJ, Ogden JC, Wourms JP (1971) Studies on the fauna associated with the deep scattering layers in the equatorial Indian Ocean, conducted on R/V Te Vega during October and November 1964. In: Farquhar GB (ed) Proceedings of an international symposium on biological sound scattering in the Ocean. pp 409–452

Brand TE, Lipps JH (1982) Foraminifera in the trophic structure of shallow-water Antarctic marine communities. J Foraminifer Res 12:96–104. doi:10.2113/gsjfr.12.2.96

Brummer GJA, Hemleben C, Spindler M (1986) Planktonic foraminiferal ontogeny and new perspectives for micropalaeontology. Nature 319:50–52. doi:10.1038/319050a0

Caron DA, Bé AWH (1984) Predicted and observed feeding rates of the spinose planktonic foraminifer *Globigerinoides sacculifer*. Bull Mar Sci 35:1–10

Caron DA, Bé AWH, Anderson OR (1982) Effects of variations in light intensity on life processes of the planktonic foraminifer *Globigerinoides sacculifer* in laboratory culture. J Mar Biol Assoc UK 62:435–451

Culver SJ, Lipps JH (2003) Predation on and by Foraminifera. In: Kelley PH, Kowalewski M, Hansen TA (eds) Predator—prey interactions in the fossil record. Kluwer Academic/Plenum Publishers, New York, pp 7–32

Darling KF, Wade CM (2008) The genetic diversity of planktic Foraminifera and the global distribution of ribosomal RNA genotypes. Mar Micropaleontol 67:216–238

Elbrächter M, Hemleben C, Spindler M (1987) On the taxonomy of the lunate *Pyrocystis* species (Dinophyta). Bot Mar 30:233–242

Erez J (1983) Calcification rates, photosynthesis and light in planktonic Foraminifera. In: Westbroek P, de Jong EW (eds) Biomineralization and biological metal accumulation. Reidel Publishing Company, Dordrecht, pp 307–312

Faber WW, Anderson OR, Caron DA (1989) Algal-foraminiferal symbiosis in the planktonic foraminifer *Globigerinella aequilateralis*: II. Effects of two symbiont species on foraminiferal growth and longevity. J Foraminifer Res 19:185–193

Faber WW, Anderson OR, Lindsey JL, Caron DA (1988) Algal-foraminiferal symbiosis in the planktonic foraminifer *Globigerinella aequilateralis*: I. Occurrence and stability of two mutually exclusive chrysophyte endosymbionts and their ultrastructure. J Foraminifer Res 18:334–343

Fairbanks RG, Wiebe PH (1980) Foraminifera and chlorophyll maximum: vertical distribution, seasonal succession, and paleoceanographic significance. Science 209:1524–1526

Gastrich MD, Bartha R (1988) Primary productivity in the planktonic foraminifer *Globigerinoides ruber* (d'Orbigny). J Foraminifer Res 18:137–142

Gast RJ, Caron DA (1996) Molecular phylogeny of symbiotic dinoflagellates from planktonic Foraminifera and Radiolaria. Mol Biol Evol 13:1192–1197

Gast RJ, McDonnell TA, Caron DA (2000) srDNA-based taxonomic affinities of algal symbionts from a planktonic foraminifer and a solitary radiolarian. J Phycol 36:172–177. doi:10.1046/j.1529-8817.2000.99133.x

Hemleben C, Auras A (1984) Variations in the calcite dissolution pattern on the Barbados Ridge Complex at Site-541 and Site-543, Deep-Sea Drilling Project Leg 78A. Initial Rep Deep Sea Drill Proj 78:471–497

Hemleben C, Bé AWH, Anderson OR, Tuntivate S (1977) Test morphology, organic layers and chamber formation of the planktonic foraminifer *Globorotalia menardii* (d'Orbigny). J Foraminifer Res 7:1–25

Hemleben C, Spindler M (1983) Recent advances in research on living planktonic Foraminifera. Utrecht Micropaleontol Bull 30:141–170

Hemleben C, Spindler M, Anderson OR (1989) Modern planktonic Foraminifera. Springer, Berlin

Hemleben C, Spindler M, Breitinger I, Deuser WG (1985) Field and laboratory studies on the ontogeny and ecology of some globorotaliid species from the Sargasso Sea off Bermuda. J Foraminifer Res 15:254–272

Hemleben C, Spindler M, Breitinger I, Ott R (1987) Morphological and physiological responses of *Globigerinoides sacculifer* (Brady) under varying laboratory conditions. Mar Micropaleontol 12:305–324

Hönisch B, Bijma J, Russell AD, Spero HJ, Palmer MR, Zeebe RE, Eisenhauer A (2003) The influence of symbiont photosynthesis on the boron isotopic composition of Foraminifera shells. Mar Micropaleontol 49:87–96

Houston RM, Huber BT (1998) Evidence of photosymbiosis in fossil taxa? Ontogenetic stable isotope trends in some Late Cretaceous planktonic Foraminifera. Mar Micropaleontol 34:29–46. doi:10.1016/S0377-8398 (97)00038-8

Houston RM, Huber BT, Spero HJ (1999) Size-related isotopic trends in some Maastrichtian planktic Foraminifera: methodological comparisons, intraspecific variability, and evidence for photosymbiosis. Mar Micropaleontol 36:169–188. doi:10.1016/S0377-8398 (99)00007-9

Huber BT, Bijma J, Darling K (1997) Cryptic speciation in the living planktonic foraminifer *Globigerinella siphonifera* (d'Orbigny). Paleobiology 23:33–62

Hulburt EM, Ryther JH, Guillard RRL (1960) The phytoplankton of the Sargasso Sea off Bermuda. J Cons 25:115–128

Itou M, Ono T, Oba T, Noriki S (2001) Isotopic composition and morphology of living *Globorotalia scitula*: a new proxy of sub-intermediate ocean carbonate chemistry? Mar Micropaleontol 42:189–210

Jørgensen BB, Erez J, Revsbech NP, Cohen Y (1985) Symbiotic photosynthesis in a planktonic foraminiferan, *Globigerinoides sacculifer* (Brady), studied with microelectrodes. Limnol Oceanogr 30: 1253–1267

King K (1977) Amino acid survey of recent calcareous and siliceous deep-sea microfossils. Micropaleontology 23:180–193

Knight R, Mantoura RFC (1985) Chlorophyll and carotenoid pigments in Foraminifera and their symbiotic algae: analysis by high performance liquid chromatography. Mar Ecol Prog Ser 23:241–249

Koeve W (2002) Upper ocean carbon fluxes in the Atlantic Ocean: the importance of the POC:PIC ratio. Glob Biogeochem Cycles. doi:10.1029/2001GB001836

Köhler-Rink S, Kühl M (2000) Microsensor studies of photosynthesis and respiration in larger symbiotic Foraminifera. I The physico-chemical microenvironment of *Marginopora vertebralis*, *Amphistegina lobifera*, and *Amphisorus hemprichii*. Mar Biol 137:473–486

Köhler-Rink S, Kühl M (2001) Microsensor studies of photosynthesis and respiration in the larger symbiont bearing Foraminifera *Amphistegina lobifera* and *Amphisorus hemprichii*. Ophelia 55:111–122

Köhler-Rink S, Kühl M (2005) The chemical microenvironment of the symbiotic planktonic foraminifer *Orbulina universa*. Mar Biol Res 1:68–78. doi:10.1080/17451000510019015

Lee JJ (1980) Nutrition and physiology of the Foraminifera. In: Levandowski M, Hutner S (eds) Biochemistry and physiology of protozoa. Academic Press, New York, pp 43–66

Lee JJ (2006) Algal symbiosis in larger Foraminifera. Symbiosis 42:63–75

Lombard F, Erez J, Michel E, Labeyrie L (2009) Temperature effect on respiration and photosynthesis of the symbiont-bearing planktonic Foraminifera *Globigerinoides ruber*, *Orbulina universa*, and *Globigerinella siphonifera*. Limnol Oceanogr 54:210–218

Meilland J (2015) Rôle des foraminifères planctoniques dans le cycle du carbone marin des hautes latitudes (Océan Indien Austral). PhD Thesis, Université d'Angers

Meilland J, Howa H, Lo Monaco C, Schiebel R (2016) Individual planktic foraminifer protein-biomass affected by trophic conditions in the Southwest Indian Ocean, 30°S–60°S. Mar Micropal 124:63–74. http://dx.doi.org/10.1016/j.marmicro.2016.02.004

Michaels AF, Caron DA, Swanberg NR, Howse FA, Michaels CM (1995) Planktonic sarcodines (Acantharia, Radiolaria, Foraminifera) in surface waters near Bermuda: abundance, biomass and vertical flux. J Plankton Res 17:131–163

Movellan A (2013) La biomasse des foraminifères planctoniques actuels et son impact sur la pompe biologique de carbone. PhD Thesis, University of Angers

Murray J (1897) On the distribution of the pelagic Foraminifera at the surface and on the floor of the ocean. Nat Scence 11:17–27

Peeters F, Ivanova E, Conan S, Brummer GJA, Ganssen G, Troelstra S, van Hinte J (1999) A size analysis of planktic Foraminifera from the Arabian Sea. Mar Micropaleontol 36:31–63

Rhumbler L (1911) Die Foraminiferen (Thalamorphoren) der Plankton-Expedition. Erster Teil: Die allgemeinen Organisations-Verhältnisse der Foraminiferen. Ergeb Plankton-Exped Humbold-Stift 1909 3:331

Rink S, Kühl M, Bijma J, Spero HJ (1998) Microsensor studies of photosynthesis and respiration in the symbiotic foraminifer Orbulina universa. Mar Biol 131:583–595

Schiebel R, Hemleben C (2000) Interannual variability of planktic foraminiferal populations and test flux in the eastern North Atlantic Ocean (JGOFS). Deep-Sea Res II 47:1809–1852

Schiebel R, Movellan A (2012) First-order estimate of the planktic foraminifer biomass in the modern ocean. Earth Syst Sci Data 4:75–89. doi:10.5194/essd-4-75-2012

Shaked Y, De Vargas C (2006) Pelagic photosymbiosis: rDNA assessment of diversity and evolution of dinoflagellate symbionts and planktonic foraminiferal hosts. Mar Ecol Prog Ser 325:59–71

Sigman DM, Haug GH (2003) The biological pump in the past. In: Treatise on geochemistry. Elsevier, pp 491–528

Sliter WV (1971) Predation on benthic foraminifers. J Foraminifer Res 1:20–28

Spero HJ (1987) Symbiosis in the planktonic foraminifer, Orbulina universa, and the isolation of its symbitic dinoflagellate, Gymnodinium béii sp. nov. J Phycol 23:307–317

Spero HJ (1986) Symbiosis, chamber formation and stable isotope incorporation in the planktonic foraminifer Orbulina universa. PhD Thesis, University of California

Spero HJ, DeNiro MJ (1987) The influence of symbiont photosynthesis on the $\delta^{18}O$ and $\delta^{13}C$ values of planktonic foraminiferal shell calcite. Symbiosis 4:213–228

Spero HJ, Lea DW (1993) Intraspecific stable isotope variability in the planktic Foraminifera Globigerinoides sacculifer: results from laboratory experiments. Mar Micropaleontol 22:221–234

Spero HJ, Parker SL (1985) Photosynthesis in the symbiotic planktonic foraminifer Orbulina universa, and its potential contribution to oceanic primary productivity. J Foraminifer Res 15:273–281

Spindler M, Hemleben C (1980) Symbionts in planktonic Foraminifera (Protozoa). In: Schwemmler W, Schenk HEA (eds) Endocytobiology, endosymbiosis and cell biology. Walter de Gruyter & Co, Berlin, New York, pp 133–140

Spindler M, Hemleben C, Salomons JB, Smit LP (1984) Feeding behavior of some planktonic foraminifers in laboratory cultures. J Foraminifer Res 14:237–249

Uhle ME, Macko SA, Spero HJ, Engel MH, Lea DW (1997) Sources of carbon and nitrogen in modern planktonic Foraminifera: the role of algal symbionts as determined by bulk and compound specific stable isotopic analyses. Org Geochem 27:103–113

Uhle ME, Macko SA, Spero HJ, Lea DW, Ruddiman WF, Engel MH (1999) The fate of nitrogen in the Orbulina universa Foraminifera: symbiont system determined by nitrogen isotope analyses of shell-bound organic matter. Limnol Oceanogr 44:1968–1977

Wiebe PH, Boyd SH, Winget CL (1976) Particulate matter sinking to the deep-sea floor at 2000 m in the Tongue of the Ocean, Bahamas, with a description of a new sedimentation trap. J Mar Res 34:341–354

Reproduction

Continuity of the species in the vastness of the deep ocean is ensured by adaptive mechanisms characteristic of pelagic organisms to promote sufficient reproductive success. Whereas wide dispersal poses no problem for monoecious (offspring produced from a single parent) organisms, gametes of sexually reproducing dioecious organisms with different parents need to fuse for successful reproduction. Therefore, dioecious organisms with a wide dispersal as assumed for planktic foraminifers need a strategy to ensure successful reproduction (cf. Hemleben et al. 1989).

The standing stock of planktic foraminifers is rather heterogeneous at an average of 10–100 individuals per m^3, i.e., one specimen per 10–100 L of seawater, or a distance of about 25–60 cm between individuals. Given an average size of a planktic foraminifer test of 250 μm, the distance between the individuals would be ~1000–4000 times their size. Assuming random (plankton-like) movement of the individuals, the distance would possibly be too long for successful reproduction in a limited time-interval of a couple of days, even at unlimited fertility. In addition, the distribution of planktic foraminifer species is patchy including temporal scales from sub-seasonal to interannual time-intervals, and spatial scales from local (kilometer scale) to meso-scale of some tens to hundred kilometres, as well as different depth habitats spanning the surface to mesobathyal depths in the water column (e.g., Schiebel and Hemleben 2000; Siccha et al. 2012).

Since the odds against gametes of the same species coming into contact in the open ocean are extremely large given the average distance between individuals, planktic foraminifers have developed adaptive strategies that help to maximize the probability of gamete fusion. These include (1) release of large numbers of gametes, (2) production of motile gametes that contain sufficient food reserves for prolonged locomotion, (3) synchronization of gamete release at distinct frequencies, and (4) establishment of a depth preference for reproduction to limit the vertical range and enhance the chance of mating. All of the four strategies have been reported for different planktic foraminifer species both from laboratory observation and field data (Spindler et al. 1978, 1979; Almogi-Labin 1984; Hemleben et al. 1989; Bijma et al. 1990, 1994; Erez et al. 1991; Bijma and Hemleben 1994; Marchant 1995; Schiebel et al. 1997).

Direct observations of the reproduction of planktic foraminifers in the laboratory, and data from natural assemblages provide statistical evidence on their reproductive behavior. Processes in reproduction also provide information on the biology of planktic foraminifers necessary to understand calcification and chemistry of their tests including stable isotope signals and chemical element ratios, and hence are relevant for the use of planktic foraminifers as proxy in paleoceanographic research.

5.1 Gametogenesis

Release of gametes in planktic foraminifers was reported as early as 1911 by Rhumbler. Le Calvez (1936) described gamete release in *Globigerinella siphonifera* and *Orbulina universa*. Details of gametogenesis and reproduction were described later from laboratory experiments and by applying electron microscopy (SEM and TEM) (e.g., Bé and Anderson 1976). Planktic foraminifers reproduce by release of flagellated cells, i.e. gametes, as observed in the spinose species *Hastigerina pelagica*, *O. universa*, *Globigerinoides conglobatus*, *Globigerinoides ruber*, *Globigerinoides sacculifer*, *Globigerina bulloides*, *Turborotalita humilis*, and *G. siphonifera*, and non-spinose *Globigerinita glutinata*, *Neogloboquadrina pachyderma*, *Neogloboquadrina dutertrei*, *Globorotalia inflata*, *Globorotalia truncatulinoides*, *Globorotalia hirsuta*, and *Globorotalia menardii* (Hemleben et al. 1989, and references therein). The vast numbers of the flagellated cells released by a single parent cell (typically 300,000–400,000) and their small size (ca. 3–5 μm) suggests that these flagellated swarmers are indeed gametes. Definitive evidence of syngamy (fusion of the swarmers) or definitive evidence for the haploid nature of the gametes still needs to be confirmed.

5.1.1 Succession of Events in Gametogenesis

As a first sign of impending gametogenesis in laboratory experiments, the normally floating spinose individuals sink to the bottom of the culture dish (Hemleben et al. 1989). Shortly after sinking, the spinose species shorten their spines by resorption from top to base (Fig. 5.1, Table 5.1) (Bé et al. 1983). Spine-fragments are discarded by rhizopodial streaming (Plate 5.1-1). In *G. sacculifer*, the formation of a final sac-like chamber is the earliest visual indication of impending gametogenesis

(Fig. 5.1). Symbiont-bearing species consume or expel their symbionts, which appear as moribund masses of yellow-brown pigmented particles around the test. The cytoplasm becomes granular and milky white, or orange to reddish due to masses of fat in many species, and withdraws to the inside of the test. Some feeble rhizopodia with granular cytoplasm may remain outside of the test and exhibit cytoplasmic streaming. Subsequently, a mass of granular cytoplasm appears in the aperture, and gradually enlarges to form a substantial bulge (Plate 5.1-2). The bulge eventually ruptures, sometimes explosively, and hundreds of thousands of flagellated gametes are released, which swim away from the parent cell with a slight undulating motion (Spindler et al. 1978). Partially expelled gametes may form string-like masses issuing from the aperture of the parental test in early stages of gamete release, then gradually spread distally, and separate into individuals or clumps of flagellated cells (Plate 5.1-6), which disperse into ambient water (Plate 5.1-3 to -5). When gamete-release is completed, only the empty parental test remains. The gross morphological and cytoplasmic events during gametogenesis of *G. sacculifer* are similar in *G. ruber*, *G. conglobatus*, *G. siphonifera*, and *O. universa*. In *G. sacculifer*, *O. universa*, *G. truncatulinoides*, *G. hirsuta*, and *H. pelagica*, remnants of fine rhizopodia may occasionally be attached to the parental test after gamete release, and exhibit rhizopodial streaming for up to 8 h before dissipating.

Due to architectural (spinose vs. non-spinose species) and autecological (symbiont-bearing *vs.* symbiont-barren species) differences, the overall pattern of reproduction varies among species. Abnormal gametogenesis is occasionally observed in individuals maintained in laboratory culture, resulting in abortive release of gametes. In some cases, the bulge forms, but the gametes are not expelled, or some gametes may be released, but the majority of the cytoplasm remains sequestered in the test and is moribund (cf. Hemleben et al. 1989).

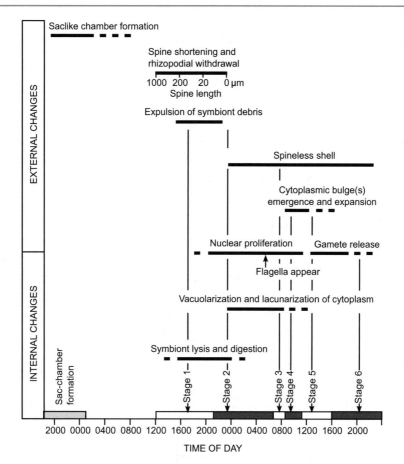

Fig. 5.1 Timetable of external and internal cellular changes associated with gametogenesis in *G. sacculifer*. Duration of the six stages of gametogenesis given by white and dark gray horizontal bars is based on numerous (i.e. hundreds of cases) observations. Arrows indicate the average time of day of each stage. Occasionally, formation of a final sac-like chamber, the earliest visual indication of impending gametogenesis, occurs in some individuals. Gradual shortening of the spines at midday and complete shedding of the spines at midnight on the day preceding gamete release clearly signal the onset of gametogenesis. Fine structural analyses indicate the onset of nuclear division, and development of large vacuoles within the cytoplasm occurs during the period from midnight until noon of the day when gametes are released. Flagella appear on the multinucleid cytoplasmic masses early in the morning, and gamete formation and release occurs in the afternoon and the early evening. Redrawn after Bé et al. (1983)

5.1.2 Fine Structural Processes During Gametogenesis

Early during gametogenesis, as exemplified by *H. pelagica*, the foraminifer descends in the water column. While sinking, prior to shedding of the bubble capsule, the cytoplasm changes from orange to bright red color. The color change commences as a small patch near the center of the cell and gradually disperses to encompass the entire cytoplasm in *H. pelagica* (Spindler et al. 1978). Upon descent of the reproducing individual (in the culture dish, and possibly also in the natural environment) early in gametogenesis, the fibrillar bodies, which are assumed to aid flotation, are reduced in abundance (cf. Chap. 3). In some specimens, fibrillar bodies persist into the late stages of gamete release, and appear as dense tubules (in TEM imagery) within an expanded vacuolar membrane (Hemleben et al. 1989). The vacuolar bodies are occasionally surrounded by a thin layer of cytoplasm.

Plate 5.1 (*1*) Spines are discarded before gamete release (GR) in *G. sacculifer* (Kage Microphotography©, with ▶ permission). When gametogenesis starts (*2*) the cytoplasmic bulge expands, and (*3*) gametes are released. Gametes are released and (*4*) are still in close vicinity to the parental tests (*N. dutertrei*). (*5*) Released gametes around parental test (*H. pelagica*). (*6*) TEM image of stained gamete of *H. pelagica* with flagella of different lengths and whip-like ends (from Spindler et al. 1978). Bars (*1,3,4*) 200 µm, (*2*) 50 µm, (*5*) 500 µm, (*6*) 2 µm

Table 5.1 Generalized schedule of gamete release in reproduction of planktic foraminifers. Gametes are released predominantly during the early afternoon. Compiled from Spindler et al. (1978), Hemleben et al. (1979), and Spindler and Hemleben (1982). After Hemleben et al. (1989)

Event	Time before gamete release	
Formation of ultimate chamber	5–1	Days
Spine shortening and shedding (in spinose species)	24–10	hours
Nuclear division	20–4	hours
Vacuolization of cytoplasm	14–6	hours
Development of flagella	9–7	hours
Cytoplasmic bulge emerges and expands	6–2	hours
Gamete release	0	hours

The lipids disperse within the cytoplasm, and droplets reduce in size upon descent of the reproducing individual (Spindler et al. 1978). The lipids will eventually be passed over to gametes as energy reserves. In symbiont-bearing species, there is increasing evidence of symbiont lysis within the perialgal vacuoles, which appear to be converted to digestive vacuoles. Excess moribund symbionts are expelled by exocytosis into the surrounding environment, and the digestive vacuoles entirely disintegrate until the late stages of gametogenesis (cf. Hemleben et al. 1989).

The nucleus commences repeated divisions producing hundreds of thousands of small daughter nuclei (Spindler et al. 1978). Each of the small nuclei are enclosed within a double membranous envelope sourced from annulate lamellae produced in quantity in the cytoplasm of *H. pelagica*, and also in other spinose species during early stages of gametogenesis prior to nuclear proliferation (Spindler and Hemleben 1982). The endoplasmic reticulum in the vicinity of the Golgi complex is transformed into flat vesicles piled up in successive layers to form the annulate lamellae, which proliferate and disperse throughout the cytoplasm. At a later stage (12–16 h before gamete release) most annulate lamellae are assembled in whorls (Plate 5.2-1). Eventually, the lamellae are arranged next to the cytoplasmic side of the membranous envelope surrounding dividing nuclei (Plate 5.2-1), and contribute to the expanding nuclear membrane during mitosis and production of daughter nuclei (Spindler and Hemleben 1982). Similarities in pore configuration within the membranes of the lamellae and those of the nuclear envelope, and the close association of lamellae with expanding and dividing nuclei of reproducing *H. pelagica* further support the conclusion that the lamellae are the origin of the massive increase in nuclear membrane during production of daughter nuclei. Similar annulate lamellae have been observed in early reproductive stages of *G. sacculifer* during spine shedding, and prior to production of the daughter nuclei (Spindler and Hemleben 1982; Bé et al. 1983).

After the nuclei are fully dispersed throughout the cytoplasm (Plate 5.2-3 and -4), the cytoplasm is separated into interconnected, multinucleated masses possessing lipid droplets, mitochondria, endoplasmic reticulum, and a full array of typical organelles found in the cytoplasm of the parent cell. Flagella begin to project from the plasma membrane surrounding the masses of multinucleated cytoplasm (cf. Hemleben et al. 1989). The interconnected network of flagellated cytoplasm becomes increasingly dispersed into individual flagellated gametes, which, upon release from the parent test, are biflagellated with flagellae of unequal length (Plate 5.1-6), similar to those found in the benthic foraminifer *Myxotheca* (Angell 1971). Each planktic foraminifer gamete consists of a dense nucleus (in TEM imagery) surrounded by an irregular zone of mitochondria,

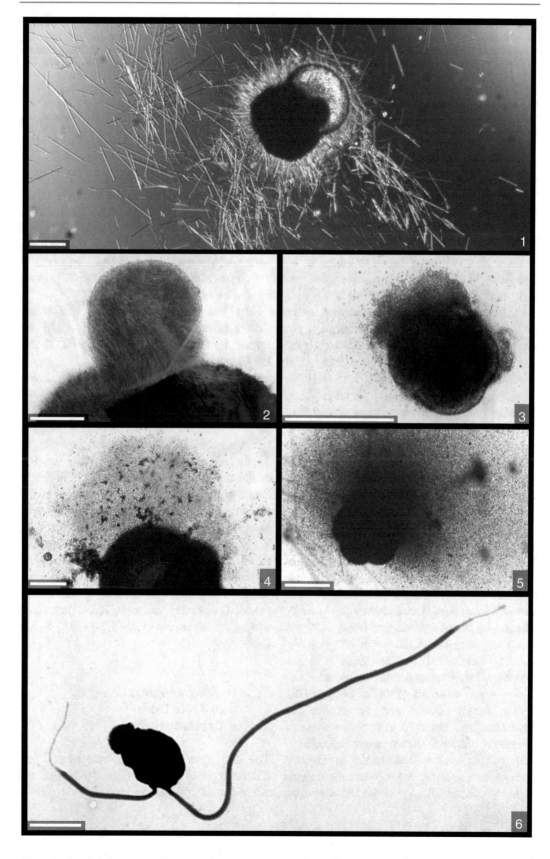

Plate 5.2 (*1*) Annulate lamellae in *H. pelagica* forming concentric aggregates when transported toward the nucleus ▶ 12–16 h before gamete release (Spindler and Hemleben 1982). (*2*) Gamete nucleus (N) of *G. ruber* with separating chromosomes (*white arrow*), and flagella in cross-section (*red arrows*). (*3,4*) Vacuolated cytoplasm with gamete nuclei (N) and flagella in longitudinal (*black arrows*) and cross-section (*red arrows*) of (*3*) *G. ruber* and (*4*) *G. sacculifer*. (*5*) Spherical bodies close to the empty shell of *H. pelagica* after gamete release, with (*6*) large central vacuole (V) including debris, and some nuclei in the surrounding cytoplasm (from Spindler et al.1978). (*7a*) Offspring of *G. truncatulinoides* with protoconch (dark) and deuteroconch (light). (*7b*) Offspring of *G. glutinata* with protoconch (*dark red*), deuteroconch (*light red*), and 3rd chamber (uncolored). (*8*) Offspring of *G. glutinata* with pustules and deuteroconch with pores (7b and 8 from K. Kimoto, with permission). Bars (*1–4,6*) 1 µm, (*5,7*) 100 µm, (*8*) 10 µm

endoplasmic reticulum, and at the periphery typical basal bodies and their flagella (Hemleben et al. 1989).

The gametes of planktic foraminifers contain a single nucleus with finely dispersed chromatin. Lipid droplets form conspicuous inclusions in the cytoplasm. Gametes are distinguished from possible motile stages of the symbionts by their nucleus with a 'foraminifer-type' fine structure (as in the parent cell cytoplasm), which is not mesokaryotic (containing persistently condensed chromatin) as in the dinoflagellate symbionts of some spinose planktic foraminifers. In addition, chloroplasts and other inclusions characteristic of symbionts are absent in the gametes, and no algal symbionts are associated with the gametes. Symbionts are present in the intratest cytoplasm of *G. sacculifer* and *G. ruber*, but not before the three-chambered ontogenetic stage of the test, and it is not known how symbiont-bearing species acquire their symbionts (see Chap. 4).

Species-specific variations in cytoplasmic fine structural processes during gametogenesis include differences in the size and shape of reproductive nuclei. Large spheroidal cytoplasmic residual bodies (spherical bodies) of some tens of micrometers in diameter are produced by *H. pelagica* during reproduction (Plate 5.2-5 and -6). They consist of a thin layer of cytoplasm with several nuclei bearing flagella at the periphery, surrounding a massive vacuole containing waste materials (Spindler et al. 1978). Similar vacuolar bodies often of smaller size have also been observed in the final mass of cytoplasm released during gametogenesis in other spinose species. Those bodies are expelled during gametogenesis, and appear to be residual digestive vacuoles. They are situated within the

protective sheath of cytoplasm, which prevents the gametes from potentially destructive effects of lytic enzymes that are isolated within the digestive vacuoles. The process of gametic nuclei proliferation is not yet entirely understood, and additional observations of the earliest stages of nuclear division would be needed to fully document the proliferation.

The small gametogenic nuclei are almost isodiametric. Additional daughter nuclei are formed by binary fission (Plate 5.2-3). During karyokinesis (nuclear fission), the nuclear envelope remains intact. Microtubular centers outside of the nuclear envelope with attached microtubules pass through it into the nucleoplasm. When the chromosomes separate (Plate 5.2-2), the nuclear envelope expands, and finally when it is in telophase two daughter nuclei are produced. The initial production of daughter nuclei from the non-reproductive nucleus appears to occur very rapidly, either by repeated budding off of smaller nuclei from the larger nucleus, or by simultaneous fragmentation of the large nucleus into many smaller nuclei (Hemleben et al. 1989). Successful reproduction results in large numbers of offspring. The earliest calcified stages of juveniles include a protoconch and a deuteroconch (Plate 5.2-7 and -8, see Chap. 6).

5.1.3 Morphological Changes of Tests During Gametogenesis

The tests remaining after gamete release bear distinct signs of gametogenesis. Resorption and shedding of spines during early stages of

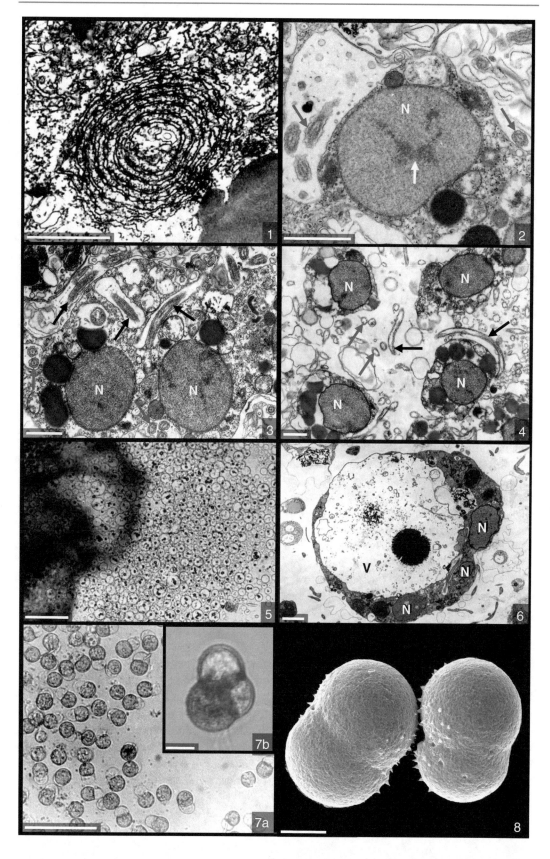

Plate 5.3 (*1–3*) Increasing GAM calcification in *G. sacculifer*. (*1*) Spine holes and remains of spines (*red arrows*) are ▶ visible, (*2*) some spines holes covered, and (*3*) all spine holes are covered by GAM calcite (*blue arrows*). (*4*) Some remains of spines (*red arrows*) are visible in *G. bulloides*. (*5,6*) Empty tests of *H. pelagica* after gametogenesis in the laboratory, with spines and septae resorbed (from Hemleben et al. 1979, 1989). (*7*) Thin-walled broken kummerform chambers in *G. sacculifer*. (*8*) *N. dutertrei* with enlarged final chamber. (*9,10*) 'Biorbulina' types of *O. universa* may indicate reproduction and/or excess food availability, e.g., when overfed in laboratory culture. (*10*) Incomplete second sphere surrounding previous sphere. Bars (*1–4*) 10 µm, (*5–10*) 200 µm

gametogenesis leave characteristic spine remnants, as exemplified in *H. pelagica* (Hemleben et al. 1979). Holes remain where spines were shed, and distinguish all modern and fossil spinose (subsequent to the C/T boundary) from non-spinose tests (Plate 5.3-1 to -4). These spine-holes may be entirely or partially covered by additional deposition of calcite during gametogenic (GAM) calcification (Plate 5.3-2 and -3) (Bé 1980; Hemleben and Spindler 1983). Resorption of internal septae, and dissolution of the test wall as in *H. pelagica* (Plate 5.3-5 and -6) is assumed to aid gamete release (Hemleben et al. 1979).

Variations in final test morphology may indicate gametogenesis in existing and fossil specimens (Hemleben and Spindler 1983). One or more (up to four) chambers may be smaller (kummerform) than the last pre-gametogenic chamber (Berger 1970). In spinose species, the kummerform chambers often lack spines, and may be either incompletely calcified or rather thick-walled, with scarce and scattered pores. In fossil specimens, kummerform chambers are often broken (Plate 5.3-7) because of their insufficiently calcified walls, and might only leave a rim where the wall was attached to the earlier chambers. In *G. sacculifer*, one polymorphous sac-like chamber, including the 'fistulose' type may be produced (Plate 5.3-7; see also Chap. 2, Plate 2.9-6 and -9). In other cases, the final chamber may be significantly larger than the previous chambers (Plate 5.3-8). In mature specimens of *O. universa*, the formation of a second sphere joined to the first one (Plate 5.3-9) or surrounding it (Plate 5.3-10) produces so-called 'Biorbulina' morphotypes (see also Chap. 4.1.5). In most cases, the final chamber may well be of normal shape and size following

the logarithmic growth phase even if reproduction had occurred.

> **Kummerform chamber:** The term kummerform (German for kümmerlich, klein, i.e. measly, small) was defined by Berger (Berger 1969, 1970), who favored the interpretation that stress, i.e. non-optimum growth conditions cause formation of kummerform chambers (see also Hecht and Savin 1972, stable isotope data of kummerform chambers). Olsson (1973) concluded that kummerform phenotypes represent mature individuals, which have achieved full adult size, and suggested that comparisons between kummerform and normal-form individuals are insignificant, and the use of the term kummerform is confusing (for summaries see Kennett 1976; Hemleben et al. 1989). Kummerform chambers are formed last in ontogeny prior to reproduction, and may be smaller or equal in size compared to the previous chamber. To speculate, the size of kummerform chambers depends on the internal calcium pool of the individual.

5.1.4 Gametogenic Calcification

An additional more or less patchy calcite layer covering the whole test after shedding of spines may be formed up to 16 h before gamete release (e.g., Bé 1980). Consequently, the test surface is thickened by additional deposition of calcite, i.e. gametogenic calcification (GAM). Spine holes are closed to varying degrees by additional calcification as observed by SEM visualization.

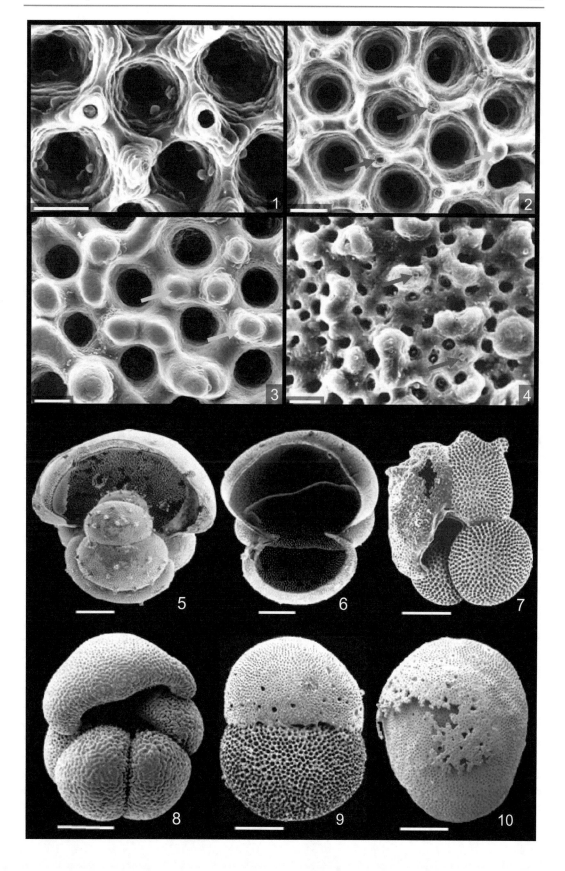

Gametogenic calcite deposited on top of the test surface (Plate 5.3-2 to -4) is an unequivocal indicator of reproduction, as found also in fossil specimens (Bé 1980). However, not all gametogenic specimens produce a substantial gametogenic calcite layer. The amount of thickening of the test wall appears to be related to the amount of excess calcium stored in the cytoplasm at the time of gametogenesis (cf. Chap. 6). It is assumed that test wall thickening in gametogenesis serves as physiological disposal of excess calcium prior to cellular changes of nuclear proliferation and gamete production (Hemleben et al. 1989).

The amount of $CaCO_3$ deposited prior to gametogenesis is specimen-specific, and may be absent or present to a varying degree in both symbiont-barren and symbiont-bearing species (cf. Bé 1980; Hemleben and Spindler 1983). Close to 100 % of adult *G. bulloides* (>150 µm in test diameter) produce a more or less complete layer of gametogenic calcite. GAM calcification in *G. bulloides* starts at structures, which are elevated above the surface of the outer shell (Plate 5.3-4), and may finally merge to form a thin veneer of calcite over most of the outer shell (Schiebel et al. 1997).

The chemical composition of the planktic foraminifer test represents a mixed signal related to the dwelling depth of the foraminifer, and is formed at the average dwelling-depth plus water depth of reproduction (Berger et al. 1978). The gametogenic calcite layer covering the test surface of *G. bulloides*, and other species, such as *Globorotalia truncatulinoides* and *Globorotalia tumida*, may be depleted in Mg relative to the pre-gametogenic test calcite, while being enriched in Mg in *G. sacculifer*, or varying in the Mg/Ca ratio as in *N. pachyderma* (Eggins et al. 2003). These differences are assumed to result from the relative position of reproduction in the water column, which is shallower in *G. truncatulinoides* and *G. tumida*, and deeper in *G. bulloides* (e.g., Hemleben et al. 1985; Schiebel et al. 2002; Schiebel and Hemleben 2005). These differences may be attributed to various factors including the presence (e.g., *G. ruber*) and absence of symbionts (e.g., *G. bulloides*), as well as regional hydrographic conditions.

GAM calcification may add about 4–20 % additional calcite to the shell that was formed during earlier ontogeny of 'thin-walled' morphotypes (and genotypes, de Vargas et al. 1999) of the symbiont-bearing species *O. universa* (Hamilton et al. 2008). In addition to differences in calcite precipitation between different species, and within the same (morpho-) species (e.g., the genotypes of *G. bulloides*, Darling and Wade 2008) GAM calcite may produce tests of differential size-normalised weight (Spero and Lea 1996). The same is true for *G. sacculifer* cultured under varying light intensity (Bé et al. 1982; Spero and Lea 1993). However, in contrast to *G. bulloides*, *G. sacculifer* has only one genotype including various morphotypes (André et al. 2013). Differences in presence, amount, and composition of gametogenic calcite within and between genotypes, plus resulting changes in dissolution susceptibility, add complexity to the interpretation of chemical data of tests from sediments (Eggins et al. 2003).

5.2   Reproduction Inferred from Population Dynamics

Shallow-dwelling species have been shown to reproduce once per month (*G. bulloides*), or twice each month (*G. ruber*), triggered by the synodic lunar cycle (Berger and Soutar 1967; Spindler et al. 1979; Reiss and Hottinger 1984; Schiebel et al. 1997). Lunar periodicity has also been inferred from population dynamics of *Globigerinella calida*, *Globigerinella siphonifera*, *Globigerinita glutinata*, *Globigerinoides sacculifer*, *Globorotalia menardii*, *Neogloboquadrina dutertrei*, *Orbulina universa*, and *Pulleniatina obliquiloculata* (e.g., Jonkers et al. 2015). However, not all specimens may reach the reproductive ontogenetic stage during one reproductive cycle and may reproduce during one of the following cycles (Spindler 1990). Intermediate to deep-dwelling species are assumed to reproduce less often than shallow-dwelling species (Hemleben et al. 1989; Schiebel and Hemleben 2005). The deep-dwelling species *G. truncatulinoides* is believed to reproduce only once per year

(Hemleben et al. 1989). For reproduction, *G. truncatulinoides* ascends from depths to the sea surface during early spring, possibly at the margins of the subtropical gyres (cf. Hemleben et al. 1985: Sargasso Sea; Schiebel et al. 2002: Azores Current). In contrast, shallow-dwelling species are assumed to descend for reproduction in the upper water column to water depths around the Deep Chlorophyll Maximum (DCM) (see also Chap. 7, Ecology, Fig. 7.7). The reproductive descent possibly marks the greatest water depth that shallow-dwelling planktic foraminifers attain during their ontogenetic cycle (Schiebel and Hemleben 2005).

Although lunar reproductive cyclicity is well known especially in the reproduction of metazoans (Richmond and Jokiel 1984), among other organisms, triggering mechanisms are still under debate, i.e., for example whether it is the affects of light and/or gravitation. The concept of lunar cyclicity (i.e. a synodic cycle) in the life cycle of some spinose planktic species is largely coherent (Schiebel and Hemleben 2005). However, as spinose species differ significantly in their biological behavior and habitat, reproductive modes may differ between species (Bijma et al. 1990; Schiebel et al. 1997). The spatial and temporal components of the population dynamics of *G. sacculifer*, *G. ruber*, and *G. siphonifera* from the Red Sea, and *G. bulloides* from the eastern North Atlantic, indicate differential lunar reproduction cyclicity from the population dynamics compared to those that are analyzed from net tow samples of the surface waters. Whereas the cohorts of the *G. sacculifer* and *G. bulloides* assemblages show quite clear synodic lunar cyclicity (Figs. 5.2 and 5.3), the distributions of *G. ruber* and *G. siphonifera* are rather less well defined and within the semi-lunar domain (Bijma et al. 1990; Jonkers et al. 2015). In *G. ruber* (white), biweekly reproduction was already suspected by Berger and Soutar (1967) and Almogi-Labin (1984). In turn, *G. siphonifera* has been assumed to reproduce on a synodic lunar frequency by Schiebel and Hemleben (2005, data from the Arabian Sea).

The ontogenetic development of species provides information on the timing of reproduction as reconstructed from large individuals (>100 μm). The diameter of the proloculus of *G. bulloides* is 20 μm on average (see Chap. 6). Juvenile specimens grow rapidly, and a test size of >100 μm can be reached in less than 10 days after reproduction (Hemleben et al. 1989; Spero and Lea 1996). Growth rates are affected by various factors, such as temperature, and quality and abundance of food. Accordingly, resulting pulses of young adult tests >125 μm were recorded between the second half of the waxing moon and the new moon of the following lunar cycle (Fig. 5.3). Those small specimens possibly resulted from the reproduction of adult (i.e. large) *G. bulloides*, mainly during the first half of the waxing moon. During the second half of the waxing moon, and during the waning moon, these individuals reach maturity, and during the first week of the following waxing moon large numbers of terminal test stages (GAM individuals) generate a new start of the ontogenetic life cycle (Fig. 5.3). All other surface dwelling species have been found (from plankton net data >100 μm) to reproduce once per fortnight or once per month (Schiebel and Hemleben 2005; Jonkers et al. 2015).

A rather blurry distribution of cohorts of *G. siphonifera* and *G. ruber*, and differences in assemblage data from different ocean basins might result from the somehow lower standing stocks than in *G. bulloides*, and hence larger standard deviations. In addition, different 'types' (morphotypes and genotypes?) of *G. ruber* and *G. siphonifera* (Huber et al. 1997; Bijma et al. 1998; Wang 2000; Darling and Wade 2008; Aurahs et al. 2011) may reproduce at different schedules, and display varying distribution patterns resulting from biological and ecological prerequisites. The analyzed samples originate from different geographic locations, from different seasons and years, and different ecologic conditions as indicated by regional differences of temperature and salinity of the surface water. However, taking into account that reproduction in *G. bulloides*, or in any other shallow-dwelling planktic foraminifer species is triggered by the synodic lunar cycle, temporal changes in the specific population structure (e.g., size

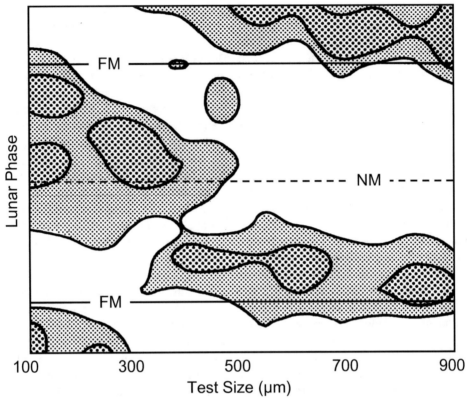

Fig. 5.2 Test-size distribution of *G. sacculifer* over an entire synodic lunar cycle from October 1 through November 17, 1984, from the Gulf of Aqaba. FM full moon, NM new moon. Surface samples were obtained every 4th day, and specimens were measured individually. Contour lines show residuals of test-size classes. Distribution of tests of any size, either under-represented, over-represented, or at maximum abundance, is given in white, fine stippled, and coarse stippled areas, respectively. Specimens grow larger over the synodic lunar cycle starting from around the first FM (*upper line*) to the following FM. The cycle is repeated as indicated by the contour lines before the 1st and after the 2nd FM. From Hemleben et al. (1989), and Bijma et al. (1990)

distribution) should be similar at equivalent longitudes, although local biotic and abiotic parameters may mask the signal. Consequently, each sample represents a transitional state of the standing stock of any species, and appears to be affected by the time of collection including season, geographic location, and day within the synodic lunar cycle.

According to data from vertical plankton tows, reproduction of shallow-dwelling planktic foraminifers at best takes place close to or within the Deep Chlorophyll Maximum (DCM), at the base of the mixed layer (i.e. thermocline, pycnocline). The DCM is thought to be the depth level where trophic conditions best support alimentation of juvenile planktic foraminifers

(Hemleben et al. 1989; Schiebel and Hemleben 2005). Reproduction depth may thus be not only biologically fixed, but also affected by variations in hydrology and ecology, i.e. food sources. After reproduction close to the base of the productive layer in the water column, the empty adult tests sink toward the seafloor. The export layer hence contains mostly post-gametogenic specimens, i.e. empty tests or tests filled with various amounts of cytoplasm remnants.

Synchronized reproduction at a narrow depth range located near the seasonal thermocline and DCM enhances the chance of successful fertilization (gamete fusion), and favors survival of the offspring by providing prey in abundance, perhaps in the form of more or less degraded

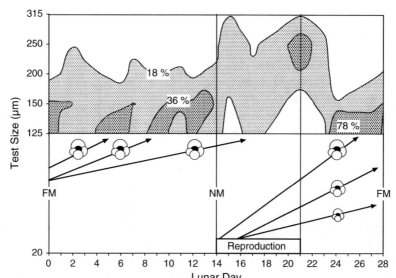

Fig. 5.3 Average test-size distribution of *G. bulloides* over a synodic lunar cycle, with FM full moon (Day 0), NM new moon (Day 14). Data in the upper panel show the relative distribution of tests >125 μm from plankton tows of the surface 60 m of the water column (linear time-scale, and 4-point interpolation). Largest individuals are assumed to reproduce mostly between Day 14 and 21. The offspring grows larger and reaches the >125-μm test-size fraction from Day 23. The lower panel shows assumed schematic growth curves of small (<125 μm) individuals starting from 20 μm (see Chap. 6), i.e. a hypothetic proloculus of *G. bulloides*. Redrawn from Schiebel et al. (1997)

organic tissues (e.g., Bijma et al. 1990; Erez et al. 1991). *Hastigerina pelagica* releases up to 500,000 gametes (Spindler et al. 1979), and other species release thousands to some tens of thousands gametes (Bé and Anderson 1976). Reproduction in *H. pelagica* occurs during the same time of day, i.e., the early hours of the afternoon (Spindler et al. 1979), which further enhances the chance of successful contact between the gametes of different parents.

Size-variations of proloculi of the same planktic foraminifer species are statistically insignificant (Parker 1962; Sverdlove and Bé 1985; Brummer et al. 1987). Therefore, alternation of different generations (i.e. sexual and asexual) is likely limited, and it is assumed reproduction is predominantly sexual. However, asexual reproduction has been reported from individually cultured *Neogloboquadrina incompta* (Kimoto and Tsuchiya 2006, and written communication K. Kimoto 2014).

In contrast to the results discussed above, *G. bulloides* is assumed to reproduce twice per month (Marchant 1995), according to data from

sediment trap samples from 2173 and 3520 m water depth off the coast of Chile. From the same samples, *N. pachyderma*, *N. dutertrei*, and *G. calida* are discussed as probably reproducing once per month (Marchant 1995; cf. also Jonkers et al. 2015 for sediment trap data from the Gulf of Mexico). From an ecologically similar sampling setup off Namibia, Lončarić et al. (2005) conclude that the only species bearing a synodic lunar reproduction strategy is *H. pelagica*, and all other 27 analyzed species bear no such lunar periodicity. Any other frequency of reproduction is reported in the 16–90 days domain, with *G. trilobus* (i.e. the trilobus type of *G. sacculifer*, Plate 2.9.1 to -3) assumed to reproduce on a 42-day cycle (Lončarić et al. 2005). However, reproduction cycles inferred from specimens from sediment trap samples are possibly affected by differential settling velocities of different species and test sizes, as well as transport of tests by currents. Those affects are increasingly difficult to account for with increasing sampling depths (cf. Berelson 2002; Von Gyldenfeldt et al. 2000; Jonkers et al. 2015). Best proof of timing and periodicity of reproduction may be derived from observations using laboratory cultures as done for *H. pelagica*. In turn, planktic foraminifers are sensitive to exogenous changes including sensitivity to ecologic conditions. Therefore, laboratory cultured planktic foraminifers might not accurately display the natural reproductive behavior of any planktic foraminifer species reproducing in the natural environment (Spindler 1990).

5.3 Deviations from the Synodic Lunar Cycle in *H. pelagica* According to Laboratory Experiments

Deviations from the synodic lunar cycle provide information to better understand the reproduction cycle, and the effect of the moon (i.e. gravitation and tides, and light) on planktic foraminifer reproduction in general. Out of a total of 848 cases of gametogenesis of *H. pelagica* observed in the laboratory, 80.7 % (97.8 %) of the specimens released their gametes between Day 3 and 7 (Day 1–9, respectively) after the full moon (Figs. 5.4 and 5.5) similar to the gametogenesis observed in the natural habitat. Only 2.2 % of all specimens reproduced earlier (up to 7 days before the full moon) or later (up to 15 days after the full moon) possibly depending on the availability of food and light (i.e., light-and-dark cycles). The closer in time before the full moon that *H. pelagica* was sampled from surface waters and transferred to the laboratory, the higher was the chance of successful reproduction (Spindler 1990).

To examine the effect of light on reproduction, specimens of *H. pelagica* were sampled one or two days before the full moon, and exposed to varying light-and-dark cycles under laboratory conditions (Spindler 1990). Depending on the number of days of prolonged light or dark periods of several days, all specimens reproduced later than the control group of specimens. Reproduction was retarded according to the number of days of both prolonged continuous (A) light or (B) dark periods (Fig. 5.6).

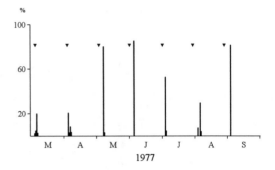

Fig. 5.4 Percentage of field-collected *H. pelagica*, which reproduced in the laboratory on the days indicated over a 7-month time-period from March through September 1977. The full moon is indicated by triangles. From Spindler et al. (1979)

Hastigerina pelagica seemed to have registered (perhaps by a physiological clock) the number of dark and light periods to synchronize reproduction, which would argue against an affect of gravitation/tides on the synodic lunar reproduction cycle. It still remains unclear, how *H. pelagica* senses the light, and in which way light affects the timing of reproduction. Experiments on sub-circadian light-and-dark cycles would add information on the trigger and synchronization of the reproduction of *H. pelagica*.

The synchonized release of gametes during the same time of day (i.e., early afternoon) would possibly indicate *H. pelagica* as being heterogamous (Spindler et al. 1979). However, the cyclic reproduction of *H. pelagica* might be an exception within the planktic foraminifers, since the genus *Hastigerina* is different from other planktic foraminifers, especially such characteristics as the mono-lamellar shell, triradial spines, and the cytoplasmic bubble capsule (Spindler et al. 1979). Cyclic reproduction triggered by the synodic lunar cycle in all other planktic foraminifer taxa (Schiebel and Hemleben 2005) has so far been inferred from statistical models on population dynamics, and hence likely affected by any one or more of the following factors: Physical (e.g., expatriation by currents), ecological (e.g., availability of food), and biological (e.g., mortality), which may occur on much shorter or longer time-scales (cf. Lončarić et al. 2005).

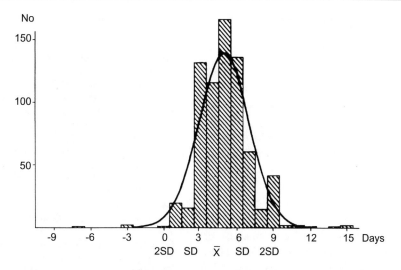

Fig. 5.5 Numbers (No) of specimens of *H. pelagica* reproducing relative to the day of the full moon (0 days). Observations started nine days before the full moon (−9), and were terminated 15 days after the full moon (15). Mean value ($\overline{X}$), standard deviation (SD). From Spindler et al. (1979)

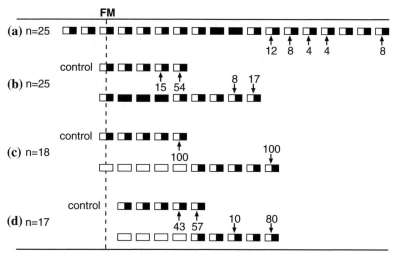

Fig. 5.6 Laboratory experiments of the effect of light-and-dark cycles on the timing of gametogenesis in *H. pelagica*. Experiments include circadian changes of 12-h light and 12-h dark (*white-black symbols*), 24 h darkness (*black symbols*), and 24 h light (*white symbols*). Day of reproduction and percentage of specimens which reproduced indicated by arrows with numbers. Full moon is given by FM and a dashed line. **a** 25 specimens, collected 2 days before the full moon were kept under a 12-h light/12-h dark cycle for 6 days after the full moon, then for 2 days in darkness, and from the morning of day 8 exposed to a 12-h/12-h light/dark cycle again. Consequently, reproduction was retarded and occurred between 9 and 15 days after the full moon. Experiment (**a**) was carried out without a control group. **b** After having been collected on the day of the full moon, part of the specimens were kept in darkness for 3 days, and reproduction occurred 3 and 4 days after the time when the specimens of the control group reproduced, which were continuously kept under 12-h/12-h light/dark cycles. **c** and **d** Same experiments as in (**b**) but part of the specimens were exposed to continuous light for several days. After Spindler (1990)

5.4   Summary and Concluding Remarks

Morphological changes, including spine shedding, and gametogenic calcification characterize specimens that undergo gametogenesis. Asexual reproduction (i.e. a haploid generation) in planktic foraminifers is assumed much less likely to be the case than sexual reproduction (cf. Hemleben et al. 1989, and references therein). Morphological alternation of microspheric (diploid generation) and macrospheric (haploid generation) generations as in benthic foraminifers has not yet been reported among planktic foraminifers. Size-variations of proloculi of the same planktic foraminifer species are statistically insignificant, supporting the assumption of predominantly sexual reproduction in planktic foraminifers. In turn, asexual reproduction has been reported from individually cultured *Neogloboquadrina incompta*.

Strong evidence shows that the reproductive cycle in spinose species living in the photic zone is linked to a synodic lunar or semi-lunar cycle, particularly well established for *H. pelagica, G. sacculifer, G. siphonifera, G. ruber*, and *G. bulloides*. Subsurface dwelling species such as *G. truncatulinoides* and *G. hirsuta* appear to have an annual or semi-annual reproduction frequency. Despite detailed knowledge of timing and cytoplasmic events during reproduction of planktic foraminifers, the complete life cycle is still not sufficiently known. More complete information from laboratory cultures and natural populations is needed to understand the events of reproduction early ontogeny of planktic foraminifers.

References

Almogi-Labin A (1984) Population dynamics of planktic Foraminifera and Pteropoda—Gulf of Aqaba, Red Sea. Proc K Ned Akad Van Wet Ser B Palaeontol Geol Phys Chem 87:481–511

André A, Weiner A, Quillévéré F, Aurahs R, Morard R, Douady CJ, de Garidel-Thoron T, Escarguel G, de Vargas C, Kucera M (2013) The cryptic and the apparent reversed: lack of genetic differentiation within the morphologically diverse plexus of the planktonic

foraminifer *Globigerinoides sacculifer*. Paleobiology 39:21–39. doi:10.1666/0094-8373-39.1.21

Angell RW (1971) Observations on gametogenesis in the Foraminifera *Myxotheca*. J Foraminifer Res 1:39–42. doi:10.2113/gsjfr.1.1.39

Aurahs R, Treis Y, Darling K, Kucera M (2011) A revised taxonomic and phylogenetic concept for the planktonic foraminifer species *Globigerinoides ruber* based on molecular and morphometric evidence. Mar Micropaleontol 79:1–14. doi:10.1016/j.marmicro.2010.12.001

Bé AWH (1980) Gametogenic calcification in a spinose planktonic foraminifer, *Globigerinoides sacculifer* (Brady). Mar Micropaleontol 5:283–310. doi:10.1016/0377-8398(80)90014-6

Bé AWH, Anderson OR (1976) Gametogenesis in planktonic Foraminifera. Science 192:890–892

Bé AWH, Anderson OR, Faber WW, Caron DA (1983) Sequence of morphological and cytoplasmic changes during gametogenesis in the planktonic foraminifer *Globigerinoides sacculifer* (Brady). Micropaleontology 29:310. doi:10.2307/1485737

Bé AWH, Spero HJ, Anderson OR (1982) Effects of symbiont elimination and reinfection on the life processes of the planktonic foraminifer *Globigerinoides sacculifer*. Mar Biol 70:73–86. doi:10.1007/BF00397298

Berelson WM (2002) Particle settling rates increase with depth in the ocean. Deep-Sea Res II 49:237–251. doi:10.1016/S0967-0645(01)00102-3

Berger WH (1969) Kummerform Foraminifera as clues to oceanic environments. Bull Am Geol Soc 53:706

Berger WH (1970) Planktonic Foraminifera: differential production and expatriation off Baja California. Limnol Oceanogr 15:183–204. doi:10.4319/lo.1970.15.2.0183

Berger WH, Killingley JS, Vincent E (1978) Stable isotopes in deep-sea carbonates: box core ERDC-92, west Equatorial Pacific. Oceanol Acta 1:203–216

Berger WH, Soutar A (1967) Planktonic Foraminifera: field experiment on production rate. Science 156:1495–1497

Bijma J, Erez J, Hemleben C (1990) Lunar and semi-lunar reproductive cycles in some spinose planktonic foraminifers. J Foraminifer Res 20:117–127

Bijma J, Hemleben C (1994) Population dynamics of the planktic foraminifer *Globigerinoides sacculifer* (Brady) from the central Red Sea. Deep-Sea Res I 41:485–510. doi:10.1016/0967-0637(94)90092-2

Bijma J, Hemleben C, Huber BT, Erlenkeuser H, Kroon D (1998) Experimental determination of the ontogenetic stable isotope variability in two morphotypes of *Globigerinella siphonifera* (d'Orbigny). Mar Micropaleontol 35:141–160

Bijma J, Hemleben C, Wellnitz K (1994) Lunar-influenced carbonate flux of the planktic foraminifer *Globigerinoides sacculifer* (Brady) from the central Red Sea. Deep-Sea Res I 41:511–530. doi:10.1016/0967-0637(94)90093-0

Brummer GJA, Hemleben C, Spindler M (1987) Ontogeny of extant spinose planktonic Foraminifera

(Globigerinidae): a concept exemplified by *Globigerinoides sacculifer* (Brady) and *G. ruber* (d'Orbigny). Mar Micropaleontol 12:357–381. doi:10.1016/0377-8398(87)90028-4

Darling KF, Wade CM (2008) The genetic diversity of planktic Foraminifera and the global distribution of ribosomal RNA genotypes. Mar Micropaleontol 67:216–238

De Vargas C, Norris R, Zaninetti L, Gibb SW, Pawlowski J (1999) Molecular evidence of cryptic speciation in planktonic foraminifers and their relation to oceanic provinces. Proc Natl Acad Sci 96:2864–2868

Eggins S, De Dekker P, Marshall J (2003) Mg/Ca variation in planktonic Foraminifera tests: implications for reconstructing palaeo-seawater temperature and habitat migration. Earth Planet Sci Lett 212:291–306

Erez J, Almogi-Labin A, Avraham S (1991) On the life history of planktonic Foraminifera: lunar reproduction cycle in *Globigerinoides sacculifer* (Brady). Paleoceanography 6:295–306

Hamilton CP, Spero HJ, Bijma J, Lea DW (2008) Geochemical investigation of gametogenic calcite addition in the planktonic Foraminifera *Orbulina universa*. Mar Micropaleontol 68:256–267. doi:10.1016/j.marmicro.2008.04.003

Hecht AD, Savin SM (1972) Phenotypic variation and oxygen isotope ratios in recent planktonic Foraminifera. J Foraminifer Res 2:55–67

Hemleben C, Bé AWH, Spindler M, Anderson OR (1979) "Dissolution" effects induced by shell resorption during gametogenesis in *Hastigerina pelagica* (d'Orbigny). J Foraminifer Res 9:118–124

Hemleben C, Spindler M (1983) Recent advances in research on living planktonic Foraminifera. Utrecht Micropaleontol Bull 30:141–170

Hemleben C, Spindler M, Anderson OR (1989) Modern planktonic Foraminifera. Springer, Berlin

Hemleben C, Spindler M, Breitinger I, Deuser WG (1985) Field and laboratory studies on the ontogeny and ecology of some globorotaliid species from the Sargasso Sea off Bermuda. J Foraminifer Res 15:254–272

Huber BT, Bijma J, Darling K (1997) Cryptic speciation in the living planktonic foraminifer *Globigerinella siphonifera* (d'Orbigny). Paleobiology 23:33–62

Jonkers L, Reynolds CE, Richey J, Hall IR (2015) Lunar periodicity in the shell flux of planktonic Foraminifera in the Gulf of Mexico. Biogeosciences 12:3061–3070. doi:10.5194/bg-12-3061-2015

Kennett JP (1976) Phenotypic variation in some Recent and late Cenozoic planktonic Foraminifera. In: Hedley RH, Adams CG (eds) Foraminifera 2. London, pp 111–170

Kimoto K, Tsuchiya M (2006) The "unusual" reproduction of planktic Foraminifera: an asexual reproductive phase of *Neogloboquadrina pachyderma* (Ehrenberg). Anuário Inst Geociências 29:461

Le Calvez J (1936) Modifications du test des foraminifères pélagiques en rapport avec la reproduction:

Orbulina universa (d'Orbigny) et *Tretomphalus bulloides* (d'Orbigny). Ann Protistol 5:125–133

Lončarić N, Brummer GJA, Kroon D (2005) Lunar cycles and seasonal variations in deposition fluxes of planktic foraminiferal shell carbonate to the deep South Atlantic (central Walvis Ridge). Deep-Sea Res I 52:1178–1188

Marchant M (1995) Die Sedimentation planktischer Foraminiferen im Auftriebsgebiet vor Chile heute und während der letzten ca. 15.000 Jahre. Berichte Fachbereich Geowiss Univ Brem 69:311–323

Olsson RK (1973) What is a kummerform planktonic foraminifer? J Paleontol 327–329

Parker FL (1962) Planktonic foraminiferal species in Pacific sediments. Micropaleontology 8:219–254

Reiss Z, Hottinger L (1984) The Gulf of Aqaba. Springer, Heidelberg

Rhumbler L (1911) Die Foraminiferen (Thalamorphoren) der Plankton-Expedition. Erster Teil: Die allgemeinen Organisations-Verhältnisse der Foraminiferen. Ergeb Plankton-Exped Humbold-Stift 1909(3):331

Richmond RH, Jokiel PL (1984) Lunar periodicity in larva release in the reef coral *Pocillopora damicornis* at Enewetak and Hawaii. Bull Mar Sci 34:280–287

Schiebel R, Bijma J, Hemleben C (1997) Population dynamics of the planktic foraminifer *Globigerina bulloides* from the Eastern North Atlantic. Deep-Sea Res I 44:1701–1713

Schiebel R, Hemleben C (2000) Interannual variability of planktic foraminiferal populations and test flux in the eastern North Atlantic Ocean (JGOFS). Deep-Sea Res II 47:1809–1852

Schiebel R, Hemleben C (2005) Modern planktic Foraminifera. Paläontol Z 79:135–148

Schiebel R, Waniek J, Zeltner A, Alves M (2002) Impact of the Azores Front on the distribution of planktic foraminifers, shelled gastropods, and coccolithophorids. Deep-Sea Res II 49:4035–4050

Siccha M, Schiebel R, Schmidt S, Howa H (2012) Short-term and small-scale variability in planktic Foraminifera test flux in the Bay of Biscay. Deep-Sea Res I 64:146–156. doi:10.1016/j.dsr.2012.02.004

Spero HJ, Lea DW (1993) Intraspecific stable isotope variability in the planktic Foraminifera *Globigerinoides sacculifer*: results from laboratory experiments. Mar Micropaleontol 22:221–234

Spero HJ, Lea DW (1996) Experimental determination of stable isotope variability in *Globigerina bulloides*: implications for paleoceanographic reconstructions. Mar Micropaleontol 28:231–246

Spindler M (1990) Reproduktionsablauf bei der planktischen Foraminifere *Hastigerina pelagica* unter besonderer Berücksichtigung morphologischer und cytologischer Vorgänge. Habilitation Thesis, University of Oldenburg

Spindler M, Anderson OR, Hemleben C, Bé AWH (1978) Light and electron microscopic observations of gametogenesis in *Hastigerina pelagica* (Foraminifera). J Protozool 25:427–433

Spindler M, Hemleben C (1982) Formation and possible function of annulate lamellae in a planktic foraminifer. J Ultrastruct Res 81:341–350

Spindler M, Hemleben C, Bayer U, Bé AWH, Anderson OR (1979) Lunar periodicity of reproduction in the planktonic foraminifer *Hastigerina pelagica*. Mar Ecol Prog Ser 1:61–64

Sverdlove MS, Bé AWH (1985) Taxonomic and ecological significance of embryonic and juvenile planktonic Foraminifera. J Foraminifer Res 15:235–241

Von Gyldenfeldt AB, Carstens J, Meincke J (2000) Estimation of the catchment area of a sediment trap by means of current meters and foraminiferal tests. Deep-Sea Res II 47:1701–1717

Wang L (2000) Isotopic signals in two morphotypes of *Globigerinoides ruber* (white) from the South China Sea: implications for monsoon climate change during the last glacial cycle. Palaeogeogr Palaeoclimatol Palaeoecol 161:381–394

Ontogeny and Test Architecture

6

Tests of varying complexity represent the most obvious and persistent visual expression of the overall planktic foraminifer morphology. Consequently, understanding the biological processes and environmental controls involved in the formation and deposition of the planktic foraminifer shell is of major interest to any related discipline, including marine biogeochemistry, sedimentology, biostratigraphy, and paleoceanography. Adult morphotypes are relatively easy to classify and have formed the basis of the taxonomic system of modern and fossil planktic foraminifers. In addition, pre-adult growth stages add important information on the biology, ecology, phylogeny, and test chemistry of planktic foraminifers. The ontogenetic development of planktic foraminifer species has hence been of interest starting from the earliest modern studies in the 20th century. Near logarithmic growth of tests during ontogeny (the rather large proloculus excluded) was recognized in *Globorotalia menardii* as early as in 1911 by Rhumbler, and subsequently confirmed for other planktic foraminifer species (e.g., Olsson 1971; Hemleben et al. 1985; Brummer 1988; Brummer et al. 1987).

The first detailed information on internal test morphology and ontogenetic development has been provided from sectioning of tests (e.g., Brönnimann 1950), X-Ray imagery of tests by Bé et al. (1969), and the comparative papers of Huang (1981), and Sverdlove and Bé (1985). However, reconstruction of ontogeny from the inner whorls of mature shells provides incomplete information on the development of tests, since additional layers of calcite are deposited at each successive addition of a new chamber, and alter the thickness and morphology of the early chambers. Calcite layers may even be deposited at a daily rate, as shown in adult *Orbulina universa* (Spero et al. 2015). In addition, resorption of previously deposited calcite, and loss of septa and spines may lead to misinterpretations of early ontogenetic stages as, for example, in *O. universa* (Hemleben et al. 1989, and references therein). Since planktic foraminifers have not yet been grown in continuous cultures in the laboratory, knowledge of the succession of growth stages from syngamy (zygote formation) to reproductive maturity is incomplete. Early ontogenetic stages, i.e. juvenile and early neanic individuals of field-collected individuals less than about 80 μm in size are particularly difficult to grow to maturity in laboratory culture.

By backtracking individual growth stages from the adult tests to the earliest detectable two-chamber stage, i.e. the proloculus and deuteroconch, Brummer et al. (1986, 1987) have reconstructed the ontogeny of various species (Plate 6.1-1 and -2). Adult tests were collected with a 10 μm net, and examined by incident light and scanning electron microscopy to establish the morphological characteristics, linking successive ontogenetic stages from old to young. The entire sequence is arranged in reverse order to represent the ontogenetic sequence from juvenile to adult. Based on sequential analysis, ontogenies of

R. Schiebel and C. Hemleben, *Planktic Foraminifers in the Modern Ocean*,
DOI 10.1007/978-3-662-50297-6_6

several globigerinid species have been developed
and conceptualized for a sequence of five successive growth stages (Brummer et al. 1987).

6.1   Morphological Development of Spinose Species

Analysis of various spinose species has resulted
in a conceptual description of ontogeny categorized in five stages (Brummer et al. 1987). The
growth stages represent a dynamic and continuous process. Transitions between stages possibly
result from ontogenetic changes in physiological
boundary conditions. The successive addition of
chambers to the planktic foraminifer test provides a means to delimit boundaries between
stages, based on individual or multiple steps in
the process of chamber addition. Ontogenetic
stages are indicated by changes in the arrangement of the chambers, onset of spine and pustule
formation, occurrence and pattern of pore distribution, structure of septae between chambers,
and morphology and orientation of the aperture.
Those features may be developed individually or
in combination. The following five growth stages
are assumed to delimit the ontogenetic development of modern spinose species (Brummer et al.
1987).

1. The prolocular stage (Plate 6.1-1 and -2) is
probably very short-lived and includes all events
from zygote formation to formation of the first
spheroidal chamber, i.e. the proloculus (or protoconch). Individual gametes are about 5 μm in
size, and initial development of the proloculus
must involve substantial cytoplasmic expansion.
The amount of expansion after gamete fusion is
presumed to be least in species with large proloculi. It is assumed that the earliest prolocular stage
is not calcified. Diameters of proloculi in spinose
species typically range between 7 and 34 μm
(Fig. 6.1). Brummer et al. (1987) report diameters
of 12 and 25 μm, respectively. Proloculi sizes
analyzed from dissected larger tests may differ
between species, and measure on average
∼ 15.7 μm in diameter (Sverdlove and Bé 1985;

Hemleben et al. 1989). Despite rather large
size-variations, size differences of proloculi in the
same   species   (Fig. 6.1)   are   statistically
non-significant (Parker 1962; Sverdlove and Bé
1985; Brummer et al. 1987). In addition, no
alternation of generations could be detected based
on size patterns of proloculi. It is hence assumed
that only one reproductive mode, i.e. sexual
reproduction   predominates   in   planktic
foraminifers.

The distinct morphological features of proloculi, such as the presence or absence of pores and
spines, are uncertain. Details of the early development   are   deduced   from   the   earliest
multi-chambered stages, because proloculi have
not yet been found (confirmed) in the water
column or sediment samples. Calcified proloculi
may exist in the surface water column (Bishop
et al. 1980; Bé et al. 1985), but their reported
occurrence has not been confirmed. In contrast to
proloculi,   calcified   two-chambered   juvenile
individuals occur in large numbers in the water
column (see Chap. 5), as confirmed from
plankton tows using 10 μm nets, and short-term
sediments traps (cf. Hemleben et al. 1989).

2. The juvenile stage (Plate 6.1-2 to -4) commences with the formation of the deuteroconch
(second chamber) as cytoplasmic growth contin-
ues. The well-calcified deuteroconch is smaller
than the weakly or non-calcified proloculus. The
deuteroconch is followed by a number of subsequent chambers of uniform morphology. Juvenile
tests of *Globigerinoides* species have a more or
less   planoconvex   (or   globorotalid)   shape
(Rhumbler 1911; Parker 1962; Huang 1981)
(Plate 6.1-3). Juvenile *Globigerina* species produce globular chambers resulting in spheroidal
tests called globigerine tests (Huang 1981).
Alternatively, a streptospiral whorl may be
formed around the proloculus (Desai and Banner
1985).

During the juvenile stage, variable and
species-specific numbers of chambers are added in
a low trochospiral whorl containing about 1.5
turns. Six chambers are added in *Globigerinella
siphonifera*, and nine in *G. ruber*, resulting in test

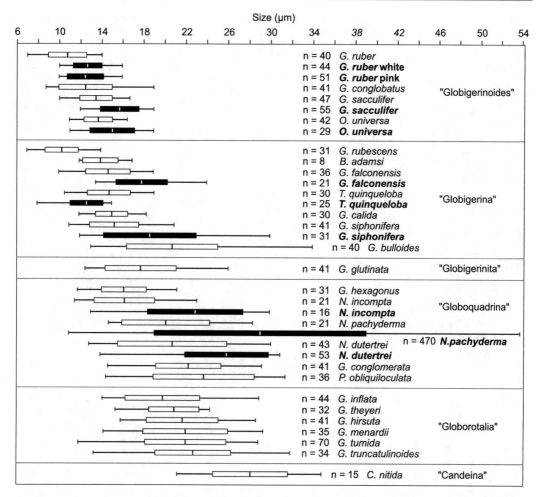

Fig. 6.1 Proloculus diameters of different planktic foraminifer species sorted by group. *Midpoints* show average diameters. *Thick bars* show standard deviations. *Whiskers* indicate the total range of data. *White bars* after data from Sverdlove and Bé (1985), and *black bars* according to data from Brummer et al. (1987). After Hemleben et al. (1989)

diameters of 80 µm and 65 µm, respectively. In *Globigerinoides sacculifer*, up to eleven chambers are added to the juvenile test, which increases in size from about 20 µm to 90 µm in maximum diameter. Juvenile tests produce a wide umbilicus and a marginal extra-umbilical aperture. Shapes of sutures and lobateness of tests varies among species. Juvenile chambers of *Globigerinoides ruber* are separated by deeply incised sutures producing a lobate test shape, whereas the sutures of juvenile *G. sacculifer* are shallow and the perimeter of the test has a smooth and shallow scalloped profile

(Plate 6.1-3 and -4). Large pores are aligned along the sutures, giving the test a typical smooth and suturally perforate morphology. Pores may be produced on both dorsal and ventral surfaces. Pores are present along both spiral and umbilical sutures of the test of juvenile *G. sacculifer* (Plate 6.1-4), and limited to the spiral side in *G. ruber*. The few thin spines at this stage lack supporting spine collars. Juveniles consume microplankton (largely phytoplankton), and may possess photosynthesizing symbionts from the three-chambered stage on.

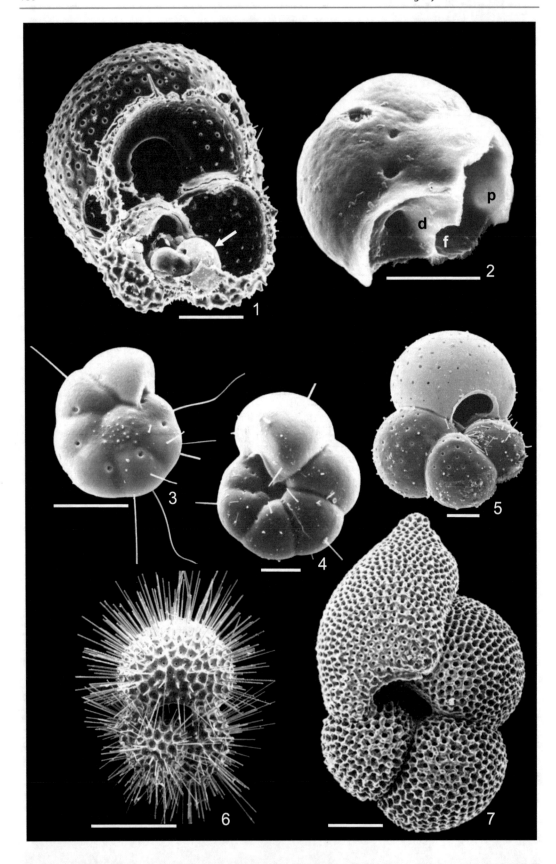

◄ **Plate 6.1** (*1*) Dissected test of adult *G. ruber* with first chambers (*arrow*). The test wall thickens on the juvenile part of the test when chambers are added. (*2–7*) Successive ontogenetic stages of *G. sacculifer*. (*2*) Dissected *G. sacculifer* with open protoconch (*p*) and deuteroconch (*d*), and foramen (*f*) connecting both chambers. (*3*) Early juvenile stage, (*4*) late juvenile stage, (*5*) late neanic stage, (*6*) adult stage and (*7*) terminal stage. Bars (*1*) 50 μm, (*2*) 10 μm, (*3–5*) 20 μm, (*6, 7*) 100 μm. (*1, 2, 3, 4, 6*) From Brummer et al. (1987)

3. The neanic stage is a transitional ontogenetic stage leading from the juvenile to the adult stage. This stage (Plate 6.1-5) is marked by substantial overall changes in test morphology. The average number of chambers per whorl decreases in comparison to the juvenile stage. During the neanic stage, typically about four chambers are added to the test in approximately one whorl. In *G. sacculifer*, the number of chambers per whorl decreases from 6.5 to 3.5 (Plate 6.1-3 to -5), and from 5 to about 3.5 chambers in *G. ruber*. Chamber shapes become increasingly globular in *Globigerinoides* species, and radially elongate in *Globigerinella* species, resulting in spheroidal or palmate tests, respectively. The umbilicus closes in most spinose taxa, and the position of the aperture changes from marginal extra-umbilical to umbilical. In other taxa, the umbilicus remains open while the aperture is situated in an extra-umbilical position as in *Turborotalita*. Pores are developed on the entire test surface. These changes in morphology occur while chambers are added, and gradually lead to the adult stage.

Characteristics of the final (adult) stage emerge in the neanic stage, including a more rugose texture of the test surface, development of numerous thick spines with spine collars, inter-spine ridges, and pore pits distributed over the entire test surface. The gradual increase in chamber volume during the neanic stage is larger than in the pre-neanic development. At the same time, the trophic demand changes from a herbivorous to a largely omnivorous diet.

4. In the adult stage (Plate 6.1-6), test features including secondary apertures are produced, which may vary between generic groups such as *Globigerinoides*, *Orbulina*, and *Sphaeroidinella* (e.g., Huang 1981; Desai and Banner 1985).

Subequatorial main apertures appear in *Globigerinella*. Chambers may deviate in shape, from the rather spherical forms of neanic chambers to forms ranging from compressed as in *Globigerinoides conglobatus* to digitate as in *Beella digitata*. Changes in chamber form may result in alterations in coiling mode, and spherical terminal chambers as observed in *O. universa*. Digitate chambers and streptospiral coiling occur in *Hastigerinella*. In *G. sacculifer*, two or three large chambers are added during the adult stage (test diameter >210 μm). The diet of adult stages is mainly carnivorous or omnivorous (see Sect. 4.1). At least one chamber, but usually two or three chambers are added before reproduction.

5. The terminal stage (Plate 6.1-7) is related to reproduction, and marked by the shedding of spines, and partial wall thickening in some species (i.e. gametogenic calcite, GAM, see Chap. 5). Terminal stages show substantial variation among some genera such as the sac-like chamber of *G. sacculifer* (cf. Plate 6.1-6 and -7). Bulla-like additions or elongated last chambers cover the umbilicus in *Turborotalita* species, and normalform or kummerform final chambers are common among other species (Berger 1970; Hemleben and Spindler 1983). The term kummerform signifies chambers, which show no size-increase, or even a size-decrease in comparison to the previous chamber. In addition, some species resorb septa (e.g., *Hastigerina pelagica*) or entire previous chamber walls (e.g., *O. universa*). The resulting final step in maturation is marked by gametogenic tests with coarsely cancellate or smooth surface texture. Open spine-holes indicate shed spines. Finally, terminal resorption of calcite and/or calcification may alter the surface texture, gross morphology, and chemical composition of tests of various species.

6.2 Morphological Development of Non-spinose Species

The ontogenetic developmental pattern observed in spinose species can be applied to the non-spinose species in most aspects. Proloculi of non-spinose species (16.3–23.9 µm) exhibit a wider size-range and are larger than proloculi of spinose species (av. 15.7 µm) on average (Fig. 6.1). Those differences are most pronounced in surface dwellers (Sverdlove and Bé 1985; Hemleben et al. 1989). *Candeina nitida* produces a particularly large proloculus of 28.3 µm in diameter on average. *Neogloboquadrina dutertrei*, *N. pachyderma*, and *Pulleniatina obliquiloculata* (sampled from the Weddell Sea) vary most markedly in proloculus sizes from 11 to 54 µm, with an average of 29 µm. Microperforate species produce proloculi, which are on average equal in diameter to those of spinose species (Brummer unpublished data; Hemleben et al. 1989).

According to Brummer (1988), juvenile tests of microperforate species (e.g., *Globigerina glutinata*, *C. nitida*, *Tenuitella iota*) are morphologically different from spinose species by being planispiral instead of trochospiral. The aperture of microperforate species has an equatorial rather than marginal position, and produces a pronounced spiral flange. The pores of microperforate species are extremely small, and evenly distributed over the entire chamber surface. In this respect, microperforate species differ from other non-spinose and spinose species, which produce larger pores along the chamber sutures (e.g., *N. dutertrei*, *Globorotalia truncatulinoides*).

In the deep-dwelling non-spinose species *G. truncatulinoides*, juvenile sutural pores are subsequently covered with additional calcite layers (Hemleben et al. 1985). Similar to *G. sacculifer*, the spiral and umbilical sides of the test of *G. truncatulinoides* exhibit marked differences in pore density during ontogeny, with more pores per unit area on the spiral side than on the umbilical side during the juvenile stage (ca. 8–11 chambers), and more pores on the umbilical side of the test thereafter (Fig. 6.2). A similar, although less marked pattern, occurs in *Globorotalia hirsuta* (Fig. 6.2). Only a few of

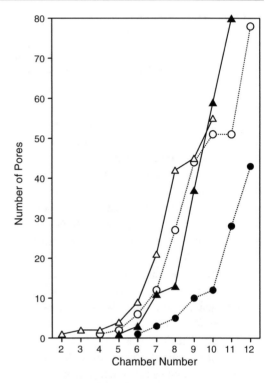

Fig. 6.2 The number of pores per chamber rapidly increases between chambers 6 and 7 (spiral side, *open triangles*), and chambers 8 and 9 (umbilical side, *filled triangles*) in *G. truncatulinoides*, and chambers 7 and 8 (spiral side, *open circles*) and chambers 10 and 11 (umbilical side, *filled circles*) in *G. hirsuta*. Modified after Hemleben et al. (1985)

the second chambers produce pores in *G. hirsuta*. The percentage of chambers with pores increases markedly after the third chamber, considering the proloculus as chamber number one. Pores are present in almost all specimens from chamber number six onward. The keel as a significant morphological feature of adult specimens of both globorotalid species develops from onset of the neanic stage. The first five to seven chambers produce no keel. In specimens larger than 100 µm, the keel is a prominent feature. The shell surface where the keel develops is initially porous in early stages, similar to the surface of newly formed chambers, and increasingly thickens with advancing test growth and calcification. Eventually, a distinctive ridge is formed along the test periphery. Keel development is similar in other globorotalid species as, for example, *Globorotalia menardii* (Hemleben et al. 1977).

During the adult stage, *C. nitida* forms its characteristic accessory apertures along sutures (Plate 2.33-12 to -17). A pronounced apertural flange develops in *Dentigloborotalia anfracta* (Plate 2.18-13). The morphology of adult stages of other non-spinose species is similar to that of the late neanic stages. Similar to spinose species, terminal features, such as kummerform chambers (e.g., Plate 2.9, Plate 5.3-7) or bullae (Plate 2.32-3 and -4) may be formed by non-spinose species. These systematic variations in the morphology of chambers and tests are not to be confused with abnormal growth such as twinned tests (e.g., Boltovskoy 1982).

6.3 Coiling Directions of Tests

Most extinct and modern planktic foraminifer species produce trochospiral tests, either coiled in clockwise (dextral, right coiling) or anti-clockwise (sinistral, left coiling) direction viewed from the spiral side. Coiling direction can usually be determined as early as the two-chambered or three-chambered stage, although coiling direction may be fixed prior to calcification of the deuteroconch (Hemleben et al. 1989 and references therein).

In modern planktic foraminifers, some species are coiled almost exclusively sinistral (e.g., *Globorotalia tumida*) or dextral (e.g., *P. obliquiloculata*). In some species like *G. truncatulinoides*, coiling direction may change over geological time periods in relation to temperature variations or other environmental factors (e.g., Ericson et al. 1954). Those changes in coiling direction have been used for stratigraphic purposes (Hemleben et al. 1989, and references therein). Whereas some non-spinose species have been analyzed for the paleoceanographic significance of coiling directions of tests, equivalent data on spinose species have so far not been published.

Molecular genetic evidence reveals that formerly distinguished left and right coiling forms of *Neogloboquadrina pachyderma* are in fact different species, i.e. *N. pachyderma* and *N. incompta*, respectively (Darling et al. 2006) (see Chapter 2). All of the *N. pachyderma* genotypes are typically left coiling, and include <3 % of right coiling individuals. In turn, typically right coiling *N. incompta* produce <3 % left coiling specimens. Coiling 'failure' of typically <3 % is realized within the same species. Such coiling failure occurs also in other trochospiral species like *Globorotalia inflata*. In turn, ratios of any coiling direction left or right >3 % possibly indicate different genotypes (cf. Darling et al. 2006).

6.4 Formation of Chambers and Pores

One of the most apparent phenomena in planktic foraminifer ontogeny is the deposition of new chambers (Table 6.1) approximately once every other day in the adult stage, and at a higher

Table 6.1 Major events during chamber formation in non-spinose species (Hemleben et al. 1986)

Time (h)	Stage	Observation
0:00–0:30	Extrusion of cytoplasmic bulge	Reduced activity and withdrawal of rhizopodia. Extrusion of cytoplasmic bulge
0:30–1:30	Organization of bulge	Fanlike rhizopodia radiate from bulge and form the outline of the outer protective envelope. Bulge gradually expands up to the protective envelope
1:20–1:40	Construction of anlage	Periphery of bulge slowly smoothens. The surface of bulge forms the position and outline of new chamber
1:30–3:00	Calcifying process	Small plaques of calcite are secreted on inner and outer surfaces of the primary organic membrane (POM). Plaques increasingly coalesce to form the first continuous bilamellar wall
3:00–6:00	Further thickening	Further thickening of completed wall, and spine development in spinose species

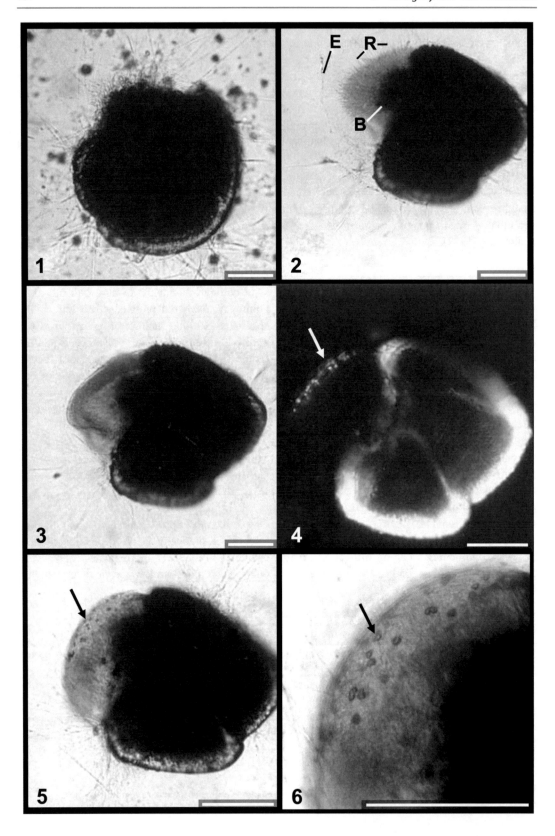

◄ **Plate 6.2** Chamber formation in *G. truncatulinoides*. (*1*) A cytoplasmic bulge with rhizopodia protrudes from the aperture, about 20 min after start (*ST*) of chamber formation. (*2*) Cytoplasm emerged from the test increases the size of the bulge (*B*). Radiating rhizopodia (*R*) are in contact with the surrounding outer protective envelope (*E*), ~30 min. after ST. (*3*) Further extension of the bulge fills the space between bulge and E, ~1:30 h after ST. The anlage of the new chamber has been formed, visible by the smooth outline of the bulge, ~1:50 h after ST. (*4*) White specks (*arrow*) indicate calcification of the chamber anlage in polarized light, ~2:15 h after ST. (*5,6*) Calcite plaques are visible as dark specks (*arrow*) on the chamber anlage in transmitted light, ~2:15 h after ST. From Bé et al. (1979). Bars 100 μm for all images

frequency in pre-adult stages. Planktic foraminifers increase their overall test size by up to 25 % per day, contributing to a geometric increase in test volume (Caron et al. 1982; Bé et al. 1982; Erez 1983; Anderson and Faber 1984). Growth of a new chamber is completed within ~6 h. Each new chamber ideally increases in volume relative to the preceding one. New chambers appear hyaline, and are only partially filled with cytoplasm, visible in low power light microscopic examination (Hemleben et al. 1986, and references therein). Major events in chamber addition and pore formation in non-spinose and spinose planktic foraminifer species are presented in the following sections.

6.4.1 Chamber Formation in Non-Spinose Species

The earliest indication of impending chamber formation in non-spinose species as, for example, in *G. truncatulinoides* and *G. hirsuta* (maintained in laboratory culture), is the withdrawal of the feeding rhizopodia, and concentration of cytoplasm within the test or near the aperture (Table 6.1, Plate 6.2) (Bé et al. 1979). A bulge of cytoplasm with a thin hyaline cover and optically dense interior emerges from the aperture (Plate 6.2-1). Initially, the bulge is delimited by a smooth membrane. Subsequently a fine halo of rhizopodia radiates from the surface of the bulge. The rhizopodia increase in density and length, producing a fan-like profile (Plate 6.2-2) about 30 min after the start (ST) of chamber formation. Those rhizopodia are different in organization and granularity from the typical feeding rhizopodia surrounding the test. The bulge is divided into an inner portion of compact and opaque

granular cytoplasm (probably cytoplasm protruding from the last chamber), and an outer translucent and less dense layer, in about half of the specimens observed during chamber-formation. The outer layer consists of a dense network of radiating rhizopodia. In the other half of observed specimens, the bulge is homogeneously translucent. About one hour after ST, the periphery of the fan-like array of rhizopodia reaches its maximum extent, and forms the outline of the outer protective envelope, thus creating a transparent region between the protective envelope and the translucent bulge. The translucent bulge expands gradually towards the outer protective envelope, while the opaque section (if present) expands only slightly, and occupies about one third to half of the bulge.

When the translucent bulge has attained maximum expansion, its periphery rapidly transforms from a rough and undulating to a smooth surface (Plate 6.2-3). The process takes only about 10 min, starting with the formation of a distinct border between the translucent and opaque regions, and is completed upon the disappearance of the rhizopodial fan. Concurrently, smoothening of the periphery starts at or near the aperture and spreads to peripheral parts of the bulge. When fully completed after about 2 h, the smooth peripheral surface of the translucent bulge forms the final position and outline of the new chamber, which is now ready for calcification (Plate 6.2-4 and -5). The new chamber wall is calcified at the periphery of the translucent section (Plate 6.2-5 and -6). Two ultrastructural elements (Plate 6.3-1 and -2) associated with the appearance of calcite deposits are the cytoplasmic envelope (CE), and the primary organic membrane (POM). CE and POM are assumed (in analogy to benthic foraminifers) to form the

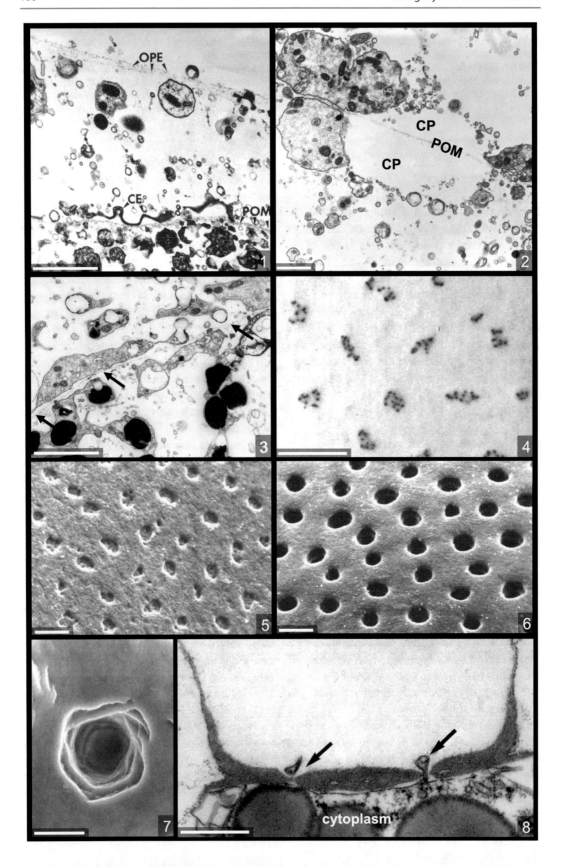

◄ **Plate 6.3** (*1*) Completion of the anlage of a new chamber in *G. truncatulinoides*, fixed ∼1.5 h after onset of chamber formation. The primary organic membrane (*POM*) is secreted to the proximal side of the cytoplasmic envelope (*CE*), and outer protective envelope (*OPE*) situated towards the distal side. (*2*) Calcite plaques (*CP*) delineated by rhizopodia are secreted at either side of the POM, in decalcified specimen. (*3*) POM (*arrows*) enveloped by cytoplasmic strands (*grey*). (*4*) New test wall of *G. menardii* (∼1.5 μm thick) with 12–18 Micropores per pore area. (*5*) Early pores, and (*6*) final pores in *G. menardii*. (*7*) Pore in *G. sacculifer* showing straight edges of calcite crystals. (*8*) Pore plate in *G. sacculifer* with cytoplasmic strands passing through micropores (*arrows*). Bars (*1–3,8*) 1 μm, (*4–7*) 10 μm. (*1, 2, 3*) From Hemleben et al. (1986, with permission from the Systematics Association). (*4, 7*) From Hemleben et al. (1977)

anlage (Angell 1967), providing a surface structure for initiating calcification, and providing the final three-dimensional shape to the new chamber (Towe and Cifelli 1967; Hemleben et al. 1989).

The CE derives from the distal rhizopodial network by differentiation and lateral extension of thin segments each about 3 μm in length, and separated by spaces, forming a patchwork of plaques interconnected by narrow bridges. The continuous and extremely thin (0.05–0.06 μm) POM forms the site of initial calcification, and probably serves as a template for calcite deposition. The POM may originate from secretory activity of the CE, assembled from a thin layer of organic fibrils on the outer surface of the membrane facing the POM. The CE is present on the distal but not on the proximal side of the POM. Calcification occurs on both sides of the POM, indicating that the CE is not exclusively involved in calcite secretion, if at all. Moreover, at advanced stages of calcification, the CE is either absent or not in direct contact with the calcareous surfaces. Alternatively, the surrounding rhizopodia may be involved in POM formation. The outer protective envelope (OPE) forms the outermost layer, probably providing a protective envelope consisting of a network of rhizopodia and organic filamentous matter, which protects the delicate anlage and rhizopodia during calcification.

Two to three hours after the onset of calcification, when the POM and CE are fully developed, a few small dark specks (in light microscopy) appear on the cytoplasm exterior in non-spinose planktic species (Plates 6.2-6 and 6.3-1). The dark specks are identified by optical interference (under polarized light) as initial secretions of calcareous particles. In specimens simultaneously fixed for transmission electron microscopy, calcite plaques occur at the POM, being lined at both the inner and outer side by cytoplasmic strands (Plate 6.3-2 and -3). These plaques are bilamellar with one lamella on the distal surface of the POM, and the other lamella on the proximal surface. The plaques grow laterally and fuse at their edges to form the first thin bilamellar test wall (Hemleben et al. 1989, and references therein).

6.4.2 Calcification of the Test Wall and Pore Formation

Calcification on the surface of the primordial test wall occurs by addition of single calcite layers each nucleated on an organic membrane (Plate 6.3-2, and Plate 6.4-1 and -2). No apparent calcification occurs at the inner organic membranes during later ontogenetic stages (after two to four additional chambers) although an inner organic membrane is deposited each time a new chamber is added. Those membranes form a dense inactive inner organic lining (IOL) covering the inner walls of previous chambers (Hemleben et al. 1989). The calcification system is primarily bilamellar, and calcification occurs on both the inner and outer side of the POM.

During early calcification, the test wall of *G. menardii* is flexible and only about 1 μm thick. Individual crystallites are surrounded by loosely packed organic matter (Plate 6.4-1), equivalent to the plaques observed in TEM images (see above). The new chamber is loosely connected to older chambers at a few points, the site of the suture being covered by a thin layer of cytoplasm. Subsequently, the new wall is more firmly attached by calcification.

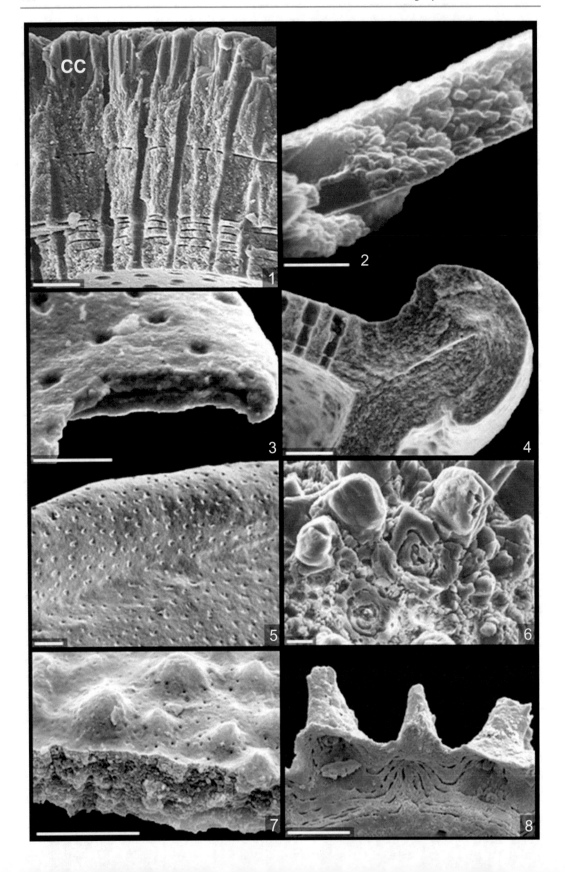

◀ **Plate 6.4** (*1*) Bilamellar test wall of *G. menardii* in cross section with one inner and nine outer calcite layers, as well as calcite crust (CC) on *top*. (*2*) Test wall of *G. menardii* during early calcification stage. (*3*) Keel development in *G. menardii*. In an early stage, the chamber wall folds upon itself and calcifies more than other parts of the wall, and (*4*) the keel is strengthened by additional calcite layers. (*5*) The smooth area of the new keel is covered by pores. (*6*) Layered pustules of *G. menardii* in cross section (etched with EDTA for 2 min). (*7*) Pustules in *Globuligerina oxfordiana* from the early Jurassic. (*8*) Layered test wall and pustules in *Morozovella aequa* (Paleocene). Bars (*2,3-8*) 10 µm, (*2*) 1 µm. (*1,2,3,5,6*) From Hemleben et al. (1977)

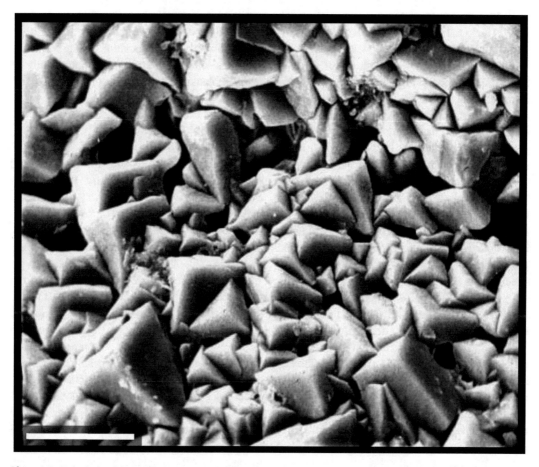

Plate 6.5 Euhedral calcite crystals on top of the shell of *G. truncatulinoides* grown under laboratory conditions, similar to calcite crust formed in the natural environment (see Plate 2.14). Bar 10 µm

In both spinose and non-spinose species, the deposition of additional layers of calcite on the surface of the new chamber as well as on the older chambers results in multiple layers of organic linings spanning the suture region between adjacent chambers (Hemleben et al. 1977; Bé and Hemleben 1970). Multiple organic layers mark the successive calcite layers (Plate 6.4-2) deposited upon the older chambers as each new chamber is added and calcified to form the bilamellar wall (Reiss 1957; Towe 1971; Oelschläger 1989; Erez 2003). Consequently, the multiple organic layers form an increasingly thick IOL within previous chambers (Oelschläger 1989). During initial and subsequent stages of test wall formation, including formation of the keel and apertural lip, microgranular (anhedral) crystals are produced. Euhedral crystals (Plate 6.5) may be formed during late calcification stages (e.g., Bé et al. 1966; Pessagno and Miyano 1968; Takayanagi et al. 1968; Hemleben 1969b).

The position of pores is already determined on the early bilamellar test wall. This wall is dissolved at multiple sites, and fine strands of cytoplasm pass through sieve-like micropores (Plate 6.3-4), and form early (Plate 6.3-5) and final pore structures by resorption (Bé et al. 1980) (Plate 6.3-6 and -7). A pore plate is produced, and spans the pore at the level of the POM (Fig. 6.3), probably by deposition of organic matter on the POM during maturation of the new chamber. The pore-wall is lined by an organic membrane, termed outer pore lining (OPL) at the distal side, and inner pore lining (IPL) at the proximal side of the shell (Fig. 6.3c, d). The pore lining is apparently thickened by repeated deposition of organic layers concurrent with each addition of plaques to the growing bilamellar wall. Micropores occur in the pore plate. In non-spinose species, the IOL of mature chambers is separated from the pore plate and positioned proximal to the cytoplasm (Fig. 6.3c, d). In spinose species (Plate 6.3-8), the IOL is fused with the pore plate (Fig. 6.3e, f), and located much closer to the chamber wall than in non-spinose species. In *H. pelagica*, different from other species, the thin and organic-rich shell exhibits (in electron micrographs) no apparent differentiation of the IOL, resulting in a monolamellar and not bilamellar test wall (Fig. 6.3a, b).

Fine rhizopodia and perhaps colloidal particles pass through the pores of the planktic foraminifer test wall. An accumulation of mitochondria at the proximal side of the pores in several benthic species indicates oxygen diffusion through the pore into the cytoplasm (Leutenegger and Hansen 1979). Symbiont-bearing *Amphistegina lobifera* take up $^{14}CO_2$ through the pores indicating gas exchange, and the support of photosynthesis of the algal symbionts by the pores. Larger molecules like ^{14}C-labelled glucose may not pass through the pores in traceable amounts (Leutenegger and Hansen 1979). In non-spinose species, the IOL seals the pore plate almost completely at least in the ultimate and penultimate chambers, and increasingly so in earlier chambers.

Porosity of the outer test wall appears to be related to temperature and oxygen concentration of ambient seawater (Fig. 6.4) (Bé 1965, 1968; Frerichs et al. 1972; Caron et al. 1987; Kuroyanagi et al. 2013). The inner walls of spiral planktic foraminifer tests sometimes exhibit pores, which connect chambers. Those pores were previously connecting to the outside of test and not closed by calcite deposition during formation of a new chamber. The entire functionality of pores in planktic foraminifers is not yet resolved.

6.4.3 Keel Development

The peripheral keel in compressed and disc-shaped planktic foraminifer tests as, for example, *G. menardii* and other globorotalids, may serve as structural reinforcement of chambers and test (Brönnimann and Brown 1956; Scott 1973a; b). Keels in modern planktic foraminifers are all of the same 'inflational fold' type, whereas four more types of keels evolved from the Cretaceous to Neogene (Norris 1991). The keel is formed by collapse of the peripheral chamber wall (Hemleben et al. 1977). Formation of the inflational fold keel commences during thickening of the chamber wall at the onset of the neanic stage. The ventral and dorsal chamber walls fold upon one another and double up to form a primordial keel (Plate 6.4-3) as an integral part of the chamber wall. The primordial keel possesses eight to ten micropores per pore similar to pores in other regions of the chamber wall including the lip. Those micropores are not functional, and form only shallow depressions of the ultimate pore area. The keel pore area is only imperfectly resorbed over the course of pore development, and more rapidly calcified than the chamber wall, and more calcite layers are formed to cover the keel than the remaining test wall (Plate 6.4-4) during the final stages of calcification (cf. Hofker 1971). Concurrently with thickening of the keel, an inner wedge (Plate 6.4-4) formed through doubling of the chamber wall

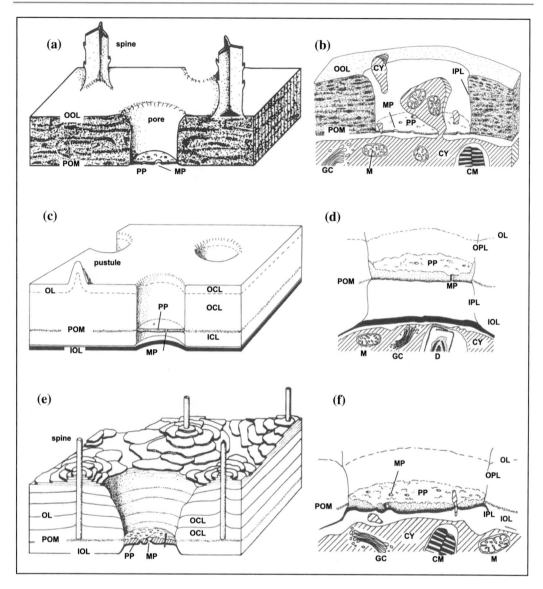

Fig. 6.3 Schematic block diagrams and cross-section of pores of different planktic foraminifer groups. **a, b** *Hastigerina pelagica* (from Hemleben et al. 1989), **c, d** non-spinose species, and **e, f** spinose species (from Bé et al. 1980). *CM* is copepod muscle tissue within a digestive vacuole; *CY* is foraminifer cytoplasm; *D* is a diatom within digestive vacuole; *GC* is a Golgi complex; *ICL* is an inner calcite layer; *IOL* is the inner organic lining; *IPL* is the inner pore lining; *M* is a mitochondrion; *MP* is a micropore; *OCL* is an outer calcite layer; *OL* is an outer organic layer; *OPL* is the outer pore lining; *POM* is the primary organic membrane; *PP* is a pore plate

during keel formation is filled in, and rounded off (Plate 6.4-5). Initially, the outer surface of the keel is relatively smooth and shows very little relief (Plate 6.4-3), except of some occasional pustules, and becomes increasingly more crystalline with the addition of calcite layers (Bé et al. 1966).

6.4.4 Pustule Formation

The surface of the test in most non-spinose species bears small conical protuberances (Fig. 6.5) called pustules. Pustules first occur during the juvenile stage, and increase in number and size during ontogeny. Pustules are larger and more

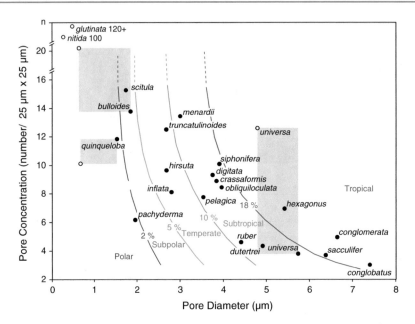

Fig. 6.4 Average pore sizes and numbers of pores per unit area, and resulting porosities (%) in 21 normal perforate planktic foraminifer species show an overall relation to latitude. Pore densities increase from the polar to tropical water masses, i.e. with ambient water temperature, and may vary (*gray boxes*) considerably as shown for *G. bulloides*, *T. quinqueloba*, and *O. universa* (dots from Bé 1968; circles from Bé et al. 1969; Hemleben et al. 2017). The two micro-perforate species *Candeina nitida* and *Globigerinita glutinata* show very high pore numbers (>100 pores per unit area, and pores <1 μm), and low pore densities <1 %, which are not related to water temperature. Deep-dwelling globorotalids (see Chap. 7) produce pore densities typical of cold subsurface water masses, and dissemble north of their actual latitudinal distribution. Redrawn after Bé (1968) and Bé et al. (1969)

densely distributed near the aperture, and smaller and less numerous (per unit area of test surface) near the periphery of the test (e.g., Plate 2.27) (Hemleben 1975). Pustules occur on the apertural lip, in the area between pores, and on the keel (e.g., Plate 2.26-13). Pustules are of structural importance and serve as anchor points for rhizopodia that radiate out from the test (see Chap. 3). Pustules start to grow when the test wall is about 4–8 μm thick, at the time of keel thickening, and soon after the onset of pore formation. New pustules are formed during each successive calcification episode of the new chamber wall. As the pustules grow laterally, they may branch (Fig. 6.5c) or merge, and cover the pores. The tips of the pustules may be rounded as in *G. menardii* (Plate 2.26-8) or pointed, and resemble shark teeth (Fig. 6.5e) as in *D. anfracta* (Plate 2.18-17).

Based on the mode of formation and morphology, pustules are not homologous with the elongate spines of spinose planktic foraminifers (Hemleben 1969a, b; Hemleben 1975; Hemleben et al. 1977). Pustules are phylogenetically old features occurring as early as the first planktic foraminifers in the early Jurassic (Plate 6.4-7). In contrast, spines are rather new developments in planktic foraminifers, and first occur during the early Tertiary, i.e. in the early Paleocene right after the Cretaceous/Paleocene boundary.

Pustules: The term pustules is synonymous to hispid wall structures, rugosities, tubercules, and pseudospines. Pustules produce the same layering as the test wall (Hemleben 1975; Benjamini and Reiss 1979), and due to the organic lamellae deposited with each increment of calcite, they are concentric in cross-section (Plate 6.4-6 and -8, Fig. 6.3).

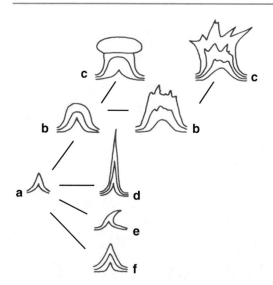

Fig. 6.5 Pustule morphologies in non-spinose planktic foraminifers originating from **a** initial stage (from Hemleben et al. 1989). **b** Types in, e.g., *G. hirsuta, G. truncatulinoides, G. menardii*, and *G. inflata*, which in the latter species may develop still further (**e**). Additional types were found in **d** *P. obliquiloculata*, **e** *D. anfracta*, and in the genus *Globigerinita* (**f**). The variations in morphology are not species specific in modern species with the exception of the pustules in *D. anfracta* (see Chap. 2)

6.4.5   Chamber Formation in Spinose Species

Chamber formation in spinose species follows the same pattern as in non-spinose species, with the exception of spine growth. Fine spines start to grow on the test surface about 2–3 h after onset of chamber formation after the earliest stage of the bilamellar wall is established (Plate 6.6-1). These spines gradually elongate and may be distinguished from the mature spines by their round cross-section. Spines originate from the POM or one of the later appearing calcifying organic layers (Fig. 6.3). Each spine is a single crystal (Plate 6.6-2) structurally supported by a calcareous spine-collar or spine-base (Fig. 6.6). Spine growth occurs at its tip through calcite deposition by a thin but continuous sheath of cytoplasm.

During the early stages of chamber formation, spines of the previously formed chambers project through the new chamber. The anlage of the new chamber is suspended on cytoplasmic strands attached to the spines. The internal array of previously formed spines is resorbed as the new chamber matures producing an open space within the new chamber (Rhumbler 1911). Involvement of the CE and the POM in early stages of chamber formation in spinose species as *G. sacculifer* is similar to non-spinose species. In contrast, plaques of calcite fuse to form the early bilamellar wall more extensively on the distal than proximal side, and cause a deeper proximal position of the POM than in non-spinose species. In spinose species, the IOL is thicker under the pore plate than under the inter-pore wall. In general, the IOL in spinose species is thinner than in non-spinose species (Fig. 6.3). In spinose species, very fine rhizopodia may occasionally pass through the micropores, and extend into the surrounding environment.

Hastigerina pelagica produces a monolamellar test wall, which lacks a visibly layered structure (Fig. 6.3a and b, Plate 6.7-1). The test wall of *H. pelagica* contains large quantities of organic matter distributed in a sponge-like way (Hemleben et al. 1989). Spines in *H. pelagica* develop within a sheath of cytoplasm at the tip of the spine. At some distance from the spine tip a thickened protective envelope of cytoplasm forms a sheath-like pellicle. More mature sections of the spine are covered by a thin layer of cytoplasm or by a rhizopodial net (Plate 6.7-2). Near the thin cytoplasmic envelope enclosing the alveoli, which form the bubble capsule in *H. pelagica*, the spine-bases are covered by a thin layer of reticulate cytoplasm, from which strands of rhizopodia protrude laterally to surround the spines. The pores in the chamber walls of *H. pelagica* produce a pore plate possibly at the level of the POM (Fig. 6.3). However, the POM is not clearly discernable in *H. pelagica* owing to the non-stratified, spongy, and organic-rich test wall. All of those differences in test wall architecture between *H. pelagica* and other modern spinose species (Fig. 6.3) suggest substantial differences also in phylogenetic histories (cf. Weiner et al. 2012).

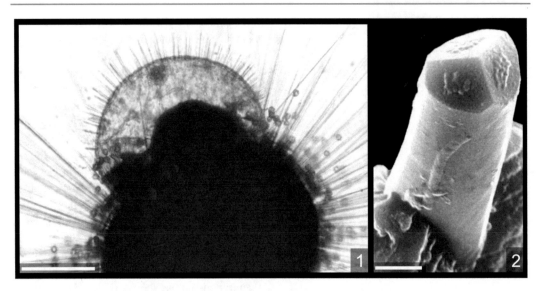

Plate 6.6 (*1*) The first spines grow on the newly formed chamber about 2–3 h after the onset of chamber formation in *G. ruber*. (2) Crystal faces of a spine populated by bacteria. Bars (*1*) 50 μm, (2) 1 μm

The spiral pre-mature test of *O. universa* is less calcified and contains more organic matter than tests of other spinose species, and, in contrast to *H. pelagica*, exhibits a stratified test wall typical of most other spinose species. Similar to *H. pelagica*, the pre-mature test of *O. universa* is extremely fragile. The earliest sign of formation of a spherical chamber is the appearance of a frothy layer of cytoplasm at the surface of the trochospiral test. The onset of chamber formation may be delayed for an uncertain period of time after the frothy layer is produced. Immediately prior to the formation of the anlage of the spherical chamber, a bulge of cytoplasm containing vesicles and mitochondria surrounds the trochospiral test, forms a distinct sphere, and expands. As the cytoplasmic layer approaches maximum expansion, its perimeter becomes more distinct. Full expansion of the spherical anlage takes about 40 to 60 min. A POM is probably produced by the vesicle-rich cytoplasm, and marks the final perimeter of the anlage. During the first 6–12 h of wall thickening, calcite may be added equally to both sides of the POM producing a typical bilamellar wall, or it may be more pronounced at the distal part of wall (Spero 1986).

Data from laboratory cultures show that wall thickening may proceed at an average rate of 0.1–0.7 μm over a 12-h period at a constant feeding rate. First, round spines are produced from the POM and appear on the surface of a 2–3-h old spherical chamber. The new spines become triangular to triradiate approximately 20 μm above their base, and attain a final length of 1.5–2.5 mm.

6.4.6 Types of Spines

The morphology of mature spines includes five general spine types (Saito et al. 1976). The *Globigerina*-type (Fig. 6.6a) and *Globigerinoides*-type (Fig. 6.6b) spines are circular in cross-section throughout their length, and occur in the two genera *Globigerina* and *Globigerinoides*. In *Globigerinoides*, the outermost spine tips may become triangular. In *Globigerina*, spines are thinner and more densely distributed on the test surface than in *Globigerinoides*. The same type of spines occurs on the spiral test of *O. universa*, but they are thinner and more densely distributed than in *Globigerinoides*. In addition to the round spines, pre-adult spiral tests

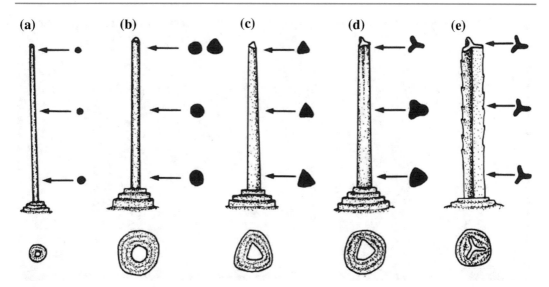

Fig. 6.6 Spines and spine bases of **a** *Globigerina* type with a small round spine base and round spine throughout. **b** *Globigerinoides* type with a round spine base and round spines, which may change to rounded triangular. **c** *Orcadia riedeli* type rounded triangular spine base and triangular spine throughout. **d** *Globigerinella* type rounded spine base and triangular spine becoming triradiate. **e** *Hastigerina* type shallow round spine base and barbed triradiate spine throughout. Compiled after Saito et al. (1976; from Hemleben et al. 1989)

of *O. universa* produce triradiate spines. *Orcadia riedeli* exhibits triangular spines throughout the entire length (Fig. 6.6c). The mature spherical tests of *Orbulina* and *Globigerinella* possess both *Globigerina*-type spines and *Globigerinella*-type spines. Spines of the latter genus have a somewhat triangular cross-section at their base, and gradually become triradiate with thin and elongate ridges (Fig. 6.6d). *Hastigerina pelagica* and *H. digitata* produce triradiate spines (Fig. 6.6e) with thin elongate ridges throughout the length of spines, and the perimeter of spines is equipped with small barbs (Rhumbler 1911; Hemleben 1969b). In contrast to the *Globigerinella* type, the ridges of the *Hastigerina* spines are overall thinner and more elongate.

Spines are different from pustules by being secondary structures, which are lodged in the test wall, and surrounded by a characteristic spine collar. Spines originate from the POM or one of the outer organic linings (OLs). In contrast, pustules are primary wall structures bearing the same layered, organic matrices as the lamellae of the wall. Spines typically do not possess internal organic matrices (Hemleben 1975).

6.5   Test Wall Thickening

Calcite precipitation during the formation of a new chamber is assumed to represent only about one-tenth of the entire test calcite mass. Most of the test $CaCO_3$ is deposited by additional calcification (e.g., Oelschläger 1989; Erez 2003). New calcite layers are actively added to the test at different intervals of maturation, i.e., during formation of new chambers, to the mature test before gametogenesis, and by gametogenic calcification. Maturation in spinose and non-spinose species is accompanied by thickening of the distal test wall (Bé and Hemleben 1970; Bé 1980). The tropical species *S. dehiscens*, for example, produces a thin and smooth calcite veneer in its mature ontogenetic stage (e.g., Bé and Hemleben 1970). After initial deposition of the bilamellar test wall, and additional lamellar thickening on the outer surface, euhedral calcite crystals are produced on the outer surface and gradually thicken to form a robust final layer with a rough textured surface, the typical '*dehiscens*' stage (see Chap. 2, Plate 2.14-4).

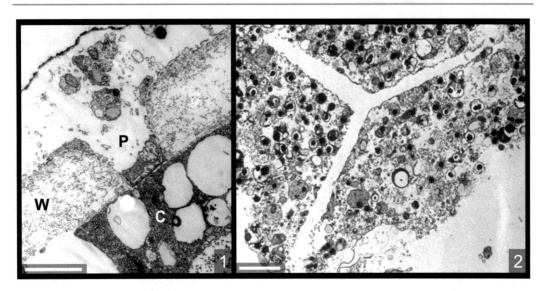

Plate 6.7 (*1*) The cytoplasm (*C*) of the ultimate chamber of *H. pelagica* with organic rich wall (*W*) and pore (*P*). (2) Cytoplasmic cover of a triradiate spine of *H. pelagica*. Bars (*1*) 2 μm, (2) 5 μm

The calcite veneer is an amorphous cortex eventually secreted at the distal part of shell (e.g., Steinhardt et al. 2015). In *O. universa*, thickening of the outer shell wall of the spherical adult test by diurnal (i.e. day and night) layers lasts over about two to nine days before gametogenesis (Spero et al. 2015).

Formation of diel calcite layers was first suspected from the presence of diurnal Mg/Ca bands in test walls of *O. universa* cultured in seawater enriched in Mg, Ca, and Li (Eggins et al. 2004; Gagnon et al. 2012). Diurnal formation of calcite layers <10 μm in thickness was confirmed in *O. universa* cultured in seawater labelled with Ba/Ca and $\delta^{18}O$, and analyzed with high (<6 μm) resolution LA-ICP-MS and SIMS (Vetter et al. 2013). Precipitation of diurnal calcite layers (Fig. 6.7) follows a light/dark cycle (Vetter et al. 2014; Spero et al. 2015). About two-thirds of the calcite is produced during the light period, and one-third during darkness (Vetter et al. 2013; Spero et al. 2015). The calcite layers precipitated periodically at a diurnal cycle produce a distinct $\delta^{18}O$, $\delta^{13}C$, Ba/Ca, $^{87}Sr/Ca$, and Mg/Ca signature. Outer layers precipitated in the daytime and at night differ in thickness and measure about 2–3 μm, and ∼1–1.5 μm, respectively (Fig. 6.7). Layers formed below the POM are only about

0.3-0.8 μm thick. The formation of high-Mg layers has been assumed to be supported by mitochondrial activity (Spero et al. 2015).

Multiple calcite layers at the submicron scale, which have been repeatedly observed in SEM analyses of cross-sections of the outer test wall of different species (Plate 6.4-2), may represent layers of diurnal calcite precipitation in symbiont-barren non-spinose species. If it turns out that diel calcite layers are a prevalent feature in modern and fossil planktic foraminifer tests, analyses of diurnal resolution would significantly improve the understanding of the paleo-environment at high temporal resolution. High-resolution analyses are feasible by, for example, LA-ICP-MS and (Nano-) SIMS (see Chap. 10) analyses of samples from a wide range of paleoceanographic settings (e.g., Vetter et al. 2014; Spero et al. 2015) (see Chap. 10.7).

A calcite crust is produced on the distal surface of the primary test wall in many spinose and non-spinose species when specimens sink into deeper and colder layers of the water column (e.g., Hemleben et al. 1989, and references therein; Lohmann 1995; Simstich et al. 2003). *Globorotalia hirsuta* and *G. truncatulinoides* (cf. Chapter 2, Plate 2.29-1 to -4) grown below 10 °C, and *N. dutertrei* grown below 15 °C produce a

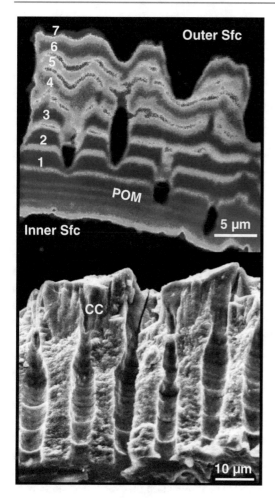

Fig. 6.7 *Upper panel* NanoSIMS image of a post-gametogenic spherical test of *O. universa*. Cross-section of test-wall showing seven high (*bright bands*) and low (*dark bands*) Mg banding pairs on the outer surface of the POM, reflecting 7 days of growth in the laboratory. Closely spaced Mg bands can be seen on the inner side of the POM. Both the low and high Mg bands near the outer test surface contain elevated Mg relative to earlier bands near the POM. Voids in the image are oblique cuts through pores. *Sfc* is the sphere surface. After Spero et al. (2015). *Lower panel* Cross section of test wall of *G. menardii* with nine calcite layers beneath calcite crust (CC).

calcite crust identical to that observed in field collected specimens (Hemleben and Spindler 1983). Those experiments confirm that changes in water temperature and accompanied changes in ambient water chemistry may lead to secretion of

calcite and formation of a secondary layer of calcite on the distal test wall (see Chap. 9). Additional overgrowth may occur on top of tests when buried in surface sediments (Lohmann 1995). Differences in stable isotope and metal/Ca (Me/Ca) ratios confirm changes between the inner and outer test wall deposited under varying taphonomic conditions including water depths and calcification temperatures (e.g., Douglas and Savin 1978; Duckworth 1977; Bé 1980; Duplessy et al. 1981; Blanc and Bé 1981; Eggins et al. 2003; Ripperger et al. 2008; Katz et al. 2010).

During gametogenesis, prior to gamete release, a substantial amount of calcite may be secreted on the outer test wall forming, for example, inter-pore ridges or smooth veneers of calcite (gametogenic calcification), and resulting in generally thicker walls than in pre-gametogenic tests. In turn, resorption of the spines and chamber wall may result in a loss of original test diameter and calcite mass (see Chap. 5). Consequently, these small and dense individuals lose buoyancy and sink towards the seafloor. This explains why thick-walled gametogenic tests of *G. sacculifer* (and other species) are rarely sampled from surface waters, but are frequently part of sedimentary assemblages (e.g., Berelson 2002; Schiebel 2002) (see Chap. 8).

6.6 Biomineralization and Test Calcite Mass

Calcite mass and original planktic foraminifer test weight (in addition to test size) is coupled to the overall marine carbonate turnover, and hence to the parameters, which affect the marine carbonate system at the individual to global scale (Zeebe and Wolf-Gladrow 2001; Schiebel 2002; Bentov et al. 2009). We here present current hypotheses on biomineralization in benthic and planktic foraminifers.

Test mass of different planktic foraminifer species is positively affected by temperature, pH, $[Ca^{2+}]$, and $[CO_3^{2-}]$, and total alkalinity of ambient seawater, as well as oxygen

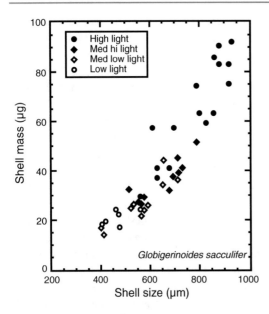

Fig. 6.8 The symbiont-bearing species *G. sacculifer* secretes larger and more massive tests when maintained under high irradiance levels than individuals grown under lower light levels. From Spero and Lea (1993)

concentration (Kuroyanagi et al. 2013), and symbiont activity (e.g., Spero and Lea 1993) (Fig. 6.8). Concentration profiles of O_2, CO_2, and the derived $[CO_3^{2-}]$, and $[HCO_3^-]$ measured within a thin layer (250–800 μm) of seawater at the outside of the test of *O. universa* show minimum and maximum *p*H during light and dark conditions, respectively (Köhler-Rink and Kühl 2005). Concentration changes within this layer are possibly affected by symbiont activity, and may change over very short time-intervals of minutes (Köhler-Rink and Kühl 2005). How symbiotic activity and changes in the chemical microenvironment affect Ca^{2+} uptake is so far not entirely understood (Köhler-Rink and Kühl 2005).

Hamilton et al. (2008) suspect that gametogenic (GAM, see Chap. 5) calcite in *O. universa* is formed through the release of Ca^{2+} or alkalinity from a not yet identified 'cytoplasmic pool' (cf. Anderson and Faber 1984; Erez 2003). In addition to active calcite production, the formation of calcite crusts, and dissolution while settling through the subsurface water column affects calcite mass of planktic foraminifer tests (see Chap. 8).

Biomineralization in planktic foraminifers is assumed to be extracellular, in spaces bounded by cytoplasmic extensions of the main cell mass, in analogy to calcite formation in the symbiont-bearing benthic foraminifer *Amphistegina lobifera* (Bentov et al. 2009). In contrast to planktic foraminifers, biomineralization in, for example, miliolid (i.e. benthic, imperforate) foraminifers and coccolithophores occurs in intracellular structures. Miliolid foraminifers precipitate needle-like calcite crystals within cytoplasmic vesicles. The crystals are then transported by pseudopodia to the site of shell formation and placed there without orientation according to crystal axis (Berthold 1976; Angell 1980; Hemleben et al. 1986). Coccoliths are formed within cell organelles called 'coccolith forming vesicles' (e.g., Brownlee and Taylor 2004). In planktic foraminifers, seawater is assumed engulfed in vacuoles by the plasma membrane that encloses the individual, a process called endocytosis (Fig. 6.9). The size of vacuoles ranges at some ten micrometers, and residence times of vacuoles are reported <1 h in the benthic foraminifer *A. lobifera* (Bentov et al. 2009). The implication that vacuolization may be a discontinuous process, and the question if and when vacuoles are open or closed systems poses important questions for the interpretation of isotope and element ratios in foraminifer test $CaCO_3$ (Elderfield et al. 1996).

Calcification in planktic foraminifers is assumed to be affected by carbon from a cytoplasmic carbon pool similar to the benthic foraminifer *A. lobifera* over short time-intervals (Ter Kuile 1991; Ter Kuile and Erez 1991). Mitochondria and acidic vesicles are assumed to increase the inorganic carbon pool (including CO_2) from which carbonate ions are obtained (Fig. 6.9). An effect of the cytoplasmic enrichment of CO_2 on $\delta^{13}C$ values of the planktic foraminifer test calcite could not be verified (Hamilton et al. 2008, for *O. universa*).

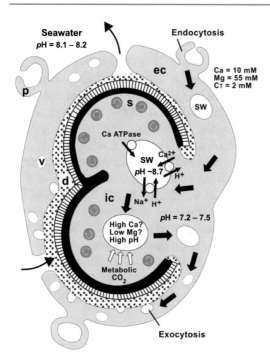

Fig. 6.9 Schematic view of seawater vacuolization, and secondary calcification (**d**, *dotted area*) in perforate foraminifers. The primary test wall (*solid black*) was produced during chamber formation. A first additional layer is given in dashed signature. Seawater (*sw*) vacuoles (*v*) are formed by endocytosis (*ec* is extralocular cytoplasm, *ic* is intralocular cytoplasm). Total carbon (C_T) including CO_2, and Ca^{2+} are exchanged between the vacuoles and the cytoplasm, and Ca^{2+} and $[CO_3^{2-}]$ are enhanced in concentration, supported by light-dependent (photosynthesis, symbionts) Ca^{2+}-ATPase. Metabolic processes and symbiotic algae (S) affect pH changes within microenvironments, as well as Mg^{2+} and Ca^{2+} concentration. Modified seawater vacuoles are exocytosed into the delimited biomineralization space, excess Ca^{2+} and CO_3^{2-} ions are supplied to the cell membrane, and $CaCO_3$ is precipitated over the existing shell. *p* are pseudopodia. From experiments with benthic foraminifers (Erez 2003; see also Bentov et al. 2009)

In contrast, an effect of the cytoplasmic Ca^{2+} pool was shown for *G. sacculifer*, and may contribute to an average increase of 11 % of the final test calcite mass (Anderson and Faber 1984).

Differences between the inorganic carbon pool in symbiont-barren and symbiont-bearing planktic foraminifers are likely, since competition for CO_2 between the foraminifer and symbionts occurs only in the latter. Calcification

mechanisms are shown to be similar in various symbiont-barren benthic foraminifer species and even at the systematic level of sub-orders (de Nooijer et al. 2008), and may be similar also in planktic foraminifers. However, observations on benthic foraminifers may not be representative for planktic foraminifers. In contrast to benthic foraminifers, planktic foraminifers bear fibrillar bodies, which are hypothesised to play a role in biomineralization, although of inconclusive functional significance (Spero 1988). Fibrillar bodies may serve as floating devices by decreasing the density of planktic foraminifer individuals, and enhancing their buoyancy (Hansen 1975; Anderson and Bé 1976a, b; Hemleben et al. 1989) (see Chap. 3).

Deposition of $CaCO_3$ starts with the formation of rather soluble high-Mg granules within cell membrane-bound spaces, so far observed in symbiont-bearing *G. sacculifer*, *G. ruber,* and *O. universa* (Erez 2003). Concentric microspherulites of 1–5 µm size then grow to 20–60 µm sized spherules. Microspherulites were observed in intracellular and extracellular locations, and are composed of platelets, which then form the primary calcite of a new chamber wall (for more information on planktic and benthic foraminifers see Erez 2003). In a second step, low-Mg $CaCO_3$ (i.e. 'secondary calcite', not to be confused with secondary calcite crusts) is precipitated and a massive test wall is formed. However, seawater vacuolization, endocytosis, exocytosis, and precipitation of calcite from those relatively small volumes of seawater of low calcium concentration does not explain the quantity of ions needed for calcification of a test wall over a time-period of several hours (de Nooijer et al. 2009, 2014; Nehrke et al. 2013).

In addition to endocytosis, transmembrane ion transport would sufficiently explain the quantity of calcium needed for test wall calcification in planktic (Lea et al. 1995) and benthic foraminifers (Toyofuku et al. 2008; Nehrke et al. 2013; de Nooijer et al. 2014). Calcium ions (Ca^{2+}) and dissolved inorganic carbon (DIC) are concentrated by transmembrane ion transport from vacuolized seawater, which forms a secluded site of calcification separated from the surrounding seawater (e.g., Erez 2003; Bentov et al. 2009). The process has been experimentally shown for the benthic foraminifer *Amphistegina* (Nehrke et al. 2013), and, by analogy, assumed feasible in planktic foraminifers. The process of transmembrane ion transport has been shown to fractionate against magnesium ions, and explains the low-Mg test calcite of planktic foraminifers (de Nooijer et al. 2014). In the model of de Nooijer et al. (2014), magnesium is assumed actively removed during the process of calcification. Ion pumps are proposed to facilitate the exchange of ions, and to transport the different ions from the seawater to the cytoplasm or vice versa, and to be responsible for the fractionation of, for example, Ca^{2+} and Mg^{2+} ions (Bentov and Erez 2006; de Nooijer et al. 2014, and references therein).

Although symbionts and metabolic processes are involved in pH regulation during calcification, CO_2 concentration of ambient seawater negatively affects calcification rates of foraminifers, which has a negative feedback on $[CO_2]$ counteracting ocean acidification (e.g., Barker and Elderfield 2002; Jansen et al. 2002; Erez 2003; de Moel et al. 2009; Moy et al. 2009; see also Glas et al. 2012). A combination of the effects and processes given above, including positive and negative feedbacks, is recorded by the chemical composition of planktic foraminifer tests, and may provide powerful proxies that can be used in paleoceanography. An additional source of calcite deposited on the surface of the test wall is produced by gametogenic (GAM) calcification, which is a specimen-specific feature, and may be absent or present to varying degrees in planktic foraminifers (see Chap. 5).

Size-normalized planktic foraminifer test-weight has been discussed as a (paleo-) proxy of the carbonate ion concentration ($[CO_3^{2-}]$) of ambient seawater (e.g., Broecker and Clark 2002; Bijma et al. 2002; Barker and Elderfield 2002; Bassinot et al. 2004). The quantitative effect (related to $[CO_3^{2-}]$) on calcite precipitation is species-specific, and may be positive or negative as shown for *G. bulloides* (i.e. heavier test at increasing $[CO_3^{2-}]$) and *G. ruber* (i.e. lighter tests), respectively, collected from the Arabian

Sea (Beer et al. 2010a) (Fig. 6.10). In *G. bulloides* from the SW Pacific Ocean, the effect of $[CO_3^{2-}]$ on test weight seems to be reversed, and parameters other than $[CO_3^{2-}]$ may affect test weight under varying environmental conditions (Gonzalez-Mora et al. 2008; Marr et al. 2011). While net calcification in *N. pachyderma* decreased under laboratory conditions at low-pH (7.8 vs. 8.1) and low-temperature (1 °C), calcification was unaffected at a temperature of 4 °C (Manno et al. 2012). In addition, juvenile and adult *N. pachyderma* were shown to differentially react to low-pH conditions with 30 and 20 % decrease in calcification, respectively.

While *G. bulloides* is symbiont-barren and *G. ruber* bears symbionts, the absence or presence of symbionts may be assumed to

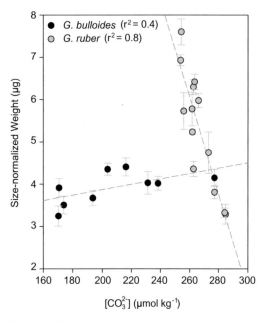

Fig. 6.10 Size-normalized test weights of *G. bulloides* (*black circles*) and *G. ruber* white (*gray circles*) show opposite relationships with carbonate ion concentration in surface waters (<60 m water depth) of Arabian Sea. Samples were obtained from transects across the upwelling area off Oman to the oligotrophic waters of the central Arabian Sea during spring and summer 1995 and 1997 (RV Meteor cruise 32-5, and RV Sonne cruise 120, respectively). Error bars are based on reciprocal numbers of specimens weighed per aliquot multiplied by mean test weight. From Beer et al. (2010a)

quantitatively affect calcification through pH changes within microenvironments in proximity to the shell surface (see Köhler-Rink and Kühl 2005). Test weight in symbiont-bearing *O. universa* has been shown by culture experiments to be related to $[CO_3^{2-}]$ (Bijma et al. 1999). Since both temperature and $[CO_3^{2-}]$ positively affect the size-normalized test weight, and occur in combination with other variables (e.g., depth habitat and pressure) affecting the marine carbonate system, the two effects are difficult to disentangle in naturally grown individuals (cf. Bijma et al. 2002; Zeebe and Wolf-Gladrow 2001).

Size-normalized   test-weight: The size-normalized test-weight (SNW) is the size-related measure of test wall thickness and density (i.e. calcite mass) of a planktic foraminifer test. Different methods employed to establish SNW are sieving (i.e. sieve size) or discrete individual test-size measurement (i.e. discrete size). Analysis of sieved tests is less time consuming than discrete individual measurement, but associated with an average analytical error of up to ± 11 % (Beer et al. 2010b). Accurate and precise morphometric data on species-specific test size and weight, produced with an automated image analysis system provides paleoceanographic information on past changes in ambient seawater $[CO_3^{2-}]$, and other environmental conditions affecting calcification of the planktic foraminifer test (Beer et al. 2010a).

Ocean acidification, OA: The term ocean acidification (OA) describes decreasing seawater pH (e.g., Intergovernmental Panel on Climate Change 2013). Modern OA from surface seawater pH 8.2 to pH 8.1 over the course of decades results from increasing atmospheric CO_2 concentration, and CO_2 uptake by the surface ocean. Modern atmospheric CO_2 increase mainly results from anthropogenic burning of fossil fuels. OA is mediated by, for example, reduced calcification of the marine calcareous plankton including planktic foraminifers, resulting in reduced shell calcite mass. Increasing CO_2 concentration and decreasing pH may threaten the production planktic foraminifers (see also Sects. 9.4 and 10.14).

6.7 Resorption and Repair Processes

Calcite resorption occurs in all planktic foraminifer species, usually during gametogenesis (see Chap. 5). Spines are shed (in spinose species), and septa may be resorbed (e.g., in *H. pelagica*, Plate 5.3 and -6). Resorption also occurs during chamber formation when the inner spine segments projecting from the surface of the previous chamber are resorbed (e.g., Rhumbler 1911). Thick masses of granular cytoplasm may form at the site of calcite resorption, and may be involved in the resorption process by locally lowering pH. Resorption of spines commences with the formation of notches on the surface of the spine. Notches (Plate 6.8-1) progressively enlarge and deepen until the shaft of the spine is fully dissolved. When a spine is sufficiently thin at the site of resorption, a rhizopodium attaches to the spine segment, rotates it away from the test, and by rhizopodial extension, carries it to the periphery of the individual where it is discarded.

Repair of the test wall after resorption or mechanical damage caused by sampling or predators, is remarkably efficient. Specimens with broken spines collected by plankton tows, produce new spines overnight if they are sufficiently cleaned from adhering debris. Spine regeneration (Plate 6.8-2 and -3) in healthy individuals is complete and leaves little or no detectable evidence of resorption.

In *O. universa*, parts of the wall of the inner spiral test and the outer spherical chamber may be entirely resorbed (Rhumbler 1911; Hemleben et al. 1979), and the spherical chamber may be

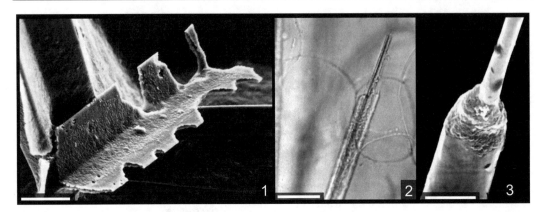

Plate 6.8 (*1*) Partially resorbed spine of *H. pelagica*. (*2*) Regrowth of broken spine in *H. pelagica*. (*3*) Regrown broken spine of *G. ruber*. Bars (*1*) 10 μm, (*2*) 50 μm, (*3*) 2 μm

successively reconstructed (Hemleben and Spindler 1983). Partially resorbed chambers can be either repaired, or more typically, completely resorbed before a new chamber is produced, leaving no evidence of the former chamber. Even if the test is partly crushed, regeneration and repair is possible if the nucleus-containing part of the cytoplasm is not harmed (Bé and Spero 1981). The pieces of a crushed test are rearranged by cytoplasmic streaming. The broken edges are fused by calcite deposition to create a repaired test, which resembles the original one, depending on the degree of damage. Ragged edges of old wall fragments may project from the surface of the repaired test where reassembly was imperfect. Tests with signs of repair, however rare, may indicate physical stress and predation in fossil specimens (Bé and Spero 1981).

6.8   Organic Composition of the Test Wall

Organic matter is included in the calcareous tests of planktic foraminifer species to different degrees (see above). The test wall of *H. pelagica*, for example, includes more organic matter than other spinose species. Higher concentrations of organic matter might result in higher dissolution susceptibilities of tests (Hemleben et al. 1989).

The amino acid composition of fossil and live planktic foraminifer tests confirms the presence of proteinaceous material, and may add information on the phylogeny and classification at the species level (King and Hare 1972a, b). Three informal groups of 16 species analyzed for their composition of amino acid (AA) are rich in different combinations of AA: (1) alanine, proline, and valine, (2) aspartic acid, and threonine, and (3) glycine, serine, and glutamic acid. Variability of amino acids is greater in spinose than in non-spinose species. The overall amino acid composition of tests resembles the classification of species according to test morphology. The amino acid composition of modern planktic foraminifers is more similar to benthic foraminifers and coccoliths than to radiolarians and diatoms. The siliceous groups produce significantly higher aspartic acid-to-glycine ratios than the calcareous groups (King 1977). The average quantity of 17 amino acids analyzed in planktic foraminifers is 2523 nmol per gram of test wall material, i.e. about 0.025 % by weight (King 1977; benthic foraminifers 2546 nm g^{-1}, radiolarians 1439 nm g^{-1}, coccoliths 1267 nm g^{-1}, diatoms 1138 nm g^{-1}).

The nature of the red pigment in tests of *G. ruber* pink and *G. rubescens* is not yet known. The pigment concentration in the *G. ruber* pink is extremely low, and could not be analyzed in one thousand tests >200 μm (G. Trommer, personal communication, 2011). Modern *G. ruber* pink occur only in the Atlantic Ocean and marginal seas, and have been absent from the Indian and Pacific Oceans for about 125 k years, i.e. following the last interglacial, MIS 5.5 (Thompson et al. 1979).

6.9 Summary and Concluding Remarks

Ontogenetic development of species under varying environmental conditions, and test architecture of the juvenile to mature individual, is of fundamental importance for the understanding and application of planktic foraminifers in ocean and climate research. Spinose and non-spinose species follow similar ontogenetic stages from (1) proloculus, to (2) juvenile, (3) neanic, (4) adult, and (5) terminal stage. Structure of the shell (i.e. calcite and organic layers), and architectural characteristics (i.e. spines, pustules, keels, pores) add information on discriminations among different taxonomic levels. From the size-distribution of proloculi, and from laboratory observation foraminifers are assumed to predominantly reproduce sexually. Coiling direction of the test is genetically determined.

Chamber formation occurs within a cytoplasmic envelope produced by rhizopodia, which also secrete the primary organic membrane (POM). The POM is the site of calcite deposition. Shell calcite deposition is affected by chemical (e.g., $[CO_3^{2-}]$) and physical (e.g., temperature) conditions of ambient seawater, as well as autecological characteristics such as the absence or presence of symbionts. Biomineralization in planktic foraminifers is assumed extracellular, by deposition in cytoplasmic-bound spaces outside of the main cell mass, and sequestered from seawater that is engulfed in vacuoles by invagination of the plasma membrane enclosing the individual (i.e. endocytosis). Calcification is possibly supported by carbon from a cytoplasmic carbon pool. Transmembrane ion transport may provide the quantity of calcium needed for calcification of the shell, beyond the calcium engulfed from seawater. Calcite precipitation during each chamber formation produces only minor amounts of the entire shell calcite mass. The major part of the test calcite is deposited by additional calcification on the test surfaces. Thickening of the adult shell may result from diurnal precipitation of micron-scale calcite layers, and gametogenic calcification, both of which processes vary between species. Diurnal calcification is assumed supported by mitochondrial activity. Euhedral calcite crystals, and a smooth calcite veneer may be produced on the outer test surface, and add substantial amounts of calcite to the test wall. In addition to active calcite precipitation, the formation of calcite crusts and dissolution during sedimentation affect the calcite mass of planktic foraminifer tests.

Damaged shells and spines can be repaired by cytoplasmic activity resulting in partial rearrangement of shell and restoration by calcification. Resorption of calcite and associated organic matter is assumed integral part of the foraminifer's metabolic cycle during gametogenesis and under unfavorable environmental conditions.

Finally, modern analytical methods facilitate detailed analyses of a wide range of chemical, structural, and morphometric characteristics of planktic foraminifer tests. However, most of the biological and environmental factors affecting calcite precipitation and planktic foraminifer test formation, and their use as proxies in paleoceanography and paleoclimate research, have not yet been fully determined.

References

Anderson OR, Bé AWH (1976a) The ultrastructure of a planktonic foraminifer, *Globigerinoides sacculifer* (Brady), and its symbiotic dinoflagellates. J Foram Res 6:1–21. doi:10.2113/gsjfr.6.1.1

Anderson OR, Bé AWH (1976b) A cytochemical fine structure study of phagotrophy in a planktonic foraminifer, *Hastigerina pelagica* (d'Orbigny). Biol Bull 151:437–449

Anderson OR, Faber WW (1984) An estimation of calcium carbonate deposition rate in a planktonic foraminifer *Globigerinoides sacculifer* using ^{45}Ca as a tracer; a recommended procedure for improved accuracy. J Foram Res 14:303–308. doi:10.2113/gsjfr.14.4.303

Angell RW (1967) The test structure and composition of the foraminifer *Rosalina floridana*. J Protozool

14:299–307. doi:10.1111/j.1550-7408.1967.tb02001.x

Angell RW (1980) Test morphogenesis (chamber formation) in the foraminifer *Spiroloculina hyalina* Schulze. J Foram Res 10:89–101. doi:10.2113/gsjfr.10.2.89

Barker S, Elderfield H (2002) Foraminiferal calcification response to glacial-interglacial changes in atmospheric CO_2. Science 297:833–836. doi:10.1126/science. 1072815

Bassinot FC, Mélières F, Gehlen M, Levi C, Labeyrie L (2004) Crystallinity of Foraminifera shells: a proxy to reconstruct past bottom water CO_3^{2-}-changes? Geochem Geophys Geosyst 5:Q08D10. doi 10.1029/2003GC000668

Bé AWH (1965) The influence of depth on shell growth in *Globigerinoides sacculifer* (Brady). Micropaleontology 11:81–97

Bé AWH (1968) Shell porosity of recent planktonic Foraminifera as a climatic index. Science 161:881–884. doi:10.1126/science.161.3844.881

Bé AWH (1980) Gametogenic calcification in a spinose planktonic foraminifer, *Globigerinoides sacculifer* (Brady). Mar Micropaleontol 5:283–310. doi:10. 1016/0377-8398(80)90014-6

Bé AWH, Hemleben C (1970) Calcification in a living planktonic foraminifer, *Globigerinoides sacculifer* (Brady). Neues Jahrb Geol Paläontol 134:221–234

Bé AWH, Spero HJ (1981) Shell regeneration and biological recovery of planktonic Foraminifera after physical injury induced in laboratory culture. Micropaleontology 27:305–316

Bé AWH, McIntyre A, Breger DL (1966) Shell microstructure of a planktonic foraminifer, *Globorotalia menardii* (d'Orbigny). Eclogae Geol Helvetiae 59:885–896

Bé AWH, Jongebloed WL, McIntyre A (1969) X-ray microscopy of recent planktonic Foraminifera. J Paleontol 43:1384–1396

Bé AWH, Hemleben C, Anderson OR, Spindler M (1979) Chamber formation in planktonic Foraminifera. Micropaleontology 294–307

Bé AWH, Hemleben C, Anderson OR, Spindler M (1980) Pore structures in planktonic Foraminifera. J Foram Res 10:117–128. doi:10.2113/gsjfr.10.2.117

Bé AWH, Spero HJ, Anderson OR (1982) Effects of symbiont elimination and reinfection on the life processes of the planktonic foraminifer *Globigerinoides sacculifer*. Mar Biol 70:73–86. doi:10.1007/BF00397298

Bé AWH, Bishop JKB, Sverdlove MS, Gardner WD (1985) Standing stock, vertical distribution and flux of planktonic Foraminifera in the Panama Basin. Mar Micropaleontol 9:307–333

Beer CJ, Schiebel R, Wilson PA (2010a) Testing planktic foraminiferal shell weight as a surface water [CO_3^{2-}] proxy using plankton net samples. Geology 38:103–106. doi:10.1130/G30150.1

Beer CJ, Schiebel R, Wilson PA (2010b) Technical note: on methodologies for determining the size-normalised weight of planktic Foraminifera. Biogeosciences 7:2193–2198. doi:10.5194/bg-7-2193-2010

Benjamini C, Reiss Z (1979) Wall-hispidity and -perforation in Eocene planktonic Foraminifera. Micropaleontology 25:141–150

Bentov S, Erez J (2006) Impact of biomineralization processes on the Mg content of foraminiferal shells: a biological perspective. Geochem Geophys Geosyst. doi:10.1029/2005GC001015

Bentov S, Brownlee C, Erez J (2009) The role of seawater endocytosis in the biomineralization process in calcareous Foraminifera. Proc Natl Acad Sci 106:21500–21504. doi:10.1073/pnas.0906636106

Berelson WM (2002) Particle settling rates increase with depth in the ocean. Deep-Sea Res II 49:237–251. doi:10.1016/S0967-0645(01)00102-3

Berger WH (1970) Planktonic Foraminifera: differential production and expatriation off Baja California. Limnol Oceanogr 15:183–204. doi:10.4319/lo.1970.15.2.0183

Berthold WU (1976) Biomineralisation bei milioliden Foraminiferen und die Matritzen-Hypothese. Naturwissenschaften 63:196–197

Bijma J, Spero HJ, Lea DW (1999) Reassessing foraminiferal stable isotope geochemistry: impact of the oceanic carbonate system (experimental results). In: Fischer G, Wefer G (eds) Use of proxies in paleoceanography. Springer, Berlin, pp 489–512

Bijma J, Hönisch B, Zeebe RE (2002) Impact of the ocean carbonate chemistry on living foraminiferal shell weight: comment on "Carbonate ion concentration in glacial-age deep waters of the Caribbean Sea" by W. S. Broecker and E Clark. Geochem Geophys Geosystems 3:1–7. doi:10.1029/2002GC000388

Bishop JK, Collier RW, Kettens DR, Edmond JM (1980) The chemistry, biology, and vertical flux of particulate matter from the upper 1500 m of the Panama Basin. Deep-Sea Res I 27:615–640

Blanc P, Bé AWH (1981) Oxygen-18 enrichment of planktonic Foraminifera due to gametogenic calcification below the euphotic zone. Science 213:1247–1250

Boltovskoy E (1982) Twinned and flattened tests in planktonic Foraminifera. J Foram Res 12:79–82

Broecker WS, Clark E (2001) Glacial-to-Holocene redistribution of carbonate ion in the deep sea. Science 294:2152–2155. doi:10.1126/science.1064171

Broecker WS, Clark E (2002) Carbonate ion concentration in glacial-age deep waters of the Caribbean Sea. Geochem Geophys Geosyst 3:1–14. doi:10.1029/2001GC000231

Brönnimann P (1950) Occurrence and ontogeny of *Globigerinatella insueta* Cushman and Stainforth from the Oligocene of Trinidad. BWI Contrib Cushman Found Foram Res 1:80–82

Brönnimann P, Brown NK (1956) Taxonomy of the Globotruncanidae. Eclogae Geol Helvetiae 48:503–561

Brownlee C, Taylor A (2004) Calcification in coccolithophores: a cellular perspective. In: Thierstein HR, Young JR (eds) coccolithophores. Springer, Berlin, pp 31–49

Brummer GJA (1988) Comparative ontogeny of modern microperforate planktonic foraminifers. In:

Brummer GJA, Kroon D (eds) Planktonic Foraminifers as tracers of ocean-climate history. Free University Press, Amsterdam, pp 77–129

Brummer GJA, Hemleben C, Spindler M (1986) Planktonic foraminiferal ontogeny and new perspectives for micropalaeontology. Nature 319:50–52. doi:10.1038/319050a0

Brummer GJA, Hemleben C, Spindler M (1987) Ontogeny of extant spinose planktonic Foraminifera (Globigerinidae): a concept exemplified by *Globigerinoides sacculifer* (Brady) and *G. ruber* (d'Orbigny). Mar Micropaleontol 12:357–381. doi:10.1016/0377-8398(87)90028-4

Caron DA, Bé AWH, Anderson OR (1982) Effects of variations in light intensity on life processes of the planktonic foraminifer *Globigerinoides sacculifer* in laboratory culture. J Mar Biol Assoc UK 62:435–451

Caron DA, Faber WW, Bé AWH (1987) Effects of temperature and salinity on the growth and survival of the planktonic foraminifer *Globigerinoides sacculifer*. J Mar Biol Assoc U K 67:323–341

Darling KF, Kucera M, Kroon D, Wade CM (2006) A resolution for the coiling direction paradox in *Neogloboquadrina pachyderma*. Paleoceanography. doi:10.1029/2005PA001189

De Moel H, Ganssen GM, Peeters FJC, Jung SJA, Kroon D, Brummer GJA, Zeebe RE (2009) Planktic foraminiferal shell thinning in the Arabian Sea due to anthropogenic ocean acidification? Biogeosciences 6:1917–1925. doi:10.5194/bg-6-1917-2009

De Nooijer LJ, Toyofuku T, Oguri K, Nomaki H, Kitazato H (2008) Intracellular pH distribution in Foraminifera determined by the fluorescent probe HPTS. Limnol Oceanogr—Methods 6:610–618

De Nooijer LJ, Langer G, Nehrke G, Bijma J (2009) Physiological controls on seawater uptake and calcification in the benthic foraminifer *Ammonia tepida*. Biogeosciences 6:2669–2675. doi:10.5194/bg-6-2669-2009

De Nooijer LJ, Spero HJ, Erez J, Bijma J, Reichart GJ (2014) Biomineralization in perforate Foraminifera. Earth-Sci Rev 135:48–58

Desai D, Banner FT (1985) The ontogeny of, and relationships between, middle Miocene and quaternary *Orbulina* (Foraminifera). J Micropalaeontol 4:81–91

Douglas RG, Savin SM (1978) Oxygen isotopic evidence for the depth stratification of tertiary and cretaceous planktic Foraminifera. Mar Micropaleontol 3:175–196

Duckworth DL (1977) Magnesium concentration in the tests of the planktonic foraminifer *Globorotalia truncatulinoides*. J Foram Res 7:304–312

Duplessy JC, Bé AWH, Blanc PL (1981) Oxygen and carbon isotopic composition and biogeographic distribution of planktonic Foraminifera in the Indian Ocean. Palaeogeogr Palaeoclimatol Palaeoecol 33:9–46

Eggins S, De Dekker P, Marshall J (2003) Mg/Ca variation in planktonic Foraminifera tests: Implications for reconstructing palaeo-seawater temperature and habitat migration. Earth Planet Sci Lett 212:291–306

Eggins S, Sadekov A, De Deckker P (2004) Modulation and daily banding of Mg/Ca in *Orbulina universa* tests by symbiont photosynthesis and respiration: a complication for seawater thermometry? Earth Planet Sci Lett 225:411–419

Elderfield H, Bertram CJ, Erez J (1996) A biomineralization model for the incorporation of trace elements into foraminiferal calcium carbonate. Earth Planet Sci Lett 142:409–423

Erez J (1983) Calcification rates, photosynthesis and light in planktonic Foraminifera. In: Westbroek P, de Jong EW (eds) Biomineralization and biological metal accumulation. Reidel Publishing Company, Dordrecht, pp 307–312

Erez J (2003) The source of ions for biomineralization in Foraminifera and their implications for paleoceanographic proxies. Rev Mineral Geochem 54:115–149

Ericson DB, Wollin G, Wollin J (1954) Coiling direction of *Globorotalia truncatulinoides* in deep-sea cores. Deep-Sea Res 2:152–158

Frerichs WE, Heiman ME, Borgman LE, Bé AWH (1972) Latitudinal variations in planktonic foraminiferal test porosity: Part 1. Optical studies. J Foram Res 2:6–13

Gagnon A, de Yoreo J, de Paolo D, Spero HJ, Russell AD, Giuffre A (2012) Me/Ca proxies and foram biomineralization: the role of cation transport. Mineral Mag 76:1730

Gehlen M, Bassinot F, Beck L, Khodja H (2004) Trace element cartography of *Globigerinoides ruber* shells using particle-induced X-ray emission. Geochem Geophys Geosyst. doi:10.1029/2004GC000822

Glas MS, Langer G, Keul N (2012) Calcification acidifies the microenvironment of a benthic foraminifer (*Ammonia* sp.). J Exp Mar Biol Ecol 424–425:53–58. doi:10.1016/j.jembe.2012.05.006

Gonzalez-Mora B, Sierro FJ, Flores JA (2008) Controls of shell calcification in planktonic foraminifers. Quat Sci Rev 27:956–961. doi:10.1016/j.quascirev.2008.01.008

Hamilton CP, Spero HJ, Bijma J, Lea DW (2008) Geochemical investigation of gametogenic calcite addition in the planktonic Foraminifera *Orbulina universa*. Mar Micropaleontol 68:256–267. doi:10.1016/j.marmicro.2008.04.003

Hansen HJ (1975) On feeding and supposed buoyancy mechanism in four recent globigerinid Foraminifera from the Gulf of Elat, Israel. Rev Esp Micropaleontol 7:325–339

Hemleben C (1969a) Ultramicroscopic shell and spine structure of some spinose planktonic Foraminifera. In: Brönniman P, Renz HH (eds) Proceedings of 1st international conference planktonic microfossils. Leiden, pp 254–256

Hemleben C (1969b) Zur Morphogenese planktonischer Foraminiferen. In: Zitteliana. pp 91–133

Hemleben C (1975) Spine and pustule relationships in some recent planktonic Foraminifera. Micropaleontology 21:334–341

Hemleben C, Spindler M (1983) Recent advances in research on living planktonic Foraminifera. Utrecht Micropaleontol Bull 30:141–170

Hemleben C, Bé AWH, Anderson OR, Tuntivate S (1977) Test morphology, organic layers and chamber formation of the planktonic foraminifer *Globorotalia menardii* (d'Orbigny). J Foram Res 7:1–25

Hemleben C, Bé AWH, Spindler M, Anderson OR (1979) "Dissolution" effects induced by shell resorption during gametogenesis in *Hastigerina pelagica* (d'Orbigny). J Foram Res 9:118–124

Hemleben C, Spindler M, Breitinger I, Deuser WG (1985) Field and laboratory studies on the ontogeny and ecology of some globorotaliid species from the Sargasso Sea off Bermuda. J Foram Res 15:254–272

Hemleben C, Anderson OR, Berthold W, Spindler M (1986) Calcification and chamber formation in Foraminifera—a brief overview. In: Riding R, Leadbeater BSC (eds) Biomineralization in lower plants and animals: The Systematics Association. Clarendon Press, Oxford, pp 237–249

Hemleben C, Spindler M, Anderson OR (1989) Modern planktonic Foraminifera. Springer, Berlin

Hemleben C, Olsson RK, Premec-Fucek V, Hernitz-Kucenjac M (2017) Wall textures of Oligocene normal perforate planktonic Foraminifera. Cushman Found Spec Publ, Chapter 3

Hofker J (1971) Wall-structure of globigerine and globorotaliid Foraminifera. Rev Espanola Micropaleontol 3:35–60

Huang CY (1981) Observations on the interior of some late Neogene planktonic Foraminifera. J Foram Res 11:173–190

Intergovernmental Panel on Climate Change (ed) (2013) Climate change 2013—The physical science basis: working group i contribution to the fifth assessment report of the intergovernmental panel on climate change. Cambridge University Press, Cambridge

Jansen H, Zeebe R, Wolf-Gladrow DA (2002) Modelling the dissolution of settling $CaCO_3$ in the ocean. Glob Biogeochem Cycles 16:1–16

Johnstone HJH, Schulz M, Barker S, Elderfield H (2010) Inside story: an X-ray computed tomography method for assessing dissolution in the tests of planktonic Foraminifera. Mar Micropaleontol 77:58–70. doi:10.1016/j.marmicro.2010.07.004

Katz ME, Cramer BS, Franzese A, Hönisch B, Miller KG, Rosenthal Y, Wright JD (2010) Traditional and emerging geochemical proxies in Foraminifera. J Foram Res 40:165–192

King K (1977) Amino acid survey of recent calcareous and siliceous deep-sea microfossils. Micropaleontology 23:180–193

King K, Hare PE (1972a) Amino acid composition of planktonic Foraminifera: a paleobiochemical approach to evolution. Science 175:1461–1463

King K, Hare PE (1972b) Amino acid composition of the test as a taxonomic character for living and fossil planktonic Foraminifera. Micropaleontology 18:285–293

Köhler-Rink S, Kühl M (2005) The chemical microenvironment of the symbiotic planktonic foraminifer *Orbulina universa*. Mar Biol Res 1:68–78. doi:10.1080/17451000510019015

Kuroyanagi A, da Rocha RE, Bijma J, Spero HJ, Russell AD, Eggins SM, Kawahata H (2013) Effect of dissolved oxygen concentration on planktonic Foraminifera through laboratory culture experiments and implications for oceanic anoxic events. Mar Micropaleontol 101:28–32. doi:10.1016/j.marmicro.2013.04.005

Lea DW, Martin PA, Chan DA, Spero HJ (1995) Calcium uptake and calcification rate in the planktonic foraminifer *Orbulina universa*. J Foram Res 25:14–23

Leutenegger S, Hansen HJ (1979) Ultrastructural and radiotracer studies of pore function in Foraminifera. Mar Biol 54:11–16

Lohmann GP (1995) A model for variation in the chemistry of planktonic Foraminifera due to secondary calcification and selective dissolution. Paleoceanography 10:445–457

Lombard F, da Rocha RE, Bijma J, Gattuso JP (2010) Effect of carbonate ion concentration and irradiance on calcification in planktonic Foraminifera. Biogeosciences 7:247–255

Manno C, Morata N, Bellerby R (2012) Effect of ocean acidification and temperature increase on the planktonic foraminifer *Neogloboquadrina pachyderma* (sinistral). Polar Biol 35:1311–1319. doi:10.1007/s00300-012-1174-7

Marr JP, Baker JA, Carter L, Allan ASR, Dunbar GB, Bostock HC (2011) Ecological and temperature controls on Mg/Ca ratios of *Globigerina bulloides* from the southwest Pacific Ocean. Paleoceanography 26:PA2209. doi 10.1029/2010PA002059

Moy AD, Howard WR, Bray SG, Trull TW (2009) Reduced calcification in modern Southern Ocean planktonic Foraminifera. Nat Geosci 2:276–280. doi:10.1038/ngeo460

Nehrke G, Keul N, Langer G, de Nooijer LJ, Bijma J, Meibom A (2013) A new model for biomineralization and trace-element signatures of Foraminifera tests. Biogeosciences 10:6759–6767. doi:10.5194/bg-10-6759-2013

Norris RD (1991) Parallel evolution in the keel structure of planktonic Foraminifera. J Foram Res 21:319–331

Oelschläger J (1989) Die Ultrastruktur der nicht-mineralisierten Schalenbestandteile bei Foraminiferen unter besonderer Berücksichtigung von *Rotorbinella rosea* (d'Orbigny)(Rotaliidae. Tübinger Mikropaläontologische Mitteilungen, Foraminifera)

Olsson RK (1971) The logarithmic spire in planktonic Foraminifera: its use in taxonomy, evolution, and paleoecology. Trans Gulf Coast Assoc Geol Soc 21st Meet 419–432

Parker FL (1962) Planktonic foraminiferal species in Pacific sediments. Micropaleontology 8:219–254

Pessagno EA, Miyano K (1968) Notes on the wall structure of the Globigerinacea. Micropaleontology 14:38–50

Reiss Z (1957) The Bilamellidea, nov. superfam., and remarks on Cretaceous Globorotaliids. Contrib Cushman Found Foram Res 8:127–145

Rhumbler L (1911) Die Foraminiferen (Thalamorphoren) der Plankton-Expedition. Erster Teil: Die allgemeinen

Organisations-Verhältnisse der Foraminiferen. Ergeb Plankton-Exped Humbold-Stift 1909 3:331

Ripperger S, Schiebel R, Rehkämper M, Halliday AN (2008) Cd/Ca ratios of in situ collected planktonic foraminiferal tests. Paleoceanography. doi:10.1029/2007PA001524

Saito T, Thompson PR, Breger DL (1976) Skeletal ultra-microstructure of some elongate-chambered planktonic Foraminifera and related species. In: Takayanagi Y, Saito T (eds) Selected papers in honor of Prof. Kiyoshi Asano. Progress in Micropaleontology Spec Pub, New York, pp 278–304

Schiebel R (2002) Planktic foraminiferal sedimentation and the marine calcite budget. Glob Biogeochem Cycles. doi:10.1029/2001GB001459

Schiebel R, Hemleben C (2000) Interannual variability of planktic foraminiferal populations and test flux in the eastern North Atlantic Ocean (JGOFS). Deep-Sea Res II 47:1809–1852

Schiebel R, Barker S, Lendt R, Thomas H, Bollmann J (2007) Planktic foraminiferal dissolution in the twilight zone. Deep-Sea Res II 54:676–686

Scott GH (1973a) Peripheral structures in chambers of *Globorotalia scitula praescitula* and some descendants. Rev Española Micropaleontol 5:235–246

Scott GH (1973b) Ontogeny and shape in *Globorotalia menardii*. J Foram Res 3:142–146

Simstich J, Sarnthein M, Erlenkeuser H (2003) Paired $\delta^{18}O$ signals of *Neogloboquadrina pachyderma* (s) and *Turborotalita quinqueloba* show thermal stratification structure in Nordic Seas. Mar Micropaleontol 48:107–125

Spero HJ (1986) Symbiosis, chamber formation and stable isotope incorporation in the planktonic foraminifer *Orbulina universa*. PhD Thesis, University of California

Spero HJ (1988) Ultrastructural examination of chamber morphogenesis and biomineralization in the planktonic foraminifer *Orbulina universa*. Mar Biol 99:9–20

Spero HJ, Bijma J, Lea DW, Bemis BE (1997) Effect of seawater carbonate concentration on foraminiferal carbon and oxygen isotopes. Nature 390:497–500

Spero HJ, Lea DW (1993) Intraspecific stable isotope variability in the planktic Foraminifera *Globigerinoides sacculifer*: results from laboratory experiments. Mar Micropaleontol 22:221–234

Spero HJ, Eggins SM, Russell AD, Vetter L, Kilburn MR, Hönisch B (2015) Timing and mechanism for intratest Mg/Ca variability in a living planktic foraminifer. Earth Planet Sci Lett 409:32–42. doi:10.1016/j.epsl.2014.10.030

Steinhardt J, de Nooijer LLJ, Brummer G-J, Reichart G-J (2015) Profiling planktonic foraminiferal crust formation. Geochem Geophys Geosyst 16:2409–2430. doi 10.1002/2015GC005752

Sverdlove MS, Bé AWH (1985) Taxonomic and ecological significance of embryonic and juvenile planktonic Foraminifera. J Foram Res 15:235–241

Takayanagi Y, Niitsuma N, Sakai T (1968) Wall microstructure of *Globorotalia truncatulinoides* (d'Orbigny). Sci Rep Tohoku Univ Second Ser Geol 40:141–170

Ter Kuile BH (1991) Mechanisms for calcification and carbon cycling in algal symbiont-bearing Foraminifera. In: Lee JJ, Anderson OR (eds) Biology of Foraminifera. Academic Press, London

Ter Kuile BH, Erez J (1991) Carbon budgets for two species of benthonic symbiont-bearing Foraminifera. Biol Bull 180:489–495

Thompson PR, Bé AWH, Duplessy J-C, Shackleton NJ (1979) Disappearance of pink-pigmented *Globigerinoides ruber* at 120,000 yr BP in the Indian and Pacific Oceans. Nature 280:554–558

Towe KM (1971) Lamellar wall construction in planktonic Foraminifera. In: Proceedings of the II Planktonic Conference, Roma 1970. Rome, pp 1213–1218

Towe KM, Cifelli R (1967) Wall ultrastructure in the calcareous Foraminifera: crystallographic aspects and a model for calcification. J Paleontol 41:742–762

Toyofuku T, De Nooijer LJ, Yamamoto H, Kitazato H (2008) Real-time visualization of calcium ion activity in shallow benthic foraminiferal cells using the fluorescent indicator Fluo-3 AM. Geochem Geophy Geosy. doi:10.1029/2007GC001772

Vetter L, Kozdon R, Mora CI, Eggins SM, Valley JW, Hönisch B, Spero HJ (2013) Micron-scale intrashell oxygen isotope variation in cultured planktic foraminifers. Geochim Cosmochim Acta 107:267–278

Vetter L, Kozdon R, Valley JW, Mora CI, Spero HJ (2014) SIMS measurements of intrashell $\delta^{13}C$ in cultured planktic foraminifer *Orbulina universa*. Geochim Cosmochim Acta 139:527–539. doi:10.1016/j.gca.2014.04.049

Weiner A, Aurahs R, Kurasawa A, Kitazato H, Kucera M (2012) Vertical niche partitioning between cryptic sibling species of a cosmopolitan marine planktonic protist. Mol Ecol 21:4063–4073. doi:10.1111/j.1365-294X.2012.05686.x

Zeebe RE, Wolf-Gladrow D (2001) CO_2 in seawater: equilibrium, kinetics, isotopes. Elsevier, Amsterdam

Ecology

## 7

"The dynamic interaction of individual organisms and populations with the physical and biotic components of the marine environment is of central importance in understanding the manifold characteristics of oceanic ecosystems. This includes the productivity of the oceans, the factors governing the distribution and range of organisms in a geographic region, the abundance and fecundity of species, the pattern of energy flow through the marine ecosystem, and the analysis of fossil remains in reconstructing ancient environments and interpreting the history of the Earth. Planktonic Foraminifera are of special significance in the study of modern and ancient marine ecosystems owing to their widespread occurrence in modern oceans, with rather clearly defined faunal provinces for many species, and the fact that they produce calcitic shells that contribute substantially to the micro-fossil faunal record" (Hemleben et al. 1989).

Most of the about 50 extant planktic foraminifer morphospecies are ubiquitous in the global ocean (e.g., Bé 1977; Hemleben et al. 1989). Single genotypes of those morphotypes are more limited to ocean basins and regions (e.g., Darling and Wade 2008). Three modern morphospecies are endemic to the Pacific and Indian Oceans, i.e. *Globigerinella adamsi*, *Globoquadrina conglomerata*, and *Globorotaloides hexagonus*. In addition, certain morphotypes (e.g., *G. sacculifer* forma *immaturus*) are limited to the Pacific and Indian Oceans

(André et al. 2013). The pink variety of *Globigerinoides ruber* has been limited to the modern Atlantic Ocean, and became extinct in the Pacific and Indian Oceans following Marine Isotope Stage (MIS) 5.5 around 125 kyrs (Thompson et al. 1979). The global distributions of some ten small-sized, rare, and dissolution-susceptible species, including *Globorotalia cavernula*, *Gallitellia vivans*, and most tenuitellid species are not well constrained due to under-sampling with plankton tows (usually >100-μm mesh-size) and dissolution during sedimentation. Best documented are the distributions of the ~35 most abundant, large-sized, and dissolution-resistant species, from plankton tows and surface sediment samples. In this chapter, general ecological demands of planktic foraminifers, the effects on shell production, and spatial and temporal distribution patterns are discussed. Particular ecological demands at the species level are discussed with their classification in Chap. 2.

Subtropical and temperate waters harbor the most diverse planktic foraminifer assemblages (e.g., Bé and Tolderlund 1971; Schmidt et al. 2004a; cf. Peters et al. 2013). Patchy distribution patterns of planktic foraminifers on various temporal and spatial scales are caused by small-scale to meso-scale hydrographic features such as fronts and eddies (Boltovskoy 1971; Beckmann et al. 1987; Siccha et al. 2012).

R. Schiebel and C. Hemleben, *Planktic Foraminifers in the Modern Ocean*,
DOI 10.1007/978-3-662-50297-6_7

Hydrology, availability of nutrients in surface waters, and primary production affect the production of planktic foraminifers. Average standing stocks of adult specimens (>100 μm) range from 10 to 100 specimens per cubic meter. Largest standing stocks of ∼190 individuals per liter are reported from Antarctic sea ice (Spindler and Dieckmann 1986), 1250 individuals (>63 μm) per cubic meter occurred in surface to subsurface waters off the ice edge in the Arctic summer (Carstens et al. 1997), and 720 individuals (>100 μm) per cubic meter in the temperate North Atlantic during spring (Schiebel and Hemleben 2000). Those large standing stocks result from high prey availability supporting the production of a wide range of opportunistic species. For example, standing stocks of opportunistic species like *G. bulloides*, *N. dutertrei*, and *N. pachyderma* are positively related to upwelling intensity and eutrophic conditions (e.g., Naidu and Malmgren 1996; Ivanova et al. 1999; Schiebel et al. 2004). In contrast, the largest overall standing stocks in tropical and subtropical waters occur rather marginal than central to major upwelling cells, caused by overall negative effects high primary production, chlorophyll concentration, and turbidity exert through light attenuation on symbiont-bearing species in central upwelling cells (Schiebel et al. 2004). However, the same morphospecies may react to overall similar ecological conditions (e.g., upwelling) in different ways, which may have various reasons. Ecological conditions may differ in detail. For example, the supply of prey may be different in quality and quantity at the spatial and temporal scale. In addition, certain morphospecies may be represented by different genotypes with different ecological adaptations. For example, *N. dutertrei* is positively related to increasing upwelling intensity (early bloom species) in the Arabian Sea (Kroon and Ganssen 1988), whereas it signifies post-upwelling conditions in the San Pedro Basin, NE Pacific (Sautter and Sancetta 1992). The wide (at least) bimodal temperature range and ecological coverage of *N. dutertrei* may indicate the presence of different genotypes (cf. Morard et al. 2015).

Average annual export production of planktic foraminifers is highest in mesotrophic waters in the temperate to subpolar ocean, caused by low average stratification of the surface water column, and frequent nutrient supply. Seasonally enhanced availability of prey during spring and fall fosters production of opportunistic species, and generalist species persist during more stratified and lower productive summer conditions (cf. Schiebel 2002; Žarić et al. 2005). Oligotrophic waters of the subtropical gyres host the lowest standing stocks due to lack of prey (e.g., Bé 1960). However, trophic conditions do not directly translate into standing stocks, and the distribution of planktic foraminifers results from a variety of factors in addition to hydrology and food (Schiebel 2002; Siccha et al. 2009).

Following the most obvious observations, sea surface temperature (SST, surface mixed layer temperature, well documented by discrete measurements and satellite imagery) may affect the distribution of species. The assumption is abundantly pursued in paleoceanography following the temperature effect on the $\delta^{18}O$ signal of planktic foraminifer tests. In turn, a direct affect of SST on the distribution of planktic foraminifer species could not yet be demonstrated, and various temperature-dependent parameters like the quality of prey (e.g., various algae) may be involved. In addition, most planktic foraminifer species are largely eurythermal (Fig. 7.1), and occur over a wide temperature range of 15–20 °C (up to 25 °C, Bé and Tolderlund 1971), with an optimum temperature range of ∼10 °C (Lombard et al. 2011). In addition to alimentation and temperature, salinity is a limiting factor to the distribution of planktic foraminifers. According to results from culture experiments, some species endure a wide salinity range of 20–45 PSU, and are most productive (reproduction rate >70 %) in waters of 33–38 PSU (e.g., Bijma et al. 1990b, 1992).

Practical salinity units, PSU: Salinity of water may be given in practical salinity units (PSU). PSU is used for practical reasons, for example, when deriving

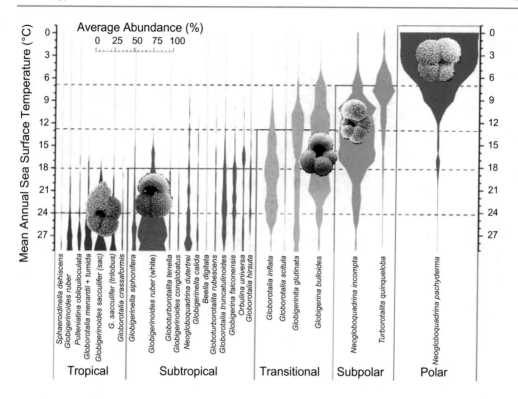

Fig. 7.1 Temperature related distribution of planktic foraminifer species in surface-sediment data from the Atlantic Ocean (Kucera et al. 2005) averaged at one degree centigrade intervals. The relation of species and sea surface temperature (SST) largely resembles the distribution in other ocean basins (Bradshaw 1959; Bé and Tolderlund 1971; Bé 1977; Bé and Hutson 1977; Žarić et al. 2005). The proportions of the major species of the respective assemblages are displayed by colored bars. Subsurface-dwelling *Globorotalia* species merely coincide with the given SSTs, and are possibly affected by ecological parameters related to SST. Modified after Kucera (2007)

seawater salinity from data on electrical conductivity. The more descriptive 'per mil' (‰) unit of seawater salinity is usually very close to PSU. Salinity of seawater typically amounts to 35 g/kg.

In contrast to surface dwelling species, subsurface dwelling species like most globorotalids (Fig. 7.1), are not exposed to sea surface conditions, and hence not affected by, for example, SST. The distribution pattern of subsurface dwellers is possibly limited by the flux of organic matter arriving at depth, as well as the distribution of subsurface water bodies (e.g., Weyl 1978;

Deuser et al. 1981; Durazzi 1981; Hemleben et al. 1985; Healy-Williams 1983; Healy-Williams et al. 1985; Itou and Noriki 2002; Schiebel et al. 2002a, b; Peeters et al. 2004).

7.1 Distribution in the Global Ocean

7.1.1 Biogeographic Provinces

Modern planktic foraminifer assemblages are attributed to five major faunal provinces at the global scale (Figs. 7.1 and 7.2): Tropical, subtropical, temperate, subpolar, and polar (e.g.,

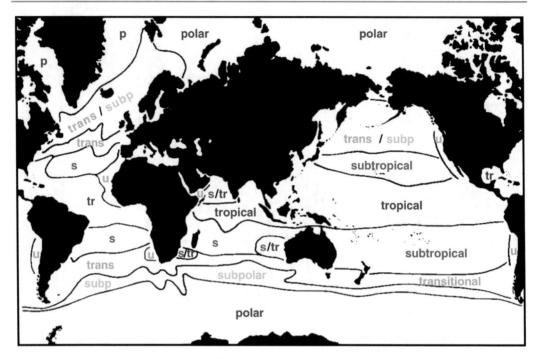

Fig. 7.2 Foraminifer provinces according to data from plankton tows and sediment samples (Hemleben et al. 1989, and references therein). Latitudinal provinces are polar (p), subpolar (subp), transitional (trans), subtropical (s), and tropical (tr). A sixth province is characterized by upwelling (u) and eutrophic conditions. Provinces in the Indian Ocean are characterized by mixing of subtropical-to-tropical (s/tr) faunal elements, and transitional-to-subpolar (trans/subp) faunal elements in the North Pacific and Atlantic Ocean. Modified after Hemleben et al. (1989)

Bradshaw 1959; Bé 1959, 1977; Hemleben et al. 1989; Kucera 2007). Those faunal provinces roughly follow zonal and areal distribution patterns, displaying water temperature and salinity (Phleger 1960; Bé and Tolderlund 1971; Tolderlund and Bé 1971; Caron et al. 1987; Bijma et al. 1990b), radiation (symbiont-bearing species; Erez 1983; Erez and Luz 1983), turbidity of ambient water (Ortiz et al. 1995), the abundance of prey, and trophic demands of planktic foraminifers at a species level (e.g., Spindler et al. 1984; Schiebel et al. 2001). To a yet unknown extend, distribution and abundance of planktic foraminifers may also follow the distribution of predators (Berger 1971). A sixth province follows the major upwelling regions, and is almost exclusively defined by eutrophic conditions, the abundance of prey, and to some extent by turbidity. Upwelling conditions are characterized by a dominance of the symbionts-barren species *G. bulloides* (e.g., Thiede 1975). In general, the biogeography of foraminifers, and foraminifer provinces are characterized by the overall distribution of species, as well as the presence of indicator species like *G. bulloides*. Depending on the genotype, *G. bulloides* (Fig. 7.3) indicates enhanced production of algal prey at temperate to high latitudes during spring, or upwelling conditions at low to mid latitudes.

Additional provinces are defined by particular ecological conditions, and mixing of different water bodies and faunas, particularly conspicuous in the Arabian Sea and northern Pacific Ocean (Fig. 7.2). Planktic foraminifer population dynamics in the Arabian Sea is affected by monsoon-induced effects in physical and biological properties of surface waters, and suboxic to anoxic conditions below the seasonal thermocline (e.g., Kroon 1988; Kroon and Ganssen 1988; Brock et al. 1992; Curry et al. 1992; Ivanova et al. 1999; Schiebel et al. 2004). The North Pacific is characterized by seasonal changes in

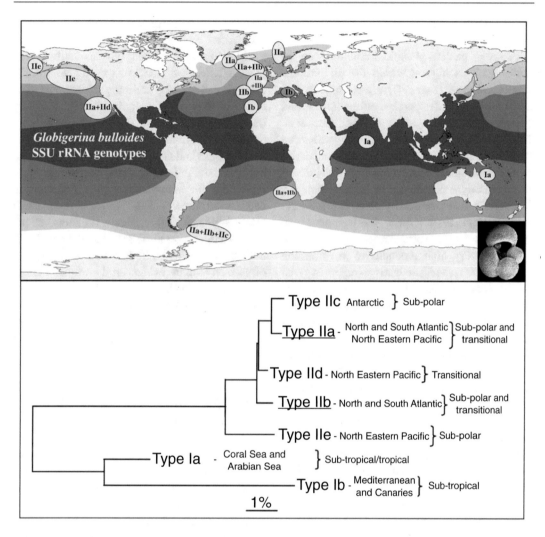

Fig. 7.3 Biogeographic distribution (*upper panel*) and evolutionary relationships (*lower panel*) of SSU rRNA genotypes isolated for the morphospecies *G. bulloides*, superimposed on the map of five major planktic foraminifer faunal provinces according to Bé and Tolderlund (1971). Genotypes isolated by Darling et al. are shown in *light grey* (1999, 2000, 2003, 2007; Stewart 2000). Mediterranean Type Ib (shown in *dark grey*) from De Vargas et al. (1997). The tree is re-drawn from Darling et al. (2007), and is rooted on the *G. bulloides* Type I genotypes at the base of the *G. bulloides* clade in the phylogenetic tree of Darling et al. (2000; see also André et al. 2014). The bipolar genotypes are underlined. From Darling and Wade (2008)

the Kuroshio-Oyashio confluence, and mixing of tropical-to-subtropical and polar-to-subpolar faunal elements (e.g., Eguchi et al. 1999; Mohiuddin et al. 2002). Faunal mixing caused by hydrodynamic features (e.g., upwelling and currents), and regional shifts of faunal provinces occurs on various temporal scales such as, for example, seasonal to glacial-interglacial time-scales (e.g., Ivanova et al. 2003; Ishikawa and Oda 2007). In addition, changing planktic foraminifer assemblages, and 'warmer' faunal elements in the eastern Pacific Ocean off California since the 1970s presumably indicate a warming trend (Field et al. 2006).

7.1.2   Diversity

Diversity of modern planktic foraminifers on the global scale is highest within the oligotrophic subtropical gyres (Fig. 7.4), as a consequence of both biological and ecological effects (Ottens and Nederbragt 1992; Brayard et al. 2005; Žarić et al. 2005; Beaugrand et al. 2013). Slightly enhanced diversity in particular at the poleward boundaries of the subtropical gyres (Fig. 7.4) may result from hydrodynamic effects, i.e. expatriation and mixing of faunal elements by currents (cf. Berger 1970a; Weyl 1978; Ottens 1991; Ottens and Nederbragt 1992). Particular ecological conditions like very high productivity in upwelling areas, and the short productive season in polar latitudes cause decreased diversity in comparison to adjacent waters, and lower latitudes, respectively (e.g., Ottens and Nederbragt 1992). Secondary effects causing decreased diversity of sediment assemblages (i.e. data used in numerical models, from, e.g., Prell et al. 1999) are differential dissolution and winnowing (e.g., Dittert et al. 1999). Reflecting the sum of parameters affecting ecological niches, the global diversity pattern is positively related to, and may be best explained (following numerical models) by absolute temperature (Rutherford et al. 1999; Beaugrand et al. 2013).

The distribution of genotypes appears geographically more restricted than the distribution of morphotypes, as for example in *G. bulloides* (Darling and Wade 2008). Primary production and the availability of prey are assumed major driving forces for regional and vertical ecological partitioning, and diversity of planktic foraminifers (Seears et al. 2012). The association of symbiont-bearing planktic foraminifer species may affect ecological partitioning by limiting those species to euphotic waters (Seears et al. 2012). Symbiont-barren species may well be depth-parapatric, as shown for *H. pelagica* Type I (above 100 m), and *H. pelagica* Type IIa (below 100 m) from the same site (Weiner et al. 2012). Both Seears's et al. (2012) and Weiner's et al. (2012) conclusions are supported by extensive genetic analyses. Gene flow and speciation are interpreted to follow ecological adaptation.

Species populate their typical depth habitat (e.g., Weiner et al. 2012) according to specific ecological demands, and may ascend and descend in the water column during ontogeny (Hemleben et al. 1989). For example, *Globorotalia truncatulinoides* spends most time of its life in subsurface and deep waters, and ascends to the sea surface during late winter/early spring to reproduce, for example, near the Azores Island and Bermuda (e.g., Durazzi 1981; Healy-Williams 1983; Healy-Williams et al. 1985; Hemleben et al. 1985; Mulitza et al. 1997; Schiebel et al. 2002a, b). The vertical separation of species is more evident in the tropics than in polar waters owing to a wider diversity of hydrographic and biotic variables from surface to depth at low latitudes compared to the more homogeneous water column at high latitudes on average (Schmidt et al. 2004a, b).

7.2   Interannual and Seasonal Distribution

Interannual variability in the production of planktic foraminifers follows variations in seasonal hydrographic and ecological changes. Consequently, standing stocks in mid latitudes may vary by more than one order of magnitude at the species to assemblage level (e.g., Schiebel and Hemleben 2000). Interannual variability of planktic foraminifer assemblages has been assumed primarily caused by trophic conditions in the productive (euphotic) surface ocean (e.g., Schiebel 2002). Regional variability may be caused by shifting fronts between water bodies due to differences in climate zones and wind patterns. In contrast, species assemblages may be (qualitatively) similar when comparing corresponding seasons. Quantitative changes in production and flux of planktic foraminifer tests may be best recorded from different latitudes and ocean basins by sediment trap samples (e.g., Žarić et al. 2005) (see Chap. 10 Methods, Table 10.1 and Fig. 10.2).

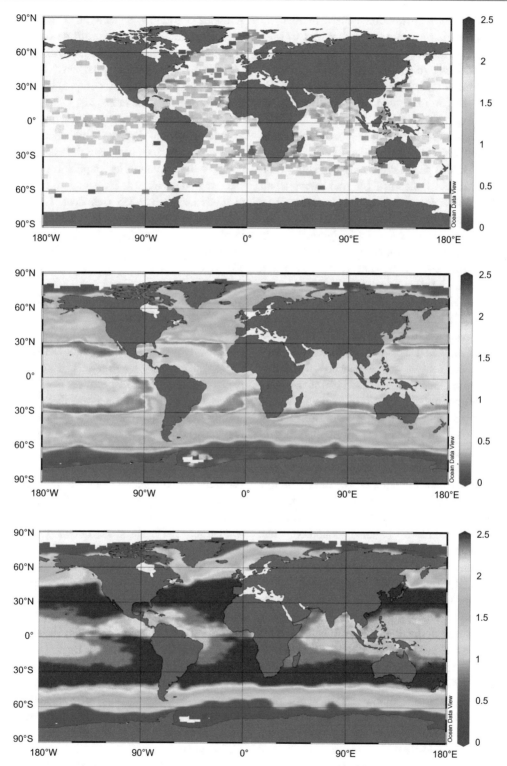

Fig. 7.4 High planktic foraminifer diversity at the global scale occurs at the poleward margins of the subtropical gyres. Diversity is lowest in polar waters. *Upper panel*: Shannon diversity is best represented in coretop assemblages according to the data of Prell et al. (1999). *Middle panel*: Modeled Shannon-Wiener diversity (*H'*, see Chap. 10). *Lower panel*: Modeled species richness (# of species) calculated from the model. *White* and *pink G. ruber* combined. Higher values correspond to higher diversity. Note different scale bars. After Žarić et al. (2005)

The seasonal distribution pattern of planktic foraminifers is most pronounced at mid to high latitudes, displaying phytoplankton succession and food chain (e.g., Bé 1960, 1977; Schiebel and Hemleben 2005; Fraile et al. 2009). In polar oceans, single maximum planktic foraminifer production occurs during the short summer, when light and temperature conditions cause enhanced primary and secondary production (Fig. 7.5). The planktic foraminifer fauna of the polar ocean is dominated by two rather small species, *Neogloboquadrina pachyderma* and *Turborotalita quinqueloba*, with *G. bulloides*, *Globigerinita glutinata*, and *Globigerinita uvula* being the most common accessory species (Carstens et al. 1997; Volkmann 2000; Pados and Spielhagen 2014). *Neogloboquadrina pachyderma* survives even in brine channels (up to 82 PSU) within the annual Antarctic sea ice (not in the Arctic!), where it feeds on diatoms (Dieckmann et al. 1991; Spindler 1996). In mid latitudes, two seasons of enhanced production during spring and fall are caused by the interplay of increased mixing depth of surface waters, nutrient recycling, and light intensity. Spring production of planktic foraminifers in mid-latitudes was shown to considerably outnumber the autumn-production (Schiebel and Hemleben 2000; Schiebel et al. 2001). In low latitudes, light intensity and temperature are high throughout the year, seasonality is low, and productivity follows regional conditions like monsoonal activity and upwelling intensity (e.g., Kroon and Ganssen 1989; Ivanova et al. 1999; Conan and Brummer 2000; Schiebel et al. 2004).

Seasonality is expressed by the co-occurrence of planktic foraminifer species, which signify different zonal distributions and hydrographic conditions (Hemleben et al. 1989; Schiebel 2002; Schiebel and Hemleben 2005; Jonkers and Kučera 2015). Seasonal changes between monsoon-driven upwelling, surface ocean mixing versus stratification, and trophic conditions result in a mix of sedimentary test assemblages. Absolute changes in water temperature (ΔT) and productivity (ΔP) may hence be reconstructed from species assemblages, as well as stable isotopes ($\delta^{18}O$ and $\delta^{13}C$) and Mg/Ca ratios (e.g.,

Williams et al. 1979; Saher et al. 2009; Wit et al. 2010; Feldmeijer 2014), and Cd/Ca ratios of tests from different species (Ripperger et al. 2008). In addition to multi-species analyses, ontogenetic changes in the chemical compositions (stable isotopes and element ratios, see Chap. 10) may provide additional information for more refined reconstructions of hydrographic changes (e.g., Katz et al. 2010). Considering the complexity of both planktic foraminifer population dynamics and regional hydrology, modern analytical methods as LA-ICP-MS (see Sect. 10.7.1) provide detailed quantitative data to achieve a higher level of understanding of paleoceanographic processes (e.g., Eggins et al. 2003; Wit et al. 2010).

7.3 Trophic Effects

The relative preference for zooplankton and phytoplankton prey by spinose and non-spinose planktic foraminifers, respectively (see Chap. 4), affects the spatial and temporal distribution of species according to the quantity and variety (i.e. quality) of available food. Most symbiont-bearing species prefer lower latitudes and less turbid (i.e. less productive) waters, whereas symbiont-barren species occur at higher relative abundance at higher latitudes and more productive (i.e. more turbid) waters (e.g., Bé and Tolderlund 1971; Bé 1977; Ottens 1992; Ortiz et al. 1995; Schiebel and Hemleben 2000). At the global scale, relative abundance of spinose species is highest in the oligotrophic central water masses in the subtropical gyres, where copepods and other zooplankton predominate (Hemleben et al. 1989 and references therein; Barnard et al. 2004; Schiebel et al. 2004; Buitenhuis et al. 2013; Moriarty and O'Brien 2013). In contrast, non-spinose species are more abundant in eutrophic waters with high phytoplankton production, such as upwelling regions, with the exception of symbiont-barren spinose *G. bulloides*.

Differential reaction of planktic foraminifer species to changing ecological conditions causes species successions, which are characteristic of

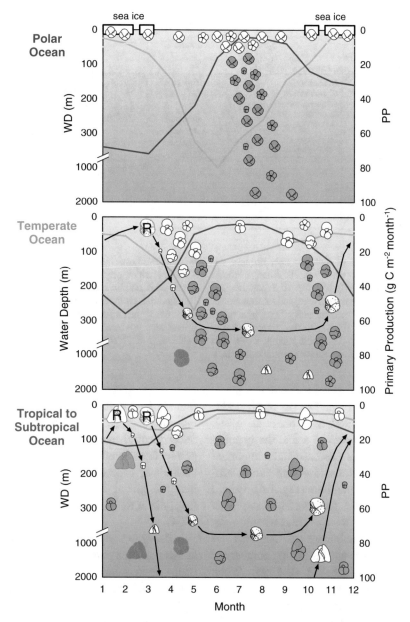

Fig. 7.5 Schematic view of seasonality, depth habitat (living specimens shown by *white* tests), and sedimentation of empty planktic foraminifer tests (*grey*), compiled from plankton-tow and sediment-trap samples (see Table 10.1, Fig. 10.2). Ecological parameters after Longhurst (1998). Mixed layer water depth (y-axis to the left, *blue line*) and photic depth (1 % isolume, *yellow line*), and integrated primary production (PP, *green line*, y-axis to the right). Biological production depends on the availability of nutrients, mixed layer depth, and light level. Seasonal succession of species according to their ecological demands (e.g., food) is exemplified by *Neogloboquadrina pachyderma* for the polar ocean (*upper panel*), and by *Globigerinita glutinata* and *Globigerina bulloides* for the temperate ocean (*middle panel*).

During winter, *N. pachyderma* lives in the lower layers of the Antarctic (not Arctic) sea ice. *Turborotalita quinqueloba* is present in the polar ocean during summer and in the temperate ocean during seasons of low water temperature. Mass flux of empty tests follows periods of major biological production. Intermediate and deep-dwelling planktic foraminifer species ascend to the sea surface to reproduce (*black* 'R'), and empty tests settle to the seafloor after reproduction. In the tropical to subtropical ocean (*lower panel*), intermediate and deep-dwelling species inhabit deeper waters than at mid-latitudes. In the tropical to subtropical ocean, production of planktic foraminifers is more balanced than at higher latitudes. From Schiebel and Hemleben (2005)

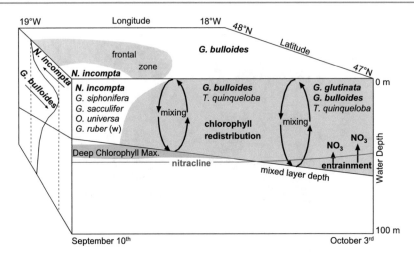

Fig. 7.6 Schematic view of hydrographic, trophic, and faunal development in the eastern North Atlantic around 47 °N, 20 °W (BIOTRANS), between 10 September and 3 October 1996. A first change in the planktic foraminifer assemblage resulted from mixing and chlorophyll redistribution in the upper 100 m of the water column. A second change due to increased mixing depth and entrainment of nutrients from below the nutricline (incl.

nutricline) followed by new phytoplankton production. As a result of chlorophyll redistribution, mainly *G. bulloides* increased in numbers. Subsequent to nutrient entrainment, *G. glutinata* proliferated (*front panel*). Depths distributions of *N. incompta* and *G. bulloides* are within the frontal area are interlocked (*side panel*). Redrawn from Schiebel et al. (2001)

different ecosystems (e.g., Deuser et al. 1981; Kroon and Ganssen 1989; Schiebel et al. 2001). At the regional and seasonal scale, the quantity and quality of food is predominantly important for the distribution of shallow- and subsurface-dwelling planktic foraminifers (Fig. 7.6). Within several days, planktic foraminifers have been shown to respond to the redistribution of chlorophyll and entrainment of nutrients by enhanced growth rates and increasing numbers of large individuals (Schiebel et al. 1995). When surface water mixing increases and the thermocline shifts to depth, for example, caused by enhanced wind stress (e.g., Schiebel et al. 1995) or induced by eddies (Kupferman et al. 1986; Beckmann et al. 1987; Fallet et al. 2011; Steinhardt et al. 2014), chlorophyll may be redistributed from the deep chlorophyll maximum and nutrients may be entrained into surface waters (Fig. 7.6). As a first consequence, the faunal portion of opportunistic species increases (e.g., *G. bulloides*). *Globigerina bulloides* is the most common morphospecies in the temperate ocean (Fig. 7.1), and has been the first planktic

foraminifer species, which has been identified as indicator of trophic conditions by Thiede (1975). Subsequently, planktic foraminifer species that prefer 'fresh' prey (e.g., *Globigerinita glutinata*) proliferate, caused by entrainment of nutrients into the mixed layer and new phytoplankton (e.g., diatoms) production (Schiebel et al. 2001, 2004).

After food sources are exhausted, opportunistic species and species specialized on particular food sources decline in numbers, and a 'background fauna' displays the average regional hydrology and biogeographic zone (Schiebel and Hemleben 2000). Consequently, opportunistic species are not characteristic of distinct depth habitats and absolute temperatures but of the quantity and quality of prey, which should be considered when interpreting the isotopic composition of their tests.

Distribution and ecological demands of intermediate- and deep-dwelling species like *Globorotalia scitula*, *Globorotalia hirsuta*, and *G. truncatulinoides*, are not as well known as those of shallow-dwelling species. Deep-dwelling species reproduce much less often (possibly as little as once per year) than shallow-dwelling species (every fortnight to once per month; Fig. 7.7). The intermediate to deep habitat is ecologically more uniform than the surface habitat, and fine scale changes in the deep planktic foraminifer distribution have not yet been sufficiently quantified. Due to their slow reaction on changing ecologic conditions, deep-living species can be used as tracers of intermediate to deep water-masses (e.g., Berger 1970b). For example, *G. truncatulinoides* probably enters the Caribbean Sea with the Subtropical Underwater through the Anegada Passage in water depths between 100 and 300 m (Schmuker and Schiebel 2002). *Globorotalia truncatulinoides* and *Globorotalia menardii* are transported within ambient water bodies by currents (e.g., Gulf Stream) over long distances (Weyl 1978), and the isotopic signature of tests is applied to the reconstruction of major current patterns, as well as life-modes of deep-living species (Mulitza et al. 1997; Spencer-Cervato and Thierstein 1997; Cléroux et al. 2007, 2009; Feldmeijer 2014).

7.4 Vertical Distribution in the Water Column

The vertical distribution (Figs. 7.6 and 7.7) of planktic foraminifers is affected by the distribution of prey in the same way as the horizontal, regional to global pattern (e.g., Bé 1960; Schiebel et al. 2001; Seears et al. 2012). Highest standing stocks of planktic foraminifers on the vertical scale are associated with the deep chlorophyll maximum usually sited around the seasonal thermocline and pycnocline in the upper 100 m of the water column (e.g., Fairbanks and Wiebe 1980; Schiebel et al. 2001; Field 2004). A comprehensive statistical analysis of the variable depth habitat of individual species in response to environmental and biological factors is exemplified for the subtropical NE Atlantic by Rebotim et al. (2016). Understanding the vertical distribution, i.e. depth habitat of planktic foraminifers in the water column is of crucial importance for reliable reconstruction of, for example, temperature and

primary productivity in paleoceanography (e.g., Phleger 1945; Wang 2000). The depth habitat of species has been directly determined from vertical plankton tows and the use of opening-closing nets (e.g., Bé 1962; Fairbanks et al. 1982; Hemleben et al. 1989; Schiebel et al. 1995), and indirectly from data on stable oxygen isotopes and Mg/Ca ratios of test calcite as temperature proxy, and hence relative measure of stratification and water depth at a regional scale (Fairbanks et al. 1980, 1982; Kohfeld et al. 1996; Mulitza et al. 1997; Field 2004; Cléroux et al. 2007, 2009; Hathorne et al. 2009; Groeneveld and Chiessi 2011).

Vertical distribution of planktic foraminifers in the water column is presumably affected by various biogenic effects such as (i) the need of sunlight of the symbiont-bearing, and independence from light by symbiont-barren species (e.g., Bé 1960; Vincent and Berger 1981; Seears et al. 2012; Weiner et al. 2012), (ii) ontogenetic vertical migration and reproduction at certain water depths (e.g., Hemleben et al. 1989; Bijma et al. 1990a; Schiebel et al. 1997), and (iii) the distribution and quality of prey (e.g., Schiebel et al. 2001). In addition, abiogenic environmental effects have been reported as affecting the depth distribution among which are surface water mixing and transportation of specimens caused by gales (Schiebel et al. 1995; Brunner and Biscaye 1997), and fresh water lenses impeding the ascent of individuals to surface waters (Deuser et al. 1988; Carstens and Wefer 1992; Carstens et al. 1997; Ufkes et al. 1998; Schmuker and Schiebel 2002).

Continent-derived matter affects the vertical distribution patterns of planktic foraminifers in hemipelagic regions along continental margins differently than in the pelagic ocean. Shelf seas are largely barren of living planktic foraminifers (e.g., Sousa et al. 2014), except where individuals have been transported onto the shelf by currents (cf. Bandy 1956; Berger 1970b). Test-size cohorts of species increasingly lack small (i.e. pre-adult) tests with decreasing water depth when approaching the continent (Retailleau et al. 2011). The lack of small test, and fragmentation of assemblages in comparison to deep marine test-size cohorts (e.g., Peeters et al. 1999; Schiebel and Hemleben 2000) is interpreted to be

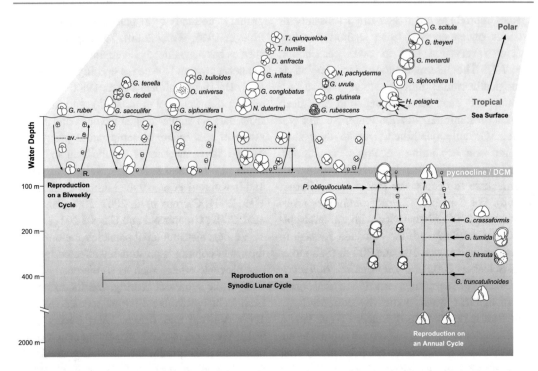

Fig. 7.7 Idealized scheme of planktic foraminifer depths habitats and life cycle in the pelagic ocean. The average water depth inhabited by planktic foraminifers (av., *stippled horizontal lines*) varies at the species level. Different foraminifer species inhabit average water depths ranging from the upper 10 m to 400 m, and *G. truncatulinoides* dwells in extreme depths down to 2000 m. Dwelling depths result from ecologic conditions and biologic prerequisites, and affect relative depths of different species rather than absolute water depths. For example, the average dwelling depth of *G. ruber* (*white*, sensu lato, s.l.) usually ranges above the pycnocline; depending on hydrographical conditions, the depth-distribution of any *G. ruber*-type may be within reach of the pycnocline. Symbiont-bearing species depend on light and live in the euphotic zone of the ocean.

Symbiont-barren species may settle in deep waters below the euphotic zone. Planktic foraminifers reproduce (R.) at species-specific depth relative to the pycnocline (i.e. seawater density), and distinct temperature and salinity conditions. Enhanced availability of prey at the deep chlorophyll maximum (DCM, associated with the pycnocline) provides trophic conditions, which support the survival of juveniles. In the *upper panel*, species are arranged according to their relative latitudinal position. *Globigerinoides ruber* is known to reproduce twice per month. *Globigerina bulloides*, *G. sacculifer*, *N. pachyderma*, *T. quinqueloba*, *H. pelagica,* and other shallow-dwelling species reproduce on a synodic lunar cycle. An annual reproduction cycle is assumed for *G. truncatulinoides*, and may be similar in other deep-dwelling species. After Schiebel and Hemleben (2005)

an indication of changing ecological conditions. River discharge from the continent affects surface salinity and trophic conditions in neritic and hemipelagic waters, which may not provide the ecological needs of planktic foraminifers (Retailleau et al. 2009). Those planktic foraminifers individuals expatriated to hemipelagic waters may still grow in size but may not reproduce. The depth-distribution of species may differ from that in pelagic waters. Subsurface dwelling *G. scitula* (Itou et al. 2001, NW Pacific;

Schiebel et al. 2002a, NE Atlantic; see also Oberhänsli et al. 1992) were found to dwell in surface waters in the hemipelagic SE Bay of Biscay (Retailleau et al. 2011), and to the NE off the Congo River mouth (R. Schiebel, unpublished data). In addition to other offshore-onshore effects, tidal currents and local upwelling over the shelf-break and submarine canyon heads are discussed as sites of enhanced primary production, and to foster the production of opportunistic planktic foraminifer species like *G. bulloides*

(Brunner and Biscaye 2003; Machain-Castillo et al. 2008; Retailleau et al. 2012).

7.5 Diurnal Vertical Migration

Diurnal changes in depth habitat have been suspected of various planktic foraminifer species (e.g., Boltovskoy 1973, and references therein; Bé 1960; Bé and Hamlin 1967; Berger 1969; Holmes 1982). Rhumbler (1911) already presumed higher abundances of planktic foraminifers in day tows than in night tows (see also Bradshaw 1959; Bé 1960). However, systematic diurnal changes in dwelling depth could not yet be deduced from assemblage data. Depth-related distribution patterns have been attributed to heterogeneity, i.e. patchiness, which is best explained by differences in the spatial rather than temporal variability (cf. Boltovskoy 1971; Siccha et al. 2012).

Diurnal changes in dwelling depth are difficult to prove because any (sub-) diurnal migration pattern could be overlain and masked by other periodic changes such as depths changes of individuals over a reproduction cycle (e.g., Schiebel et al. 1997), local episodic events like storms (Schiebel et al. 1995), and transportation of planktic foraminifers within surface water masses by currents (Kupferman et al. 1986; Schiebel and Hemleben 2000). In addition, relations between grazers and prey, as well as parameters, which affect the absolute abundance of species during reproduction, both of which potentially affecting the depth distribution of species, so far remain unanswered.

Planktic foraminifers may be capable of limited active vertical migration by changing the quantity of lipids in their cytoplasm, and through activity of fibrillar bodies (see Chap. 3), to a yet unknown degree (Hansen 1975; Anderson and Bé 1976a, b). Individuals are presumed to migrate up and down the water column to occupy species-specific depth habitats predominantly for reproduction and alimentation at a synodic lunar cycle (i.e. two to four weeks on average, see Sect. 5.2), they can possibly not undertake active vertical diurnal migration over tens of meters like

other zooplankton and phytoplankton (cf. Boltovskoy 1973; Riley 1976; Holmes 1982; Ralston et al. 2007).

Evidence of systematic though passive diurnal change in the depth habitat of planktic foraminifers is provided by analyses of floating sediment traps (Siccha et al. 2012). The kilometer-scale and sub-diurnal variability of planktic foraminifer distribution in the surface water column in the central Bay of Biscay was sampled in spring 2009, using drifting sediment traps deployed at 200 m depth for three consecutive intervals between April 7 and 19, 2009. The hydrodynamic bias and its effects on the sampling efficiency, trap track, and sample composition (incl. species-specific size distributions) were carefully checked for sampling artefacts, and autocorrelation of the planktic foraminifer flux at distances <2 km could not be attributed to the temporal domain. Significant negative autocorrelation of the distribution of the total live foraminifer assemblage, as well as of living *G. scitula*, was detected for intervals of 2 km and 6 h, following the temporal signal of the internal tide in the Bay of Biscay. *Globorotalia scitula* is particularly well suited to detect depth changes in this study, because its average depth-habitat between 100 and 300 m (e.g., Erez and Honjo 1981; Ortiz et al. 1995) is bracketing the deployment depth of the sediment traps. Accordingly, the distribution of *G. scitula* indicates passive (non-selective for size!) diurnal displacement of assemblages by internal tidal waves rather than active individual depths migration (Siccha et al. 2012).

7.6 Test Size

Planktic foraminifer test size provides information on (paleo-) ecological conditions of the ocean (Figs. 7.8 and 7.9). Test-size analyses have been pursued since the early works of Ericson (1959) and Hecht (1976), following the ideas of Bergmann's (1847) rule relating body size to temperature, and hence ecogeography. Whereas ecological effects on body size are obvious (e.g., Bergmann 1847), the multiple factors that may affect foraminifer test size are difficult to

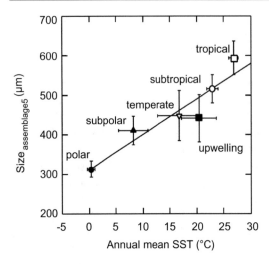

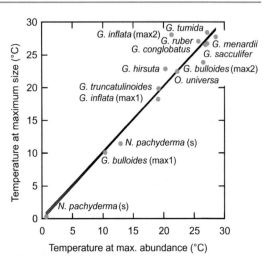

Fig. 7.8 Average test size (μm) of the largest 5 % of specimens from samples (Size$_{assemblage5}$) per biogeographic area, plotted against annual average sea surface temperature (SST, data from Levitus et al. 1994). Error bars give the 95 %-confidence intervals. Linear regression, r = 0.938, p = 0.006). From Schmidt et al. (2004a)

Fig. 7.9 Relationship of sea surface temperature (data from Levitus et al. 1994), maximum test size, and maximum (max.) relative abundance of single taxa ($r^2 = 0.928$, $p = 0.001$) in surface sediments (data from Prell et al. 1999). Note that *G. inflata* and *G. bulloides* have two optima both in size and abundance, possibly displaying varying ecological demands of different genotypes at the regional scale (cf. Darling and Wade 2008; Morard et al. 2011; André et al. 2014). *N. pachyderma* (sinistral coiled test) is signified by (*s*), and dextral *N. pachyderma* (i.e. *N. incompta*) by (*d*). From Schmidt et al. (2004a)

disentangle (Schmidt et al. 2006, and references therein). Over long time-intervals, evolutionary effects should be considered. The test size of species may increase over evolutionary time following Cope's rule (Stanley 1973; Schmidt et al. 2006). Mixing of fossil assemblages may result in test-size changes, which cannot be explained by evolution and ecological effects alone. When interpreting fossil assemblages, taphonomic effects including differential sedimentation and preservation of tests need to be taken into account (see Chap. 8). Fortunately, planktic foraminifers occur in large standing stocks and usually at sufficient numbers of ubiquitous species in above-CCD sediments over the past 100 million years, and serve as model organisms (among others) in deciphering relationships of body size, environment, and evolution (Schmidt et al. 2004b).

The modern ocean hosts some of the largest planktic foraminifers of all times (Schmidt et al. 2004b), resulting in high modern calcite flux and burial rates of foraminifer CaCO$_3$ (Schiebel 2002). Climate warming since the 1970s is assumed to still enhance planktic foraminifer calcite production (Field et al. 2006). Largest

assemblage test-size in the modern ocean occurs at tropical and subtropical latitudes, and smallest test assemblages characterize high-latitude waters (Fig. 7.8). Given that most planktic foraminifer species occur over wide temperature and salinity ranges, and associated environmental parameters (Bé and Tolderlund 1971; Hemleben et al. 1987; Lombard et al. 2009, 2011), the positive correlation of maximum average test size and abundance with surface water temperature at the global scale is possibly significant (Fig. 7.9).

The latitudinal distribution pattern of planktic foraminifers is disrupted by currents and hydrographic fronts (Fig. 7.10a), including regional hydrographic features such as upwelling cells (e.g., Schiebel et al. 2001; Schmidt et al. 2004a, b). Hydrographic fronts presumably negatively affect test size, in addition to an overall negative affect on planktic foraminifer diversity (Ottens and Nederbragt 1992). Upwelled waters are colder than surrounding surface waters, comprise more macronutrients, and hence produce more food for

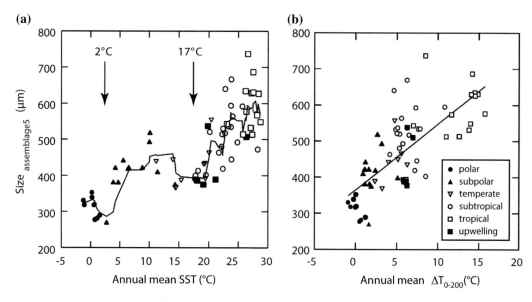

Fig. 7.10 Average test size (μm) of the largest 5 % of specimens (Size$_{assemblage5}$) from Holocene sediment samples plotted against (*a*) mean annual sea surface temperature (SST [°C]). The areas of minimum size (*arrows*) at 2 °C and 17 °C correspond to the polar and the subtropical fronts, respectively. (*b*) Surface water stratification, given as the difference between mean annual SST and temperature at 200 m water depth (ΔT$_{0-200}$). Small and large assemblage test sizes occur at weak (e.g., upwelling) and strong (e.g., central subtropical gyres) annual mean surface water stratification, respectively. Accordingly, planktic foraminifer test size indicates regional surface ocean stratification, and consequently of trophic conditions. The *black line* represents the five-point moving average in panel (*a*) and the regression line in panel (*b*). The legend relates to both panels (*a*) and (*b*). Temperature data from Levitus et al. (1994). Modified after Schmidt et al. (2004a)

planktic foraminifers. Due to enhanced biological productivity, upwelled waters are more turbid than lower productive waters, which favors small-sized symbiont-barren opportunists like G. *bulloides*, whereas larger symbiont-bearing generalist species like G. *sacculifer* are disadvantaged and hence less frequent, which results in an overall decreased test size and diversity. In contrast, low productivity in well-stratified surface waters, for example, in the subtropical gyres favors large-sized symbiont-bearing species (Figs. 7.8, 7.9 and 7.10b). Whereas primary production and the availability of freshly produced food (see Sect. 4.1) in surface waters affect the size of surface dwelling planktic foraminifer species and assemblages, subsurface dwelling species are affected by the flux of more or less degraded organic matter arriving at depth (e.g., Itou et al. 2001). Subsurface dwelling foraminifers, including predominantly globorotalid species, produce rather large-sized tests (at rather low water

temperature), which might in part be caused by their longer life cycle in comparison to shallow-dwelling species, as well as by their generalist (K-selected) behavior.

The effect of ecology on test size is applied as a proxy of a variety of physical and chemical marine parameters as well as alimentation at the regional scale, and over the recent geological past. The largest sized tests of G. *ruber* in the eastern Mediterranean during the Holocene occurred during the formation of Sapropel S1, and far from optimum ecological conditions (Mojtahid et al. 2015). Decreased surface water salinity during S1 apparently caused a descent of the symbiont-bearing G. *ruber* to deeper waters (Deuser et al. 1988; Schmuker and Schiebel 2002; Rohling et al. 2004). Less favorable light-conditions and hence decreased symbiont activity at depth, in combination with decreased salinity, may have caused the production of smaller tests (Hemleben et al. 1989, and references therein). The opposite reaction, i.e. the production of larger tests may be explained by delayed reproduction and prolonged maturity, resulting in longer individual growth and larger tests (Mojtahid et al. 2015).

The life of adult planktic foraminifer individuals is most often terminated by reproduction (see Chap. 5), after which the empty tests settle to depth through the water column, and, if not dissolved, form part of the sedimentary assemblage. Accordingly, sediment assemblages are biased towards the largest test-size attained by any species. In addition to individuals that have completed their life cycle by reproduction, smaller prematurely deceased individuals contribute to the sediment assemblages. Taking into account the usually log-normal size-distribution of planktic foraminifer species assemblages (Peeters et al. 1999; Schiebel and Hemleben 2000; Schmidt et al. 2006, and references therein), about half of the adult individuals are lost between the smaller size-class and the next larger size-class. Premature death may be caused by horizontal or vertical expatriation by currents (Berger 1970b; Vincent and Berger 1981; Schiebel et al. 1995) to waters, which provide suboptimal ecological conditions, for example, concerning the quality and quantity of food (cf. Schiebel et al. 2001), light attenuation (Bé et al. 1982, only in symbiont-bearing species), and salinity (cf. Bijma et al. 1994). Consequently, only specimens, which have completed their ontogenetic development (see Chap. 6) count towards the 'maximum size' proxy in paleo-ecological analyses (Schmidt 2002). In contrast, growth rate, survival rate and premature mortality, and the ratio of pre-adult tests of a certain species in sediment assemblages could serve as measure of expatriation and ecological change during the life of a planktic foraminifer.

7.7 Summary and Concluding Remarks

Ecological parameters affect production and distribution of planktic foraminifers (e.g., test calcite and biomass) at the species and assemblage level. Consequently, foraminifer tests are indicators of modern and past environmental change and carbon turnover. Unfortunately, the understanding of planktic foraminifer ecology remains fragmentary although first ideas date back to the late 19th and early 20th century (Murray 1897; Rhumbler 1911), and first targeted programs have been conducted from the 1950s (e.g., Bradshaw 1959; Bé 1960). The understanding is fragmentary by nature, since plankton tow samples of living specimens, and sediment trap samples of the test flux represent only temporal and local snap-shots of the population dynamics, Continuous Plankton Recorder (CPR) hauls only include a narrow depth-layer of the ocean. Laboratory cultures facilitate continuous observation over short intervals of time, but cannot entirely simulate ecological conditions of the natural habitat of planktic foraminifers, which embraces at least the upper 50 m of the water column, and includes a natural composition of prey, which may not be provided artificially. In addition, climate constantly changes at the regional to global scale, including ecological conditions at their entity. Consequently, a combination of the above given approaches needs to be pursued to better understand the production of planktic foraminifers. More complete monitoring programs, and reinforced concerted efforts of the international community of data producers (i.e., sampling, culturing, and modeling) should lead to a better understanding of foraminifers as qualitative and quantitative proxies of the changing climate and ocean.

References

Anderson OR, Bé AWH (1976a) A cytochemical fine structure study of phagotrophy in a planktonic foraminifer, *Hastigerina pelagica* (d'Orbigny). Biol Bull 151:437–449

Anderson OR, Bé AWH (1976b) The ultrastructure of a planktonic foraminifer, *Globigerinoides sacculifer* (Brady), and its symbiotic dinoflagellates. J Foraminifer Res 6:1–21. doi:10.2113/gsjfr.6.1.1

André A, Quillévéré F, Morard R, Ujiié Y, Escarguel G, de Vargas C, de Garidel-Thoron T, Douady CJ (2014) SSU rDNA divergence in planktonic Foraminifera: molecular taxonomy and biogeographic implications. PLoS ONE. doi:10.1371/journal.pone.0104641

André A, Weiner A, Quillévéré F, Aurahs R, Morard R, Douady CJ, de Garidel-Thoron T, Escarguel G, de Vargas C, Kucera M (2013) The cryptic and the apparent reversed: lack of genetic differentiation within the morphologically diverse plexus of the planktonic foraminifer *Globigerinoides sacculifer*. Paleobiology 39:21–39. doi:10.1666/0094-8373-39.1.21

Bandy OL (1956) Ecology of Foraminifera in northeastern Gulf of Mexico. Geol Surv Prof Pap US 274:179–204

Barnard RT, Batten S, Beaugrand G, Buckland C, Conway DVP, Edwards M, Finlayson J, Gregory LW, Halliday NC, John AWG, Johns DG, Johnson AD, Jonas TD, Lindley JA, Nyman J, Pritchard P, Reid PC, Richardson AJ, Saxby MA, Sidey J, Smith MA, Stevens DP, Taylor CM, Tranter PRG, Walne AW, Wootton M, Wotton COM, Wright JC (2004) Continuous plankton records: plankton atlas of the North Atlantic Ocean (1958–1999) II. Biogeographical charts. Mar Ecol-Prog Ser Suppl 11–75

Beaugrand G, Rombouts I, Kirby RR (2013) Towards an understanding of the pattern of biodiversity in the oceans. Glob Ecol Biogeogr 22:440–449

Bé AWH (1977) An ecological, zoogeographic and taxonomic review of recent planktonic Foraminifera. In: Ramsay ATS (ed) Oceanic micropaleontology. Academic Press, London, pp 1–100

Bé AWH (1960) Ecology of recent planktonic Foraminifera: Part 2: Bathymetric and seasonal distributions in the Sargasso Sea off Bermuda. Micropaleontology 6:373–392. doi:10.2307/1484218

Bé AWH (1959) A method for rapid sorting of Foraminifera from marine plankton samples. J Paleontol 33:846–848

Bé AWH (1962) Quantitative multiple opening-and-closing plankton samplers. Deep-Sea Res 9:144–151. doi:10.1016/0011-7471(62)90007-4

Bé AWH, Hamlin WH (1967) Ecology of recent planktonic Foraminifera: Part 3: distribution in the North Atlantic during the summer of 1962. Micropaleontology 13:87–106. doi:10.2307/1484808

Bé AWH, Hutson WH (1977) Ecology of planktonic Foraminifera and biogeographic patterns of life and fossil assemblages in the Indian Ocean. Micropaleontology 23:369–414. doi:10.2307/1485406

Bé AWH, Spero HJ, Anderson OR (1982) Effects of symbiont elimination and reinfection on the life processes of the planktonic foraminifer *Globigerinoides sacculifer*. Mar Biol 70:73–86. doi:10.1007/BF00397298

Bé AWH, Tolderlund DS (1971) Distribution and ecology of living planktonic Foraminifera in surface waters of the Atlantic and Indian Oceans. In: Funell BM, Riedel WR (eds) The micropalaeontology of oceans. University Press, Cambridge, pp 105–149

Beckmann W, Auras A, Hemleben C (1987) Cyclonic cold-core eddy in the eastern North Atlantic. III. Zooplankton. Mar Ecol Prog Ser 39:165–173

Berger WH (1971) Sedimentation of planktonic Foraminifera. Mar Geol 11:325–358. doi:10.1016/0025-3227(71)90035-1

Berger WH (1970a) Planktonic Foraminifera: selective solution and the lysocline. Mar Geol 8:111–138. doi:10.1016/0025-3227(70)90001-0

Berger WH (1970b) Planktonic Foraminifera: differential production and expatriation off Baja California. Limnol Oceanogr 15:183–204. doi:10.4319/lo.1970.15.2.0183

Berger WH (1969) Ecologic patterns of living planktonic Foraminifera. Deep-Sea Res 16:1–24. doi:10.1016/0011-7471(69)90047-3

Bergmann C (1847) Über die Verhältnisse der Wärmeökonomie der Thiere zu ihrer Grösse. Gött Stud 3:595–708

Bijma J, Erez J, Hemleben C (1990a) Lunar and semi-lunar reproductive cycles in some spinose planktonic foraminifers. J Foraminifer Res 20:117–127

Bijma J, Faber WW, Hemleben C (1990b) Temperature and salinity limits for growth and survival of some planktonic foraminifers in laboratory cultures. J Foraminifer Res 20:95–116. doi:10.2113/gsjfr.20.2.95

Bijma J, Hemleben C, Oberhänsli H, Spindler M (1992) The effects of increased water fertility on tropical spinose planktonic foraminifers in laboratory cultures. J Foraminifer Res 22:242–256

Bijma J, Hemleben C, Wellnitz K (1994) Lunar-influenced carbonate flux of the planktic foraminifer *Globigerinoides sacculifer* (Brady) from the central Red Sea. Deep-Sea Res I 41:511–530. doi:10.1016/0967-0637(94)90093-0

Boltovskoy E (1971) Patchiness in the distribution of planktonic Foraminifera. In: Farinacci A (ed) Proceedings of the second planktonic conference. Rome 1970, pp 107–155

Boltovskoy E (1973) Daily vertical migration and absolute abundance of living planktonic Foraminifera. J Foraminifer Res 3:89–94. doi:10.2113/gsjfr.3.2.89

Bradshaw JS (1959) Ecology of living planktonic Foraminifera in the North and Equatorial Pacific Ocean. Contrib Cushman Found Foraminifer Res 10:25–64

Brayard A, Escarguel G, Bucher H (2005) Latitudinal gradient of taxonomic richness: combined outcome of temperature and geographic mid-domains effects? J Zool Syst Evol Res 43:178–188. doi:10.1111/j.1439-0469.2005.00311.x

Brock JC, McClain CR, Hay WW (1992) A southwest monsoon hydrographic climatology for the northwestern Arabian Sea. J Geophys Res 97:9455. doi:10.1029/92JC00813

Brunner CA, Biscaye PE (1997) Storm-driven transport of foraminifers from the shelf to the upper slope, southern Middle Atlantic Bight. Cont Shelf Res 17:491–508

Brunner CA, Biscaye PE (2003) Production and resuspension of planktonic foraminifers at the shelf break of the Southern Middle Atlantic Bight. Deep-Sea Res I 50:247–268

Buitenhuis ET, Vogt M, Moriarty R, Bednaršek N, Doney SC, Leblanc K, Le Quéré C, Luo YW, O'Brien C, O'Brien T, Peloquin J, Schiebel R, Swan C (2013)

MAREDAT: towards a world atlas of MARine ecosystem DATa. Earth Syst Sci Data 5:227–239. doi:10.5194/essd-5-227-2013

Caron DA, Faber WW, Bé AWH (1987) Effects of temperature and salinity on the growth and survival of the planktonic foraminifer *Globigerinoides sacculifer*. J Mar Biol Assoc UK 67:323–341

Carstens J, Hebbeln D, Wefer G (1997) Distribution of planktic Foraminifera at the ice margin in the Arctic (Fram Strait). Mar Micropaleontol 29:257–269

Carstens J, Wefer G (1992) Recent distribution of planktonic Foraminifera in the Nansen Basin, Arctic Ocean. Deep-Sea Res I 39:507–524

Cléroux C, Cortijo E, Duplessy JC, Zahn R (2007) Deep-dwelling Foraminifera as thermocline temperature recorders. Geochem Geophys Geosyst. doi:10.1029/2006GC001474

Cléroux C, Lynch-Stieglitz J, Schmidt MW, Cortijo E, Duplessy JC (2009) Evidence for calcification depth change of *Globorotalia truncatulinoides* between deglaciation and Holocene in the western Atlantic Ocean. Mar Micropaleontol 73:57–61

Conan SMH, Brummer GJA (2000) Fluxes of planktic Foraminifera in response to monsoonal upwelling on the Somalia Basin margin. Deep-Sea Res II 47:2207–2227

Curry WB, Ostermann DR, Guptha MVS, Ittekkot V (1992) Foraminiferal production and monsoonal upwelling in the Arabian Sea: evidence from sediment traps. In: Summerhays CP, Prell WS, Emeis KC (eds) Upwelling systems: evolution since the early Miocene. Geological society. Special Publications, London, pp 93–106

Darling KF, Kucera M, Wade CM (2007) Global molecular phylogeography reveals persistent Arctic circumpolar isolation in a marine planktonic protist. Proc Natl Acad Sci 104:5002–5007

Darling KF, Kucera M, Wade CM, von Langen P, Pak D (2003) Seasonal distribution of genetic types of planktonic foraminifer morphospecies in the Santa Barbara Channel and its paleoceanographic implications. Paleoceanography. doi:10.1029/2001PA000723

Darling KF, Wade CM (2008) The genetic diversity of planktic Foraminifera and the global distribution of ribosomal RNA genotypes. Mar Micropaleontol 67:216–238

Darling KF, Wade CM, Kroon D, Brown AJL, Bijma J (1999) The diversity and distribution of modern planktic foraminiferal small subunit ribosomal RNA genotypes and their potential as tracers of present and past ocean circulations. Paleoceanography 14:3–12

Darling KF, Wade CM, Stewart IA, Kroon D, Dingle R, Brown AJL (2000) Molecular evidence for genetic mixing of Arctic and Antarctic subpolar populations of planktonic foraminifers. Nature 405:43–47

Deuser WG, Ross EH, Hemleben C, Spindler M (1981) Seasonal changes in species composition, numbers, mass, size, and isotopic composition of planktonic Foraminifera settling into the deep Sargasso Sea. Palaeogeogr Palaeoclimatol Palaeoecol 33:103–127

Deuser WG, Muller-Karger FE, Hemleben C (1988) Temporal variations of particle fluxes in the deep subtropical and tropical North Atlantic: Eulerian versus Lagrangian effects. J Geophys Res Oceans 1978–2012(93):6857–6862

De Vargas C, Zaninetti L, Hilbrecht H, Pawlowski J (1997) Phylogeny and rates of molecular evolution of planktonic Foraminifera: SSU rDNA sequences compared to the fossil record. J Mol Evol 45:285–294

Dieckmann G, Spindler M, Lange MA, Ackley SF, Eicken H (1991) Antarctic sea ice: a habitat for the foraminifer *Neogloboquadrina pachyderma*. J Foraminifer Res 21:182–189

Dittert N, Baumann KH, Bickert T, Henrich R, Huber R, Kinkel H, Meggers H (1999) Carbonate dissolution in the deep-sea: methods, quantification and paleoceanographic application. In: Fischer G, Wefer G (eds) Use of proxies in paleoceanography. Springer, Berlin, pp 255–284

Durazzi JT (1981) Stable-isotope studies of planktonic Foraminifera in North Atlantic core tops. Palaeogeogr Palaeoclimatol Palaeoecol 33:157–172

Eggins S, De Dekker P, Marshall J (2003) Mg/Ca variation in planktonic Foraminifera tests: implications for reconstructing palaeo-seawater temperature and habitat migration. Earth Planet Sci Lett 212:291–306

Eguchi NO, Kawahata H, Taira A (1999) Seasonal response of planktonic Foraminifera to surface ocean condition: sediment trap results from the central North Pacific Ocean. J Oceanogr 55:681–691

Erez J (1983) Calcification rates, photosynthesis and light in planktonic Foraminifera. In: Westbroek P, de Jong EW (eds) Biomineralization and biological metal accumulation. Reidel Publishing Company, Dordrecht, pp 307–312

Erez J, Honjo S (1981) Comparison of isotopic composition of planktonic Foraminifera in plankton tows, sediment traps and sediments. Palaeogeogr Palaeoclimatol Palaeoecol 33:129–156

Erez J, Luz B (1983) Experimental paleotemperature equation for planktonic Foraminifera. Geochim Cosmochim Ac 47:1025–1031

Ericson DB (1959) Coiling direction of *Globigerina pachyderma* as a climatic index. Science 130:219–220

Fairbanks RG, Sverdlove M, Free R, Wiebe PH, Bé AWH (1982) Vertical distribution and isotopic fractionation of living planktonic Foraminifera from the Panama Basin. Nature 298:841–844

Fairbanks RG, Wiebe PH (1980) Foraminifera and chlorophyll maximum: vertical distribution, seasonal succession, and paleoceanographic significance. Science 209:1524–1526

Fairbanks RG, Wiebe PH, Be AWH (1980) Vertical distribution and isotopic composition of living planktonic Foraminifera in the Western North Atlantic. Science 207:61–63. doi:10.1126/science.207.4426.61

Fallet U, Ullgren JE, Castañeda IS, van Aken HM, Schouten S, Ridderinkhof H, Brummer GJA (2011) Contrasting variability in foraminiferal and organic paleotemperature proxies in sedimenting particles of the Mozambique Channel (SW Indian Ocean). Geochim Cosmochim Acta 75:5834–5848. doi:10.1016/j.gca.2011.08.009

Feldmeijer W (2014) Sensing seasonality by planktonic Foraminifera. PhD Thesis, Vrije Universiteit Amsterdam

Field DB (2004) Variability in vertical distributions of planktonic Foraminifera in the California Current: relationships to vertical ocean structure. Paleoceanography. doi:10.1029/2003PA000970

Field DB, Baumgartner TR, Charles CD, Ferreira-Bartrina V, Ohman MD (2006) Planktonic Foraminifera of the California Current reflect 20th-century warming. Science 311:63–66

Fraile I, Schulz M, Mulitza S, Merkel U, Prange M, Paul A (2009) Modeling the seasonal distribution of planktonic Foraminifera during the Last Glacial Maximum. Paleoceanography. doi:10.1029/2008PA001686

Groeneveld J, Chiessi CM (2011) Mg/Ca of *Globorotalia inflata* as a recorder of permanent thermocline temperatures in the South Atlantic. Paleoceanography 26:PA2203. doi:10.1029/2010PA001940

Hansen HJ (1975) On feeding and supposed buoyancy mechanism in four recent globigerinid Foraminifera from the Gulf of Elat, Israel. Rev Esp Micropaleontol 7:325–339

Hathorne EC, James RH, Lampitt RS (2009) Environmental versus biomineralization controls on the intratest variation in the trace element composition of the planktonic Foraminifera *G. inflata* and *G. scitula*. Paleoceanography 24:PA4204. doi:10.1029/2009PA001742

Healy-Williams N (1983) Fourier shape analysis of *Globorotalia truncatulinoides* from late Quaternary sediments in the southern Indian Ocean. Mar Micropaleontol 8:1–15

Healy-Williams N, Ehrlich R, Williams DF (1985) Morphometric and stable isotopic evidence for subpopulations of *Globorotalia truncatulinoides*. J Foraminifer Res 15:242–253

Hecht AD (1976) An ecologic model for test size variation in recent planktonic Foraminifera: applications to the fossil record. J Foraminifer Res 6:295–311

Hemleben C, Spindler M, Anderson OR (1989) Modern planktonic Foraminifera. Springer, Berlin

Hemleben C, Spindler M, Breitinger I, Deuser WG (1985) Field and laboratory studies on the ontogeny and ecology of some globorotaliid species from the Sargasso Sea off Bermuda. J Foraminifer Res 15:254–272

Hemleben C, Spindler M, Breitinger I, Ott R (1987) Morphological and physiological responses of *Globigerinoides sacculifer* (Brady) under varying laboratory conditions. Mar Micropaleontol 12:305–324

Holmes NA (1982) Diel vertical variations in abundance of some planktonic Foraminifera from the Rockall Trough, northeastern Atlantic Ocean. J Foraminifer Res 12:145–150

Ishikawa S, Oda M (2007) Reconstruction of Indian monsoon variability over the past 230,000 years: Planktic foraminiferal evidence from the NW Arabian

Sea open-ocean upwelling area. Mar Micropaleontol 63:143–154. doi:10.1016/j.marmicro.2006.11.004

Itou M, Noriki S (2002) Shell fluxes of solution-resistant planktonic foraminifers as a proxy for mixed-layer depth. Geophys Res Lett. doi:10.1029/2002GL014693

Itou M, Ono T, Oba T, Noriki S (2001) Isotopic composition and morphology of living *Globorotalia scitula*: a new proxy of sub-intermediate ocean carbonate chemistry? Mar Micropaleontol 42:189–210

Ivanova E, Conan SMH, Peeters FJ, Troelstra SR (1999) Living *Neogloboquadrina pachyderma* sin and its distribution in the sediments from Oman and Somalia upwelling areas. Mar Micropaleontol 36:91–107

Ivanova E, Schiebel R, Singh AD, Schmiedl G, Niebler HS, Hemleben C (2003) Primary production in the Arabian Sea during the last 135 000 years. Palaeogeogr Palaeoclimatol Palaeoecol 197:61–82

Jonkers L, Kučera M (2015) Global analysis of seasonality in the shell flux of extant planktonic Foraminifera. Biogeosciences Discuss 12:1327–1372. doi:10.5194/bgd-12-1327-2015

Katz ME, Cramer BS, Franzese A, Hönisch B, Miller KG, Rosenthal Y, Wright JD (2010) Traditional and emerging geochemical proxies in Foraminifera. J Foraminifer Res 40:165–192

Kohfeld KE, Fairbanks RG, Smith SL, Walsh ID (1996) *Neogloboquadrina pachyderma* (sinistral coiling) as paleoceanographic tracers in polar oceans: evidence from Northeast Water Polynya plankton tows, sediment traps, and surface sediments. Paleoceanography 11:679–699

Kroon D (1988) Distribution of extant plankic foraminiferal assemblages in Red Sea and northern Indian Ocean surface waters. In: Brummer GJA, Kroon D (eds) Planktonic foraminifers as tracers of ocean climate history. Free University Press, Amsterdam, pp 37–74

Kroon D, Ganssen G (1988) Northern Indian Ocean upwelling cells and the stable isotope composition of living planktic Foraminifera. In: Brummer GJA, Kroon D (eds) Planktonic foraminifers as tracers of ocean-climate history. Free University Press, Amsterdam, pp 219–238

Kroon D, Ganssen G (1989) Northern Indian Ocean upwelling cells and the stable isotope composition of living planktonic foraminifers. Deep-Sea Res I 36:1219–1236

Kucera M (2007) Chapter six: Planktonic Foraminifera as tracers of past oceanic environments. In: Hillaire-Marce C, de Vernal A (eds) Developments in marine geology. Elsevier, pp 213–262

Kucera M, Rosell-Melé A, Schneider R, Waelbroeck C, Weinelt M (2005) Multiproxy approach for the reconstruction of the glacial ocean surface (MARGO). Quat Sci Rev 24:813–819. doi:10.1016/j.quascirev.2004.07.017

Kupferman SL, Becker GA, Simmons WF, Schauer U, Marietta MG, Nies H (1986) An intense cold core eddy in the North-East Atlantic. Nature 319:474–477

Levitus S, Boyer T, Burgett R, Conkright M (1994) World Ocean Atlas-CD-ROM data set. In: National oceanographic data center. Washinton DC, p 604

Lombard F, Labeyrie L, Michel E, Bopp L, Cortijo E, Retailleau S, Howa H, Jorissen F (2011) Modelling planktic foraminifer growth and distribution using an ecophysiological multi-species approach. Biogeosciences 8:853–873. doi:10.5194/bg-8-853-2011

Lombard F, Labeyrie L, Michel E, Spero HJ, Lea DW (2009) Modelling the temperature dependent growth rates of planktic Foraminifera. Mar Micropaleontol 70:1–7. doi:10.1016/j.marmicro.2008.09.004

Longhurst AR (1998) Ecological geography of the sea. Academic Press, San Diego

Machain-Castillo ML, Monreal-Gómez MA, Arellano-Torres E, Merino-Ibarra M, González-Chávez G (2008) Recent planktonic foraminiferal distribution patterns and their relation to hydrographic conditions of the Gulf of Tehuantepec, Mexican Pacific. Mar Micropaleontol 66:103–119. doi:10.1016/j.marmicro.2007.08.003

Mohiuddin MM, Nishimura A, Tanaka Y, Shimamoto A (2002) Regional and interannual productivity of biogenic components and planktonic foraminiferal fluxes in the northwestern Pacific Basin. Mar Micropaleontol 45:57–82

Mojtahid M, Manceau R, Schiebel R, Hennekam R, de Lange GJ (2015) Thirteen thousand years of southeastern Mediterranean climate variability inferred from an integrative planktic foraminiferal-based approach: Holocene climate in the SE Mediterranean. Paleoceanography 30:402–422. doi:10.1002/2014PA002705

Morard R, Darling KF, Mahé F, Audic S, Ujiié Y, Weiner AKM, André A, Seears HA, Wade CM, Quillévéré F, Douady CJ, Escarguel G, de Garidel-Thoron T, Siccha M, Kucera M, de Vargas C (2015) PFR^2: a curated database of planktonic Foraminifera 18S ribosomal DNA as a resource for studies of plankton ecology, biogeography and evolution. Mol Ecol Resour. doi:10.1111/1755-0998.12410

Morard R, Quillévéré F, Douady CJ, de Vargas C, de Garidel-Thoron T, Escarguel G (2011) Worldwide genotyping in the planktonic foraminifer Globoconella inflata: implications for life history and paleoceanography. PLoS ONE. doi:10.1371/journal.pone.0026665

Moriarty R, O'Brien TD (2013) Distribution of mesozooplankton biomass in the global ocean. Earth Syst Sci Data 5:45–55. doi:10.5194/essd-5-45-2013

Mulitza S, Dürkoop A, Hale W, Wefer G, Niebler HS (1997) Planktonic Foraminifera as recorders of past surface-water stratification. Geology 25:335–338

Murray J (1897) On the distribution of the pelagic Foraminifera at the surface and on the floor of the ocean. Nat Science 11:17–27

Naidu PD, Malmgren BA (1996) A high-resolution record of late Quaternary upwelling along the Oman Margin, Arabian Sea based on planktonic Foraminifera. Paleoceanography 11:129–140

Oberhänsli H, Bénier C, Meinecke G, Schmidt H, Schneider R, Wefer G (1992) Planktonic foraminifers as tracers of ocean currents in the eastern South Atlantic. Paleoceanography 7(5): 607–632. doi: 10.1029/92PA01236

Ortiz JD, Mix AC, Collier RW (1995) Environmental control of living symbiotic and asymbiotic Foraminifera of the California Current. Paleoceanography 10:987–1009

Ottens JJ (1992) Planktic Foraminifera as indicators of ocean environments in the northeast Atlantic. PhD Thesis, Free University, Amsterdam

Ottens JJ (1991) Planktic Foraminifera as North-Atlantic water mass indicators. Oceanol Acta 14:123–140

Ottens JJ, Nederbragt AJ (1992) Planktic foraminiferal diversity as indicator of ocean environments. Mar Micropaleontol 19:13–28

Pados T, Spielhagen RF (2014) Species distribution and depth habitat of recent planktic Foraminifera in Fram Strait, Arctic Ocean, Polar Research 33. http://dx.doi.org/10.3402/polar.v33.22483

Peeters F, Ivanova E, Conan S, Brummer GJA, Ganssen G, Troelstra S, van Hinte J (1999) A size analysis of planktic Foraminifera from the Arabian Sea. Mar Micropaleontol 36:31–63

Peeters FJC, Acheson R, Brummer GJA, de Ruijter WPM, Schneider RR, Ganssen GM, Ufkes E, Kroon D (2004) Vigorous exchange between the Indian and Atlantic oceans at the end of the past five glacial periods. Nature 430:661–665. doi:10.1038/nature02785

Peters SE, Kelly DC, Fraass AJ (2013) Oceanographic controls on the diversity and extinction of planktonic Foraminifera. Nature 493:398–401. doi:10.1038/nature11815

Phleger FB (1960) Ecology and distribution of recent Foraminifera. The Johns Hopkins Press, Baltimore

Phleger FB (1945) Vertical distribution of pelagic Foraminifera. Am J Sci 243:377–383

Prell WL, Martin A, Cullen JL, Trend M (1999) The Brown University Foraminiferal Data Base, IGBP PAGES/World) Data Center-A for Paleoclimatology Data Contribution Series # 1999-027. NOAA/NGDC Paleoclimatology Program, Boulder CO, USA

Ralston DK, McGillicuddy DJ, Townsend DW (2007) Asynchronous vertical migration and bimodal distribution of motile phytoplankton. J Plankton Res 29:803–821

Rebotim A, Voelker AHL, Jonkers L, Waniek JJ, Meggers H, Schiebel R, Fraile I, Schulz M, Kucera M (2016) Factors controlling the depth habitat of planktonic foraminifera in the subtropical eastern North Atlantic. Biogeosciences Discuss. doi:10.5194/bg-2016-348

Retailleau S, Eynaud F, Mary Y, Abdallah V, Schiebel R, Howa H (2012) Canyon heads and river plumes: how might they influence neritic planktonic Foraminifera communities in the SE Bay of Biscay? J Foraminifer Res 42:257–269

Retailleau S, Howa H, Schiebel R, Lombard F, Eynaud F, Schmidt S, Jorissen F, Labeyrie L (2009) Planktic foraminiferal production along an offshore-onshore transect in the south-eastern Bay of Biscay. Cont Shelf Res 29:1123–1135

Retailleau S, Schiebel R, Howa H (2011) Population dynamics of living planktic foraminifers in the hemipelagic southeastern Bay of Biscay. Mar Micropaleontol 80:89–100

Rhumbler L (1911) Die Foraminiferen (Thalamorphoren) der Plankton-Expedition. Erster Teil: Die allgemeinen Organisations-Verhältnisse der Foraminiferen. Ergeb Plankton-Exped Humbold-Stift 1909 3:331

Riley GA (1976) A model of plankton patchiness. Limnol Oceanogr 21:873–880

Ripperger S, Schiebel R, Rehkämper M, Halliday AN (2008) Cd/Ca ratios of in situ collected planktonic foraminiferal tests. Paleoceanography. doi:10.1029/2007PA001524

Rohling EJ, Sprovieri M, Cane T, Casford JSL, Cooke S, Bouloubassi I, Emeis KC, Schiebel R, Rogerson M, Hayes A, Jorissen FJ, Kroon D (2004) Reconstructing past planktic foraminiferal habitats using stable isotope data: a case history for Mediterranean sapropel S5. Mar Micropaleontol 50:89–123. doi:10.1016/S0377-8398(03)00068-9

Rutherford S, d' Hondt S, Prell W (1999) Environmental controls on the geographic distribution of zooplankton diversity. Nature 400:749–753

Saher MH, Rostek F, Jung SJA, Bard E, Schneider RR, Greaves M, Ganssen GM, Elderfield H, Kroon D (2009) Western Arabian Sea SST during the penultimate interglacial: a comparison of $U_{37}^{K'}$ and Mg/Ca paleothermometry. Paleoceanography. doi:10.1029/2007PA001557

Sautter LR, Sancetta C (1992) Seasonal associations of phytoplankton and planktic Foraminifera in an upwelling region and their contribution to the seafloor. Mar Micropaleontol 18:263–278

Schiebel R (2002) Planktic foraminiferal sedimentation and the marine calcite budget. Glob Biogeochem Cycles. doi:10.1029/2001GB001459

Schiebel R, Bijma J, Hemleben C (1997) Population dynamics of the planktic foraminifer *Globigerina bulloides* from the eastern North Atlantic. Deep-Sea Res I 44:1701–1713

Schiebel R, Hemleben C (2000) Interannual variability of planktic foraminiferal populations and test flux in the eastern North Atlantic Ocean (JGOFS). Deep-Sea Res II 47:1809–1852

Schiebel R, Hemleben C (2005) Modern planktic Foraminifera. Paläontol Z 79:135–148

Schiebel R, Hiller B, Hemleben C (1995) Impacts of storms on recent planktic foraminiferal test production and CaCO$_3$ flux in the North Atlantic at 47 °N, 20 °W (JGOFS). Mar Micropaleontol 26:115–129

Schiebel R, Schmuker B, Alves M, Hemleben C (2002a) Tracking the recent and late Pleistocene Azores front by the distribution of planktic foraminifers. J Mar Syst 37:213–227

Schiebel R, Waniek J, Bork M, Hemleben C (2001) Planktic foraminiferal production stimulated by chlorophyll redistribution and entrainment of nutrients. Deep-Sea Res I 48:721–740

Schiebel R, Waniek J, Zeltner A, Alves M (2002b) Impact of the Azores Front on the distribution of planktic foraminifers, shelled gastropods, and coccolithophorids. Deep-Sea Res II 49:4035–4050

Schiebel R, Zeltner A, Treppke UF, Waniek JJ, Bollmann J, Rixen T, Hemleben C (2004) Distribution of diatoms, coccolithophores and planktic foraminifers along a trophic gradient during SW monsoon in the Arabian Sea. Mar Micropaleontol 51:345–371

Schmidt DN (2002) Size variability in planktic foraminifers. PhD Thesis, ETH Zürich

Schmidt DN, Lazarus D, Young JR, Kucera M (2006) Biogeography and evolution of body size in marine plankton. Earth-Sci Rev 78:239–266

Schmidt DN, Renaud S, Bollmann J, Schiebel R, Thierstein HR (2004a) Size distribution of Holocene planktic foraminifer assemblages: biogeography, ecology and adaptation. Mar Micropaleontol 50:319–338

Schmidt DN, Thierstein HR, Bollmann J, Schiebel R (2004b) Abiotic forcing of plankton evolution in the Cenozoic. Science 303:207–210

Schmuker B, Schiebel R (2002) Planktic foraminifers and hydrography of the eastern and northern Caribbean Sea. Mar Micropaleontol 46:387–403

Seears HA, Darling KF, Wade CM (2012) Ecological partitioning and diversity in tropical planktonic Foraminifera. BMC Evol Biol 12:54. doi:10.1186/1471-2148-12-54

Siccha M, Schiebel R, Schmidt S, Howa H (2012) Short-term and small-scale variability in planktic Foraminifera test flux in the Bay of Biscay. Deep-Sea Res I 64:146–156. doi:10.1016/j.dsr.2012.02.004

Siccha M, Trommer G, Schulz H, Hemleben C, Kucera M (2009) Factors controlling the distribution of planktonic Foraminifera in the Red Sea and implications for the development of transfer functions. Mar Micropaleontol 72:146–156. doi:10.1016/j.marmicro.2009.04.002

Sousa SHM, de Godoi SS, Amaral PGC, Vicente TM, Martins MVA, Sorano MRGS, Gaeta SA, Passos RF, Mahiques MM (2014) Distribution of living planktonic Foraminifera in relation to oceanic processes on the southeastern continental Brazilian margin (23 °S–25 °S and 40 °W–44 °W). Cont Shelf Res 89:76–87. doi:10.1016/j.csr.2013.11.027

Spencer-Cervato C, Thierstein HR (1997) First appearance of *Globorotalia truncatulinoides*: cladogenesis and immigration. Mar Micropaleontol 30:267–291

Spindler M (1996) On the salinity tolerance of the planktonic foraminifer *Neogloboquadrina pachyderma* from Antarctic sea ice. In: Proceedings of the NIPR symposium on polar biology, pp 85–91

Spindler M, Dieckmann GS (1986) Distribution and abundance of the planktic foraminifer *Neogloboquadrina pachyderma* in sea ice of the Weddell Sea (Antarctica). Polar Biol 5:185–191

Spindler M, Hemleben C, Salomons JB, Smit LP (1984) Feeding behavior of some planktonic foraminifers in laboratory cultures. J Foraminifer Res 14:237–249

Stanley SM (1973) An explanation for Cope's rule. Int J Org Evol 27:1–26

Steinhardt J, Cléroux C, Ullgren J, de Nooijer L, Durgadoo JV, Brummer GJ, Reichart GJ (2014) Anti-cyclonic eddy imprint on calcite geochemistry of several planktonic foraminiferal species in the Mozambique Channel. Mar Micropaleontol 113:20–33. doi:10.1016/j.marmicro.2014.09.001

Stewart IA (2000) The molecular evolution of planktonic Foraminifera and its implications for the fossil record. PhD Thesis, University of Edinburgh

Thiede J (1975) Distribution of Foraminifera in surface waters of a coastal upwelling area. Nature 253:712–714

Thompson PR, Bé AWH, Duplessy J-C, Shackleton NJ (1979) Disappearance of pink-pigmented *Globigerinoides ruber* at 120,000 yr BP in the Indian and Pacific Oceans. Nature 280:554–558

Tolderlund DS, Bé AWH (1971) Seasonal distribution of planktonic Foraminifera in the western North Atlantic. Micropaleontology 17:297–329

Ufkes E, Jansen JHF, Brummer GJA (1998) Living planktonic Foraminifera in the eastern South Atlantic during spring: indicators of water masses, upwelling and the Congo (Zaire) River plume. Mar Micropaleontol 33:27–53

Vincent E, Berger WH (1981) Planktonic Foraminifera and their use in paleoceanography. Ocean Lithosphere Sea 7:1025–1119

Volkmann R (2000) Planktic foraminifer ecology and stable isotope geochemistry in the Arctic Ocean: implications from water column and sediment surface studies for quantitative reconstructions of oceanic parameters. Reports on Polar Research. PhD Thesis, AWI Bremerhaven

Wang L (2000) Isotopic signals in two morphotypes of *Globigerinoides ruber* (white) from the South China Sea: implications for monsoon climate change during the last glacial cycle. Palaeogeogr Palaeoclimatol Palaeoecol 161:381–394

Weiner A, Aurahs R, Kurasawa A, Kitazato H, Kucera M (2012) Vertical niche partitioning between cryptic sibling species of a cosmopolitan marine planktonic protist. Mol Ecol 21:4063–4073. doi:10.1111/j.1365-294X.2012.05686.x

Weyl PK (1978) Micropaleontology and ocean surface climate. Science 202:475–481

Williams DF, Be AWH, Fairbanks RG (1979) Seasonal oxygen isotopic variations in living planktonic Foraminifera off Bermuda. Science 206:447–449. doi:10.1126/science.206.4417.447

Wit JC, Reichart GJ, A Jung SJ, Kroon D (2010) Approaches to unravel seasonality in sea surface temperatures using paired single-specimen foraminiferal $\delta^{18}O$ and Mg/Ca analyses. Paleoceanography. doi:10.1029/2009PA001857

Žarić S, Donner B, Fischer G, Mulitza S, Wefer G (2005) Sensitivity of planktic Foraminifera to sea surface temperature and export production as derived from sediment trap data. Mar Micropaleontol 55:75–105

Sedimentation and Carbon Turnover

8

Calcification and dissolution of test CaCO₃ cause changes in the surface water carbonate system. Deep-water chemistry affects and is affected by the dissolution of tests (e.g., Berger and Piper 1972; Dittert et al. 1999). Thermodynamic dissolution of tests is evident below the calcite lysocline. Below the calcite compensation depth (CCD, Fig. 8.1) only minor proportions of calcite are preserved (Broecker and Peng 1982). However, dissolution of calcareous tests may occur even some distance above the calcite lysocline (Anderson and Sarmiento 1994; Schiebel 2002; Schiebel et al. 2007), possibly caused by the remineralization of organic matter, and decreasing pH within microenvironments (Milliman et al. 1999). On the other hand, tests may settle below the CCD because they sink faster than they are dissolved. Consequently, well-preserved calcareous tests may occur in sediments deposited below the CCD, although the quantitative composition of the thanatocoenosis (fossil tests contained in sea floor sediments over geological time periods) below the CCD may not display the original fauna. Supra-lysoclinal dissolution of tests in surface sediments may be caused by remineralization of organic matter and chemical conditions at the fluffy (Fig. 8.2) sediment-water interface (De Villiers 2005). In addition to changes in the faunal composition and test calcite budget, encrustation and dissolution affect the chemical composition, i.e. isotope and element ratios of

planktic foraminifer tests (e.g., Lohmann 1995; Van Raden et al. 2011).

In this chapter, an overview is given on how foraminifer tests and assemblages are affected by transportation and expatriation, dissolution, and encrustation during sedimentation, i.e. when settling through the water column and being embedded in the surface sediment. Upon the arrival of planktic foraminifer tests at surface sediments taphonomic processes take over (e.g., Berger 1971; Lončarić et al. 2007).

8.1 Test Flux Dynamics

Planktic foraminifer assemblages in the water column and sea floor sediments represent the sum of production (i.e. population dynamics) and preservation (i.e. taphonomic processes) of tests, including transportation, dissolution and encrustation during sedimentation (e.g., Berger 1971; Vincent and Berger 1981; Schiebel 2002). In general, increased numbers of empty and sinking tests result from increased growth rates mostly in the surface waters. Maximum numbers of empty tests within the water column occur following time-intervals of maximum production such as, for example, in spring at mid-latitudes, and upwelling seasons in monsoon climates (Schiebel 2002). Following seasons of enhanced biological production in surface waters, a vast number of large and fast settling tests occur in

8

© Springer-Verlag Berlin Heidelberg 2017
R. Schiebel and C. Hemleben, *Planktic Foraminifers in the Modern Ocean*,
DOI 10.1007/978-3-662-50297-6_8

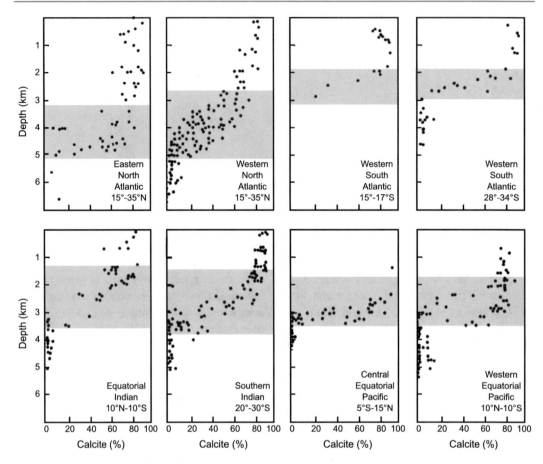

Fig. 8.1 Calcite content in sea floor sediments indicating varying depths of calcite lysocline and calcite compensation depth (CCD) at the ocean-basin scale. Maximum calcite dissolution occurs between the lysocline (about 80 % calcite preservation) and the CCD (<20 % calcite preservation). The upper limit of the depth-ranges shaded in *blue* gives the average depth of the lysocline, the lower limit gives the CCD. Deviations from the average result from regional effects. Redrawn from Broecker and Peng (1982). See also Sarmiento and Gruber (2006)

Fig. 8.2 Planktic foraminifer tests in the fluffy sea floor sediment surface are exposed to physical and chemical processes, which may cause alteration of the shell. Fluffy layer on top of fine-grained sediment contained within multicorer tube (left panel, width 4 cm). SEM image of fluffy sediment surface with planktic foraminifer tests (*arrows*) in the upper left corner (right panel, width 0.5 mm). Photo Ch. Hemleben

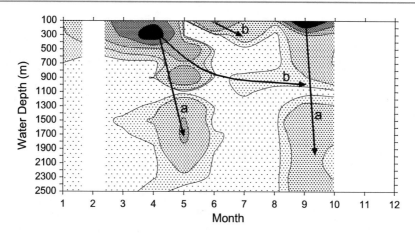

Fig. 8.3 Schematic view of average monthly planktic foraminifer test calcite flux between 100 and 2500 m in the eastern North Atlantic around 47°N, 20°W. Settling tracks of fast (**a**) and slow (**b**) settling tests are indicated by *arrows*. The month of May was sampled over five years. Black and gray levels correspond to $CaCO_3$ flux >60, >30–60, >10–30, >3–10, >1–3, and <1 mg $m^{-2}d^{-1}$. Time-intervals without data coverage are shown in *white*. From Schiebel (2002)

the deep-water column (Schiebel 2002). At the same time, small tests become attached to larger particles and settle to depth. Differential settling velocities of planktic foraminifer tests result in different settling tracks (Fig. 8.3), which can be traced through the water column by repetitive sampling at the same location (time-series station).

Regional qualitative and quantitative discrepancies between fluxes are caused by differential production and preservation of tests. Discrepancies may be due to the better preservation of settling tests under low-oxygen conditions as, for example, within the oxygen minimum zone (OMZ) of the Arabian Sea (cf. Hermelin et al. 1992) in comparison to the well-oxygenated water column of the eastern North Atlantic. Although standing stocks in surface waters are similar at both sites during the SW monsoon and during spring, respectively, the number of empty tests in the deep water column in the Arabian Sea is much higher than in the eastern temperate North Atlantic (Schiebel 2002).

Assuming an average life expectancy of surface dwelling species of one month, assemblages of empty planktic foraminifer tests follow the seasonal production and sedimentation pattern (see Chap. 7). The settling community is composed mostly of empty tests, and includes specimens that have undergone reproduction, and specimens, which died without having reproduced, mostly in juvenile and neanic stages. Following reproduction or death, planktic foraminifer tests sink out of their habitat and settle towards the sea floor (e.g., Berger 1971; Schiebel and Hemleben 2005). Planktic foraminifer tests settle to depth mostly individually in contrast to other particles like coccoliths, which are usually transported to depth within aggregates (e.g., Bishop et al. 1977; Thiel et al. 1989; Schiebel 2002; De La Rocha and Passow 2007; Schmidt et al. 2014). Settling velocities of individual tests depend on size, shape, and thickness of the shell, i.e., calcite mass (Fok-Pun and Komar 1983; Takahashi and Bé 1984). Small, thin-walled, and discoidal tests settle slower than large, heavy, and spherical tests (Table 8.1).

In addition to the empty test assemblage, cytoplasm-filled individuals, which may be alive but have lost buoyancy, are dragged to depth and contribute to the settling assemblage (Boltovskoy and Lena 1970; Takahashi and Bé 1984; Schiebel and Movellan 2012). Remaining cytoplasm within these tests has possibly no effect on their settling velocities (Table 8.2, Fig. 8.4). Spines decelerate the tests on their way to the sea floor by decreasing their weight-to-size ratio (Takahashi and Bé 1984; Furbish and Arnold 1997). In the

Table 8.1 Settling velocity (m day^{-1}) of empty planktic foraminifer tests calculated after Takahashi and Bé (1984)

Species	Sieve size (µm)							
	>100–125	>125–150	>150–200	>200–250	>250–315	>315–400	>400–500	>500
T. quinqueloba	100	115	142	–	–	–	–	–
G. siphonifera	107	167	196	334	361	373	639	–
G. falconensis	83	122	206	322	351	733	–	–
G. bulloides	83	115	237	328	434	597	885	1031
G. scitula	129	179	265	326	425	–	–	–
G. glutinata	100	185	247	359	476	–	–	–
N. incompta	91	129	222	408	493	–	–	–
G. inflata	100	173	232	350	515	738	1082	1534
G. hirsuta	122	167	296	441	691	986	1205	1551
Average	102	150	227	358	493	685	953	1339

From Schiebel and Hemleben (2000)

Table 8.2 Average diameter, weight, and sinking speed determined from settling experiments of tests of selected non-spinose (first five species) and spinose planktic foraminifer species

P-specimens	n	Test diameter		Test weight		Sinking speed	
		(µm)	±	(µg)	±	(m day^{-1})	±
G. menardii	12	658	117	27	11	1104	228
G. inflata	10	234	45	4	2	504	91
P. obliquiloculata	13	391	72	14	8	796	362
N. dutertrei	12	458	93	16`	8	842	290
G. glutinata	11	221	31	3	1	328	26
O. universa	8	573	74	8	6	277	144
G. ruber	10	314	49	6	3	198	94
G. sacculifer	11	328	99	10	6	274	143
G. siphonifera	10	377	96	5	4	271	191
G. bulloides	12	299	44	4	1	328	174
A-specimens							
G. menardii	15	572	174	22	15	918	322
G. inflata	11	309	101	9	8	753	311
P. obliquiloculata	9	399	71	13	11	769	333
N. dutertrei	14	374	94	11	7	766	359
G. glutinata	7	229	96	4	5	259	45
O. universa	8	521	52	17	7	701	219
G. ruber	22	289	82	8	6	723	321
G. sacculifer	16	430	170	25	20	1054	531
G. siphonifera	15`	347	130	5	3	364	160
G. bulloides	11	211	28	1	1	208	46

All specimens are from plankton tow samples (P-specimens). A-specimens were treated in low temperature asher to remove organic matter, and some specimens did not lose all spines in settling experiments. After Takahashi and Bé (1984)

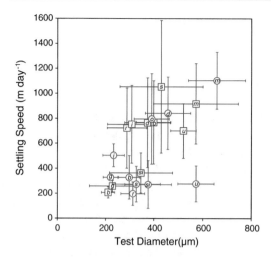

Fig. 8.4 Test size (µm) related settling speed (m day^{-1}) of empty tests (*blue*) and cytoplasm bearing tests (*red*). Non-spinose species: *G. menardii* (m), *G. inflata* (i), *P. obliquiloculata* (o), *N. dutertrei* (d) and *G. glutinata* (g). Spinose species: *O. universa* (u), *G. ruber* (r), *G. sacculifer* (s), *G. siphonifera* (si), and *G. bulloides* (b). Settling velocities between originally spinose and non-spinose species, and cytoplasm bearing versus empty tests are not systematically different. Data from Takahashi and Bé (1984), see Table 8.2

general case where adult specimens had undergone gametogenesis, the spines would have been shed in surface waters. In addition, spines are particularly prone to dissolution due to their high surface-to-volume ratio, and rapidly dissolved after being shed or after death of the individual. Consequently, few spine-bearing tests occur in the subsurface water column. Subsurface dwelling species may increasingly contribute to the settling assemblage in the deeper water column. However, deep-dwelling species reproduce mostly in surface waters, and are included in the assemblage of empty tests within surface waters (Hemleben et al. 1987; Schiebel et al. 2002).

8.1.1 Accumulation of Tests Within the Water Column

Small tests settle through the water column more slowly than assumed from test size, weight, and drag coefficient alone (Fok-Pun and Komar 1983;

Takahashi and Bé 1984). In particular, tests of small and thin-shelled species like *T. quinqueloba* settle at low velocity and decelerate with depth due to increasing seawater viscosity. These light tests may come to a halt within the water column, and accumulate at interfaces of changing viscosity and density between different water bodies. Small and slow-sinking tests are particularly prone to accumulate in the mid water column. Mesobathyal assemblages of *T. quinqueloba* form, for example, during summer in the mid-latitude eastern North Atlantic, when low-productive conditions in surface waters allow accumulation of particles at density-interfaces in the mid water column (Fig. 8.3). Enhanced plankton production and consequently enhanced flux of settling matter during spring and fall causes scavenging of tests (cf. Honjo and Manganini 1993). Small tests, which had accumulated over the low productive summer season would be cleared out from the water column, and dragged to depths in spring and fall (Fig. 8.3).

Viscosity and density: The viscosity of a liquid may also be expressed as 'fluidity' or 'thickness'. Less fluid means thicker and more viscous. The viscosity of seawater increases with salinity and decreases with temperature. In addition, viscosity increases with increasing pressure and depth. Temperature, salinity, and pressure affect (sea-) water density in the same sense as they affect viscosity, and buoyancy increases with increasing density. Since seawater is thicker and denser than freshwater, swimming in the sea is much easier for us (humans) than swimming in a lake. Hypersaline lakes like the Dead Sea even allow humans to float without much physical effort. Likewise, the viscosity and density of normal saline seawater supports buoyancy of live planktic foraminifers. In turn, active vertical displacement of planktic foraminifers by changing their buoyancy, via the amount of lipids embedded in vacuoles and fibrillar bodies (i.e. cell organelles, which may act as floating bodies; Sect. 3.2.5), is impeded by the viscosity of ambient seawater.

Changes of viscosity and density of the mesobathyal water column in the subtropical to temperate North Atlantic are caused by the Mediterranean outflow water (MOW), in addition to the effect resulting from decreasing temperature. MOW is more saline than the adjacent Atlantic water bodies and spreads from Gibraltar to mid depths mostly to the west and north (cf. Van Aken 2000). MOW was repeatedly detected by CTD recording during sampling campaigns in summer at the same depths as test 'clouds' of small and thin-shelled *T. quinqueloba* (Fig. 8.3). One of those mesobathyal assemblages of empty tests composed mainly of small (100–125 μm) sized *T. quinqueloba* was sampled at 1000–1500 m water depth in the eastern North Atlantic in July and August 1992 at the upper limit of the MOW. A disproportionately large number of *T. quinqueloba* tests were also observed during September when the species constituted about 50 % of the assemblage sampled with a sediment trap (Schiebel 2002). In contrast, *T. quinqueloba* constitutes only 5–15 % of the live fauna in summer and fall in the eastern North Atlantic around 47°N, 20°W (Schiebel and Hemleben 2000) and about 10 % of the sea floor sediment assemblage at the same location (Prell et al. 1999). Ratios of *T. quinqueloba* tests much higher than 10 % are explained by an accumulation mechanism such as viscosity driven density separation.

Preservation of *T. quinqueloba* tests trapped at horizons of sharp density-changes in the mesobathyal water column has generally been good. Most of the tests show no signs of dissolution although being highly susceptible to dissolution (Table 8.3), despite long exposure times to ambient seawater of several weeks or months (Fig. 8.3). Above average preservation of tests is facilitated by the high calcite saturation state ($\Omega >1$) at 500–1500 m water depth in the North Atlantic (Schiebel et al. 2007), and even higher

calcite super-saturation of waters sourced from the western Mediterranean (Millero et al. 1979).

8.1.2 Pulsed Test Flux

Sedimentation of planktic foraminifer tests occurs in complex intermittent pulses (Fig. 8.5) rather than a steady flow (e.g., Sautter and Thunell 1989; Bijma et al. 1994; Peeters et al. 1999), resulting from ecological (e.g., food availability) and biological prerequisites (e.g., reproductive cycles). Pulsed flux events cause mass dumps of fast settling particles, and yield a major contribution of tests to the formation of deep-sea sediments (Sect. 8.5.2, Fig. 8.18). Mass dumps take place at regions and during periods of high biological productivity such as, for example, seasonal upwelling, spring blooms, and during fall (Kemp et al. 2000; Kawahata 2002; Schiebel 2002). During low productive periods, a steady rain of slow-sinking tests contributes only small amounts of tests to deep marine sediments (Schiebel 2002). In addition to seasonal test flux pulses, interannual changes affect different species to varying degrees, depending on their adaptation to regional ecologic conditions. Comparatively balanced flux occurs in the tropical ocean, and displays less distinct seasonality than at high latitudes. However, pulsed foraminifer test flux at low latitudes follows the same processes as at higher latitudes (Schiebel and Hemleben 2005; Buesseler et al. 2007). Seasonal and Interannual variability of species-specific flux patterns, and their relation to ecological conditions are determined by time-series analyses of samples (e.g., Lončarić et al. 2007; Wejnert et al. 2010; Kuhnt et al. 2013) (Sect. 10.1.8).

Quantitative differences between test (i.e. test numbers) and calcite (i.e. $CaCO_3$ mass) fluxes result from differential production of species (environmental conditions and biological prerequisites), as well as differential flux modes. High settling velocities (Tables 8.1 and 8.2) of relatively large tests dominate the maximum $CaCO_3$ flux pulses following maximum production of planktic foraminifers. In contrast, large numbers of small tests settle through the water

Table 8.3 Ranking of average (Av., plus standard deviation, ±) dissolution susceptibility

Genus	Species	Av.	±	[1]	[2]	[3]
Turborotalita	*humilis*	**96**	6	100	100	89
Globorotalia	*tumida*	**90**	10	95	97	79
Berggrenia	*pumilio*	**89**	–	–	89	–
Neogloboquadrina	*pachyderma*	**88**	12	77	86	100
Sphaeroidinella	*dehiscens*	**87**	10	91	94	75
Globorotalia	*crassaformis*	**82**	5	86	77	82
Pulleniatina	*obliquiloculata*	**81**	10	82	91	71
Neogloboquadrina	*dutertrei*	**81**	7	73	83	86
Globorotalia	*inflata*	**78**	16	64	74	96
Globorotalia	*truncatulinoides*	**74**	17	59	71	93
Globorotalia	*menardii*	**68**	–	–	–	68
Globorotalia	*cavernula*	–	–	–	–	–
Globorotalia	*theyeri*	–	–	–	–	–
Globorotalia	*ungulata*	–	–	–	–	–
Globoquadrina	*conglomerata*	**65**	1	–	66	64
Beella	*digitata*	**62**	1	–	63	61
Beella	*megastoma*	–	–	–	–	–
Globorotalia	*hirsuta*	**62**	10	55	69	–
Globorotaloides	*hexagonus*	**60**	–	–	60	–
Globorotalia	*scitula*	**57**	0	–	57	57
Candeina	*nitida*	**51**	1	50	51	–
Tenuitella	*iota*	**50**	6	–	46	54
Tenuitella	*compressa*	–	–	–	–	–
Tenuitella	*fleisheri*	–	–	–	–	–
Tenuitella	*parkerae*	–	–	–	–	–
Globigerina	*falconensis*	**46**	4	–	49	43
Globigerinita	*glutinata*	**46**	4	45	43	50
Globigerinita	*minuta*	–	–	v	–	–
Globigerinella	*calida*	**40**	8	–	34	46
Globigerinita	*uvula*	**40**	–	–	40	–
Neogloboquadrina	*incompta*	**37**[*]	–	–	–	–
Orbulina	*universa*	**34**	23	9	54	39
Turborotalita	*quinqueloba*	**32**	8	41	29	25
Turborotalita	*clarkei*	–	–	–	–	–
Globigerinoides	*conglobatus*	**30**	3	32	26	32
Globigerina	*bulloides*	**29**	8	36	31	21
Globigerinoides	*sacculifer*	**25**	3	23	23	29
Globigerinella	*siphonifera*	**22**	12	14	17	36
Bolliella	*adamsi*	**20**	–	–	20	–
Globoturborotalita	*tenella*	**20**	7	27	14	18

(continued)

Table 8.3 (continued)

Genus	Species	Av.	±	[1]	[2]	[3]
Globoturborotalita	rubescens	**13**	4	18	11	11
Globigerinoides	ruber (white)	**9**	5	5	9	14
Globigerinoides	ruber (pink)	–	–	–	–	–
Hastigerina	pelagica	**4**	1	–	3	4
Hastigerina	digitata	–	–	–	–	–

High numbers correspond to high preservation potentials (after Dittert et al. 1999). Averages are calculated from data given by [1] Berger (1970), [2] Parker and Berger (1971), and [3] Berger (1979). Data[*] on *N. incompta* (i.e. *N. pachyderma* subantarctic variety) are from Malmgren (1983). High standard deviation signifies species, which may produce shells of varying structure (e.g., pore density, thickness, GAM calcification). In case of missing data (e.g., *B. megastoma*), the respective species are related to other species of the same genus. No information is given on the ranking of *Dentigloborotalia anfracta, Gallitellia vivans, Orcadia riedeli, Streptochilus globigerus,* and *tenuitellids*

column at comparatively low velocity, and constitute low $CaCO_3$ flux during times of low production (cf. Deuser 1987). Both large and small tests predominantly result from time intervals of maximum production (Sect. 8.1.1). For example, maximum test flux at 2000 m water depth during summer in the NE Atlantic is mainly caused by small (20–125 µm in diameter) and thin-shelled tests of *T. quinqueloba* of relatively low calcite mass (Schiebel 2002), which were mainly produced in spring (cf. Schiebel and Hemleben 2000). Maximum test flux may hence cause only moderate $CaCO_3$ flux, and vice versa (Fig. 8.6).

8.1.3   Mass Sedimentation of Tests

Disproportionally high test-calcite flux of >1 g $CaCO_3$ m^{-2}day^{-1} at 1000–2500 m water depth occurred during March 1995 around new moon in the Arabian Sea (Schiebel 2002) (Sect. 8.5.1). The $CaCO_3$ flux pulse was mainly caused by large tests of *G. siphonifera* and *G. sacculifer* (>315–700 µm). Although *G. sacculifer* is a frequent faunal element in the Arabian Sea (Auras-Schudnagies et al. 1989; Conan and Brummer 2000; Naidu and Malmgren 1996), mass sedimentation of large tests of *G. sacculifer* was only observed in three (consecutive samples) out of 285 samples obtained in eight sampling campaigns (Schiebel 2002). Production and flux of *G. sacculifer* are related to the synodic lunar

cycle (Almogi-Labin 1984; Bijma et al. 1990a; Erez et al. 1991; Bijma et al. 1994). Consequently, the observed mass flux event does not represent the average sedimentation scenario, but is a monthly recurrent (though under-sampled) feature of deep marine sedimentation (cf. Anderson and Sarmiento 1994). Above-average ratios of large tests in low latitude sea floor sediments may be proof of these mass flux events (cf. Peeters et al. 1999). The temporal resolution of most deep sediment traps (>1000 m water depth) is too low to record single mass flux events. Shallow sediments traps (<1000 m depth) generally sample at low trapping efficiency (Scholten et al. 2001), and may miss mass flux events. Mass flux events may not be detected by most moored sediment traps, because mass flux over very short time-intervals of a couple of hours or days have been part of samples, which integrate over longer time-intervals of typically one or two weeks (see Sect. 10.1.7, Table 10.1). In contrast, the ratio of large to small tests obtained by multinet samples during mass flux events is much larger than in trap samples. Variations in standing stocks and fluxes are statistically significant for shorter time-intervals of hours rather than days. Unfortunately, mass flux events have rarely been sampled by plankton net-haul because their exact occurrence is unpredictable depending on ecological, biological, and hydrographical conditions, and wide dispersal of tests in the vastness of the deep ocean.

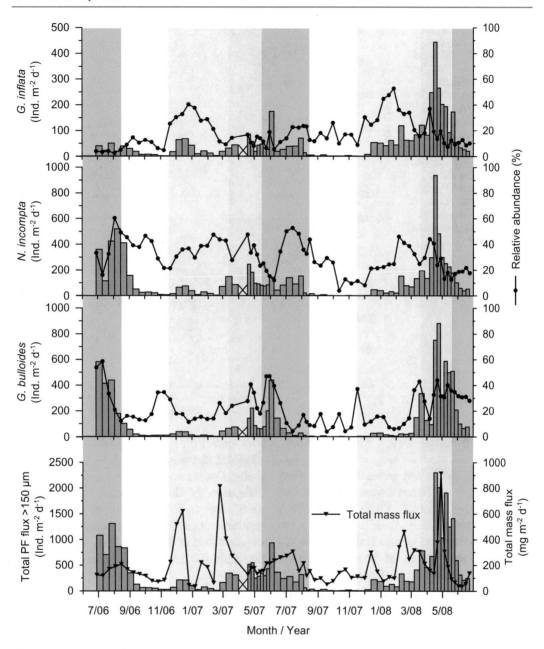

Fig. 8.5 Total mass flux and planktic foraminifer (PF) test flux (*gray bars*) at the SE Bay of Biscay sampled with a sediment trap moored at 1700 m water depth from June 2006 to June 2008 (x-axis shows beginnings of the months). Fluxes of the most abundant species *G. bulloides*, *N. incompta*, and *G. inflata* are given as absolute and relative abundances (mean binomial standard errors of 3.8, 4.8 and 5 %, respectively). Maximum production and flux of planktic foraminifer relatively occurs with a time lag of some weeks after high primary production (*green*) in spring. Time lags depend on species-specific settling modes and velocities (see Tables 8.1 and 8.2). Production in summer (*red*), fall (*white*), and winter (*blue*) is lower than in spring, and test flux occurs with a time lag of several weeks. Relative changes in seasonal production according to changes in sverdrup's cristical depth (CRD) after Obata et al. (1996). 'X' indicates sampling gap. After Kuhnt et al. (2013)

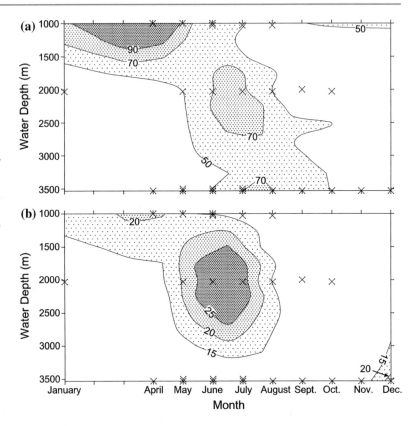

Fig. 8.6 a Planktic foraminifer CaCO$_3$ flux (mg m^{-2}day^{-1}), and **b** test flux (10^3 tests m^{-2}day^{-1}) sampled with sediment traps moored at 47°N, 20° W, between 1000 m and 3530 m water depth. **a** Maximum CaCO$_3$ mass flux in spring results from relatively few tests. In turn, **b** maximum test flux in summer is caused by high numbers of tests of low calcite mass. Crosses indicate sampling intervals. From Schiebel (2002)

8.1.4 The 'Large Tests' Phenomenon

Planktic foraminifer test assemblages deposited at the sea floor are biased towards large and fast sinking (up to 1500 m d^{-1}) tests (e.g., Berelson 2002). In particular, sea floor sediments of the tropical to subtropical oceans contain disproportionately high numbers of large tests, when compared to the live fauna (Peeters et al. 1999). Those large tests may result from flux events like the mass sedimentation described above (Sect. 8.1.3). Pulsed flux events are a major contribution to deep-sea sediment accumulation and remove shell-bound bicarbonate over long time-scales from the upper ocean by transferring and burying it in deep-sea sediments (cf. Berger and Wefer 1990; Wefer 1989). Mass sedimentation requires the presence of species that have the biological prerequisites to form large tests and large numbers of specimens (Brummer et al. 1987; Hemleben et al. 1987; Caron et al. 1990) and are adapted to specific environmental conditions (Bijma et al. 1990b; Huber et al. 2000). In

addition to biological prerequisites, hydrographic conditions need to support mass dumps of large shells. One example is the fall dump of large diatoms in the Guaymas Basin, Gulf of California (Kemp et al. 2000), due to breakdown of stratification in the surface water column and sedimentation of an accumulated 'shade flora' in fall. A similar scenario is imaginable for planktic foraminifers.

In contrast to large tests, small tests are exposed to dissolution in the water column much longer than large tests due to their low sinking velocity (around 100 m d^{-1}) and are preferentially removed from the assemblage when settling freely. The majority of small tests settle through the water column much slower than assessed from their test morphology (cf. Takahashi and Bé 1984). In turn, small tests can be quantitatively transported with good preservation from surface waters to depth during mass dump events along with other particles (Schiebel 2002), as observed from sediment trap samples from 2000 and 3000 m water depth below a naturally fertilized high-nutrient low-chlorophyll (HNLC) region at Crozet seamount in the southern Indian Ocean (Salter et al. 2014). However, small tests are prone to winnowing, particularly by bottom water currents along bathymetric features like continental slopes, seamounts, and canyons (cf. Stow et al. 2002). To conclude, test assemblages result from a complex combination of biological, ecological, oceanographic, sedimentological, and taphonomic effects at the local to regional scale, favoring sedimentation of large tests over small tests with increasing water depth (cf. Berelson 2002).

8.2 Transportation and Expatriation

Horizontal transport and expatriation of live planktic foraminifers and empty tests by surface and subsurface currents adds complexity to the ecological and paleoceanographic analysis of faunas and assemblages (e.g., Weyl 1978). Ecologic conditions may change over the individual ontogenetic development and biogeochemical conditions, which affect calcite precipitation. Stable isotopes and Me/Ca ratios of the test calcite provide a mixed signal and do not necessarily display ecological conditions of the sampling location (Van Sebille et al. 2015). Transportation and expatriation of foraminifers within their 'original water body' (e.g., within an eddy) would not make analysis easier, because conditions at the sampling site would still not be displayed. The same is true for the transport of dead individuals and empty tests.

Depending on the velocity of surface and subsurface currents, planktic foraminifer tests are horizontally transported by up to several hundred kilometers. 'Statistical funnels' of a radius of up to ~ 500 km result from a modeling study of Siegel and Deuser (1997), using input data typical of small-sized and slow-sinking foraminifer tests (50–200 m day^{-1}, see Tables 8.1 and 8.2), being affected by the Gulf Stream recirculation in the Sargasso Sea near Bermuda. For a sediment trap deployed at 1125 m water depth in the West Spitzbergen Current, average trajectory lengths of

25–50 km for *N. pachyderma* and 50–100 km for *T. quinqueloba* were calculated (von Gyldenfeldt et al. 2000). The reconstructed catchment areas are up to 230 km long and 140 km wide, i.e. catchment areas of 23,900 km^2 for tests of *N. pachyderma,* and 33,300 km^2 for tests of *T. quinqueloba* (Fig. 8.7). Both species are small sized and have long residence times in the water column. At current velocities of up to 40 cm s^{-1} in the West Spitzbergen Current tests are transported over long distances and short time-intervals (von Gyldenfeldt et al. 2000). Since larger and heavier planktic foraminifer tests settle through the water column at higher velocity (Tables 8.1. and 8.2) they are transported over shorter distances of some tens of kilometers before arrival at depth. Therefore, larger tests provide results of higher regional accuracy, which is of particular importance when working in areas of high spatial variability such as hydrographic fronts and in regions which are characterized by high current velocities (cf. Caromel et al. 2013).

8.3 Dissolution

Dissolution of shells increases with decreasing carbonate ion concentration ([CO$_3^{2-}$]) and calcite saturation state (Ω) at increasing water depth, and may continue at the sea floor sediment surface (e.g., Berger 1971; Henrich and Wefer 1986; Broecker and Clark 2001; De Villiers 2005; Schiebel et al. 2007). Increasing excess alkalinity (TA*) below 3500–5000 m water depth may contribute to benthic carbonate dissolution (Berelson et al. 2007). In addition to carbonate chemistry of ambient seawater, the degree of dissolution is related to structure (i.e. dissolution susceptibility) of the foraminifer shell (Plate 8.1), as well as settling velocity and exposure time of tests (Schiebel et al. 2007).

In general, dissolution susceptibility is species-specific (Table 8.3), with *Hastigerina pelagica, Globigerinoides ruber,* and *Globoturborotalita rubescens* being most susceptible, and *Turborotalita humilis, Berggrenia pumilio,* encrusted *Neogloboquadrina pachyderma,* and

Plate 8.1 Dissolution of planktic foraminifer shells illustrated with SEM images. (*1*) Well preserved assemblage with ▶
some mechanical damage only, and containing pteropod (P) shells made of aragonite. (*2*) Assemblage with well preserved
planktic foraminifer shells, (*3*) moderate dissolution, and (*4*) heavy dissolution (S is *S. dehiscens*). (*5*) Assemblage with
near complete dissolution of calcareous shells, being dominated by agglutinated benthic foraminifer tests. (*6*) Loosening
of layered shell structure, and peeling off of the inner calcite layers. (*7*) Labyrinth structures on the outer shell (*left side*),
and peeling off of outer calcite layers (*upper right corner*). (*8*) Close-up of labyrinth structures on the outer shell. Bars of
assemblages (*1–5*) 1 mm, close-ups (*6, 7*) 5 μm, (*8*) 2 μm

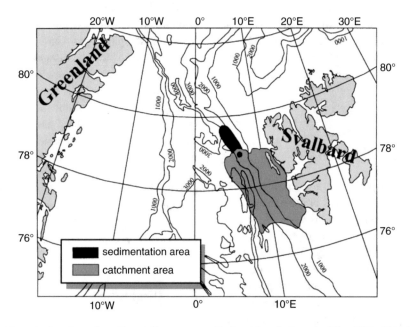

Fig. 8.7 Catchment (*gray*) and sedimentation areas
(*black*) calculated from assemblages of *T. quinqueloba*
and *N. pachyderma*, sampled by a sediment trap moored
in the West Spitzbergen current west of Svalbard. The
solid *black circle* within the catchment area indicates the
position of the mooring. The planktic foraminifer tests are
displaced largely from the SE to NW within the catchment
area, and would (if not sampled by the sediment trap) be
displaced further to the NW to finally be embedded within
the surface sediment within the sedimentation area. After
von Gyldenfeldt et al. (2000)

Sphaeroidinella dehiscens, as well as some
globorotalid species being most resistant to dissolution (see Dittert et al. 1999). Dissolution and
destruction of the shell ultrastructure renders
tests increasingly prone to fragmentation also by
physical force. A fragmentation index is hence
applied to the reconstruction of past [CO_3^{2-}] and
Ω at the basin scale (Berger 1973; Broecker and
Clark 1999; Dittert and Henrich 2000; Conan
et al. 2002; Volbers and Henrich 2002). In
addition, preservation of planktic foraminifer
tests varies at the regional and temporal scale
and might be affected by biogeochemistry of
ambient water and *p*H within microenvironments
(Milliman et al. 1999). On a global average, one
fourth of the initially produced planktic foraminifer calcite is assumed to settle on the sea
floor above the lysocline (Schiebel 2002). Dissolution–resistant species (e.g., *S. dehiscens*, see
Plate 8.1) increasingly dominate towards depth
and may eventually constitute the residual test
assemblages in deep basins (e.g., Ivanova et al.
2003). Starting at the lysocline to calcite compensation depth (CCD), the predominance of
foraminifer calcite mass increasingly shifts
towards a predominance of coccolithophore calcite mass in sediments below the CCD in subtropical gyres (Frenz et al. 2005), while coarser

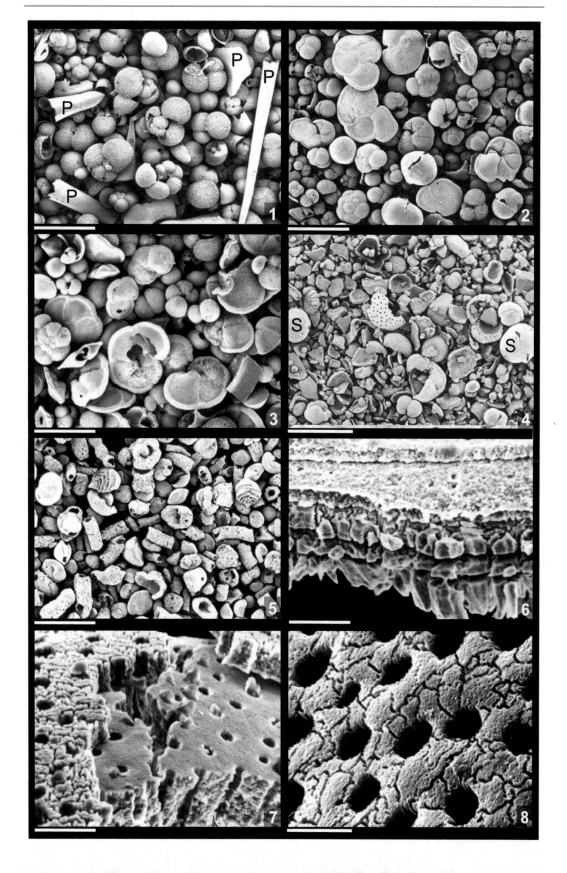

particles (i.e. foraminifer tests and fragments) may dominate elsewhere (Paull et al. 1988). The phenomenon is caused by the fact that coccoliths are enveloped in organic matter and often embedded in fecal pellets, and are composed of purer and hence more dissolution resistant calcite than planktic foraminifers.

The most significant decrease in planktic foraminifer test flux occurs in the ocean's twilight zone at water depths between 100 and 1000 m (Fig. 8.8), and hence at depths of calcite supersaturation (Fig. 8.9) where thermodynamic calcite dissolution is unlikely (cf. Broecker and Peng 1982; Zeebe and Wolf-Gladrow 2001; Sarmiento

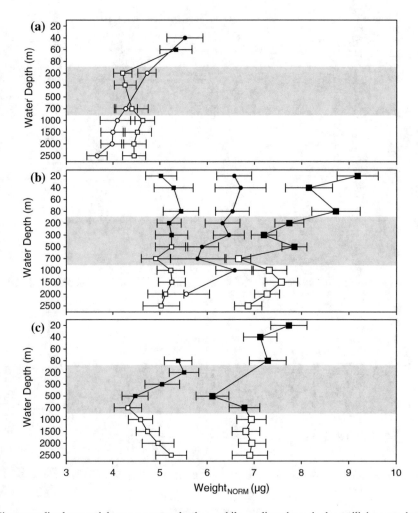

Fig. 8.8 Size normalized test weight versus water depth, in the eastern North Atlantic **a** in spring 1992, and **b** in fall 1996 (47°N, 20°W), and **c** in the Arabian Sea (16°N, 60°E) during SW monsoon 1995. Although absolute test weight largely differs between species and ocean basins, maximum decrease in test weight uniformly occurs in the twilight zone (*shaded*) between about 100 and 1000 m water depth. Individual tests of *G. bulloides* and *G. glutinata* lose on average about one-fifth in weight while settling through the twilight zone between about 100 and 1000 m water depth. Cytoplasm bearing tests are indicated by filled symbols, empty tests by open symbols. *Large squares* represent large *G. bulloides* (300 μm minimum test diameter), *small squares* represent small *G. bulloides* (250 μm), and *circles* represent small *G. glutinata* (250 μm). Standard deviation is given as error bars. From Schiebel et al. (2007)

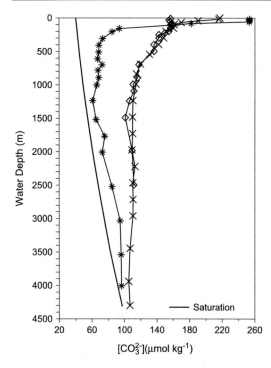

Fig. 8.9 Carbonate ion concentration [CO$_3^{2-}$] in the North Atlantic (47°N, 20°W) during spring (*diamonds*) and fall (*crosses*), and in the Arabian Sea (16°N, 60°E) during SW monsoon (*asterisks*) indicates calcite supersaturation throughout the analyzed water depths. From Schiebel et al. (2007)

and Gruber 2006; Friis et al. 2006; Schiebel et al. 2007). However, dissolution of planktic foraminifer calcite may not only be caused by the Δ [CO$_3^{2-}$] of ambient seawater at the outside of tests, and also takes place at the inside of tests. Dissolution at the inside of tests possibly results from bacterially mediated decomposition of cytoplasm and decreasing *p*H in microenvironments (Schiebel et al. 1997; Milliman et al. 1999; cf. also Boltovskoy and Lena 1970; Turley and Stutt 2000; Jansen et al. 2002). Dissolution of settling tests has been observed to be stronger in well-oxygenated waters than in low-oxygen environments where bacterial activity is limited by the availability of oxygen (e.g., Schiebel 2002). Accordingly, calcite preservation is better in the prominent oxygen minimum zone (OMZ) of the Arabian Sea than in well-oxygenated waters of the North Atlantic

(Fig. 8.9). In addition, better preservation of foraminifer tests in the Arabian Sea than in the NE Atlantic is probably due to higher average settling velocities of larger tests, and hence shorter exposure times of tests to ambient seawater in particular during seasonal (e.g., SW monsoon) mass flux events (Schiebel et al. 2007).

Below the twilight zone, at 1000–2500 m water depth, bacterially mediated dissolution has largely ceased [CO$_3^{2-}$] and planktic foraminifer shell flux may increase to values higher than above (Figs. 8.8 and 8.9). Since mostly large and dissolution-resistant tests arrive at depths below the twilight zone, the average weight (i.e. calcite mass) and settling velocity of the remaining test assemblage increases with depth (Berelson 2002; Schiebel et al. 2007).

In addition to thermodynamic and bacterially mediated processes calcite dissolution far above the lysocline may take place within the guts of grazers (Hemleben et al. 1989; Jansen and Wolf-Gladrow 2001), but which is minor part of the global planktic foraminifer carbon turnover and calcite budgets (see Sect. 4.8 Predation). Quantitative dissolution of tests within fast sinking aggregates of marine snow is unlikely since planktic foraminifer tests are only occasionally contained within organic-rich and microbe-rich aggregates (cf. Ransom et al. 1998; Schmidt et al. 2014). To conclude, dissolution of planktic foraminifer tests in supersaturated waters with respect to calcite ($\Omega > 1$) is hitherto not sufficiently explained and 'the global carbonate budget is far from resolved' (Berelson et al. 2007).

Dissolution and overgrowth of tests during sedimentation affect the composition of faunal assemblages through the presence or absence of tests of more or less dissolution-resistant species (e.g., Dittert et al. 1999). In the case of uncertain degrees of dissolution and overgrowth, care must be taken when analyzing biogeochemical data (i.e., stable isotopes, and element ratios) measured on planktic foraminifer tests (e.g., Pearson and Palmer 2000; Van Raden et al. 2011; Pearson 2012). Quantification of dissolution and its effect on the faunal composition of planktic foraminifer assemblages is difficult. Dissolution of selected specimens can be visualized by scanning electron microscopy (SEM) or other high-resolution technology like computed tomography (CT; Johnstone et al. 2010). Since analyses of entire assemblages using SEM or CT would be too laborious, fragmentation indexes are employed for quantitative assessment of dissolution of samples (Plate 8.1). A fragmentation index for the evaluation of the effect of dissolution proposed by Ivanova (1988) relates the number of fragments (F) to the number of fragments plus entire tests (TE):

$$F = (F + TE) * 100 \qquad (8.1)$$

The solution index (SI) of Berger (1973) relates the number of tests of resistant species (SR) to the total number of tests of common low latitude species (ST). When using the SI index of Berger (1973), the respective resistant and susceptible species need to be defined for a given region.

$$SI = SR/ST \qquad (8.2)$$

8.4   Overgrowth

Overgrowth is due to non-biogenic processes. Dissolution and overgrowth of empty planktic foraminifer tests may affect tests settling through the water column (Deuser et al. 1981) and exposed to ambient seawater over days or weeks depending on their species-specific (test shape, shell surface and thickness) and size-related (i.e. test mass) settling velocity. Laboratory experiments have shown that various species form calcite crusts during their ontogeny (Hemleben et al. 1985). A massive calcite crust of disputed origin may cover tests of N. pachyderma from polar and sub-polar waters (Simstich et al. 2003) as well as other species at lower latitudes (Lohmann 1995). Deduced from the $\delta^{18}O$ signal of N. pachyderma tests, Simstich et al. (2003) suggest active crust formation by the live foraminifer at sub-thermocline water depths between 70 and 250 m in the subpolar North Atlantic, whereas T. quinqueloba hardly forms any crust at the same time. In contrast to the North Atlantic, calcite

crusts have not been observed to the same extent in N. pachyderma tests from sediment-trap samples from 2000 and 3000 m water depth at subpolar waters near Crozet Seamount in the southern Indian Ocean (Salter et al. 2014). The questions arise as to what degree encrustation of N. pachyderma is affected by the chemistry (e.g., $[CO_3^{2-}]$) of ambient seawater at subsurface depths and to what degree calcite crusts of fossil tests of N. pachyderma in sea floor sediments are of biogenic or non-biogenic origin.

Calcite crust: Some uncertainty concerning the origin of calcite crusts covering planktic foraminifer shells may be due to inconsequent and confusing use of terminology. The formation of calcite crusts is sometimes attributed to 'deep growth'. 'Deep growth' has been identified in species like N. pachyderma and is absent in other species as, for example, G. ruber. The term 'deep growth' might have been deduced from 'overgrowth', and does not refer to any biogenic process. Overgrowth is not engaged in active calcification of the individual shell and signifies purely thermodynamic calcite precipitation. Such calcite overgrowth on top of fossil shells is assumed precipitated in equilibrium with sediment chemistry and the chemistry of interstitial pore waters. Model calculations on the ratio of dissolution to encrustation of planktic foraminifer shells by Lohmann (1995) provide a theoretical explanation and quantification of the chemical composition, including stable isotopes, of planktic foraminifer tests from sediment samples (Fig. 8.10).

A positive ΔCO_3^{2-} of surface and subsurface waters in the subpolar to temperate North Atlantic (Fig. 8.11) would foster thermodynamic calcite precipitation and encrustation of tests during sedimentation (cf. Simstich et al. 2003; Sarmiento and Gruber 2006, and references therein). In contrast, decreasing $[CO_3^{2-}]$, and negative ΔCO_3^{2-} in the subsurface to deep water

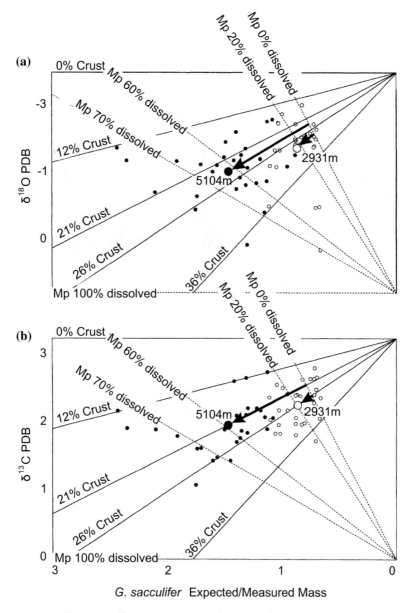

Fig. 8.10 Changes in $\delta^{18}O$ and $\delta^{13}C$ of fossil *G. sacculifer* tests from surface sediment samples from the Sierra Leone Rise. *Solid dots* and *open circles* indicate samples from 5104 m and 2931 m water depth, respectively (large symbols indicate mean values). Degrees of assumed encrustation and dissolution between the hypothetical end-members of '0 % Crust' and '100 % dissolved' are indicated by *solid* and *stippled lines*, respectively. M_P signifies the mass of primary chamber calcite. From Lohmann (1995)

column would impede calcite precipitation and may cause dissolution of planktic foraminifer shells, which is the case in the tropical to temperate Atlantic (e.g., Broecker and Clark 1999; Broecker and Clark 2001) and Southern Ocean (cf. Salter et al. 2014). In addition, low *p*H within microenvironments produced by degradation of organic tissues presumably causes considerable dissolution at the inside of tests (Milliman et al. 1999; Schiebel 2002; Schiebel et al. 2007; Johnstone et al. 2010; Constandache et al. 2013).

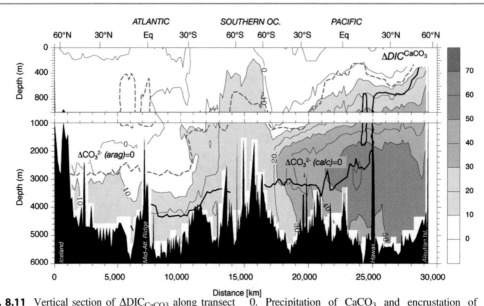

Fig. 8.11 Vertical section of ΔDIC_{CaCO3} along transect from Iceland in the North Atlantic to the left, to the South Atlantic, along 60°S off the East Antarctic into the South Pacific, and along ~160°E into the North Pacific off the Aleutian Islands. Super-saturation with respect to the mineral phases of aragonite (*hatched line*) and calcite (*solid line*) occurs above the saturation horizons, ΔCO_3^{2-} = 0. Precipitation of $CaCO_3$ and encrustation of tests fostered above and impeded below saturation horizon. Significant dissolution tends to coincide with the saturation horizon of aragonite. From Ocean Biogeochemical Dynamics by Jorge L. Sarmiento and Nicolas Gruber. Copyright (C) 2006 by Princeton University Press. Reprinted by permission

Dissolution at the sediment-water interface including the benthic fluff layer (see Fig. 8.2) above and within surface sediments depends on the residence time of tests and small-scale chemical conditions (cf. Lohmann 1995; De Villiers 2005; Feely et al. 2008). Dissolution or overgrowth of fossils tests within surface sediments act over much longer time scales, i.e. years to millennia, compared to days and weeks within the water column. As a result of long-term exposure to either carbonate super-saturation or under-saturation, overgrowth or dissolution of tests embedded in the sediment may appear less selective than in short-term processes within the water column. Precipitation of crusts affects tests embedded in calcareous surface sediments (Boussetta et al. 2011) and is a frequent phenomenon in supra-lysoclinal sediments (cf. Lohmann 1995; Van Raden et al. 2011). In contrast, little or no overgrowth occurs in clay–rich sediments and produces perfectly well preserved ('glassy') tests which provide ideal (un-contaminated) carriers of paleoceanographic proxies (Sexton et al. 2006). The fact that the same species may or may not be encrusted in calcareous and clayey sediments, respectively, indicates potential formation of late sedimentary to early diagenetic overgrowth on top of fossil planktic foraminifer tests.

The combination of both dissolution and crust formation adds considerable uncertainty to the paleoceanographic interpretation of stable isotope data from fossil foraminifer calcite, which is difficult to disentangle and quantify. The long-known (e.g., Bouvier-Soumagnac et al. 1986) but often ignored combined signal of biogenic plus taphonomic effects needs to be quantified for detailed reconstruction of biological, ecological, and sedimentological processes, rather than information on average conditions resulting from analysis of the bulk test calcite. Data on bulk test calcite often foster the misleading idea that isotope and element ratios of bulk test calcite display the ecology, i.e. depth habitat and seasonal occurrence of any extinct species. When differentiating between primary (i.e. ontogenetic growth) and secondary (i.e. non-biogenic) calcite, the ecology and taphonomy of a planktic foraminifer should be reconstructed in detail to provide accurate data on the paleoenvironment.

8.5 Carbon Turnover

Planktic foraminifers affect the regional to global carbon budget by sequestration of calcareous tests mainly from bicarbonate (HCO_3^-, $\sim 90\%$), carbonate (CO_3^{2-}, $\sim 10\%$), carbonic acid (H_2CO_3, $\sim 1\%$), and carbon dioxide (CO_2, $<1\%$) depending on pH, temperature, salinity, and concentration of dissolved inorganic carbon (DIC) of ambient seawater (e.g., Zeebe and Wolf-Gladrow 2001, Bjerrum plot). When precipitating their shell, planktic foraminifers fix half of the CO_2 sourced from the different carbonate-species within their test calcite, and

release the other half to the environment (Fig. 8.12). Stoichiometrically, calcification of planktic foraminifer tests follows the equation $Ca^{2+} + 2HCO_3^- \rightarrow CaCO_3 + CO_2 + H_2O$ (for bicarbonate only). Calcification of planktic foraminifer tests is hence a source of CO_2 to surface waters and atmosphere, called carbonate counter pump, which acts on short time-intervals of days to seasons. On long geological time-scales of millions of years, sedimentation and burial of planktic foraminifer tests is a sink of CO_2 (Zeebe 2012). Opposite to the carbon of planktic foraminifer tests, the carbon ingested with their food and stored in the foraminifer cytoplasm is quantitatively removed from the surface water carbon pool during sedimentation, and constitutes a sink of CO_2 (Fig. 8.12).

Sedimentation of planktic foraminifers removes and transfers carbon from the surface to

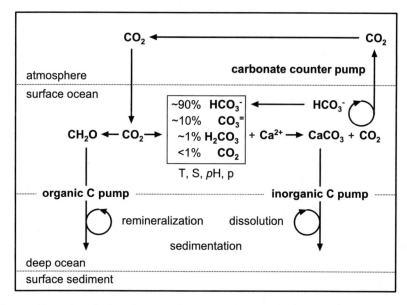

Fig. 8.12 Non-stoichiometric scheme of the planktic foraminifer carbon pump, including the organic (CH_2O, i.e. cytoplasm carbon) and inorganic (test $CaCO_3$) carbon mass. At a regional to global scale, the planktic foraminifer C_{INORG} to C_{ORG} ratio of settling assemblages ranges at about 5:1 to 10:1, whereas the C_{INORG} to C_{ORG} ratio of live individuals is $\sim 1{:}3$ (Schiebel and Movellan 2012). For one mole of $CaCO_3$-bound CO_2, one mole of CO_2 is released into ambient seawater, and is recycled (*round arrow*) to HCO_3^- or released to the environment, i.e. the carbonate counter pump (e.g., Zeebe and

Wolf-Gladrow 2001). The ratio between the different carbonate species HCO_3^-, $CO_3^=$, and H_2CO_3 involved in the formation of planktic foraminifer shell calcite depends on temperature (T), salinity (S), pH, pressure (p), and DIC concentration. Production of organic and inorganic matter occurs mainly in the surface mixed ocean. Remineralization and dissolution (*round arrows*) occurs primarily at mesobathyal depths and quantitatively affects sedimentation and CO_2 burial. For absolute numbers on the organic and inorganic carbon pump see Schiebel and Movellan (2012) and Schiebel (2002), respectively

the deep water column where the carbon is temporarily withdrawn from ocean-to-atmosphere exchange for intermediate time-scales of decades to centuries, depending on upwelling dynamics and turnover rates of the global marine current systems (cf. Broecker 1987; Archer and Maier-Reimer 1994). When arriving at the sea floor, planktic foraminifer $CaCO_3$ including the captured CO_2 may be stored over long time-scales of millions of years depending on diagenetic effects and tectonic processes, and the preservation and dissolution of tests (e.g., Dittert et al. 1999; Broecker and Clark 1999; Broecker and Clark 2003).

The biological carbon pump in the ocean is the sum of processes, which affect the production, transportation, and remineralization of organic (e.g., cytoplasm) and dissolution of inorganic carbon (e.g., planktic foraminifer shell calcite). Planktic foraminifers affect the marine carbonate pump mainly through shell $CaCO_3$ flux from surface waters towards surface sediments. The soft tissue pump (i.e. cytoplasm) concerns organic carbon, and has so far been regarded virtually non-existent in planktic foraminifers, because sedimentary planktic foraminifer tests are in general produced through reproduction, and settle to depth after having redistributed most cytoplasm to their offspring. In addition, planktic foraminifers have little effect on the efficiency of the biological carbon pump, since they are only minor part of 'ballast' in aggregates of particulate organic carbon (cf. De La Rocha and Passow 2007). However, planktic foraminifer soft tissue is systematically exported from surface waters to the sub-surface water column (cf. Boltovskoy and Lena 1970; Schiebel and Movellan 2012; Salter et al. 2014). Therefore, quantitative data on fossil planktic foraminifer assemblages could complement $\delta^{13}C$ data as a proxy of the biological carbon pump of the

ancient oceans (e.g., Broecker 1971; Hilting et al. 2008).

Planktic foraminifer standing stocks and carbon turnover are highest in the surface mixed layer of the ocean, where CO_2 exchange of ambient seawater is closely coupled to the atmospheric CO_2 pool through diffusion at sub-seasonal time-scales. Live, i.e. cytoplasm bearing foraminifer individuals, which grow in surface waters may be mixed to depth by currents and surface water mixing, for example by eddies and during storms (Beckmann et al. 1987; Schiebel et al. 1995, cf. Koeve et al. 2002). On a global average, convection removes both calcite-carbon and cytoplasm-bound carbon at a ratio of $\sim 10{:}1$, respectively, from the atmosphere-coupled surface ocean to sub-surface depth (Schiebel and Movellan 2012). Planktic foraminifers thus contribute, although to a minor degree, to the marine biological carbon pump. Because their test size and calcite mass (Beer et al. 2010) are closely related to biomass (see Chap. 5, Fig. 5.4 on biomass) their calcite carbon to soft tissue carbon ratio may be used as proxy of the marine biological carbon pump (Movellan 2013).

8.5.1 Regional Calcite Budgets

Regional planktic foraminifer calcite flux ranges between <0.001 and >2000 mg m^{-2} d^{-1} in oligotrophic, eutrophic, and mesotrophic waters (Fig. 8.13) of the global ocean (Schiebel 2002). Highest test calcite flux occurs at mid latitudes (Fig. 8.14, see also Žarić et al. 2006) caused by seasonally enhanced primary production and production of planktic foraminifers (Fig. 8.15). Data on shell $CaCO_3$ flux span more than four orders of magnitude within water depth intervals between the surface ocean and 2500 m water depth (Fig. 8.13). Export production and flux of tests starts at the base of the surface mixed layer at about 100 m water depth (e.g., Koeve 2002).

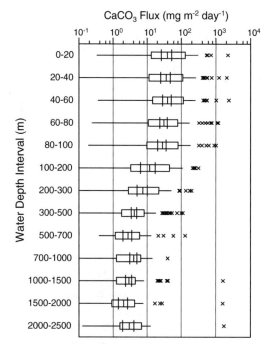

Fig. 8.13 Average planktic foraminifer CaCO₃ flux deduced from multinet samples from the North Atlantic, Arabian Sea, and Caribbean Sea (n = 1777). *Boxes* cover the upper and lower quartile, with horizontal lines for the upper and lower adjacent values. Whereas the CaCO₃ flux varies over five orders of magnitude, the average flux between the surface and mesobathyal water column decreases by only one order of magnitude over the twilight zone between 100 and 700 m water depth. The upper to lower quartile of fluxes between the sea surface and 200 m water depth exceeds one order of magnitude, indicating large variations in export production (e.g., Koeve 2002). Outliers (x) result from pulsed flux events following time-periods of enhanced production. Note that no outliers exist to the left of average distributions, indicating relatively constant 'background' (off-peak) test flux. Three extreme outliers between 1000 and 2500 m water depth result from mass flux events of *G. siphonifera* and *G. sacculifer* tests in the Arabian Sea. From Schiebel (2002)

The most significant decrease in flux takes place between 100 and 700 m depth. Changes in test CaCO₃ flux between 700 and 2500 m are of minor amplitude (Figs. 8.13, 8.14 and 8.15).

Exceptionally high planktic foraminifer CaCO₃ flux results from mass sedimentation of tests (see Sect. 8.1.3). Since mass flux events are episodic and rapid, they are rarely sampled by plankton net tows, and not detectable from sediment trap samples or surface sediments because of too low temporal sampling resolution.

Presence and absence of species with different biological prerequisites and ecological demands may exert a considerable effect on the flux of tests and CaCO₃. For example, very high planktic foraminifer CaCO₃ flux >1000 mg m^{-2} d^{-1}, between 1000 and 2500 m (Fig. 8.13), was mainly caused by large specimens of *G. siphonifera* and *G. sacculifer* in the Arabian Sea in March 1995. The same is true for other seasonally pulsed CaCO₃ flux peaks observed in the Arabian Sea in April and during August–September (see Sect. 8.1) (Schiebel 2002). Those flux peaks were caused by opportunistic species (*N. dutertrei, G. bulloides*), which proliferate during the late stages of the NE and SW monsoons, respectively. Test flux pulses (Tables. 8.1 and 8.2) arrive at depths with the typical delay resulting from test-size related settling-velocity (cf. Takahashi and Bé 1984; Kroon and Ganssen 1989; Rixen et al. 2000; Schiebel and Hemleben 2000).

Moderate to low production and flux of planktic foraminifer tests and CaCO₃ occurs in mesotrophic to oligotrophic waters of the temperate ocean and subtropical gyres (Fig. 8.14) and may be dominated by seasonal mass flux events in the same way as in eutrophic waters (e.g., Thunell and Honjo 1987) (see Sect. 8.1). Following mass flux events such as, for example, during the spring bloom in the NE Atlantic export flux decreases and flux pulses occur in the deeper water column. Maximum seasonality and sharp test flux peaks at high latitudes are caused by productivity during the short euphotic time-interval in summer (e.g., Fischer et al. 1988). In contrast, relatively balanced export flux, and steady sedimentation of tests in the tropical to subtropical ocean (Fig. 8.14, Caribbean; see also Sect. 7.2, Fig. 7.5) results from low seasonality compared to higher latitudes and year-round production and flux of foraminifer test calcite.

Low production and flux of planktic foraminifer tests occur in oligotrophic regions such as subtropical gyres (Fig. 8.14, Azores). However, seasonal test and CaCO₃ flux peaks may also occur in oligotrophic waters. Distinct CaCO₃ flux pulses at subsurface water depths in the subtropical gyre of the North Atlantic are, for

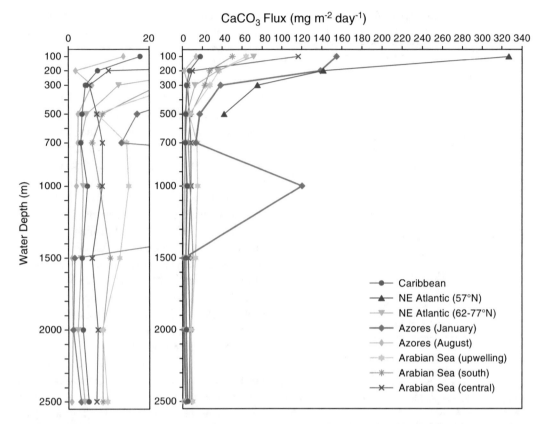

Fig. 8.14 Monthly averages of regional planktic foraminifer CaCO$_3$ flux (mg m^{-2}d^{-1}) between 100 (export layer) and 2500 m water depth, calculated from plankton net samples (Schiebel 2002). High export flux occurs in mesotrophic and eutrophic regions such as in the NE Atlantic and Arabian Sea. Low export flux occurs in oligotrophic regions such as in the Caribbean Sea. The most significant decrease in flux takes place between 100 and 700 m water depth. Exceptionally high CaCO$_3$ flux at 100 m water depth in January in waters south of the Azores Islands was caused by reproduction and flux of large tests of *G. truncatulinoides*. Data are given for the lower level of each water depth interval. Left panel shows an enlarged view of low fluxes <20 mg m^{-2}d^{-1} given in right panel. From Schiebel (2002)

example, caused by *G. truncatulinoides*. After reproduction in surface waters during winter to early spring, the empty tests of adult *G. truncatulinoides* form a confined test-cloud settling through the subsurface water column (e.g., Deuser et al. 1981; Hemleben et al. 1987; Schiebel et al. 2002). Opposite to the flux of empty tests, live individuals of subsurface to deep-dwelling species in general contribute only a minor part to the foraminifer assemblage at subsurface depths, resulting from small standing stocks, which are dispersed over the vast expanses of the deep ocean (e.g., Lončarić et al. 2006).

8.5.2   Global Calcite Budget

The global planktic foraminifer calcite flux at 100 m water depth (F$_{100}$) is estimated at 1.3–3.2 Gigatons (Gt, 10^9 tons) year^{-1} (Fig. 8.16), equivalent to 23–56 % of the total open marine CaCO$_3$ particulate inorganic carbon (PIC) flux (Schiebel 2002). Test and calcite fluxes are calculated from the regional distribution of species obtained from net-tow (tests >100 μm in minimum diameter) and sediment trap samples and are assumed to cover most of the entire modern range of marine biogenic PIC flux (Schiebel 2002). During most of the year (off-peak periods), a large

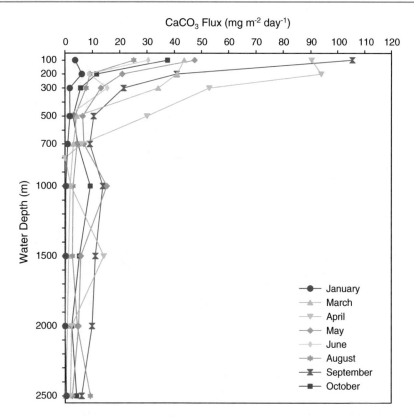

Fig. 8.15 Average monthly planktic foraminifer CaCO₃ flux in the temperate eastern North Atlantic around 47°N, 20°W (BIOTRANS). Maximum decrease in CaCO₃ flux occurs above 700 m water depths. Maximum export flux occurs in spring and fall. Test flux increase in deep waters below 700 m results from enhanced spring production and pulsed mass sedimentation in April and May. In addition, CaCO₃ flux is increasingly dominated by large and fast settling tests with high calcite mass at increasing water depth (Berelson 2002). In summer, small and slow settling tests cause low CaCO₃ flux. Winter is characterized by low production and flux of planktic foraminifer CaCO₃. From Schiebel (2002)

part of the test calcite is dissolved while settling through the mesobathyal water column between 100–1000 m depth (Fig. 8.14).

As little as 1–3 % of the test CaCO₃ initially exported from the surface mixed layer to sub-pycnocline waters may reach the above-lysocline seafloor on average (Schiebel 2002). Pulsed flux events, i.e. mass dumps of fast settling particles, yield a major contribution of tests to the formation of deep-sea sediments above the CCD. Highest flux and sedimentation rates of tests and calcite occur at latitudes between about 30–70° (Fig. 8.17), where high and pointed spring production (spring bloom), and food supply coincides with high planktic

foraminifer diversity. The same applies to other regions of pointed seasonal production in the tropical to temperate ocean. On a global scale, about a quarter of the initially produced planktic foraminifer test CaCO₃ settles on the sea floor and forms a major portion of sediment calcite above the calcite compensation depth, CCD (e.g., Berger 1971; Vincent and Berger 1981; Dittert et al. 1999; Schiebel 2002; Frenz et al. 2005).

The total planktic foraminifer contribution of CaCO₃ to sediments above the CCD in the modern global ocean is estimated at 0.36–0.88 Gt yr^{-1} (Fig. 8.18), which amounts to 32–80 % of the total marine sedimentary calcite budget (see also Archer 1996; Schiebel 2002;

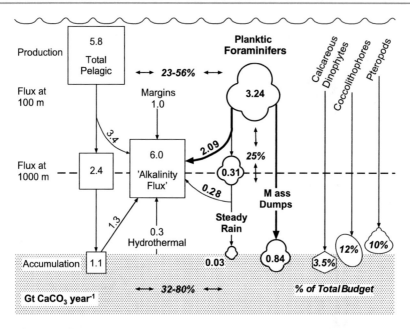

Fig. 8.16 Global planktic foraminifer CaCO$_3$ flux budget (*center*) in comparison with CaCO$_3$ budgets given by Milliman et al. (1999, *squares* to the *left*). Planktic foraminifer shell calcite flux at 100 m depth is assumed equivalent to 23–56 % of the total open marine CaCO$_3$ flux according to Milliman et al. (1999). An average of about 25 % of the initially produced planktic foraminifer CaCO$_3$ is assumed to settle on the seafloor. The global planktic foraminifer contribution of CaCO$_3$ to marine sediments amounts to 32–80 % of the total above-CCD budget and is assumed ~ 1.1 Gt CaCO$_3$ yr^{-1}. An estimate of the coccolithophore, pteropod, and calcareous dinophyte contribution to the global open marine above-ACD (pteropods) and above-CCD CaCO$_3$ accumulation is given to right. From Schiebel (2002)

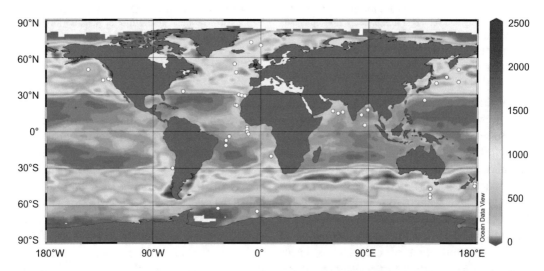

Fig. 8.17 Modeled annual total foraminifer test flux (10^3 individuals m^{-2}, for 18 species included in the empirical model) is highest from the subtropical to subpolar ocean of the northern and southern hemisphere. *Circles* mark positions of sediment traps comprised in the calibration data set. The general relative pattern of the global test flux is well represented by the model results. Many regional patterns are not properly reproduced due to insufficient forcing by environmental parameters, and the correlation between primary production and planktic foraminifer test CaCO$_3$ flux is weak (cf. Schiebel 2002). Absolute fluxes are assumed significantly underestimated in most cases. From Žarić et al. (2006)

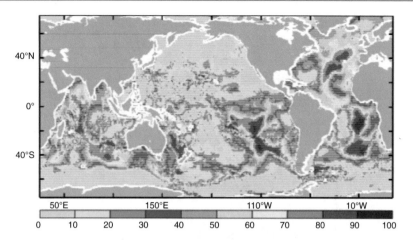

Fig. 8.18 Map of the $CaCO_3$ weight percent in the surface sediments. Low partial pressure of CO_2 of young deep-water bodies in the Atlantic Ocean causes deep $CaCO_3$ lysoclines and compensations depths, and results in well-preserved calcareous sediments of wide distribution. Oldest deep-water bodies of high pCO_2 in the North Pacific cause shallow lysoclines and compensations depths and ample carbonate dissolution. Accumulation of particles other than $CaCO_3$ in the Southern Ocean results in dilution and carbonate-poor sediments, at relatively well preservation of calcareous shells. From Dunne et al. (2012). See also Sarmiento and Gruber (2006)

Sarmiento and Gruber 2006). The total accumulation of $CaCO_3$ in the modern ocean is estimated at 1.1 Gt yr^{-1} (e.g., Milliman 1974; Milliman 1993; Milliman and Droxler 1996). In addition to planktic foraminifer $CaCO_3$ flux, three major groups of calcareous plankton, i.e. coccolithophores, pteropods, and calcareous dinophytes add to the deep marine $CaCO_3$ flux. Aragonite shells of pteropods are largely dissolved at the aragonite lysocline to aragonite compensation depth (ACD) above the calcite lysocline and CCD, respectively. Below the CCD, coccolithophore calcite takes over and increasingly constitutes the calcite fraction of abyssal sediment with increasing water depth and pCO_2, i.e. increasing pH (Frenz et al. 2005).

8.5.3 Global Biomass

In addition to the calcite-bound carbon of the tests, the global biomass (i.e. cytoplasm) of planktic foraminifers is estimated at $\sim$ 8.5–32.7 Teragrams (Tg, i.e. 10^{12} g) C yr^{-1} including specimens >125 μm in diameter (Schiebel and Movellan 2012). When adding juvenile and neanic specimens (<125 μm in tests size), the total planktic foraminifer biomass production is assumed as high as $\sim$ 25–100 Tg C yr^{-1} (i.e., 0.025–0.1 Gigatons, Gt). The average global biomass-bound planktic foraminifer carbon would hence be four to six times less than the $CaCO_3$ bound carbon of their test assemblages. The 25–100 Tg are estimated for a global ocean area of $322 * 10^6$ km^2 assumed to support planktic foraminifer production over nine months per year, accounting for three aphotic (winter) months without any significant production on a global average (Obata et al. 1996; Schiebel and Movellan 2012).

Assemblage biomass of planktic foraminifers varies by up to five orders of magnitude at intermediate water depth (100–700 m) and on average decreases by three orders of magnitude over 13 distinct water-depth intervals (see Methods Chap. 12) analyzed between the surface and deep water column at 2500 m depth (Fig. 8.19). Highest assemblage biomass in surface waters in the temperate North Atlantic and Arabian Sea is possibly biased by data from high-productive

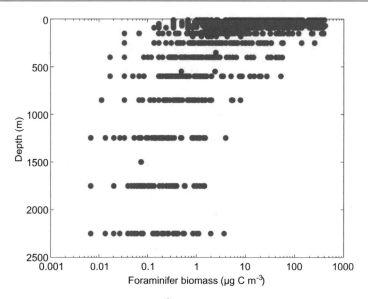

Fig. 8.19 Log-normalised carbon-biomass (Log$_{10}$ µg m^{-3}) given for the total planktic foraminifer assemblage >125 µm (data available from http://dx.doi.org/10.1594/PANGAEA. 777386). Data are calculated from average individual protein-biomass data and faunal counts from the eastern North Atlantic Ocean, Caribbean Sea, and Arabian Sea (n = 1087, without zero values). All data given for the mid-points of the sampled water depth intervals. From Schiebel and Movellan (2012)

seasons, i.e. spring and SW monsoon in the Atlantic Ocean and Arabian Sea, respectively. Enhanced planktic foraminifer biomass in the Caribbean Sea, off Japan, and Oregon might be affected by land-derived input to the hemi-pelagic ocean, and effects on the primary and secondary production including planktic foraminifers. Low biomass in Red Sea waters is caused by oligotrophic conditions (Fig. 8.20). Biological production (including planktic foraminifers) in both Artic and Antarctic waters has been assumed for time-intervals of only three months per year, and long aphotic polar seasons of nine months (Obata et al. 1996). However, Boetius et al. (2013) report significant under-ice primary production in the Arctic Ocean, which might also enhance the availability of food to planktic foraminifers. In conjunction with decreasing sea ice cover in the Arctic Ocean (e.g., Intergovernmental Panel on Climate Change 2007; Intergovernmental Panel on Climate Change 2013) primary production and secondary production, including planktic foraminifers, may increase over the 21st century and beyond.

8.6   Summary and Concluding Remarks

The value of planktic foraminifers as proxy in paleoceanography and as part of the marine carbon turnover critically depends on the understanding of production and sedimentation of tests. Temporal scales of days to seasons, and regional sedimentation to basin scale transportation control the production of test assemblages (thanatocoenoses). Preservation and dissolution depend on thermodynamic (ΔCO_3^{2-}, Ω) and biological (often bacterially mediated processes) conditions.

Production and flux of planktic foraminifer test calcite affects, and is affected by, regional to global ocean carbon turnover. On short time-scales, test production is a source of CO_2 to the ocean surface and lower atmosphere and acts

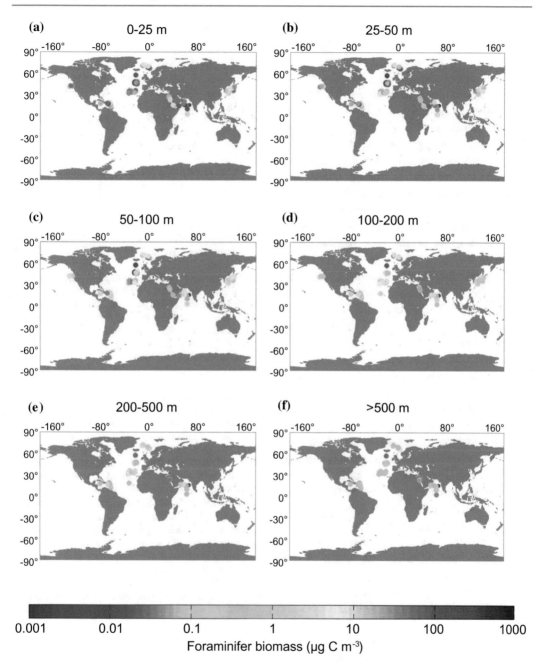

Fig. 8.20 Log-normalised (Log$_{10}$ µg C m^{-3}) average depth related (a to f) planktic foraminifer assemblage biomass (PFAB) binned on a 3° × 3° grid, comprising the North Atlantic Ocean, Caribbean, Arabian Sea, Gulf of Aden, Red Sea. Data on the eastern and western North Pacific Ocean off Oregon and Japan, respectively, are only on the upper 200 m of the water column. From Schiebel and Movellan (2012)

as a sink of CO_2 on long geological time scales. Sedimentation of organic carbon within foraminifer shells adds to the biological carbon pump. Increasing $[CO_2]$ and decreasing pH (ocean acidification, OA) are assumed to negatively affect calcification of planktic foraminifers.

In analogy to past acidification events, planktic foraminifers are assumed to buffer changes in pH. The buffering capacity of foraminifer shell formation on increasing $[CO_2]$ and OA has not yet been quantified at the global scale. To date, calcification of planktic foraminifer tests is neither well constrained for the biological processes, nor for the quantitative effects on the marine carbon turnover. Future studies on the natural environment, as well as culture experiments (laboratory and mesocosm) should help to better understand planktic foraminifer calcification in a changing ocean and to calibrate new proxies for the reconstruction of the past marine carbon turnover.

Processes that affect sedimentation and dissolution are still insufficiently understood. This is partly owing to the fact that experimental approaches are limited by technological constraints. Smart chemostat experiments are needed to attain a better systematic understanding and quantification of small-scale to global processes, and to facilitate new modeling approaches. In addition, information on the natural environment is limited by methodological constraints like trapping efficiency of sediment traps. Concerted programs and sampling campaigns like the Joint Global Ocean Flux Study (JGOFS) was core project of the International Geosphere-Biosphere Programme (IGBP) (e.g., Ducklow and Harris 1993) and mesocosm experiments (e.g., Riebesell et al. 2013) enhance the systematic and quantitative understanding of mass fluxes. A combination of methods may provide a better qualitative and quantitative understanding of processes, which determine sedimentation and preservation versus dissolution of the planktic foraminifer shell calcite, i.e. the interplay of chemical, physical, and biological factors. Modern analytical methods will provide detailed information on ontogenetic shell calcite, and different kinds of calcite layers covering the pre-gametogenic planktic foraminifer shell. Those data are indispensable for a better understanding of carbon budgets and the use of planktic foraminifers as proxies in paleoceanography discussed in other chapters of this book.

References

Almogi-Labin A (1984) Population dynamics of planktic Foraminifera and Pteropoda—Gulf of Aqaba, Red Sea. Proc K Ned Akad Van Wet Ser B Palaeontol Geol Phys Chem 87:481–511

Anderson LA, Sarmiento JL (1994) Redfield ratios of remineralization determined by nutrient data analysis. Glob Biogeochem Cycles 8:65–80. doi:10.1029/93GB03318

Archer DE (1996) An atlas of the distribution of calcium carbonate in sediments of the deep sea. Glob Biogeochem Cycles 10:159–174. doi:10.1029/95GB03016

Archer D, Maier-Reimer E (1994) Effect of deep-sea sedimentary calcite preservation on atmospheric CO_2 concentration. Nature 367:260–263. doi:10.1038/367260a0

Auras-Schudnagies A, Kroon D, Ganssen G, Hemleben C, van Hinte JE (1989) Distributional pattern of planktonic foraminifers and pteropods in surface waters and top core sediments of the Red Sea, and adjacent areas controlled by the monsoonal regime and other ecological factors. Deep-Sea Res I 36:1515–1533. doi:10.1016/0198-0149(89)90055-1

Beckmann W, Auras A, Hemleben C (1987) Cyclonic cold-core eddy in the eastern North Atlantic. III Zooplankton Mar Ecol Prog Ser 39:165–173

Beer CJ, Schiebel R, Wilson PA (2010) Technical note: on methodologies for determining the size-normalised weight of planktic Foraminifera. Biogeosciences 7:2193–2198. doi:10.5194/bg-7-2193-2010

Berelson WM (2002) Particle settling rates increase with depth in the ocean. Deep-Sea Res II 49:237–251. doi:10.1016/S0967-0645(01)00102-3

Berelson WM, Balch WM, Najjar R, Feely RA, Sabine C, Lee K (2007) Relating estimates of $CaCO_3$ production, export, and dissolution in the water column to measurements of $CaCO_3$ rain into sediment traps and dissolution on the sea floor: a revised global carbonate budget. Glob Biogeochem Cycles. doi:10.1029/2006GB002803

Berger WH (1970) Planktonic Foraminifera: selective solution and the lysocline. Mar Geol 8:111–138. doi:10.1016/0025-3227(70)90001-0

Berger WH (1971) Sedimentation of planktonic Foraminifera. Mar Geol 11:325–358. doi:10.1016/0025-3227(71)90035-1

Berger WH (1973) Deep-sea carbonates: Pleistocene dissolution cycles. J Foraminifer Res 3:187–195. doi:10.2113/gsjfr.3.4.187

Berger WH (1979) Preservation of Foraminifera. In: Lipps JH, Berger WH, Buzas MA, Douglas RG, Ross EH (eds) Foraminiferal ecology and paleoecology: Houston, Texas, Society of Economic Paleontologists and Mineralogists Short Course No 6, pp 105–155

Berger WH, Piper DJW (1972) Planktonic Foraminifera: differential settling, dissolution, and redeposition.

Limnol Oceanogr 17:275–287. doi:10.4319/lo.1972.17.2.0275

Berger WH, Wefer G (1990) Export production: seasonality and intermittency, and paleoceanographic implications. Palaeogeogr Palaeoclimatol Palaeoecol 89:245–254. doi:10.1016/0031-0182(90)90065-F

Bijma J, Erez J, Hemleben C (1990a) Lunar and semi-lunar reproductive cycles in some spinose planktonic foraminifers. J Foraminifer Res 20:117–127

Bijma J, Faber WW, Hemleben C (1990b) Temperature and salinity limits for growth and survival of some planktonic foraminifers in laboratory cultures. J Foraminifer Res 20:95–116. doi:10.2113/gsjfr.20.2.95

Bijma J, Hemleben C, Wellnitz K (1994) Lunar-influenced carbonate flux of the planktic foraminifer Globigerinoides sacculifer (Brady) from the central Red Sea. Deep-Sea Res I 41:511–530. doi:10.1016/0967-0637(94)90093-0

Bishop JKB, Edmond JM, Ketten DR, Bacon MP, Silker WB (1977) The chemistry, biology, and vertical flux of particulate matter from the upper 400 m of the equatorial Atlantic Ocean. Deep-Sea Res 24:511–548. doi:10.1016/0146-6291(77)90526-4

Boetius A, Albrecht S, Bakker K, Bienhold C, Felden J, Fernández-Méndez M, Hendricks S, Katlein C, Lalande C, Krumpen T, Nicolaus M, Peeken I, Rabe B, Rogacheva A, Rybakova E, Somavilla R, Wenzhöfer F (2013) Export of algal biomass from the melting Arctic Sea ice. Science 339:1430–1432. doi:10.1126/science.1231346

Boltovskoy E, Lena H (1970) On the decomposition of the protoplasm and the sinking velocity of the planktonic foraminifers. Int Rev Gesamten Hydrobiol Hydrogr 55:797–804. doi:10.1002/iroh.19700550507

Boussetta S, Bassinot F, Sabbatini A, Caillon N, Nouet J, Kallel N, Rebaubier H, Klinkhammer G, Labeyrie L (2011) Diagenetic Mg-rich calcite in Mediterranean sediments: quantification and impact on foraminiferal Mg/Ca thermometry. Mar Geol 280:195–204. doi:10.1016/j.margeo.2010.12.011

Bouvier-Soumagnac Y, Duplessy JC, Bé AWH (1986) Isotopic composition of a laboratory cultured planktonic foraminifer O. universa—implications for paleoclimatic reconstructions. Oceanol Acta 9:519–522

Broecker WS (1971) A kinetic model for the chemical composition of sea water. Quat Res 1:188–207. doi:10.1016/0033-5894(71)90041-X

Broecker WS (1987) The biggest chill. Nat Hist 97:74–82

Broecker WS, Clark E (1999) CaCO₃ size distribution: a paleocarbonate ion proxy? Paleoceanography 14:596–604. doi:10.1029/1999PA900016

Broecker WS, Clark E (2001) An evaluation of Lohmann's Foraminifera weight dissolution index. Paleoceanography 16:531–534. doi:10.1029/2000PA000600

Broecker WS, Clark E (2003) CaCO₃ dissolution in the deep sea: paced by insolation cycles. Geochem Geophys Geosystems 4:1059. doi:10.1029/2002GC000450

Broecker WS, Peng TH (1982) Tracers in the sea. Eldigio Press, New York

Brummer GJA, Hemleben C, Spindler M (1987) Ontogeny of extant spinose planktonic Foraminifera (Globigerinidae): a concept exemplified by Globigerinoides sacculifer (Brady) and G. ruber (d'Orbigny). Mar Micropaleontol 12:357–381. doi:10.1016/0377-8398(87)90028-4

Buesseler KO, Lamborg CH, Boyd PW, Lam PJ, Trull TW, Bidigare RR, Bishop JKB, Casciotti KL, Dehairs F, Elskens M, Honda M, Karl DM, Siegel DA, Silver MW, Steinberg DK, Valdes J, Mooy BV, Wilson S (2007) Revisiting carbon flux through the ocean's twilight zone. Science 316:567–570. doi:10.1126/science.1137959

Caromel AGM, Schmidt DN, Phillips JC (2013) Repercussions of differential settling on sediment assemblages and multi-proxy palaeo-reconstructions. Biogeosciences Discuss 10:6763–6781. doi:10.5194/bgd-10-6763-2013

Caron DA, Roger Anderson O, Lindsey JL, Faber WW, Lin Lim EE (1990) Effects of gametogenesis on test structure and dissolution of some spinose planktonic Foraminifera and implications for test preservation. Mar Micropaleontol 16:93–116

Conan SMH, Brummer GJA (2000) Fluxes of planktic Foraminifera in response to monsoonal upwelling on the Somalia basin margin. Deep-Sea Res II 47:2207–2227

Conan SMH, Ivanova EM, Brummer GJA (2002) Quantifying carbonate dissolution and calibration of foraminiferal dissolution indices in the Somali basin. Mar Geol 182:325–349. doi:10.1016/S0025-3227(01)00238-9

Constandache M, Yerly F, Spezzaferri S (2013) Internal pore measurements on macroperforate planktonic Foraminifera as an alternative morphometric approach. Swiss J Geosci 106:179–186

De La Rocha CL, Passow U (2007) Factors influencing the sinking of POC and the efficiency of the biological carbon pump. Deep-Sea Res II 54:639–658. doi:10.1016/j.dsr2.2007.01.004

De Villiers S (2005) Foraminiferal shell-weight evidence for sedimentary calcite dissolution above the lysocline. Deep-Sea Res I 52:671–680. doi:10.1016/j.dsr.2004.11.014

Deuser WG (1987) Seasonal variations in isotopic composition and deep-water fluxes of the tests of perennially abundant planktonic Foraminifera of the Sargasso Sea: results from sediment-trap collections and their paleoceanographic significance. J Foraminifer Res 17:14–27

Deuser WG, Ross EH, Hemleben C, Spindler M (1981) Seasonal changes in species composition, numbers, mass, size, and isotopic composition of planktonic Foraminifera settling into the deep Sargasso Sea. Palaeogeogr Palaeoclimatol Palaeoecol 33:103–127

Dittert N, Henrich R (2000) Carbonate dissolution in the South Atlantic Ocean: evidence from ultrastructure breakdown in Globigerina bulloides. Deep-Sea Res I 47:603–620

Dittert N, Baumann KH, Bickert T, Henrich R, Huber R, Kinkel H, Meggers H (1999) Carbonate dissolution in the deep-sea: methods, quantification and paleoceanographic application. In: Fischer G, Wefer G (eds) Use of proxies in paleoceanography. Springer, Berlin, pp 255–284

Ducklow HW, Harris RP (1993) Introduction to the JGOFS North Atlantic bloom experiment. Deep-Sea Res II 40:1–8. doi:10.1016/0967-0645(93)90003-6

Dunne JP, Hales B, Toggweiler JR (2012) Global calcite cycling constrained by sediment preservation controls. Glob Biogeochem Cycles 26. doi:10.1029/2010GB003935

Erez J, Almogi-Labin A, Avraham S (1991) On the life history of planktonic Foraminifera: lunar reproduction cycle in *Globigerinoides sacculifer* (Brady). Paleoceanography 6:295–306

Feely RA, Sabine CL, Hernandez-Ayon JM, Ianson D, Hales B (2008) Evidence for upwelling of corrosive "acidified" water onto the continental shelf. Science 320:1490–1492

Fischer G, Fütterer D, Gersonde R, Honjo S, Ostermann D, Wefer G (1988) Seasonal variability of particle flux in the Weddell Sea and its relation to ice cover. Nature 335:426–428

Fok-Pun L, Komar PD (1983) Settling velocities of planktonic Foraminifera: density variations and shape effects. J Foraminifer Res 13:60–68

Frenz M, Baumann KH, Boeckel B, Höppner R, Henrich R (2005) Quantification of foraminifer and coccolith carbonate in South Atlantic surface sediments by means of carbonate grain-size distributions. J Sediment Res 75:464–475

Friis K, Najjar RG, Follows MJ, Dutkiewicz S (2006) Possible overestimation of shallow-depth calcium carbonate dissolution in the ocean. Glob Biogeochem Cycles. doi:10.1029/2006GB002727

Furbish DJ, Arnold AJ (1997) Hydrodynamic strategies in the morphological evolution of spinose planktonic Foraminifera. Geol Soc Am Bull 109:1055–1072. doi:10.1130/0016-7606(1997)109<1055:HSITME>2.3.CO;2

Hemleben C, Spindler M, Breitinger I, Deuser WG (1985) Field and laboratory studies on the ontogeny and ecology of some globorotaliid species from the Sargasso Sea off Bermuda. J Foraminifer Res 15:254–272

Hemleben C, Spindler M, Breitinger I, Ott R (1987) Morphological and physiological responses of *Globigerinoides sacculifer* (Brady) under varying laboratory conditions. Mar Micropaleontol 12:305–324

Hemleben C, Spindler M, Anderson OR (1989) Modern planktonic Foraminifera. Springer, Berlin

Henrich R, Wefer G (1986) Dissolution of biogenic carbonates: effects of skeletal structure. Mar Geol 71:341–362

Hermelin JOR, Summerhays CP, Prell WS, Emeis KC (1992) Variations in the benthic foraminiferal fauna of the Arabian Sea: a response to changes in upwelling intensity? Upwelling systems: evolution since the early miocene. Geological Society, Special Publications, London, pp 151–166

Hilting AK, Kump LR, Bralower TJ (2008) Variations in the oceanic vertical carbon isotope gradient and their implications for the Paleocene-Eocene biological pump. Paleoceanography. doi:10.1029/2007PA001458

Honjo S, Manganini SJ (1993) Annual biogenic particle fluxes to the interior of the North Atlantic Ocean; studied at 34°N 21°W and 48°N 21°W. Deep-Sea Res II 40:587–607

Huber R, Meggers H, Baumann KH, Henrich R (2000) Recent and Pleistocene carbonate dissolution in sediments of the Norwegian-Greenland Sea. Mar Geol 165:123–136. doi:10.1016/S0025-3227(99)00138-3

Intergovernmental Panel on Climate Change (2007) Climate change 2007: synthesis report. Contribution of working groups I, II, and III to the fourth assessment report of the intergovernmental panel on climate change. IPCC, Geneva Switzerland

Intergovernmental Panel on Climate Change (ed) (2013) Climate change 2013—The physical science basis: working group I contribution to the fifth assessment report of the intergovernmental panel on climate change. Cambridge University Press, Cambridge

Ivanova EV (1988) Late quaternary paleoceanology of the Indian Ocean (based on planktonic foraminifers and pteropods). PP Shirshov Institute of Oceanology USSR Academy of Sciences, Moscow (in Russian)

Ivanova E, Schiebel R, Singh AD, Schmiedl G, Niebler HS, Hemleben C (2003) Primary production in the Arabian Sea during the last 135,000 years. Palaeogeogr Palaeoclimatol Palaeoecol 197:61–82

Jansen H, Wolf-Gladrow DA (2001) Carbonate dissolution in copepod guts: a numerical model. Mar Ecol Prog Ser 221:199–207

Jansen H, Zeebe R, Wolf-Gladrow DA (2002) Modelling the dissolution of settling $CaCO_3$ in the ocean. Glob Biogeochem Cycles 16:1–16

Johnstone HJH, Schulz M, Barker S, Elderfield H (2010) Inside story: an X-ray computed tomography method for assessing dissolution in the tests of planktonic Foraminifera. Mar Micropaleontol 77:58–70. doi:10.1016/j.marmicro.2010.07.004

Kawahata H (2002) Suspended and settling particles in the Pacific. Deep-Sea Res II 49:5647–5664. doi:10.1016/S0967-0645(02)00216-3

Kemp AES, Pike J, Pearce RB, Lange CB (2000) The "Fall dump"—a new perspective on the role of a "shade flora" in the annual cycle of diatom production and export flux. Deep-Sea Res II 47:2129–2154

Koeve W (2002) Upper ocean carbon fluxes in the Atlantic Ocean: the importance of the POC:PIC ratio. Glob Biogeochem Cycles. doi:10.1029/2001GB001836

Koeve W, Pollehne F, Oschlies A, Zeitzschel B (2002) Storm-induced convective export of organic matter during spring in the northeast Atlantic Ocean. Deep-Sea Res I 49:1431–1444. doi:10.1016/S0967-0637(02)00022-5

Kroon D, Ganssen G (1989) Northern Indian Ocean upwelling cells and the stable isotope composition of living planktonic foraminifers. Deep-Sea Res I 36:1219–1236

Kuhnt T, Howa H, Schmidt S, Marié L, Schiebel R (2013) Flux dynamics of planktic foraminiferal tests in the south-eastern Bay of Biscay (northeast Atlantic margin). J Mar Syst 109–110:169–181. doi:10.1016/j.jmarsys.2011.11.026

Lohmann GP (1995) A model for variation in the chemistry of planktonic Foraminifera due to secondary calcification and selective dissolution. Paleoceanography 10:445–457

Lončarić N, Peeters FJC, Kroon D, Brummer GJA (2006) Oxygen isotope ecology of recent planktic Foraminifera at the central Walvis Ridge (SE Atlantic). Paleoceanography. doi:10.1029/2005PA001207

Lončarić N, van Iperen J, Kroon D, Brummer GJA (2007) Seasonal export and sediment preservation of diatomaceous, foraminiferal and organic matter mass fluxes in a trophic gradient across the SE Atlantic. Prog Oceanogr 73:27–59

Malmgren BA (1983) Ranking of dissolution susceptibility of planktonic Foraminifera at high latitudes of the South Atlantic Ocean. Mar Micropaleontol 8:183–191. doi:10.1016/0377-8398(83)90023-3

Millero FJ, Morse J, Chen CT (1979) The carbonate system in the western Mediterranean Sea. Deep-Sea Res I 26:1395–1404

Milliman JD (1974) Marine carbonates. Springer, New York

Milliman JD (1993) Production and accumulation of calcium carbonate in the ocean: budget of a nonsteady state. Glob Biogeochem Cycles 7:927–957

Milliman JD, Droxler AW (1996) Neritic and pelagic carbonate sedimentation in the marine environment: ignorance is not bliss. Geol Rundsch 85:496–504

Milliman JD, Troy PJ, Balch WM, Adams AK, Li YH, Mackenzie FT (1999) Biologically mediated dissolution of calcium carbonate above the chemical lysocline? Deep-Sea Res I 46:1653–1669

Movellan A (2013) La biomasse des foraminifères planctoniques actuels et son impact sur la pompe biologique de carbone. PhD. Thesis, University of Angers

Naidu PD, Malmgren BA (1996) A high-resolution record of late quaternary upwelling along the Oman margin, Arabian Sea based on planktonic Foraminifera. Paleoceanography 11:129–140

Obata A, Ishizaka J, Endoh M (1996) Global verification of critical depth theory for phytoplankton bloom with climatological in situ temperature and satellite ocean color data. J Geophys Res 101:20657–20667

Parker FL, Berger WH (1971) Faunal and solution patterns of planktonic Foraminifera in surface sediments of the South Pacific. Deep-Sea Res 18:73–107. doi:10.1016/0011-7471(71)90017-9

Paull CK, Hills SJ, Thierstein HR (1988) Progressive dissolution of fine carbonate particles in pelagic sediments. Mar Geol 81:27–40. doi:10.1016/0025-3227(88)90015-1

Pearson PN (2012) Oxygen isotopes in Foraminifera: overview and historical review. In: Ivany LC, Huber BT (eds) Reconstructing earth's deep-time climate—the state of the art in 2012. Paleontological Society Short Course. Paleontological Society Papers, pp 1–38

Pearson PN, Palmer MR (2000) Atmospheric carbon dioxide concentrations over the past 60 million years. Nature 406:695–699

Peeters F, Ivanova E, Conan S, Brummer GJA, Ganssen G, Troelstra S, van Hinte J (1999) A size analysis of planktic Foraminifera from the Arabian Sea. Mar Micropaleontol 36:31–63

Prell WL, Martin A, Cullen JL, Trend M (1999) The Brown University Foraminiferal Data Base (BFD)

Ransom B, Shea KF, Burkett PJ, Bennett RH, Baerwald R (1998) Comparison of pelagic and nepheloid layer marine snow: Implications for carbon cycling. Mar Geol 150:39–50

Riebesell U, Gattuso JP, Thingstad TF, Middelburg JJ (2013) Arctic ocean acidification: pelagic ecosystem and biogeochemical responses during a mesocosm study. Biogeosciences 10:5619–5626. doi:10.5194/bg-10-5619-2013

Rixen T, Haake B, Ittekkot V (2000) Sedimentation in the western Arabian Sea the role of coastal and open-ocean upwelling. Deep-Sea Res II 47:2155–2178

Salter I, Schiebel R, Ziveri P, Movellan A, Lampitt R, Wolff GA (2014) Carbonate counter pump stimulated by natural iron fertilization in the polar frontal zone. Nat Geosci 7:885–889. doi:10.1038/ngeo2285

Sarmiento JL, Gruber N (2006) Ocean biogeochemical dynamics. Princeton University Press, Princeton and Oxford

Sautter LR, Thunell RC (1989) Seasonal succession of planktonic Foraminifera: results from a four-year time-series sediment trap experiment in the Northeast Pacific. J Foraminifer Res 19:253–267

Schiebel R (2002) Planktic foraminiferal sedimentation and the marine calcite budget. Glob Biogeochem Cycles. doi:10.1029/2001GB001459

Schiebel R, Hemleben C (2000) Interannual variability of planktic foraminiferal populations and test flux in the eastern North Atlantic Ocean (JGOFS). Deep-Sea Res II 47:1809–1852

Schiebel R, Hemleben C (2005) Modern planktic Foraminifera. Paläontol Z 79:135–148

Schiebel R, Movellan A (2012) First-order estimate of the planktic foraminifer biomass in the modern ocean. Earth Syst Sci Data 4:75–89. doi:10.5194/essd-4-75-2012

Schiebel R, Hiller B, Hemleben C (1995) Impacts of storms on recent planktic foraminiferal test production and $CaCO_3$ flux in the North Atlantic at 47°N, 20°W (JGOFS). Mar Micropaleontol 26:115–129

Schiebel R, Bijma J, Hemleben C (1997) Population dynamics of the planktic foraminifer *Globigerina bulloides* from the eastern North Atlantic. Deep-Sea Res I 44:1701–1713

Schiebel R, Waniek J, Zeltner A, Alves M (2002) Impact of the Azores front on the distribution of planktic foraminifers, shelled gastropods, and coccolithophorids. Deep-Sea Res II 49:4035–4050

Schiebel R, Barker S, Lendt R, Thomas H, Bollmann J (2007) Planktic foraminiferal dissolution in the twilight zone. Deep-Sea Res II 54:676–686

Schmidt K, De La Rocha CL, Gallinari M, Cortese G (2014) Not all calcite ballast is created equal: differing effects of foraminiferan and coccolith calcite on the formation and sinking of aggregates. Biogeosciences 11:135–145. doi:10.5194/bg-11-135-2014

Scholten JC, Fietzke J, Vogler S, Rutgers van der Loeff MM, Mangini A, Koeve W, Waniek J, Stoffers P, Antia A, Kuss J (2001) Trapping efficiencies of sediment traps from the deep Eastern North Atlantic: the ^{230}Th calibration. Deep-Sea Res II 48:2383–2408. doi:10.1016/S0967-0645(00)00176-4

Sexton PF, Wilson PA, Pearson PN (2006) Microstructural and geochemical perspectives on planktic foraminiferal preservation: "Glassy" versus "Frosty". Geochem Geophys Geosystems. doi:10.1029/2006GC001291

Siegel DA, Deuser WG (1997) Trajectories of sinking particles in the Sargasso Sea: modeling of statistical funnels above deep-ocean sediment traps. Deep-Sea Res I 44:1519–1541

Simstich J, Sarnthein M, Erlenkeuser H (2003) Paired $\delta^{18}O$ signals of *Neogloboquadrina pachyderma* (s) and *Turborotalita quinqueloba* show thermal stratification structure in Nordic Seas. Mar Micropaleontol 48:107–125

Stow DAV, Pudsey CJ, Howe JA, Faugeres JC, Viana AR (2002) Deep-water contourite systems: modern drifts and ancient series, seismic and sedimentary characteristics. In: Geological Society Memoir 22. Geological Society of London, London, p 464

Takahashi K, Bé AWH (1984) Planktonic Foraminifera: factors controlling sinking speeds. Deep-Sea Res Part Oceanogr Res Pap 31:1477–1500

Thiel H, Pfannkuche O, Schriever G, Lochte K, Gooday AJ, Hemleben C, Mantoura RFG, Turley CM, Patching JW, Riemann F (1989) Phytodetritus on the deep-sea floor in a central oceanic region of the Northeast Atlantic. Biol Oceanogr 6:203–239. doi:10.1080/01965581.1988.10749527

Thunell RC, Honjo S (1987) Seasonal and interannual changes in planktonic foraminiferal production in the North Pacific. Nature 328:335–337

Turley CM, Stutt ED (2000) Depth-related cell-specific bacterial leucine incorporation rates on particles and its biogeochemical significance in the Northwest Mediterranean. Limnol Oceanogr 45:419–425

Van Aken HM (2000) The hydrography of the mid-latitude Northeast Atlantic Ocean: II: the intermediate water masses. Deep-Sea Res I 47:789–824

Van Raden UJ, Groeneveld J, Raitzsch M, Kucera M (2011) Mg/Ca in the planktonic Foraminifera *Globorotalia inflata* and *Globigerinoides bulloides* from Western Mediterranean plankton tow and core top samples. Mar Micropaleontol 78:101–112. doi:10.1016/j.marmicro.2010.11.002

Van Sebille E, Scussolini P, Durgadoo JV, Peeters FJC, Biastoch A, Weijer W, Turney C, Paris CB, Zahn R (2015) Ocean currents generate large footprints in marine palaeoclimate proxies. Nat Commun 6:6521. doi:10.1038/ncomms7521

Vincent E, Berger WH (1981) Planktonic Foraminifera and their use in paleoceanography. Ocean Lithosphere Sea 7:1025–1119

Volbers ANA, Henrich R (2002) Present water mass calcium carbonate corrosiveness in the eastern South Atlantic inferred from ultrastructural breakdown of *Globigerina bulloides* in surface sediments. Mar Geol 186:471–486. doi:10.1016/S0025-3227(02)00333-X

Von Gyldenfeldt AB, Carstens J, Meincke J (2000) Estimation of the catchment area of a sediment trap by means of current meters and foraminiferal tests. Deep-Sea Res II 47:1701–1717

Wefer G (1989) Particle flux in the ocean: effects of episodic production. In: Berger WH, Smetacek VS, Weger G (eds) Productivity of the ocean: present and past. Dahlem Workshop Proceedings, pp 139–154

Wejnert KE, Pride CJ, Thunell RC (2010) The oxygen isotope composition of planktonic Foraminifera from the Guaymas Basin, Gulf of California: seasonal, annual, and interspecies variability. Mar Micropaleontol 74:29–37

Weyl PK (1978) Micropaleontology and ocean surface climate. Science 202:475–481

Žarić S, Schulz M, Mulitza S (2006) Global prediction of planktic foraminiferal fluxes from hydrographic and productivity data. Biogeosciences 3:187–207

Zeebe RE (2012) History of seawater carbonate chemistry, atmospheric CO_2, and ocean acidification. Annu Rev Earth Planet Sci 40:141–165

Zeebe RE, Wolf-Gladrow D (2001) CO_2 in Seawater: equilibrium, Kinetics. Isotopes, Elsevier, Amsterdam

Biogeochemistry

9

The calcareous planktic foraminifer shell has been analyzed for its chemical composition, and assumed proxy of the chemical composition of seawater since the pioneering works of Samuel Epstein and Cesare Emiliani in the 1950s (e.g., Epstein et al. 1951; Emiliani 1955). Seawater is a natural pool of chemical elements and isotopes mainly resulting from erosion and hydrothermalism, and being subjected to environmental change (e.g., Stein et al. 2007; Derry 2009). Carrying environmental signals, the isotopic composition and element ratios are used as proxies in paleoceanography (see, e.g., the reviews of Fischer and Wefer 1999; Katz et al. 2010). The biogeochemistry of the planktic foraminifer test has been used in numerous studies reconstructing past marine conditions. The large majority of these analyses have utilized the entire foraminiferal test, i.e. the actively precipitated calcite during test formation over different ontogenetic stages, including organic tissues on top and between the calcite layers, the latter only in the case of multilayered species. In addition, those analyses may include additional calcite occasionally covering the 'original' shell precipitated by the foraminifer (e.g., King and Hare 1972; Hemleben et al. 1977, 1989; Bé 1980; Lohmann 1995). The biogeochemistry of entire tests consequently contains data from more or less wide water-depths and time-intervals, and is often over-interpreted or misinterpreted when ecological demands at the species level and regional scale, as well as remineralisation processes during sedimentation, are not considered.

The biogeochemistry is here discussed with a biological and ecological perspective of the planktic foraminifer, differentiating between primary (production) and secondary (remineralization and encrustation) effects of proxy formation. Along with the technological development of mass spectrometers, it has been possible to analyze a wide range of trace elements in planktic foraminifer calcite in addition to major and minor elements. The composition of the shell calcite, including effects of secondary overgrowth and remineralization are discussed in relation to environmental parameters, which affect the organism's habitat, i.e. physical and biological parameters like surface mixed layer depth, seasonality, and food availability. Detailed reviews on the geochemistry of the planktic foraminifer shell with a strong paleoceanographic perspective are given, for example, by Rohling and Cooke (1999), Lea (1999), and Katz et al. (2010).

> **Trace element:** A trace element is defined as average concentration of $<10^{-6}$ µg per gram. Cd, V, and U would consequently be trace elements in planktic foraminifer shell calcite. Other elements such as, for example Sr, B, and Zn would classify minor elements (see Lea 1999).

© Springer-Verlag Berlin Heidelberg 2017
R. Schiebel and C. Hemleben, *Planktic Foraminifers in the Modern Ocean*,
DOI 10.1007/978-3-662-50297-6_9

9.1   Stable Isotopes and Element Ratios

Shell formation of planktic foraminifers is generally assumed to be coupled to the carbonate equilibrium of ambient seawater (see, e.g., Rohling and Cooke 1999, and references therein). Therefore, stable oxygen ($^{18}O/^{16}O$ ratio, i.e. $\delta^{18}O$), carbon isotopes ($^{13}C/^{12}C$ ratio, i.e. $\delta^{13}C$), and other isotope and element ratios of shell carbonate are widely applied paleoceanographic proxies to reconstruct temperature, salinity, primary productivity, carbon dioxide concentration, and carbonate ion concentration of ancient oceans from deep-sea sediments (e.g., Urey 1947; Epstein et al. 1951; Emiliani 1955; Rohling and Cooke 1999; Broecker and Clark 1999; Ren et al. 2012a). From the combination of stable isotope records of different species that live at different water depths (Emiliani 1954), and often in combination with other proxies like data from benthic foraminifer tests (Fig. 9.1), a detailed reconstruction of biogeochemical state and hydrography of an ancient ocean is obtained at the regional scale (Fig. 9.2). Quaternary glacial-to-interglacial changes in global ice volume and climatically induced cycles in the terrestrial biosphere largely coincide with changes in the stable isotope record in benthic and planktic foraminifer tests, and in addition to ecological analyses provide a detailed relative (in comparison to absolute data from radioactive isotopes) stratigraphic time scale (Fig. 9.1) of the past ~5.3 Million years (e.g., Emiliani 1955; Shackleton and Opdyke 1973; Imbrie et al. 1984; Martinson et al. 1987; Lisiecki and Raymo 2005). Radiocarbon incorporated in the planktic foraminifer test calcite allows direct determination of absolute age (i.e. radiocarbon data) over the past ~50 kyrs (e.g., Voelker et al. 2000; Voelker 2002; Mollenhauer et al. 2005; Barker et al. 2007).

> **Isotope:** Atoms of the same chemical element but with different numbers of neutrons are called isotopes. Isotopes hence differ in weight. Isotopes with lower numbers of neutrons are lighter than isotopes with more neutrons. For example, the nucleus of the light oxygen isotope ^{16}O contains 8 neutrons and 8 protons, whereas the heavy ^{18}O isotope contains 10 neutrons and 8 protons.

Nitrogen isotopes ($^{15}N/^{14}N$, i.e. $\delta^{15}N$) of the planktic foraminifer test calcite display the $\delta^{15}N$ composition of the ambient seawater nitrate (Ren et al. 2009, 2012a). Systematic differences in the $\delta^{15}N$ between species and ontogenetic stages (i.e. test sizes) are assumed to display regional variations in trophic conditions, hydrology of surface waters (i.e. mixing vs. stratification), and depth habitats of planktic foraminifers (Ren et al. 2009). In addition to changes of the nitrogen–pool and $\delta^{15}N$ of ambient seawater, recycling of ammonium (NH_4^+) affects the $\delta^{15}N$ of the test in symbiont-bearing planktic foraminifer species to a varying degree depending on the availability of nitrogen from ambient seawater (Uhle et al. 1997, 1999). Taking the sum of biotic and abiotic effects in planktic foraminifer stable nitrogen isotope composition into account, $\delta^{15}N$ data add important information on the metabolism, ecology, and habitat of shallow and deep-dwelling species from sedimentary archives (Ren et al. 2012a). In turn, past (glacial-to-interglacial) changes in foraminifer test $\delta^{15}N$ are assumed to result from changing nitrogen fixation at complete nutrient consumption in low latitudes (Ren et al. 2009). Consequently, $\delta^{15}N$ data indicate changes in nutrient concentration and primary production, which finally affects the biological carbon pump, atmospheric CO_2, and climate (Ren et al. 2009). Test $\delta^{15}N$ is therefore proxy of planktic foraminifer paleoecology, as well as regional to global nutrient and carbon turnover.

Most of the about 50 extant planktic foraminifer morphospecies are largely eurythermal and euryhaline within the limits of the global open marine temperature and salinity range, and hence are ubiquitous in the global ocean (Bé and Tolderlund 1971; Hemleben et al. 1989). Consequently, the entire latitudinal range of the ocean can be covered by analyzing a relatively small

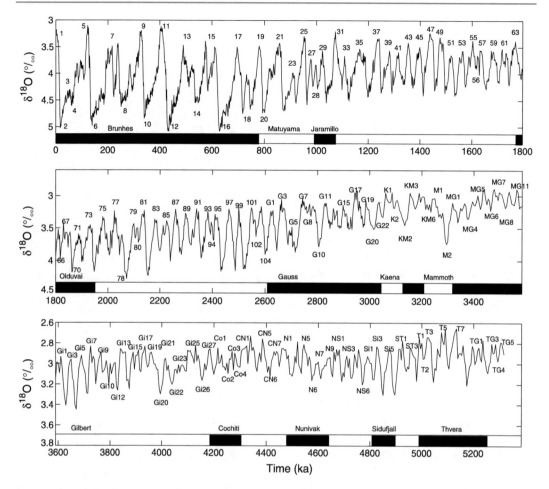

Fig. 9.1 The 'LR04 stack' of Lisiecki and Raymo (2005) constructed by graphic correlation of 57 globally distributed benthic $\delta^{18}O$ records is the standard stack applied for correlation of $\delta^{18}O$ data. Down-core $\delta^{18}O$ data of planktic foraminifers are correlated to the LR04 stack for graphic construction of an age model. Differences between the absolute $\delta^{18}O$ data of the benthic and planktic foraminifer data result from biotic and abiotic effects, such as the absence and presence of symbionts, and varying isotope distributions between surface and deep water bodies, respectively. Those effects may vary between regions (see Fig. 9.3). Note that the scale of the vertical axis changes across panels. From Lisiecki and Raymo (2005)

number of planktic foraminifer species, i.e. *Globigerinoides sacculifer*, *Globigerinoides ruber*, *Globigerina bulloides*, and *Neogloboquadrina pachyderma*, and their overlapping distribution patterns from tropical to polar waters, respectively (see Chap. 7, Ecology, Fig. 7.1). Biogeochemical proxies of the tests of those four species, hence are most applied in paleoceanography to reconstruct surface hydrology and trophic conditions of the open ocean (Fig. 9.2). Various other species may be used in addition to the four ubiquitous species for further information, and to improve paleoceanographic interpretations of surface and deep waters over the course of seasons, and between regions of varying hydrology (e.g., Fischer and Wefer 1999; Henderson 2002; Cléroux et al. 2009; Richey et al. 2012).

9.1.1 δ^{13}C and δ^{18}O

Planktic foraminifer tests are ideal recorders of stable isotopes and minor-element ratios of surface seawater, because the metabolism of planktic foraminifer individuals is probably not affected by variations in the same isotopes and element ratios. Stable oxygen ($^{18/16}$O) and carbon ($^{13/12}$C) isotopes (in delta notation, given in per mil [‰]) of the planktic foraminifer test are standard proxies in paleoceangraphy, and the formation of the isotopic signal needs to be understood in the modern species. Even if assuming that both δ^{18}O and δ^{13}C were recorded within the shell calcite in equilibrium with ambient seawater (Erez and Luz 1983; Lea et al. 1995; Bemis et al. 1998), the signal would still be affected by a variety of autecological (e.g., dwelling depth), regional, and chemical effects. Both O and C isotopes are affected by the salinity of ambient seawater, which is affected by evaporation, sea ice freezing and melting, precipitation, and terrestrial freshwater input (see, e.g., Fischer and Wefer 1999). Ambient water temperature affects the δ^{18}O in a direct way, and δ^{13}C through metabolic effects such as, for example CO_2 incorporation at a rate of 8–15 % (Spero and Lea 1996). On top of those rather small-scale and short-term effects, both O and C isotopes are affected by global effects, which change on longer time-scales. Global effects, like the 'ice effect' and global carbon turnover affect the isotopic composition on land and in the ocean (Fig. 9.3). The interplay of various regional to global effects determines the isotopic composition of seawater and dissolved inorganic carbon (DIC) pool, both of which affect the isotope composition of the planktic foraminifer test calcite (see Rohling and Cooke 1999 for a review). Regional ecological conditions, which affect the isotopic composition of planktic foraminifer test calcite, are temperature and salinity of ambient seawater. Between those two effects, temperature is predominant for most species due to regional sea surface temperature (SST) changes on a much wider range than salinity of open marine waters.

Delta notation: The delta (δ) notation in stable oxygen and carbon isotopes is calculated as:

$$\delta^{18}O_{SAMPLE} = 1000 \times \left[\left(^{18}O/^{16}O\right)_{SAMPLE} - \left(^{18}O/^{16}O\right)_{STANDARD} \right] / \left(^{18}O/^{16}O\right)_{STANDARD}$$

The same formula works for the $^{13/12}$C ratio. In our case, the SAMPLE is a number of planktic foraminifer tests, i.e. calcite, large enough to allow for reproducible analysis with a mass spectrometer. The measured data need to be compared to an oxygen and carbon isotope standard, which is measured together with each batch of original planktic foraminifer sample. Calcite standards have changed over time. The first standard to be used was calcite from the Pee Dee Belemnite (PDB standard), a cephalopod from the Cretaceous Pee Dee Formation outcropping in North and South Carolina, U.S.A. When the Pee Dee Belemnites were exploited, a new standard came into use, called the Vienna PDB standard (VPDB), a synthetic standard related to the original PDB standard (Rohling and Cooke 1999; Coplen 1994; Brand et al. 2014 for further reading). VSMOW, i.e. Vienna Standard Mean Ocean Water is a recalibration from SMOW (Standard Mean Ocean Water), and is mainly used as standard in the analyses of water samples (Coplen 1994). Calcite standard material is not available for the time being (2016)

The δ^{18}O ratio decreases by $0.2 - 0.25‰ \, °C^{-1}$. The δ^{13}C ratio is less, if at all, affected by temperature than δ^{18}O. Salinity increases due to evaporation and sea ice freezing. Precipitation, terrestrial freshwater input, and sea ice melting decrease salinity of the seawater. Those environmental factors are affected by currents, changes in air humidity, and sea surface roughness. Increasing salinity causes enhanced δ^{18}O (0.2–$0.4 \, ‰ \, PSU^{-1}$) and δ^{13}C ratios, and may amount to 1–2 ‰ at the

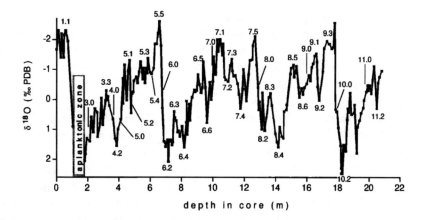

Fig. 9.2 $\delta^{18}O$ curve of *G. ruber* (white) from the Red Sea sediment core KL11 (Hemleben et al. 1996) for comparison with the LR04 standard stack (Fig. 9.1; Lisiecki and Raymo 2005). Please note the differences in absolute $\delta^{18}O$ values between the two curves, resulting from differences between benthic (LR04) and planktic foraminifers (Hemleben et al. 1996), as well as global to regional differences between the two data sets. Absence of planktic foraminifers (aplanktonic zone) during the last glacial maximum (LGM, core depth of 0.9-1.7 m) were caused by extremely high salinities. Numbering of MISs according to Imbrie et al. (1984). From Hemleben et al. (1996)

regional to global scale (e.g., Rohling and Cooke 1999).

Planktic foraminifer calcite production and flux have an impact on, and are affected by, carbonate ion concentration $[CO_3^{2-}]$, and pH of ambient seawater (Wolf-Gladrow et al. 1999a, b; Beer et al. 2010; Marr et al. 2011; Marshall et al. 2013). $[CO_3^{2-}]$ and pH negatively affect $\delta^{18}O$ and $\delta^{13}C$ ratios (Spero et al. 1997; Russell and Spero 2000; Ziveri et al. 2012), while CO_2 concentration has a positive effect (Spero 1992). $[CO_3^{2-}]$ and pH of seawater are positively affected by photosynthesis and negatively affected by respiration (Fig. 9.3). When the planktic foraminifer sequesters calcite from bicarbonate (HCO_3^-) or carbonate (CO_3^{2-}), CO_2 is released to the ambient water (e.g., Zeebe and Wolf-Gladrow 2001). Formation of planktic foraminifer test calcite and release of CO_2 ($2HCO_3^- + Ca^{2+} \rightarrow CaCO_3 + CO_2 + H_2O$) hence decreases pH, $[CO_3^{2-}]$, and total alkalinity. As CO_2 is the second most important greenhouse gas after water vapor, the production of calcareous plankton may affect climate on decadal to geological time scales of hundreds of thousands to millions of years (e.g., Bramlette 1958; Hay 1985; Archer et al. 2000; Zeebe 2012).

9.1.2 Vital Effects on Stable Isotopes and Element Ratios

Biological effects are often attributed to vital effects, and have an impact on the isotope composition of planktic foraminifer tests, as well as on calcification rates, in addition to global, regional, and chemical effects on the stable isotope composition of ambient seawater (e.g., Billups and Spero 1995). Species-specific effects of dwelling-depths and trophic demands may affect seasonal and regional isotope signals (Duplessy et al. 1981; Ganssen 1983). Symbionts increase the $\delta^{13}C$ and $\delta^{15}N$, and decrease the $\delta^{18}O$ of planktic foraminifer tests according to the level of irradiance and photosynthesis typically by up to 1 ‰ (Spero and DeNiro 1987; Uhle et al. 1997; Zeebe et al. 1999; Ezard et al. 2015; see also Bemis et al. 2000). In turn, $\delta^{15}N$ decrease in some symbiont-bearing planktic foraminifer species is suspected to result from ammonium recycling by symbionts (Ren et al. 2012b). The metabolic incorporation of carbon and oxygen in general decreases the $\delta^{13}C$ of the shell according to the isotopic composition of prey (e.g., Berger 1971; Berger et al. 1978; Uhle et al. 1997). This effect is possibly masked by the change of metabolic activity during ontogeny (Hemleben and Bijma 1994). Juvenile individuals grow more rapidly and have a systematically higher metabolic activity than adult specimens (Berger et al. 1978; Bouvier-Soumagnac and Duplessy 1985). In addition, planktic foraminifers change their depth-habitat during ontogeny for various reasons (e.g., food availability), and may dwell in water masses of different temperature, and carbon and oxygen isotope composition. Accordingly, stable carbon and oxygen isotope ratios increase during ontogeny by up to 1 ‰. Finally, gametogenic (GAM) calcite may add an effect on the $\delta^{18}O$ of up to 1 ‰ to the pre-gametogenic shell in *G. sacculifer* (Duplessy et al. 1981). The effect may vary between species depending on the proportion of GAM calcite added to the shell. Size and weight-dependent effects of up to 2 ‰ are reported for *N. pachyderma* from the Arctic Ocean (Hillaire-Marcel et al. 2004). Vital effects, which affect the chemical composition of planktic foraminifer test calcite, hence need to be evaluated at the level of species and for each ontogenetic stage, i.e. for any similar test size increment. To minimize uncertainty in paleoceanographic interpretation caused by vital effects, adult specimens selected from a narrow size range are analyzed for their chemical composition. For an overview of vital effects and $\delta^{18}O$ see the review of Niebler et al. (1999).

Metabolic effect changes during ontogeny (Fig. 9.3) are stronger in juvenile planktic foraminifers with high metabolic activity and rapid chamber formation than in adult individuals with rather slow metabolic activity (Berger et al. 1978; Spero and Lea 1996). At the same time, planktic foraminifers are assumed to change their depth habitat (Chap. 7, Ecology), and in addition to metabolic effects, changing environmental conditions (e.g., temperature) affect the isotope composition of the shell (e.g., Spero and Lea 1996; Mulitza et al. 1997). Consequently, narrow test-size ranges of adult specimens are analyzed for stable isotope ratios, which usually span not more than ∼50 μm in test diameter (Spero and Lea 1996 suggest size ranges ±1 s.d.). Correction factors are applied to the isotope values of different species to account species to account for

the so-called vital effects, which supposedly are the sum of ecological and biological effects on planktic foraminifer isotope ratios (see Niebler et al. 1999 for a summary).

9.1.3 Effect of Photosynthesis on Stable Isotopes

Activity of planktic foraminifer symbionts (see Chap. 4.3) and changes in the chemical microenvironment of various carbonate species (e.g., CO_2, $H_2CO_3{}^-$, CO_3^{2-}) and pH cause carbon fractionation (Fig. 9.4). Enhanced photosynthetic $^{12}CO_2$ uptake and ^{13}C enrichment in inorganic carbon may occur during shell calcification, as exemplified for symbiont-bearing *Orbulina universa* versus symbiont-barren *G. bulloides* (Spero and DeNiro 1987; Bemis et al. 1998, 2000; Köhler-Rink and Kühl 2005; Lombard et al. 2009; see also the review of Bijma et al. 1999) (Fig. 9.5). Changes of the chemical microenvironment, which are caused by the light environment (between 99 and 365 μmol photons $m^{-2} s^{-1}$), and symbiont activity occur over time periods of minutes (Rink et al. 1998; Köhler-Rink and Kühl 2005). Symbiont activity enhances pH and CO_3^{2-}, and lowers CO_2 near the planktic foraminifer test wall (shell) (Fig. 9.4), and hence negatively affects $\delta^{13}C$ and $\delta^{18}O$ by up to 1 ‰ depending on a (symbiont-bearing) species (Fig. 9.3) in the water column (e.g., Bijma et al. 1999; Rohling and Cooke 1999; Zeebe et al. 1999). Note that the effect of photosynthesis on $\delta^{18}O$ in shell calcite of *O. universa* is independent of temperature, for an absolute temperature range between 15 and 25 °C (Fig. 9.5).

9.1.4 Effect of Carbonate Ion Concentration on Stable Isotopes

Carbonate ion concentration ($[CO_3^{2-}]$) affects both $\delta^{13}C$ and $\delta^{18}O$ in natural environments to the same direction, i.e. both $\delta^{13}C$ and $\delta^{18}O$ decrease at increasing $[CO_3^{2-}]$ given that alkalinity and total

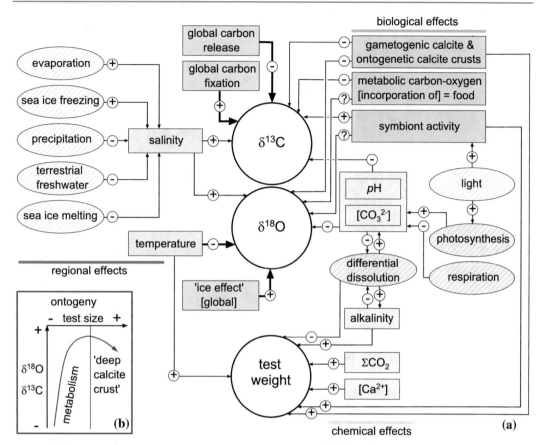

Fig. 9.3 Effects on the stable carbon and oxygen isotope incorporation, and weight of planktic foraminifer tests produced in the average surface pelagic ocean at pH >8.1. **a** Stable isotope composition of planktic foraminifer tests, and test weight is formed by a combination of global (*orange*), regional (*blue*), chemical (*yellow*), and biological effects (*green*). Differential dissolution is both biologically and chemically (thermodynamically) affected (Dittert et al. 1999; Milliman et al. 1999; Schiebel et al. 2007). Predominant effects are indicated with bold arrows. Positive and negative coupling is marked by (+) and (–), respectively. 'ΣCO_2' includes all carbon species dominated by HCO_3^- and CO_3^{2-} at pH 8.1–8.2 (Zeebe and Wolf-Gladrow 2001). **b** Ontogenetic shifts of the isotopic composition (Berger 1971; Berger et al. 1978) of the shell calcite are complex, and are attributed to metabolism, and the formation of additional ontogenetic calcite layers in some species such as, for example, *N. pachyderma* (e.g., Simstich et al. 2003; Hillaire-Marcel et al. 2004) in subpolar waters. After Schiebel and Hemleben (2005)

CO_2 (ΣCO_2) remain constant (Bijma et al. 1999). The 'carbonate-ion effect' is assumed to affect symbiont-barren (e.g., *G. bulloides*) and symbiont-bearing (e.g., *O. universa*) species to about the same degree (Zeebe et al. 1999; Bijma et al. 1999), and result from an internal (cytoplasmic) inorganic carbon pool, by analogy to observations on benthic foraminifers (ter Kuile and Erez 1991). These vital effects, i.e. non-equilibrium test-calcite compositions in isotopes and element ratios were suspected by Parker (1958), and experimentally assessed for the uptake of ^{14}C and ^{45}Ca in hermatypic corals and benthic foraminifers by Erez (1978), and ^{45}Ca in *G. sacculifer* by Anderson and Faber (1984). In paleoceanographic analyses, stable isotope data from planktic foraminifer tests consequently need to be analyzed on the basis of temporally varying pH and $[CO_3^{2-}]$ of past seawater on the regional scale (Lea et al. 1999a). $[CO_3^{2-}]$ and consequently pH may be reconstructed to some degree from size-normalized shell weight (calcite mass) along

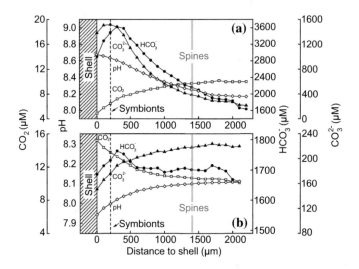

Fig. 9.4 Calculated HCO$_3^-$ and CO$_3^{2-}$ profiles and measured CO$_2$ and pH profiles under (**a**) light and (**b**) dark conditions in *Orbulina universa* showing the effect of photosynthesis on the carbonate chemistry near the foraminifer shell. Symbiont activity (**a**) enhances pH, CO$_3^{2-}$, and H$_2$CO$_3^-$, and lowers CO$_2$ near the planktic foraminifer shell. Note the different scales in (**a**) and (**b**). The vertical *dashed line* indicates the start of the symbiont swarm. The *gray vertical line* indicates the outer extension of spines. After Köhler-Rink and Kühl (2005)

with stable isotope data to account for effects caused by environmental change and stratigraphic control in down-core analyses (cf. Barker and Elderfield 2002; Broecker and Clark 2003; Beer et al. 2010).

Precision of modern mass spectrometers can analyze at sufficient reproducibility single large chambers, single tests of most adult planktic foraminifer species, or some specimens of small-sized and thin-shelled species (e.g., Kozdon et al. 2011). Technological progress in mass spectrometry allows reproducibility of single-specimen analyses and comparison of single data points. Since ecological conditions under which a planktic foraminifer precipitates test calcite are subject to statistical variability, multi-specimen analyses are performed to enhance reproducibility of results. The number of specimens needed to achieve a certain level of statistical reproducibility is exemplified for *Pulleniatina obliquiloculata* and *G. sacculifer* by Schiffelbein and Hills (1984) (Fig. 9.6). Those results are applicable for any planktic foraminifer species, with an analytical precision depending on the variability of regional ecological conditions and the species'

autecological prerequisites, which are determined empirically.

Carbonate ion effect: Carbonate ion concentration [CO$_3^{2-}$] has a positive effect (among other effects induced, e.g., by pH, CO$_2$, T, salinity, food) on the calcification of planktic foraminifer (among other calcifiers) tests. The higher the [CO$_3^{2-}$], the higher the amount of calcite produced by planktic foraminifers (e.g., Bijma et al. 1999; Bijma et al. 2002). The carbonate ion effect seems to be species-specific, and depends on light (higher at stronger irradiation) in symbiont-bearing species. In addition, the effect has been shown to be lower in *G. sacculifer* than in *O. universa* both being symbiont-bearing species (Lombard et al. 2010). It may be speculated that the carbonate ion effect is stronger in symbiont-barren species (e.g., *G. bulloides*) than symbiont-bearing species, the former of which lack the buffering effect of CO$_2$ consumption by symbiont activity. In turn,

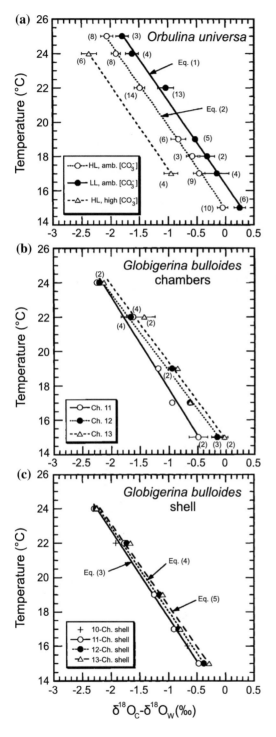

Fig. 9.5 Effect of temperature on $\delta^{18}O_{CARBON}$-δ^{18} O_{WATER} values ($\pm1\sigma$) in cultured individuals of two planktic foraminifer species. **a** *Orbulina universa* kept under high light (HL) >380 µEinstein m^{-2} s^{-1}, and low light (LL) 20–30 µEinstein m^{-2} s^{-1} conditions. Ambient $[CO_3^{2-}]$ of 171 µmol kg^{-1}, and high $[CO_3^{2-}]$ of 458 µmol kg^{-1}. Slopes of the regressions are −4.8 (0.21 ‰ °C^{-1}). **b** *Globigerina bulloides* chambers 11, 12, and 13, and **c** reconstructed whole *G. bulloides* shells consisting of 11, 12, and 13 chambers. Data on 10-chambered shells result from experiments carried out at 16 °C (Spero and Lea 1996), and 22 °C (Bemis et al. 1998). Note the effect of photosynthesis (*O. universa* only), i.e. the offset between the curves is independent of temperature. Absolute numbers of test for each experimental group are given in parentheses. From Bemis et al. (1998)

dissolution of planktic foraminifer test calcite increases with decreasing $[CO_3^{2-}]$, increasing pH (and $[CO_2]$), and decreasing saturation state (Ω) at increasing water depth (Schiebel et al. 2007). Consequently, dissolution of planktic foraminifer tests changes with atmospheric $[CO_2]$, and hence seawater $[CO_2]$, for example, over glacial-interglacial cycles (Broecker and Clark 2001). In addition to the effect on calcification rate, $[CO_3^{2-}]$ affects the stable carbon and oxygen isotope ratio through kinetic fractionation processes, and the consumption of metabolic CO_2 by symbionts (Spero et al. 1997; Bijma et al. 1999).

9.1.5   Paleotemperature Equations

The chemical composition of the planktic foraminifer shell calcite including stable oxygen isotopes ($^{18}O/^{16}O$) represents the sum of biotic and abiotic effects, i.e. differences in the species-specific biological prerequisites and environmental requirements (see Fig. 9.3). Consequently, equations for paleotemperature reconstruction are ideally calibrated at the

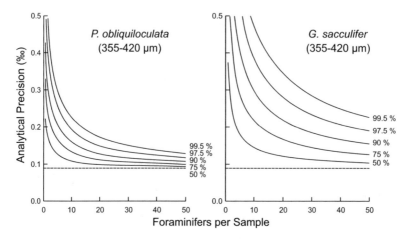

Fig. 9.6 Precision-reliability sampling curves for $\delta^{18}O$ multi-shell isotope analyses are species-specific. Whereas about ten tests per analysis are sufficient to increase precision by 50 % (99.5 % level) in *P. obliquiloculata*, about 20 tests are needed in *G. sacculifer*. One-sided confidence limits for sample standard deviation (σ_T) are generated using the jackknife-based estimates of σ.

Dashed line represents machine precision. *P. obliquiloculata* (355–420 μm), thirty isotope analyses of three specimens each (*left panel*). *G. sacculifer* (355–420 μm), thirty isotope analyses of four specimens each (*right panel*). Data are from Holocene sediment samples. After Schiffelbein and Hills (1984)

highest possible taxonomic level (i.e. species, morphotypes, ecophenotypes) to account for regional variability in environmental conditions (e.g., Bemis et al. 1998, 2002; Peeters et al. 2002; Mulitza et al. 2003). Temperature equations follow the second order equation

$$T = K1 - K2(\Delta - A) + K3(\Delta - A)^2 \quad (9.1)$$

relating temperature (T[°C]) and isotopic composition of the planktic foraminifer test calcite (Δ) and ambient seawater (A) (see, e.g., Epstein et al. 1951, 1953; Epstein and Mayeda 1953; Erez and Luz 1982, 1983; see also Kim and O'Neil 1997). The term 'A' in (9.1) may be directly measured from the ambient seawater when working on modern systems, and is estimated for past conditions. K1 and K2 are empirically determined coefficients of slope and intersection, respectively. Shackleton (1974) provides a general temperature equation, which yields reasonable results independent of species and region analyzed:

$$T(°C) = 16.9 - 4.38\,(\delta^{18}O_C - \delta^{18}O_W) + 0.10\,(\delta^{18}O_C - \delta^{18}O_W)^2 \quad (9.2)$$

Species-specific and regionally calibrated equations provide more reasonable results, but are, in turn, less applicable on a larger scale. Bemis et al. (2002) provide temperature equations on samples from the Southern California Bight, which account for differences between species, test size, light level in case of symbiont-bearing species, and dwelling depth (Eqs. 9.3, 9.4, and 9.5). Mulitza et al. (2003) provide equations from samples obtained from the eastern Equatorial and South Atlantic (Eqs. 9.6, 9.7, 9.8, and 9.9).

N. incompta (*N. pachyderma*)
$$T(°C) = 17.3 - 6.07\,(\delta^{18}O_C - \delta^{18}O_W) \quad (9.3)$$
G. bulloides
$$T(°C) = 13.4 - 4.48\,(\delta^{18}O_C - \delta^{18}O_W) \quad (9.4)$$

O. universa

$$T(^{\circ}C) = 15.7 - 4.46\,(\delta^{18}O_C$$
$$- \delta^{18}O_W) + 0.35\,(\delta^{18}O_C - \delta^{18}O_W)^2$$
$$(9.5)$$

N. pachyderma

$$T(^{\circ}C) = 12.69 - 3.55\,(\delta^{18}O_C - \delta^{18}O_W) \quad (9.6)$$

G. bulloides

$$T(^{\circ}C) = 14.62 - 4.70(\delta^{18}O_C - \delta^{18}O_W) \quad (9.7)$$

G. ruber

$$T(^{\circ}C) = 14.20 - 4.44\,(\delta^{18}O_C - \delta^{18}O_W) \quad (9.8)$$

G. sacculifer

$$T(^{\circ}C) = 14.91 - 4.35(\delta^{18}O_C - \delta^{18}O_W) \quad (9.9)$$

Temperature-to-$\delta^{18}O$ relationships from surface seawater samples of the eastern Equatorial and South Atlantic from Mulitza et al. (2003) are different from those reported from the Southern California Bight (Bemis et al. 2002) even for the same species *G. bulloides* (see Eqs. 9.4 and 9.7).

Regional differences possibly result from biological and ecological differences between morphotypes and ecophenotypes of the same species. Systematic differences in the stable isotope composition of shell calcite result from habitat-specific effects like dwelling depth, region, season, and hence food availability (Fig. 9.7) highlight the need for species-specific paleotemperature equations (e.g., Mortyn and Charles 2003; Birch et al. 2013).

9.2 Clumped Isotopes

A rather new approach on clumped isotopes has been developed for planktic foraminifer analyses since the early 2000s. Clumped isotopes provide quantitative information for paleo-environmental reconstruction (e.g., Ghosh et al. 2006; Eiler 2007; Dennis et al. 2011). The clumped-isotope geochemistry utilizes the extent to which the rare species of the respective isotopes (e.g., ^{17}O, ^{18}O, ^{13}C, ^{15}N, D) bond with each other and not with the light isotopes (i.e. ^{16}O, ^{12}C, ^{14}N, ^{1}H), and the deviation from their stochastic distribution. Bonds of the rare isotopes, called isotopologues (e.g., $^{18}O^{13}C^{16}O$ and $^{17}O^{13}C^{16}O$) are very rare.

Therefore, a larger volume of sample is required for a single measurement, i.e. 5–10 mg of planktic foraminifer calcite provided by $\sim$1000 medium-sized tests ($\sim$250 µm) of *G. bulloides*. A mass spectrometer designed for high-precision measurement is needed (Ghosh et al. 2006; Eiler 2007; Schmid and Bernasconi 2010).

For clumped isotope analyses of planktic foraminifer test calcite, the carbonate-bound CO_2 is released by adding phosphoric acid, and the excess abundance in ^{13}C-^{18}O is defined as

$$\Delta_{47} = \left(R^{47}_{SAMPLE} / R^{47}_{STOCHASTIC} - 1\right) \times 1000$$
$$(9.10)$$

with R^{47} being the 47/44 ratio of the analyzed CO_2.

Clumped isotope analysis has been successfully applied to planktic foraminifers, and may add important information on the ecology and paleoecology at the species level (Tripati et al. 2010; Wacker et al. 2014). However, most of the existing data on planktic foraminifers are indistinguishable from equilibrium (Tripati et al. 2010).

9.3 Mg/Ca Ratio and $\delta^{44}Ca$

The Mg/Ca ratio of planktic foraminifer test calcite (Fig. 9.8) is utilized as a proxy of seawater (paleo-) temperature (e.g., Cronblad and Malmgren 1981; Nürnberg et al. 1996; Hastings et al. 1998; Russell et al. 2004; Martínez-Botí et al. 2011). The Mg/Ca ratio and $\delta^{44}Ca$ (Fig. 9.9) are often analyzed in combination with other metal-to-Ca (Me/Ca) ratios and stable isotopes, such as, for example, Mn/Ca, Ba/Ca, Zn/Ca, Sr/Ca, ^{88}Sr, and ^{138}Ba (e.g., Rosenthal et al. 1997; Eggins et al. 2003; Gehlen et al. 2004; Kunioka et al. 2006; Marr et al. 2013). Systematic changes of the sum of Me/Ca ratios are interpreted for their biological, ecological, and paleoceanographic information, ideally using planktic foraminifer species from different ecological niches and with different dissolution susceptibilities

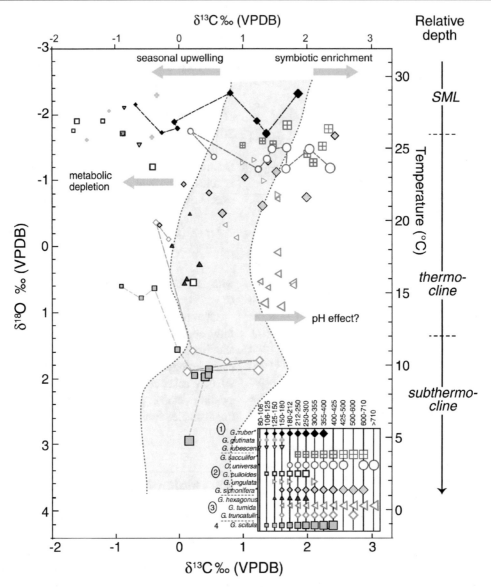

Fig. 9.7 Carbon (δ^{13}C) versus oxygen (δ^{18}O) isotopes of 12 extant planktic foraminifer species across test-size spectra from core-top samples (GLOW3 Box-Core), showing systematic changes in stable isotope composition of shell calcite. The changes result from habitat-specific effects like dwelling depth, trophic conditions (e.g., upwelling), symbiont activity, and metabolic effects. Each symbol corresponds to a single species. Symbol size is scaled to the test size-fraction analyzed. The taxa selected span the range of open ocean planktic foraminifer depth habitats, sampling surface mixed layer to subthermocline zones. *Dashed lines/gray* shading indicates target water column δ^{13}C DIC envelope. *Inset* at the lower *right* shows habitat groupings: (*1*) surface mixed layer (SML), (*2*) deeper surface mixed layer/upper thermocline, (*3*) thermocline, and (*4*) sub-thermocline. Size-ordered data arrays of four species (one from each of the eco-groups) are connected by lines to illustrate isotopic trajectories. From Birch et al. (2013)

(Dekens et al. 2002; Gussone et al. 2009). Multispecies and multiproxy analyses provide a comprehensive understanding of paleoenvironment from the surface and subsurface water column (e.g., Brown and Elderfield

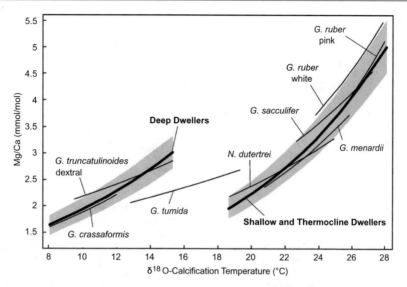

Fig. 9.8 Planktic foraminifer Mg/Ca versus annual temperature [T(°C)] calibrations from δ^{18}O data. Mg/Ca ratios increase with ambient water temperature. Thick curves illustrate the multispecies calibrations. Thin curves give the species-specific calibrations. The uncertainties of Mg/Ca temperatures calculated with the multispecies calibrations (± 1.0 °C for shallow and thermocline dwellers, ± 1.3 °C for deep dwellers) are represented by the *shaded areas*. Note the parallel offset of the multispecies calibrations by ~ 8 °C. From Regenberg et al. (2009)

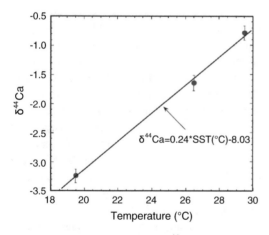

Fig. 9.9 Temperature effect on δ^{44}Ca values of *G. sacculifer* cultured under controlled conditions. *Data points* and *error bars* represent weighted means of two to three independent analyses and statistical uncertainties, respectively. The *bold line* gives the absolute temperature calibration. From Nägler et al. (2000)

1996; Sadekov et al. 2010; Hönisch et al. 2013; Jonkers et al. 2013; Regenberg et al. 2014).

Calibration of the Mg/Ca paleothermometer from cultured planktic foraminifers (e.g., von Langen et al. 2005), and empty tests from

sediment trap samples reveals an increase of the Mg/Ca ratio of 8.5–10.2 % °C^{-1}, which is a ~ 0.3 mmol mol^{-1} Mg/Ca change per °C (Lea 2003; Anand et al. 2003). Mg/Ca ratios of planktic foraminifer calcite are positively correlated to ^{44}Ca isotopes (Fig. 9.10), the latter of which provides another independent proxy of seawater temperature, which is probably not affected by diagenetic alteration (Nägler et al. 2000; Gussone et al. 2003). Small differential effects of $[CO_3^{2-}]$ on δ^{44}Ca in the symbiont-bearing species *G. ruber* and *Globigerinella siphonifera* (Kisakürek et al. 2011) may be explained by species-specific differences in the depth habitat and hence calcification temperature.

The fact that the Mg/Ca ratio of planktic foraminifer tests is a more or less reliable paleothermometer may be explained by a variety of temperature-dependent physiological processes (Bentov and Erez 2006; Jonkers et al. 2013). Branson et al. (2013) provide evidence for consistent thermodynamically related Mg/Ca uptake from ambient seawater during precipitation of the shell calcite. The preference of Ca over Mg, and production of low-Mg planktic foraminifer

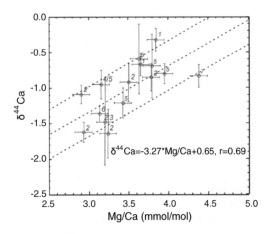

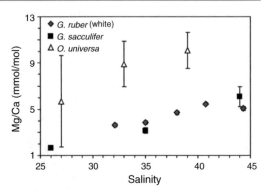

Fig. 9.10 δ^{44}Ca and Mg/Ca ratios of *G. sacculifer* are positively correlated. Tests from interglacial periods MIS 1, 5, and 7 (*red numbers*), tend to have higher δ^{44}Ca and Mg/Ca ratios than tests from glacial periods MIS 2, 4, and 6 (*blue numbers*). From Nägler et al. (2000)

Fig. 9.11 Salinity dependence of Mg/Ca in tests of cultured planktic foraminifers. Mg/Ca of *G. ruber* (24 °C, *diamonds*) and of *O. universa* (22 °C, *triangles*) are bulk analyses. Mg/Ca of *G. sacculifer* (26.5 °C, *squares*) are in situ analyses of consecutive chambers of one specimen. *Error bars* give 1σ for *G. ruber* and *O. universa*, and 1SEM for *G. sacculifer*. From Kisakürek et al. (2008)

calcite is due to active incorporation of Mg (Mg-pump) counteracting a large electrochemical gradient (cf. Zeebe and Sanyal 2002; Bentov and Erez 2006; De Nooijer et al. 2014).

While the Mg/Ca ratio of planktic foraminifer tests results from regional hydrographic conditions, δ^{18}O-derived temperature results from both regional and global effects. Consequently, the difference of Mg/Ca to δ^{18}O-derived temperatures is a measure of global ice volume, and to some degree of ambient seawater salinity (e.g., Elderfield and Ganssen 2000; Weldeab et al. 2006; Kisakürek et al. 2008; Dueñas-Bohórquez et al. 2009; Mathien-Blard and Bassinot 2009; Hönisch et al. 2013; Arbuszewski et al. 2010) (Fig. 9.11). Species-specific relations of Mg/Ca versus δ^{18}O ratios are interpreted as a result of differences in dwelling depth and hence mostly temperature (Regenberg et al. 2009; Bolton et al. 2011; Hönisch et al. 2013) (Fig. 9.12). The effect of ambient modern surface ocean pH (8.0-8.3) and $[CO_3^{2-}]$ on Mg/Ca ratios is assumed negligible, as exemplified for the symbiont-bearing *G. ruber* (white) by Kisakürek et al. (2008) (Fig. 9.12). Small differential effects of $[CO_3^{2-}]$ on δ^{44}Ca in the symbiont-bearing species *G. ruber* and *G. siphonifera* (Kisakürek et al. 2011) are explained by species-specific differences in the depth habitat and hence calcification temperature.

In addition to the effects discussed above, the Mg/Ca ratio is related to the ontogenetic development of planktic foraminifer species, i.e. test size and dwelling depth, as well as the presence of symbionts (e.g., Elderfield et al. 2002; von Langen et al. 2005; Friedrich et al. 2012). Mg/Ca ratios of most large-sized planktic foraminifer species rapidly decrease over juvenile and neanic development, i.e. over a test size interval of about 100–200 μm (Fig. 9.13). Mg/Ca ratios of small-sized species change less over the 100–200 μm size fraction than those of large-sized species, and may even increase as in *G. glutinata* (Friedrich et al. 2012). In general, Mg/Ca ratios of adult individuals decrease less, if at all, than in pre-adult individuals, and might even increase as in *O. universa* (Fig. 9.13).

Most ontogenetic Mg/Ca changes are confirmed by δ^{18}O and δ^{13}C data (Fig. 9.14) measured on samples from the subtropical (Elderfield et al. 2002, 19°N, 20°W) and temperate North Atlantic (Friedrich et al. 2012, 47°N, 20°W). Opposite relationships of δ^{13}C to Mg/Ca ratios in *O. universa* in comparison to other symbiont-bearing (e.g., *G. ruber*) and symbiont-barren species, are explained by photosynthetic effects on calcification (cf. Hamilton et al. 2008), and may in addition be affected by the concomitant formation of additional calcite layers ('calcite crusts') in *O. universa* (but not in *G. ruber*; e.g.,

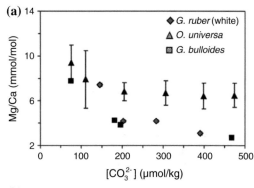

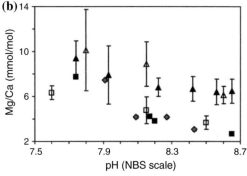

Fig. 9.12 Dependence of planktic foraminifer test Mg/Ca ratios on **a** carbonate ion concentration, and **b** pH as deduced from cultured specimens. *G. ruber* was grown at 27 °C, whereas the other species were grown at 24 °C. All experiments were conducted at oceanic salinity. Grey diamonds *G. ruber*, filled triangles *O. universa* (Russell et al. 2004), empty triangles *O. universa* (Lea et al. 1999b), filled squares *G. bulloides* (Russell et al. 2004), empty squares *G. bulloides* (Lea et al. 1999b). *Error bars* represent 1σ. NBS is the National Bureau of Standards. From Kisakürek et al. (2008)

Bolton et al. 2011). The outer shell of *O. universa* has been shown to comprise low and high Mg growth bands (Eggins et al. 2004). Both high-Mg night bands and low-Mg day bands were produced by *O. universa* under laboratory conditions, and are interpreted to result from changes in mitochondrial uptake of Mg rather than changes in ambient water temperature (Spero et al. 2015). Mg/Ca ratios of both night and day bands hence represent seawater temperature to the same degree. Those diurnal calcite-bands, formed over 2–9 days, add about 10–30 μm to the absolute thickness of adult shell with about 30 % of the calcite being produced at night (see Fig. 6.7; Spero et al. 2015).

In addition to pre-gametogenic (pre-GAM) Mg/Ca ratios of the shell, low-Mg calcite layers may form on the proximal part of shells of different species, and to different degrees, ranging between 20–50 % of the wall thickness in *G. sacculifer*, and up to 70 % in *Neogloboquadrina dutertrei*, and may even vary between different chambers of the same specimen such as in *G. ruber* and *N. dutertrei* (Eggins et al. 2003; Bolton et al. 2011; Jonkers et al. 2012). Those low-Mg calcite layers provide low Mg/Ca calcification temperatures. In contrast, Mg-rich calcite layers may form up to 20 % of the total calcite in fossil tests during early diagenesis in Mg-rich sediments (Boussetta et al. 2011; van Raden et al. 2011). When being analyzed at high resolution, i.e. chamber by chamber, and calcite layer by calcite layer using LA-ICP-MS or NanoSIMS (see Chap. 10), ontogenetic changes of the individual depth habitat provide valuable information on the ecology of planktic foraminifer species, and the regional paleoceanography at high temporal resolution (Eggins et al. 2003; Kunioka et al. 2006; Bolton et al. 2011; Groeneveld and Chiessi 2011; Marr et al. 2011; Branson et al. 2013).

An effect of dissolution on Mg/Ca ratios of the planktic foraminifer tests settling through the water column may vary according regional hydrology, and temperature differences between surface and deep waters, i.e. stratification of the water column (Friedrich et al. 2012). In addition, post-depositional dissolution may affect Mg/Ca ratios (Brown and Elderfield 1996; Regenberg et al. 2006, 2007).

9.4 Boron Isotopes and B/Ca Ratio

The uptake of Boron (^{11}B and ^{10}B) isotopes in planktic foraminifer shells is related to pH, and hence is a proxy of past seawater pH (Sanyal et al. 1996; Spero et al. 1997; Kasemann et al. 2009; Allen et al. 2011). In addition, surface ocean pH affects calcification of planktic foraminifer tests (Moy et al. 2009; de Moel et al. 2009), and the skeletons of other major marine calcifiers such as, for example, coccolithophores (calcite; e.g.,

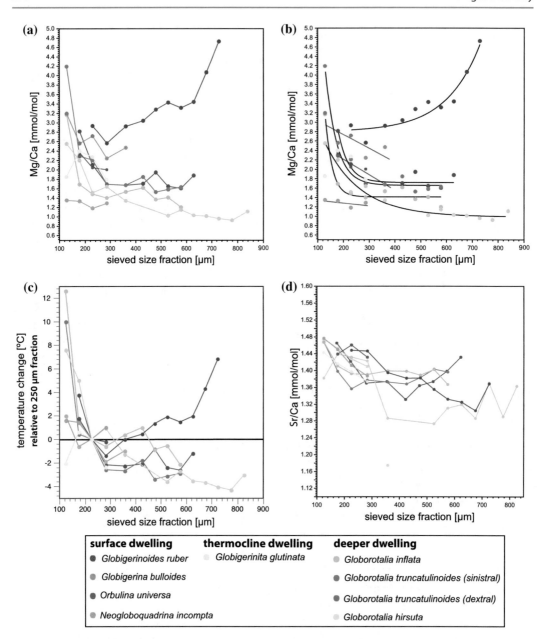

Fig. 9.13 Mg/Ca ratio in relation to the ontogenetic development of planktic foraminifer species. **a** Mg/Ca ratio versus sieved test size fraction species from a core-top sample in the NE Atlantic at 47°N, 20°W, 4577 m water depth (MC575, GeoTü Archive). **b** Same as **a** but including best-fit regression lines (black lines $r^2 > 0.9$; red lines $r^2 < 0.9$). **c** Relative temperature change estimated based on measured Mg/Ca ratios shown in relation to the 200–250-μm size fraction (*horizontal black line*) sensitivity of 10 % per 1 °C. **d** Sr/Ca versus test size. *G. inflata* is categorized deeper dwelling at this location, since its variable habitat occasionally includes sub-thermocline waters. From Friedrich et al. (2012)

Riebesell et al. 2000), and pteropods (aragonite; e.g., Orr et al. 2005; Bednaršek et al. 2012). Rapid ocean acidification (OA) events have been recurrent in Earth's history (e.g., Paleocene-Eocene Thermal Maximum, PETM) along with major calcite dissolution cycles, CO_2

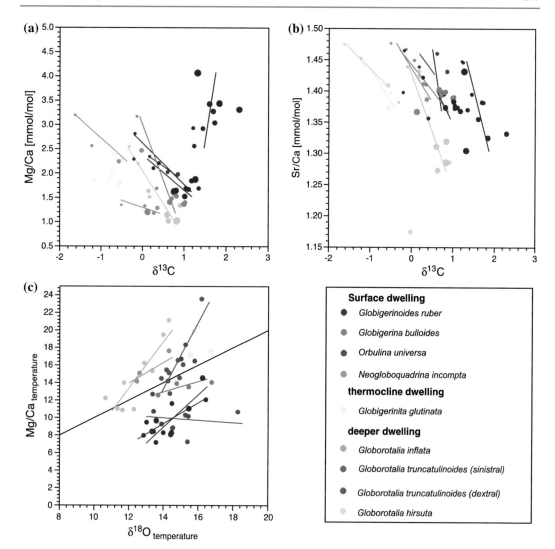

Fig. 9.14 Mg/Ca and Sr/Ca plotted against stable isotope values. **a** Comparison between Mg/Ca and $\delta^{13}C$, and **b** Sr/Ca and $\delta^{13}C$ values. **c** Mg/Ca-based temperature compared with $\delta^{18}O$ calcification temperature using a constant $\delta^{18}O_{SEAWATER}$ of 0.7‰. *Black line* represents a 1:1 relation. Equations for $\delta^{18}O$ temperature calculation of *G. bulloides*, *N. incompta*, and *O. universa* are from Bemis et al. (2002), *G. ruber* from Mulitza et al. (2003), and all other species from Shackleton (1974). Mg/Ca calibration for *G. bulloides* from Elderfield and Ganssen (2000), *G. ruber* from Anand et al. (2003), *G. truncatulinoides* from Regenberg et al. (2009), *O. universa* from Lea et al. (1999b), and all other species from multi-species calibration from Anand et al. (2003). *G. inflata* is categorized deeper dwelling at this location, since its variable habitat occasionally includes sub-thermocline waters. *Symbol size* indicates sieved size fractions. From Friedrich et al. (2012)

release, and climate warming (e.g., Zachos et al. 2005; Uchikawa and Zeebe 2010). In contrast, slow increase in atmospheric CO_2 during the upper Cretaceous ($\sim$100–66 Ma) allowed sufficient time for equilibration of surface to deep ocean carbon species (CO_2, CO_3^{2-}, HCO_3^-, etc.), as well as to the marine calcifiers to adjust to the changing conditions, which resulted in massive calcite (mainly coccolithophores) sedimentation despite very high atmospheric $[CO_2]$ around 2000 ppm (Caldeira and Wickett 2003; Lohbeck et al. 2012). Good examples of the upper Cretaceous paleoenvironment and massive calcite sedimentation are the cliffs of northern France

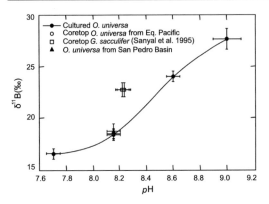

Fig. 9.15 Empirical relationship between boron isotopic composition of cultured *O. universa* and pH of culture media. The pH *error bars* give maximum variations of measured pH of the culture solutions during the experiments. The B isotopic composition of *O. universa* from core-top samples from the eastern equatorial Pacific (V19-28), and that of live adult *O. universa* from the San Pedro Basin off Santa Catalina Island (CA) are plotted against the modern surface ocean pH at the respective locations. Data of Holocene *G. sacculifer* from the equatorial W Pacific and tropical Atlantic are shown for comparison. From Sanyal et al. (1996)

(e.g., Étretat), southern England (e.g., Dover), Møn (Denmark), and Rügen (Germany).

Changes in seawater pH exerts a quantitative effect on a variety of calcifying marine organisms (see Zeebe and Wolf-Gladrow 2001, for a comprehensive review) and may result in a significant decrease in the number of taxa. Apart from a loss in biodiversity and biological niches, the ocean would lose some of its capacity to take up CO_2, the latter of which is the reason for modern ocean acidification (e.g., Feely et al. 2004; Orr et al. 2005). Due to the complexity of the marine carbonate system and the relatively short interval of time over which pH and its effects on marine calcification have been monitored (since the early 1980s at Bermuda, Gruber et al. 2002) the entire effect of OA is not yet completely understood. An enormous average decrease of 0.1 pH units, from pH 8.2 to 8.1, i.e. a 30 % decrease in pH of the surface ocean is assumed since the beginning of industrialization (Sabine et al. 2004). Analogous situations to the modern OA are analyzed from the sedimentary record of past climate and OA change using $\delta^{11}B$ (‰) as proxy.

The control of $\delta^{11}B$ by the pH of ambient seawater was first analyzed from live planktic foraminifers (*O. universa* and *G. sacculifer*) from cultures by Sanyal et al. (1996; 2001) (Fig. 9.15). Earlier data from Spivack et al. (1993) had proposed a pH of 7.4 for Miocene (21 Ma) seawater analyzed from (unspecified) foraminifers from the western Pacific Ocean Drilling Program (ODP) Hole 803. The ^{11}B paleo-pH proxy utilizes the pH dependent fractionation of $B(OH)_4^-$ and $B(OH)_3$, the latter being 20 ‰ heavier in ^{11}B than the $B(OH)^{4-}$ (Sanyal et al. 1996). It is then assumed that only $B(OH)^{4-}$ is utilized by the foraminifer for substitution of CO_3^{2-} when precipitating the test-calcite (Hemming and Hanson 1992). The measured $\delta^{11}B$ values are then converted to pH by applying a partition coefficient, K_D (Sanyal et al. 1996).

Although pH dependency of $\delta^{11}B$ values in planktic foraminifer calcite is evident, symbiont activity reduces the sensitivity of the test calcite to pH driven $\delta^{11}B$ incorporation in living (cultured and tow-sampled) *G. ruber*$_w$ (Henehan et al. 2013). Symbiont-barren species probably record the boron isotope composition of ambient seawater more directly than symbiont-bearing species. In addition, $\delta^{11}B$ of *G. ruber*$_w$ seems to result from effects related to the ontogenetic development of individuals, and to increase from the 250–355-μm to 400–455-μm test-size fraction (Henehan et al. 2013). A similar ontogenetic effect is reported for *G. sacculifer* from surface sediment samples, concluding that increased $\delta^{11}B$ incorporation into the shell is caused by enhanced symbiont activity, and an effect of light intensity in larger specimens (i.e. adults), which live at shallower water depth than smaller (i.e. juvenile) specimens (Hönisch et al. 2003; Hönisch and Hemming 2004). In contrast, Ni et al. (2007) suggest dissolution as a primary cause of test-size dependent $\delta^{11}B$ variation. However, comparing the results of Hönisch and co-workers (e.g., 2003) on the $\delta^{11}B$ signal of planktic foraminifer tests from surface sediments with those of Henehan et al. (2013) on live specimens from surface waters and culture experiments shows that an ontogenetic (i.e. test-size related) effect can be amplified by dissolution. Smaller (lighter) tests are more affected by dissolution than larger (heavier) tests due to their lower settling velocity and longer exposure time to deep and more calcite-aggressive (low Ω) waters (Berelson 2002; Schiebel 2002; Schiebel et al. 2007).

The boron-to-calcium (B/Ca) ratio increases at increasing pH as indicated by culture-grown *O. universa* at pH 6.2–8.4 (Sanyal et al. 1996). In addition to the findings of Sanyal et al. (1996), Allen et al. (2011) point toward the controls of different carbonate species including CO_3^{2-} (measurable), temperature (insignificant at a range of 17.7–26.5 °C), and salinity and boron concentration (increases) on B/Ca ratios. Allen et al. (2011) also report increasing and inconsistent B/Ca ratios from the inside to the outside of the planktic foraminifer shell, which are negatively correlated to Mg/Ca ratios. These data need to be confirmed by data from other locations and different species to allow a more solid interpretation.

9.5 Cd/Ca Ratio

Cadmium (Cd) and phosphate (PO_4) concentrations in seawater are closely correlated (e.g., the pioneering work of Boyle et al. 1976; Broecker and Peng 1982; Zahn and Keir 1994). Both Cd and the macronutrient PO_4 are depleted through consumption by phytoplankton growth in surface waters. Both [Cd] and [PO_4] rapidly increase in mesobathyal waters below the surface mixed layer, and are relatively constant at high concentration in the water column below ~ 1000 m depth (e.g., Broecker and Peng 1982; Rickaby et al. 2000; Sarmiento and Gruber 2006). Cd concentration is indicative of different water masses on the basin scale, and the cadmium-to-calcium (Cd/Ca) ratio is used as a tracer of paleo-circulation of deep-water masses, often in combination with other element ratios, and $\delta^{13}C$ of benthic foraminifer tests (e.g., Boyle 1981, 1988; Oppo and Horowitz 2010).

The Cd/Ca ratio of planktic foraminifer shell calcite is a proxy of seawater phosphate concentration (e.g., Boyle 2006), and hence of the macronutrient concentration in general when assuming a constant Redfield ratio (cf. Boyle et al. 1976). In contrast to PO_4, Cd is not quantitatively consumed by marine biota and fractionation of Cd is hence assumed negligible (cf. Boyle 1981). Cadmium incorporation into planktic foraminifer calcite is assumed to be generally higher in symbiont-barren than in symbiont-bearing species (Delaney 1989; Mashiotta et al. 1997; Lea 1999). Cadmium is taken up in preference to phosphorus at an average fractionation factor $\alpha_{Cd/P} = 2$ (Elderfield and Rickaby 2000). In addition, uptake of Cd in the planktic foraminifer shell is temperature-dependent, following a partition coefficient D_{Cd} (Rickaby and Elderfield 1999) (Fig. 9.16). When used in combination with independent temperature proxies like the Mg/Ca ratio and $\delta^{18}O$, and when accounting for the different partition coefficients, the Cd/Ca ratio provides an independent measure of phytoplankton productivity (e.g., Boyle et al. 1976; Rickaby and Elderfield 1999; Ripperger et al. 2008).

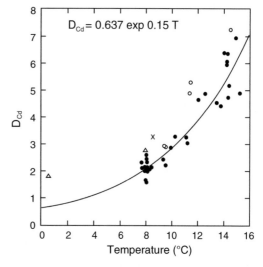

Fig. 9.16 Average temperature-related partition coefficient D_{Cd} of cadmium incorporation in *G. bulloides* shell calcite analyzed from surface sediment sampled from the North Atlantic (*dots*), interglacial Southern Ocean (*circles*), glacial (LGM) Southern Ocean (*cross*), and *N. pachyderma* from the Southern Ocean (*triangle*). From Rickaby and Elderfield (1999)

Redfield ratio: The Redfield ratio (e.g., Redfield et al. 1963) describes element ratios in phytoplankton, and is classically assumed at 106:16:1 for C:N:P, respectively, but which might deviate (e.g., Tett et al. 1985). The C:N:P ratio in seawater would hence be indicative of the availability of nutrients to primary productivity. The lack of either nutrient is identified in case C:N:P ratios would deviate from the Redfield ratio. It has more recently been shown that C:N:P ratios may largely deviate from the classical Redfield ratio (for a review see Sarmiento and Gruber 2006).

The Cd/Ca ratio of tow-sampled planktic foraminifers from the surface water column (0–2500 m water depth) of the North Atlantic and Arabian Sea is correlated to mean phosphate concentration of ambient seawater (Fig. 9.17), depending on region, season, and species-specific average dwelling depth over the individual ontogenetic development (Ripperger et al. 2008). Accordingly, Cd/Ca ratios are higher in specimens from eutrophic waters, i.e. SW monsoonal upwelling off the Arabian Peninsula than in specimens produced during the low-productive intermonsoon in the Arabian Sea. Shallow dwelling *G. ruber* from upwelled water bear higher Cd/Ca ratios than *G. sacculifer* (both symbiont-bearing) and *G. bulloides* (symbiont-barren) from oligotrophic waters.

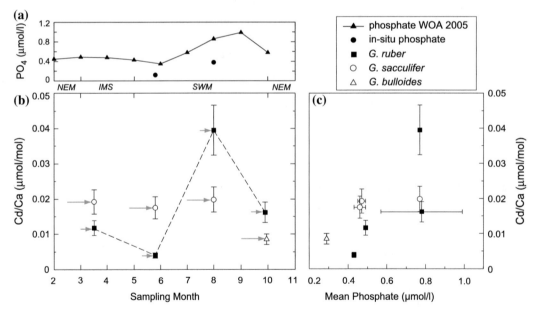

Fig. 9.17 a Seawater phosphate concentration (*triangles*) in the uppermost 75 m of the water column in the Arabian Sea at 16°N, 60°E (from Garcia et al. 2006), and seawater phosphate concentrations determined at the time of sampling (*circles*). **b** Cd/Ca ratios of live *G. ruber* (*squares*) and *G. sacculifer* (*circles*). The *arrows* denote the life span (i.e. test calcification) of the two species, which is a fortnight in *G. ruber* and a full synodic lunar cycle in *G. sacculifer*. The different monsoon seasons are indicated in panel (**a**), i.e., northeast (NEM), inter (IMS), southwest (SWM) monsoon. A single in situ data point from *G. bulloides* (*triangle*) from the North Atlantic is shown for comparison. (**c**) Cd/Ca ratios of live *G. ruber* (*squares*), *G. sacculifer* (*open circles*), and *G. bulloides* (*open triangles*) versus mean seawater phosphate concentration for the upper 75 m of the water column. The *horizontal error* bars denote the range of seawater phosphate concentration that prevailed during the lifespan of the foraminifers. The uncertainty of the Cd/Ca data is given by the relative standard deviation value of ±18 % obtained for multiple analyses of *G. ruber* sampled from the upper 200 m of the water column. From Ripperger et al. (2008)

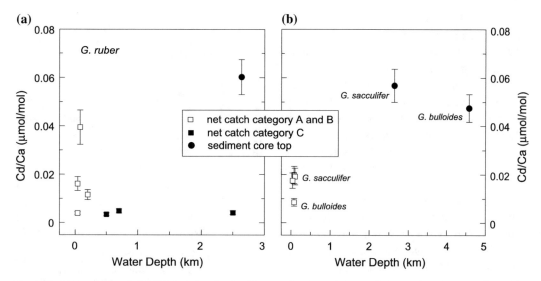

Fig. 9.18 Water depth related Cd/Ca ratios obtained for different species from plankton net tows (*squares*) and sediment core tops (*circles*). In situ collected samples and corresponding sediment core tops are from approximately the same location. **a** Cd/Ca ratios of *G. ruber* tests from plankton net tows in the surface (category **A**), subsurface (category **B**), and deep (category **C**) water column from different months, as well as from surface sediments.

b Water depth related Cd/Ca ratios of *G. sacculifer* and *G. bulloides*. Cd/Ca ratios of specimens from surface sediments are systematically higher than those of specimens from the water column since sediment assemblages are mainly sourced from high-productive seasons (e.g., Schiebel 2002), and hence Cd-rich waters. From Ripperger et al. (2008)

Specimens from sea floor sediment samples in general have higher Cd/Ca ratios than those sampled from the water column (Fig. 9.18). The phenomenon is explained by their ecological niche and test production, and not by alteration of tests during sedimentation (Ripperger et al. 2008). Since most of the planktic foraminifer test calcite is produced during times of high productivity, at high PO_4 and Cd concentrations, these tests constitute the major portion of sediment assemblages produced over short time-periods of mass flux events (Schiebel 2002). In contrast, planktic foraminifers from net-tow samples analyzed here (Fig. 9.18) mostly represent tests, which were produced at low-PO_4 and low-Cd concentration of ambient seawater (Ripperger et al. 2008).

Original Cd/Ca ratios are probably retained during sedimentation, as indicated by the distinct Cd/Ca ratios of deep dwelling *Globorotalia truncatulinoides*, shallow dwelling eutrophic *G. bulloides*, and oligo-to-mesotrophic symbiont-bearing *O. universa* from sea floor sediments of the North Atlantic (Fig. 9.19). However, it cannot

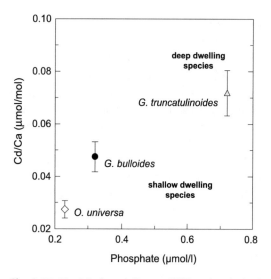

Fig. 9.19 Planktic foraminifer test Cd/Ca ratios obtained for three different species analyzed from the sediment core top from the NE Atlantic at ~47°N, 20°W (MC575), versus seawater phosphate concentration (from Garcia et al. 2006). The latter corresponds to the mean phosphate concentration prevailing during the species-specific bloom month and at the species-specific habitat. From Ripperger et al. (2008)

be excluded that dissolution may alter the Cd/Ca ratio of planktic foraminifer tests (McCorkle et al. 1995), and further analyses should add to a better understanding of Cd/Ca proxy.

Unfortunately, Cd concentration in planktic foraminifer calcite is rather low (10^{-7}–10^{-9} mol mol^{-1} Ca, Lea 1999), and large sample volumes are required for analyses of the Cd/Ca ratio and Cd isotopes, usually not available from tow-samples (Ripperger and Rehkämper 2007; Ripperger et al. 2008). In addition, the Cd/Ca proxy for paleo-productivity needs to be calibrated at the species level to account for differences between symbiont-barren and symbiont-bearing species (cf. Mashiotta et al. 1997). The relation of Cd-uptake in planktic foraminifer test calcite needs to be analyzed from cultured specimens grown under controlled concentrations of Fe and Zn, and $p\mathrm{CO}_2$ of ambient seawater (Cullen et al. 1999; Cullen 2006).

9.6 Other Isotope and Element Ratios

Isotope and element ratios other than those discussed above provide additional information on the biology and ecology of modern planktic foraminifers. Sr/Ca and U/Ca ratios in planktic foraminifer test calcite have been shown to depend on the carbonate ion concentration ($[\mathrm{CO}_3^{2-}]$) and $p\mathrm{H}$ of ambient seawater, and to change at a species-specific rate (Russell et al. 2004). Current and future work on those isotopes and metal-to-Ca ratios may add more data and a better understanding of planktic foraminifer biology and ecology, and provide new paleo-proxies.

9.6.1 Sr/Ca Ratio

Strontium may substitute calcium in foraminifer test CaCO$_3$, and the Sr/Ca ratio can be easily measured along with other element ratios such as, for example, the Mg/Ca ratio (e.g., Lea et al. 1999b; Elderfield et al. 2002; Rosenthal et al. 2004; Friedrich et al. 2012). Sr/Ca ratios seem to be related to ontogenetic changes in growth rates

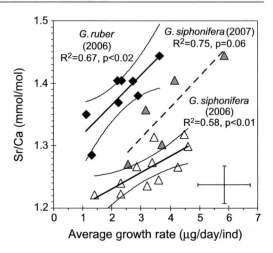

Fig. 9.20 Dependence of Sr/Ca ratios on growth rates in *G. ruber* and *G. siphonifera*. The *error bars* denote average values for 2σ on Sr/Ca (± 0.03 mmol mol^{-1}) and 2σ of the mean on average growth rate (± 0.9 µg/day/individual). 95 % confidence intervals are given for regressions with a significance level better than 5 %. The average Sr/Ca ratios of the dataset are 1.38 ± 0.08 mmol mol^{-1} (2σ for *G. ruber*), 1.26 ± 0.07 mmol mol^{-1} for *G. siphonifera* from the year 2006 cultures, and 1.36 ± 0.14 for *G. siphonifera* from the year 2007 cultures. From Kisakürek et al. (2011)

of planktic foraminifers, as shown for *G. ruber* (white) by Kisakürek et al. (2008), and for *G. ruber* and *G. siphonifera* by Kisakürek et al. (2011) (Fig. 9.20). As such, the Sr/Ca ratio is used in combination with the Mg/Ca ratio to calibrate ontogenetic effects in the uptake of the latter (Friedrich et al. 2012). The Sr/Ca incorporation in test calcite of cultured *G. bulloides, O. universa,* and *G. siphonifera* is mostly affected by temperature, and less affected by salinity and $p\mathrm{H}$ over the natural range of open marine surface waters (Lea et al. 1999b; Kisakürek et al. 2011). Data from cultured *G. ruber* (white) indicate similar effects of temperature and salinity on the Sr/Ca ratio, whereas a $p\mathrm{H}$-effect seems to be negligible in modern surface waters (Kisakürek et al. 2008).

9.6.2 Ba/Ca, U/Ca, Nd/Ca, and SO$_4$/Ca Ratios

Barium-to-calcium (Ba/Ca) ratios of ambient seawater seem to be quantitatively incorporated

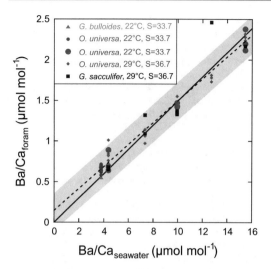

Fig. 9.21 Ba/Ca ratios in shells of cultured *O. universa*, *G. sacculifer*, and *G. bulloides* relative to the Ba/Ca ratio of experimental seawater. The symbiont-bearing and symbiont-barren foraminifers were grown at different salinities and temperatures, and yet all species and experiments fall on the same partitioning line. The linear regression through this data set is Ba/Ca$_{shell}$ = 0.13 + 0.14 * Ba/Ca$_{seawater}$ (*dashed line*). When forced through the origin, the Ba/Ca partitioning for this data set can be described as Ba/Ca$_{shell}$ = 0.15 ($\pm$ 0.05) * Ba/Ca$_{seawater}$ (*solid line*), and the linear and forced regression agree on a 95 % confidence level (*gray bar*). From Hönisch et al. (2011)

into planktic foraminifer calcite, and are a proxy of hydrographic parameters including stratification of water masses, and fresh water river discharge into the ocean (Lea and Spero 1994; Weldeab et al. 2007). Intra-shell Ba/Ca ratios around 0.7 μmol mol^{-1} on average probably display the Ba/Ca concentration of ambient seawater, and are not significantly affected by temperature, salinity (Fig. 9.21), and *p*H (Hönisch et al. 2011). A partition coefficient of D$_{Ba}$ = 0.15 $\pm$ 0.05 (95 % confidence interval) has been reported for spinose planktic foraminifer species, and might be different in some non-spinose species (Hönisch et al. 2011).

The uranium-to-calcium (U/Ca) ratio of planktic foraminifer test calcite does possibly indicate the U/Ca ratio of ambient seawater, and is highly susceptible to dissolution, i.e. decreases with dissolution (Russell et al. 1994; Hayes et al. 2014). The vanadium-to-calcium ratio in planktic

foraminifer calcite is species dependent and decreases, like the U/Ca ratio, with the dissolution of tests (Hastings et al. 1996). Neodymium isotopes (^{143}Nd/^{144}Nd) and Nd/Ca ratios in planktic foraminifer tests from surface waters are assumed to correctly record the distribution of neodymium in ambient seawater (Vance et al. 2004; Katz et al. 2010; Tachikawa et al. 2013). However, preservation of the Nd isotope signal may be affected by redox conditions within sediments (Roberts et al. 2012). Sulfate-to-calcium ratios (SO$_4^{2-}$/Ca^{2+}) in test calcite of *O. universa* are possibly related to ratios of ambient seawater, and may provide a proxy of the past marine state of oxygenation and carbon concentration (Paris et al. 2014). The ^{7}Li/^{6}Li ratio (δ^7Li) is a proxy of silica concentration in seawater, driven mainly by continental weathering and hydrothermal activity (Misra and Froelich 2009).

9.7 Summary and Concluding Remarks

The use of stable oxygen (δ^{18}O) and carbon (δ^{13}C) isotopes as classical proxies of water temperature, salinity, and productivity has been optimized over the past 40 years. Vital effects have been taken into account, and effects of photosynthesis and carbonate ion concentration [CO$_3^{2-}$] have been quantified. In addition to stable oxygen and carbon isotopes, new proxies have been developed to assess the biogeochemistry and paleoceanographic use of the planktic foraminifer shell at high precision and accuracy. In modern approaches, δ^{18}O and δ^{13}C data are combined with other geochemical data in multi-proxy and multi-species analyses. Stable nitrogen isotopes (δ^{15}N) are assessed as proxy of (paleo-) ecology and productivity, and add important information on the regional and global carbon turnover. The use of clumped isotopes is still in its infancy, and may open new perspectives in the near future. Boron isotopes (δ^{11}B) provide data on the *p*H of ambient seawater, and may be combined with morphometric data of the planktic foraminifer such as, for example, shell thickness, and size-normalized test weight. Calcium isotopes ($\delta^{44}Ca$) and Mg/Ca ratios are temperature proxies. Other Me/Ca ratios are proxy of primary productivity (Cd/Ca), temperature (Sr/Ca) and salinity (Ba/Ca), dissolution (U/Ca), and seawater oxygenation (SO_4^{2-}/Ca^{2+}). The sum of those approaches and proxies, in combination with faunal data, facilitate more reliable quantitative reconstructions of the past ocean, climate, and environment change.

References

Allen KA, Hönisch B, Eggins SM, Yu J, Spero HJ, Elderfield H (2011) Controls on boron incorporation in cultured tests of the planktic foraminifer *Orbulina universa*. Earth Planet Sci Lett 309:291–301. doi:10.1016/j.epsl.2011.07.010

Anand P, Elderfield H, Conte MH (2003) Calibration of Mg/Ca thermometry in planktonic Foraminifera from a sediment trap time series. Paleoceanography 18:1050. doi:10.1029/2002PA000846

Anderson OR, Faber WW (1984) An estimation of calcium carbonate deposition rate in a planktonic foraminifer *Globigerinoides sacculifer* using ^{45}Ca as a tracer; a recommended procedure for improved accuracy. J foraminifer Res 14:303–308. doi:10.2113/gsjfr.14.4.303

Arbuszewski J, deMenocal P, Kaplan A, Farmer EC (2010) On the fidelity of shell-derived $\delta^{18}O_{seawater}$ estimates. Earth Planet Sci Lett 300:185–196. doi:10.1016/j.epsl.2010.10.035

Archer DE, Winguth A, Lea D, Mahowald N (2000) What caused the glacial/interglacial atmospheric pCO_2 cycles? Rev Geophys 38:159–189. doi:10.1029/1999RG000066

Barker S, Broecker W, Clark E, Hajdas I (2007) Radiocarbon age offsets of Foraminifera resulting from differential dissolution and fragmentation within the sedimentary bioturbated zone. Paleoceanography. doi:10.1029/2006PA001354

Barker S, Elderfield H (2002) foraminiferal calcification response to glacial-interglacial changes in atmospheric CO_2. Science 297:833–836. doi:10.1126/science.1072815

Bé AWH (1980) Gametogenic calcification in a spinose planktonic foraminifer, *Globigerinoides sacculifer* (Brady). Mar Micropaleontol 5:283–310. doi:10.1016/0377-8398(80)90014-6

Bé AWH, Tolderlund DS (1971) Distribution and ecology of living planktonic Foraminifera in surface waters of the Atlantic and Indian Oceans. In: Funnell BM, Riedel WR (eds) The micropalaeontology of oceans. University Press, Cambridge, pp 105–149

Bednaršek N, Tarling GA, Bakker DCE, Fielding S, Jones EM, Venables HJ, Ward P, Kuzirian A, Lézé B, Feely RA, Murphy EJ (2012) Extensive dissolution of live pteropods in the Southern Ocean. Nat Geosci 5:881–885. doi:10.1038/ngeo1635

Beer CJ, Schiebel R, Wilson PA (2010) Testing planktic foraminiferal shell weight as a surface water $[CO_3^{2-}]$ proxy using plankton net samples. Geology 38:103–106. doi:10.1130/G30150.1

Bemis BE, Spero HJ, Bijma J, Lea DW (1998) Reevaluation of the oxygen isotopic composition of planktonic Foraminifera: experimental results and revised paleotemperature equations. Paleoceanography 13:150–160

Bemis BE, Spero HJ, Lea DW, Bijma J (2000) Temperature influence on the carbon isotopic composition of *Globigerina bulloides* and *Orbulina universa* (planktonic Foraminifera). Mar Micropaleontol 38:213–228

Bemis BE, Spero HJ, Thunell RC (2002) Using species-specific paleotemperature equations with Foraminifera: a case study in the Southern California Bight. Mar Micropaleontol 46:405–430. doi:10.1016/S0377-8398(02)00083-X

Bentov S, Erez J (2006) Impact of biomineralization processes on the Mg content of foraminiferal shells: a biological perspective. Geochem Geophys Geosyst. doi:10.1029/2005GC001015

Berelson WM (2002) Particle settling rates increase with depth in the ocean. Deep-Sea Res II 49:237–251. doi:10.1016/S0967-0645(01)00102-3

Berger WH (1971) Sedimentation of planktonic Foraminifera. Mar Geol 11:325–358. doi:10.1016/0025-3227(71)90035-1

Berger WH, Killingley JS, Vincent E (1978) Stable isotopes in deep-sea carbonates: box core ERDC-92, west equatorial Pacific. Oceanol Acta 1:203–216

Bijma J, Spero HJ, Lea DW (1999) Reassessing foraminiferal stable isotope geochemistry: impact of the oceanic carbonate system (experimental results). In: Fischer G, Wefer G (eds) Use of proxies in paleoceanography. Springer, Berlin, Heidelberg, pp 489–512

Billups K, Spero HJ (1995) Relationship between shell size, thickness and stable isotopes in individual planktonic Foraminifera from two equatorial Atlantic cores. J foraminifer Res 25:24–37

Birch H, Coxall HK, Pearson PN, Kroon D, O'Regan M (2013) Planktonic Foraminifera stable isotopes and water column structure: disentangling ecological signals. Mar Micropaleontol 101:127–145

Bolton A, Baker JA, Dunbar GB, Carter L, Smith EGC, Neil HL (2011) Environmental versus biological controls on Mg/Ca variability in *Globigerinoides ruber* (white) from core top and plankton tow samples in the southwest Pacific Ocean. Paleoceanography 26: PA2219. doi:10.1029/2010PA001924

Boussetta S, Bassinot F, Sabbatini A, Caillon N, Nouet J, Kallel N, Rebaubier H, Klinkhammer G, Labeyrie L (2011) Diagenetic Mg-rich calcite in Mediterranean sediments: quantification and impact on foraminiferal

Mg/Ca thermometry. Mar Geol 280:195–204. doi:10.1016/j.margeo.2010.12.011

Bouvier-Soumagnac Y, Duplessy JC (1985) Carbon and oxygen isotopic composition of planktonic Foraminifera from laboratory culture, plankton tows and Recent sediment; implications for the reconstruction of paleoclimatic conditions and of the global carbon cycle. J foraminifer Res 15:302–320. doi:10.2113/gsjfr.15.4.302

Boyle EA (1981) Cadmium, zinc, copper, and barium in Foraminifera tests. Earth Planet Sci Lett 53:11–35. doi:10.1016/0012-821X(81)90022-4

Boyle EA (1988) Cadmium: chemical tracer of deepwater paleoceanography. Paleoceanography 3:471–489. doi:10.1029/PA003i004p00471

Boyle EA (2006) A direct proxy for oceanic phosphorus? Science 312:1758–1759

Boyle EA, Sclater F, Edmond JM (1976) On the marine geochemistry of cadmium. Nature 263:42–44. doi:10.1038/263042a0

Bramlette MN (1958) Significance of coccolithophorids in calcium-carbonate deposition. Geol Soc Am Bull 69:121. doi:10.1130/0016-7606(1958)69[121:SOCICD]2.0.CO;2

Brand WA, Coplen TB, Vogl J, Rosner M, Prohaska T (2014) Assessment of international reference materials for isotope-ratio analysis (IUPAC Technical Report). Pure Appl Chem 86(3): 425–467. doi:10.1515/pac-2013-1023

Branson O, Redfern SAT, Tyliszczak T, Sadekov A, Langer G, Kimoto K, Elderfield H (2013) The coordination of Mg in foraminiferal calcite. Earth Planet Sci Lett 383:134–141. doi:10.1016/j.epsl.2013.09.037

Broecker WS, Clark E (2003) $CaCO_3$ dissolution in the deep sea: paced by insolation cycles. Geochem Geophys Geosyst 4:1059. doi:10.1029/2002GC000450

Broecker WS, Clark E (1999) $CaCO_3$ size distribution: a paleocarbonate ion proxy? Paleoceanography 14:596–604. doi:10.1029/1999PA900016

Broecker WS, Peng TH (1982) Tracers in the Sea. Eldigio Press, New York

Brown SJ, Elderfield H (1996) Variations in Mg/Ca and Sr/Ca ratios of planktonic Foraminifera caused by postdepositional dissolution: evidence of shallow Mg-dependent dissolution. Paleoceanography 11:543–551. doi:10.1029/96PA01491

Caldeira K, Wickett ME (2003) Oceanography: anthropogenic carbon and ocean pH. Nature 425:365. doi:10.1038/425365a

Cléroux C, Lynch-Stieglitz J, Schmidt MW, Cortijo E, Duplessy JC (2009) Evidence for calcification depth change of *Globorotalia truncatulinoides* between deglaciation and Holocene in the western Atlantic Ocean. Mar Micropaleontol 73:57–61

Coplen TB (1994) Reporting of stable hydrogen, carbon, and oxygen isotopic abundances (Technical Report). Pure Appl Chem. doi:10.1351/pac199466020273

Cronblad HG, Malmgren BA (1981) Climatically controlled variation of Sr and Mg in Quaternary planktonic Foraminifera. Nature 291:61–64

Cullen JT (2006) On the nonlinear relationship between dissolved cadmium and phosphate in the modern global ocean: Could chronic iron limitation of phytoplankton growth cause the kink? Limnol Oceanogr 51:1369–1380

Cullen JT, Lane TW, Morel FMM, Sherrell RM (1999) Modulation of cadmium uptake in phytoplankton by seawater CO_2 concentration. Nature 402:165–167

Dekens PS, Lea DW, Pak DK, Spero HJ (2002) Core top calibration of Mg/Ca in tropical Foraminifera: refining paleotemperature estimation. Geochem Geophys Geosyst 3:1–29. doi:10.1029/2001GC000200

Delaney ML (1989) Uptake of cadmium into calcite shells by planktonic Foraminifera. Chem Geol 78:159–165

De Moel H, Ganssen GM, Peeters FJC, Jung SJA, Kroon D, Brummer GJA, Zeebe RE (2009) Planktic foraminiferal shell thinning in the Arabian Sea due to anthropogenic ocean acidification? Biogeosciences 6:1917–1925. doi:10.5194/bg-6-1917-2009

Dennis KJ, Affek HP, Passey BH, Schrag DP, Eiler JM (2011) Defining an absolute reference frame for "clumped" isotope studies of CO_2. Geochim Cosmochim Acta 75:7117–7131

De Nooijer LJ, Spero HJ, Erez J, Bijma J, Reichart GJ (2014) Biomineralization in perforate Foraminifera. Earth-Sci Rev 135:48–58

Derry LA (2009) Geochemistry: a glacial hangover. Nature 458:417–418. doi:10.1038/458417a

Dittert N, Baumann KH, Bickert T, Henrich R, Huber R, Kinkel H, Meggers H (1999) Carbonate dissolution in the deep-sea: methods, quantification and paleoceanographic application. In: Fischer G, Wefer G (eds) Use of proxies in paleoceanography. Springer, Berlin, pp 255–284

Dueñas-Bohórquez A, da Rocha RE, Kuroyanagi A, Bijma J, Reichart GJ (2009) Effect of salinity and seawater calcite saturation state on Mg and Sr incorporation in cultured planktonic Foraminifera. Mar Micropaleontol 73:178–189

Duplessy JC, Bé AWH, Blanc PL (1981) Oxygen and carbon isotopic composition and biogeographic distribution of planktonic Foraminifera in the Indian Ocean. Palaeogeogr Palaeoclimatol Palaeoecol 33:9–46

Eggins S, De Dekker P, Marshall J (2003) Mg/Ca variation in planktonic Foraminifera tests: implications for reconstructing palaeo-seawater temperature and habitat migration. Earth Planet Sci Lett 212:291–306

Eggins S, Sadekov A, De Deckker P (2004) Modulation and daily banding of Mg/Ca in *Orbulina universa* tests by symbiont photosynthesis and respiration: a complication for seawater thermometry? Earth Planet Sci Lett 225:411–419

Eiler JM (2007) "Clumped-isotope" geochemistry—The study of naturally-occurring, multiply-substituted isotopologues. Earth Planet Sci Lett 262:309–327

Elderfield H, Ganssen G (2000) Past temperature and $\delta^{18}O$ of surface ocean waters inferred from foraminiferal Mg/Ca ratios. Nature 405:442–445

Elderfield H, Rickaby REM (2000) Oceanic Cd/P ratio and nutrient utilization in the glacial Southern Ocean. Nature 405:305–310

Elderfield H, Vautravers M, Cooper M (2002) The relationship between shell size and Mg/Ca, Sr/Ca, $\delta^{18}O$, and $\delta^{13}C$ of species of planktonic Foraminifera. Geochem Geophys Geosyst 3:1–13. doi:10.1029/2001GC000194

Emiliani C (1955) Pleistocene temperatures. J Geol 63:538–578

Emiliani C (1954) Depth habitats of some species of pelagic Foraminifera as indicated by oxygen isotope ratios. Am J Sci 252:149–158

Epstein S, Buchsbaum R, Lowenstam HA, Urey HC (1953) Revised carbonate-water isotopic temperature scale. Geol Soc Am Bull 64:1315–1326

Epstein S, Buchsbaum R, Lowenstam H, Urey HC (1951) Carbonate-water isotopic temperature scale. Geol Soc Am Bull 62:417–426

Epstein S, Mayeda T (1953) Variation of ^{18}O content of waters from natural sources. Geochim Cosmochim Acta 4:213–224

Erez J (1978) Vital effect on stable-isotope composition seen in Foraminifera and coral skeletons. Nature 273:199–202

Erez J, Luz B (1983) Experimental paleotemperature equation for planktonic Foraminifera. Geochim Cosmochim Acta 47:1025–1031

Erez J, Luz B (1982) Temperature control of oxygen-isotope fractionation of cultured planktonic Foraminifera. Nature 297:220–222

Ezard THG, Edgar KM, Hull PM (2015) Environmental and biological controls on size-specific $\delta^{13}C$ and $\delta^{18}O$ in recent planktonic Foraminifera. Paleoceanography 30. doi:10.1002/2014PA002735

Feely RA, Sabine CL, Lee K, Berelson W, Kleypas J, Fabry VJ, Millero FJ (2004) Impact of anthropogenic CO_2 on the $CaCO_3$ system in the oceans. Science 305:362–366

Fischer G, Wefer G (1999) Use of proxies in paleoceanography: examples from the South Atlantic. Springer, Berlin, Heidelberg

Friedrich O, Schiebel R, Wilson PA, Weldeab S, Beer CJ, Cooper MJ, Fiebig J (2012) Influence of test size, water depth, and ecology on Mg/Ca, Sr/Ca, $\delta^{18}O$ and $\delta^{13}C$ in nine modern species of planktic foraminifers. Earth Planet Sci Lett 319–320:133–145. doi:10.1016/j.epsl.2011.12.002

Ganssen G (1983) Dokumentation von küstennahem Auftrieb anhand stabiler Isotope in Rezenten Foraminiferen vor Nordwest-Afrika. Meteor Forschungsergebnisse Reihe C 1–46

Garcia HE, Locarnini RA, Boyer TP, Antonov JI (2006) World Ocean Atlas 2005. Vol. 4, Nutrients (phosphate, nitrate, silicate). In: Levitus S (ed) NOAA Atlas NESDIS 64. NOAA, Silver Spring

Gehlen M, Bassinot F, Beck L, Khodja H (2004) Trace element cartography of *Globigerinoides ruber* shells using particle-induced X-ray emission. Geochem Geophys Geosyst. doi:10.1029/2004GC000822

Ghosh P, Adkins J, Affek H, Balta B, Guo W, Schauble EA, Schrag D, Eiler JM (2006) $^{13}C-^{18}O$ bonds in carbonate minerals: a new kind of paleothermometer. Geochim Cosmochim Acta 70:1439–1456

Groeneveld J, Chiessi CM (2011) Mg/Ca of *Globorotalia inflata* as a recorder of permanent thermocline temperatures in the South Atlantic. Paleoceanography 26:PA2203. doi:10.1029/2010PA001940

Gruber N, Keeling CD, Bates NR (2002) Interannual variability in the North Atlantic Ocean carbon sink. Science 298:2374–2378

Gussone N, Eisenhauer A, Heuser A, Dietzel M, Bock B, Böhm F, Spero HJ, Lea DW, Bijma J, Nägler TF (2003) Model for kinetic effects on calcium isotope fractionation ($\delta^{44}Ca$) in inorganic aragonite and cultured planktonic Foraminifera. Geochim Cosmochim Acta 67:1375–1382. doi:10.1016/S0016-7037(02)01296-6

Gussone N, Hönisch B, Heuser A, Eisenhauer A, Spindler M, Hemleben C (2009) A critical evaluation of calcium isotope ratios in tests of planktonic foraminifers. Geochim Cosmochim Acta 73: 7241–7255. doi:10.1016/j.gca.2009.08.035

Hamilton CP, Spero HJ, Bijma J, Lea DW (2008) Geochemical investigation of gametogenic calcite addition in the planktonic Foraminifera *Orbulina universa*. Mar Micropaleontol 68:256–267. doi:10.1016/j.marmicro.2008.04.003

Hastings DW, Emerson SR, Erez J, Nelson BK (1996) Vanadium in foraminiferal calcite: Evaluation of a method to determine paleo-seawater vanadium concentrations. Geochim Cosmochim Acta 60:3701–3715

Hastings DW, Russell AD, Emerson SR (1998) foraminiferal magnesium in *Globeriginoides sacculifer* as a paleotemperature proxy. Paleoceanography 13:161–169

Hayes CT, Martínez-García A, Hasenfratz AP, Jaccard SL, Hodell DA, Sigman DM, Haug GH, Anderson RF (2014) A stagnation event in the deep South Atlantic during the last interglacial period. Science 346:1514–1517. doi:10.1126/science.1256620

Hay WW (1985) Potential errors in estimates of carbonate rock accumulating through geologic time. The carbon cycle and atmospheric CO_2: natural variations Archean to Present. Geoph Monogr Series 32:573–583

Hemleben C, Bijma J (1994) foraminiferal population dynamics and stable carbon isotopes. In: Zahn R, Kaminski MA, Labeyrie L, Pedersen TF (eds) Carbon cycling in the glacial ocean: Constraints on the ocean's role in global change. NATO ASI Series I 17, pp 145-166

Hemleben C, Bé AWH, Anderson OR, Tuntivate S (1977) Test morphology, organic layers and chamber formation of the planktonic foraminifer *Globorotalia menardii* (d'Orbigny). J Foraminifer Res 7:1–25

Hemleben C, Spindler M, Anderson OR (1989) Modern planktonic Foraminifera. Springer, Berlin

Hemleben C, Meischner D, Zahn R, Almogi-Labin A, Erlenkeuser H, Hiller B (1996) Three hundred eighty thousand year long stable isotope and faunal records from the Red Sea: Influence of global sea level change on hydrography. Paleoceanography 11(2): 147-156. doi:10.1029/95PA03838

Hemming NG, Hanson GN (1992) Boron isotopic composition and concentration in modern marine carbonates. Geochim Cosmochim Acta 56:537–543

Henderson GM (2002) New oceanic proxies for paleoclimate. Earth Planet Sci Lett 203:1–13

Henehan MJ, Rae J, Foster GL, Erez J, Prentice KC, Kucera M, Bostock HC, Martinez-Boti MA, Milton JA, Wilson PA, Marshal BJ, Elliott T (2013) Calibration of the boron isotope proxy in the planktonic Foraminifera *Globigerinoides ruber* for use in palaeo-CO$_2$ reconstruction. Earth Planet Sci Lett 364:111–122

Hillaire-Marcel C, de Vernal A, Polyak L, Darby D (2004) Size-dependent isotopic composition of planktic foraminifers from Chukchi Sea vs. NW Atlantic sediments-implications for the Holocene paleoceanography of the western Arctic. Quat Sci Rev 23:245–260

Hönisch B, Allen KA, Lea DW, Spero HJ, Eggins SM, Arbuszewski J, deMenocal P, Rosenthal Y, Russell AD, Elderfield H (2013) The influence of salinity on Mg/Ca in planktic foraminifers—evidence from cultures, core-top sediments and complementary δ^{18}O. Geochim Cosmochim Acta 121:196–213

Hönisch B, Allen KA, Russell AD, Eggins SM, Bijma J, Spero HJ, Lea DW, Yu J (2011) Planktic foraminifers as recorders of seawater Ba/Ca. Mar Micropaleontol 79:52–57

Hönisch B, Bijma J, Russell AD, Spero HJ, Palmer MR, Zeebe RE, Eisenhauer A (2003) The influence of symbiont photosynthesis on the boron isotopic composition of Foraminifera shells. Mar Micropaleontol 49:87–96

Hönisch B, Hemming NG (2004) Ground-truthing the boron isotope-paleo-pH proxy in planktonic Foraminifera shells: Partial dissolution and shell size effects. Paleoceanography. doi:10.1029/2004PA001026

Imbrie J, Hays JD, Martinson DG, McIntyre A, Mix AC, Morley JJ, Pisias NG, Prell WL, Shackleton NJ (1984) The orbital theory of Pleistocene climate: Support from a revised chronology of the marine δ^{18}O record. In: Berger A (ed) Milankovitch and climate. Part I. Reidel Publishing Company, Dordrecht, p 269

Jonkers L, de Nooijer LJ, Reichart GJ, Zahn R, Brummer GJA (2012) Encrustation and trace element composition of *Neogloboquadrina dutertrei* assessed from single chamber analyses—implications for paleotemperature estimates. Biogeosciences 9:4851–4860. doi:10.5194/bg-9-4851-2012

Jonkers L, Jiménez-Amat P, Mortyn PG, Brummer G-JA (2013) Seasonal Mg/Ca variability of *N. pachyderma* (s) and *G. bulloides*: Implications for seawater temperature reconstruction. Earth Planet Sci Lett 376:137–144. doi:10.1016/j.epsl.2013.06.019

Katz ME, Cramer BS, Franzese A, Hönisch B, Miller KG, Rosenthal Y, Wright JD (2010) Traditional and emerging geochemical proxies in Foraminifera. J foraminifer Res 40:165–192

Kasemann SA, Schmidt DN, Bijma J, Foster GL (2009) In situ boron isotope analysis in marine carbonates and its application for Foraminifera and palaeo-pH. Chemical Geology 260: 138–147. doi:10.1016/j.chemgeo.2008.12.015

Kim ST, O'Neil JR (1997) Equilibrium and nonequilibrium oxygen isotope effects in synthetic carbonates. Geochim Cosmochim Acta 61:3461–3475. doi:10.1016/S0016-7037(97)00169-5

King K, Hare PE (1972) Amino acid composition of planktonic Foraminifera: a paleobiochemical approach to evolution. Science 175:1461–1463

Kisakürek B, Eisenhauer A, Böhm F, Garbe-Schönberg D, Erez J (2008) Controls on shell Mg/Ca and Sr/Ca in cultured planktonic foraminiferan, *Globigerinoides ruber* (white). Earth Planet Sci Lett 273:260–269

Kisakürek B, Eisenhauer A, Böhm F, Hathorne EC, Erez J (2011) Controls on calcium isotope fractionation in cultured planktic Foraminifera, *Globigerinoides ruber* and *Globigerinella siphonifera*. Geochim Cosmochim Acta 75:427–443. doi:10.1016/j.gca.2010.10.015

Köhler-Rink S, Kühl M (2005) The chemical microenvironment of the symbiotic planktonic foraminifer *Orbulina universa*. Mar Biol Res 1:68–78. doi:10.1080/17451000510019015

Kozdon R, Kelly DC, Kita NT, Fournelle JH, Valley JW (2011) Planktonic foraminiferal oxygen isotope analysis by ion microprobe technique suggests warm tropical sea surface temperatures during the Early Paleogene. Paleoceanography. doi:10.1029/2010PA002056

Kunioka D, Shirai K, Takahata N, Sano Y, Toyofuku T, Ujiie Y (2006) Microdistribution of Mg/Ca, Sr/Ca, and Ba/Ca ratios in *Pulleniatina obliquiloculata* test by using a NanoSIMS: implication for the vital effect mechanism. Geochem Geophys Geosystems 7: Q12P20. doi:10.1029/2006GC001280

Lea DW (1999) Trance elements in foraminiferal calcite. In: Sen Gupta B (ed) Modern Foraminifera. Kluwer Academic Publishers, Dordrecht, pp 259–277

Lea DW (2003) Elemental and isotopic proxies of past ocean temperatures. In: Holland HD, Turekian KK (eds) Treatise on geochemistry. Elsevier-Pergamon, Oxford, pp 365–390

Lea DW, Bijma J, Spero HJ, Archer D (1999a) Implications of a carbonate ion effect on shell carbon and oxygen isotopes for glacial ocean conditions. In: Fischer G, Wefer G (eds) Use of proxies in paleoceanography: examples from the South Atlantic. Springer, Berlin, Heidelberg, pp 513–522

Lea DW, Martin PA, Chan DA, Spero HJ (1995) Calcium uptake and calcification rate in the planktonic foraminifer *Orbulina universa*. J foraminifer Res 25:14–23

Lea DW, Mashiotta TA, Spero HJ (1999b) Controls on magnesium and strontium uptake in planktonic Foraminifera determined by live culturing. Geochim Cosmochim Acta 63:2369–2379

Lea DW, Spero HJ (1994) Assessing the reliability of paleochemical tracers: barium uptake in the shells of planktonic Foraminifera. Paleoceanography 9:445–452

Lisiecki LE, Raymo ME (2005) A Pliocene-Pleistocene stack of 57 globally distributed benthic $\delta^{18}O$ records. Paleoceanography. doi:10.1029/2004PA001071

Lohbeck KT, Riebesell U, Reusch TBH (2012) Adaptive evolution of a key phytoplankton species to ocean acidification. Nat Geosci 5:346–351. doi:10.1038/ngeo1441

Lohmann GP (1995) A model for variation in the chemistry of planktonic Foraminifera due to secondary calcification and selective dissolution. Paleoceanography 10:445–457

Lombard F, Erez J, Michel E, Labeyrie L (2009) Temperature effect on respiration and photosynthesis of the symbiont-bearing planktonic Foraminifera *Globigerinoides ruber*, *Orbulina universa*, and *Globigerinella siphonifera*. Limnol Oceanogr 54:210–218

Marr JP, Baker JA, Carter L, Allan ASR, Dunbar GB, Bostock HC (2011) Ecological and temperature controls on Mg/Ca ratios of *Globigerina bulloides* from the southwest Pacific Ocean. Paleoceanography 26: PA2209. doi:10.1029/2010PA002059

Marr JP, Carter L, Bostock HC, Bolton A, Smith E (2013) Southwest Pacific Ocean response to a warming world: using Mg/Ca, Zn/Ca, and Mn/Ca in Foraminifera to track surface ocean water masses during the last deglaciation. Paleoceanography 28:347–362. doi:10.1002/palo.20032

Marshall BJ, Thunell RC, Henehan MJ, Astor Y, Wejnert KE (2013) Planktonic foraminiferal area density as a proxy for carbonate ion concentration: a calibration study using the Cariaco Basin ocean time series. Paleoceanography 28:363–376. doi:10.1002/palo.20034

Martínez-Botí MA, Mortyn PG, Schmidt DN, Vance D, Field DB (2011) Mg/Ca in Foraminifera from plankton tows: evaluation of proxy controls and comparison with core tops. Earth Planet Sci Lett 307:113–125. doi:10.1016/j.epsl.2011.04.019

Martinson DG, Pisias NG, Hays JD, Imbrie J, Moore TC Jr, Shackleton NJ (1987) Age dating and the orbital theory of the ice ages: development of a high-resolution 0 to 300,000-year chronostratigraphy. Quat Res 27:1–29

Mashiotta TA, Lea DW, Spero HJ (1997) Experimental determination of cadmium uptake in shells of the planktonic Foraminifera *Orbulina universa* and *Globigerina bulloides*: Implications for surface water paleoreconstructions. Geochim Cosmochim Acta 61:4053–4065. doi:10.1016/S0016-7037(97)00206-8

Mathien-Blard E, Bassinot F (2009) Salinity bias on the Foraminifera Mg/Ca thermometry: correction procedure and implications for past ocean hydrographic reconstructions. Geochem Geophys Geosystems 10: Q12011. doi:10.1029/2008GC002353

McCorkle DC, Martin PA, Lea DW, Klinkhammer GP (1995) Evidence of a dissolution effect on benthic foraminiferal shell chemistry: $\delta^{13}C$, Cd/Ca, Ba/Ca, and Sr/Ca results from the Ontong Java Plateau. Paleoceanography 10:699–714. doi:10.1029/95PA01427

Milliman JD, Troy PJ, Balch WM, Adams AK, Li YH, Mackenzie FT (1999) Biologically mediated dissolution of calcium carbonate above the chemical lysocline? Deep-Sea Res I 46:1653–1669

Misra S, Froelich PN (2009) Measurement of lithium isotope ratios by quadrupole-ICP-MS: application to seawater and natural carbonates. J Anal At Spectrom 24:1524. doi:10.1039/b907122a

Mollenhauer G, Kienast M, Lamy F, Meggers H, Schneider RR, Hayes JM, Eglinton TI (2005) An evaluation of ^{14}C age relationships between co-occurring Foraminifera, alkenones, and total organic carbon in continental margin sediments. Paleoceanography. doi:10.1029/2004PA001103

Mortyn PG, Charles CD (2003) Planktonic foraminiferal depth habitat and $\delta^{18}O$ calibrations: plankton tow results from the Atlantic sector of the Southern Ocean. Paleoceanography 18:1037. doi:10.1029/2001PA000637

Moy AD, Howard WR, Bray SG, Trull TW (2009) Reduced calcification in modern Southern Ocean planktonic Foraminifera. Nat Geosci 2:276–280. doi:10.1038/ngeo460

Mulitza S, Boltovskoy D, Donner B, Meggers H, Paul A, Wefer G (2003) Temperature: $\delta^{18}O$ relationships of planktonic Foraminifera collected from surface waters. Palaeogeogr Palaeoclimatol Palaeoecol 202:143–152

Mulitza S, Dürkoop A, Hale W, Wefer G, Niebler HS (1997) Planktonic Foraminifera as recorders of past surface-water stratification. Geology 25:335–338

Nägler TF, Eisenhauer A, Müller A, Hemleben C, Kramers J (2000) The $\delta^{44}Ca$-temperature calibration on fossil and cultured *Globigerinoides sacculifer*: new tool for reconstruction of past sea surface temperatures. Geochem Geophy Geosy. doi:10.1029/2000GC000091

Niebler HS, Hubberten HW, Gersonde R (1999) Oxygen isotope values of planktic Foraminifera: A tool for the reconstruction of surface water stratification. In: Fischer G, Wefer G (eds) Use of proxies in paleoceanography: examples from the South Atlantic. Springer, Berlin, Heidelberg, pp 165–189

Ni Y, Foster GL, Bailey T, Elliott T, Schmidt DN, Pearson P, Haley B, Coath C (2007) A core top assessment of proxies for the ocean carbonate system in surface-dwelling foraminifers. Paleoceanography. doi:10.1029/2006PA001337

Nürnberg D, Bijma J, Hemleben C (1996) Assessing the reliability of magnesium in foraminiferal calcite as a proxy for water mass temperatures. Geochim Cosmochim Acta 60:803–814

Oppo DW, Horowitz M (2010) Glacial deep water geometry: South Atlantic benthic foraminiferal Cd/Ca and δ^{13}C evidence. Paleoceanography 15:147–160

Orr JC, Fabry VJ, Aumont O, Bopp L, Doney SC, Feely RA, Gnanadesikan A, Gruber N, Ishida A, Joos F, Key RM, Lindsay K, Maier-Reimer E, Matear R, Monfray P, Mouchet A, Najjar RG, Plattner G-K, Rodgers KB, Sabine CL, Sarmiento JL, Schlitzer R, Slater RD, Totterdell IJ, Weirig M-F, Yamanaka Y, Yool A (2005) Anthropogenic ocean acidification over the twenty-first century and its impact on calcifying organisms. Nature 437:681–686. doi:10.1038/nature04095

Paris G, Fehrenbacher JS, Sessions AL, Spero HJ, Adkins JF (2014) Experimental determination of carbonate-associated sulfate δ^{34}S in planktonic Foraminifera shells. Geochem Geophys Geosyst 15:1452–1461. doi:10.1002/2014GC005295

Parker FL (1958) Eastern Mediterranean Foraminifera, sediment cores from the Mediterranean Sea and Red Sea. Rep Swed Deep-Sea Exped 1947–1948(8): 219–285

Peeters FJC, Brummer GJA, Ganssen G (2002) The effect of upwelling on the distribution and stable isotope composition of *Globigerina bulloides* and *Globigerinoides ruber* (planktic Foraminifera) in modern surface waters of the NW Arabian Sea. Glob Planet Change 34:269–291. doi:10.1016/S0921-8181(02)00120-0

Redfield AC, Ketchum BH, Richards FA (1963) The influence of organisms on the composition of sea-water. In: Hill MN (ed) The Sea, vol 2. Wiley Interscience, New York, pp 26–77

Regenberg M, Nürnberg D, Schönfeld J, Reichart GJ (2007) Early diagenetic overprint in Caribbean sediment cores and its effect on the geochemical composition of planktonic Foraminifera. Biogeosciences 4:957–973

Regenberg M, Nürnberg D, Steph S, Groeneveld J, Garbe-Schönberg D, Tiedemann R, Dullo WC (2006) Assessing the effect of dissolution on planktonic foraminiferal Mg/Ca ratios: evidence from Caribbean core tops. Geochem Geophys Geosyst. doi:10.1029/2005GC001019

Regenberg M, Regenberg A, Garbe-Schönberg D, Lea DW (2014) Global dissolution effects on planktonic foraminiferal Mg/Ca ratios controlled by the calcite-saturation state of bottom waters. Paleoceanography 29:127–142. doi:10.1002/2013PA002492

Regenberg M, Steph S, Nürnberg D, Tiedemann R, Garbe-Schönberg D (2009) Calibrating Mg/Ca ratios of multiple planktonic foraminiferal species with δ^{18}O-calcification temperatures: paleothermometry for the upper water column. Earth Planet Sci Lett 278:324–336

Ren H, Sigman DM, Meckler AN, Plessen B, Robinson RS, Rosenthal Y, Haug GH (2009) foraminiferal isotope evidence of reduced nitrogen fixation in the ice age Atlantic Ocean. Science 323:244–248. doi:10. 1126/science.1165787

Ren H, Sigman DM, Thunell RC, Prokopenko MG (2012a) Nitrogen isotopic composition of planktonic Foraminifera from the modern ocean and recent sediments. Limnol Oceanogr 57:1011–1024. doi:10. 4319/lo.2012.57.4.1011

Ren H, Sigman DM, Thunell RC, Prokopenko MG (2012b) Nitrogen isotopic composition of planktonic Foraminifera from the modern ocean and recent sediments. Limnol Oceanogr 57(4): 1011–1024. doi:10.4319/lo.2012.57.4.1011

Richey JN, Poore RZ, Flower BP, Hollander DJ (2012) Ecological controls on the shell geochemistry of pink and white *Globigerinoides ruber* in the northern Gulf of Mexico: Implications for paleoceanographic reconstruction. Mar Micropaleontol 82–83:28–37. doi:10. 1016/j.marmicro.2011.10.002

Rickaby REM, Elderfield H (1999) Planktonic foraminiferal Cd/Ca: paleonutrients or paleotemperature? Paleoceanography 14:293–303

Rickaby REM, Greaves MJ, Elderfield H (2000) Cd in planktonic and benthic foraminiferal shells determined by thermal ionisation mass spectrometry. Geochim Cosmochim Acta 64:1229–1236

Riebesell U, Zondervan I, Rost B, Tortell PD, Zeebe RE, Morel FMM (2000) Reduced calcification of marine plankton in response to increased atmospheric CO_2. Nature 407:364–367

Rink S, Kühl M, Bijma J, Spero HJ (1998) Microsensor studies of photosynthesis and respiration in the symbiotic foraminifer *Orbulina universa*. Mar Biol 131:583–595

Ripperger S, Rehkämper M (2007) A highly sensitive MC-ICPMS method for Cd/Ca analyses of foraminiferal tests. J Anal At Spectrom 22:1275–1283

Ripperger S, Schiebel R, Rehkämper M, Halliday AN (2008) Cd/Ca ratios of in situ collected planktonic foraminiferal tests. Paleoceanography. doi:10.1029/ 2007PA001524

Roberts NL, Piotrowski AM, Elderfield H, Eglinton TI, Lomas MW (2012) Rare earth element association with Foraminifera. Geochim Cosmochim Acta 94:57–71. doi:10.1016/j.gca.2012.07.009

Rohling EJ, Cooke S (1999) Stable oxygen and carbon isotopes in foraminiferal carbonate shells. In: Sen Gupta BS (ed) Modern Foraminifera. Kluwer Academic Publishers, Dordrecht, pp 239–258

Rosenthal Y, Boyle EA, Slowey N (1997) Temperature control on the incorporation of magnesium, strontium, fluorine, and cadmium into benthic foraminiferal shells from Little Bahama Bank: prospects for thermocline paleoceanography. Geochim Cosmochim Acta 61:3633–3643

Rosenthal Y, Perron-Cashman S, Lear CH, Bard E, Barker S, Billups K, Bryan M, Delaney ML, deMenocal PB, Dwyer GS, Elderfield H, German CR, Greaves M, Lea DW, Marchitto TM, Pak DK, Paradis GL, Russell AD, Schneider RR, Scheiderich K, Stott L, Tachikawa K, Tappa E, Thunell R, Wara M, Weldeab S, Wilson PA (2004) Interlaboratory comparison study of Mg/Ca and Sr/Ca measurements in planktonic Foraminifera for paleoceanographic research. Geochem Geophys Geosyst. doi:10.1029/2003GC000650

Russell AD, Emerson S, Nelson BK, Erez J, Lea DW (1994) Uranium in foraminiferal calcite as a recorder of seawater uranium concentrations. Geochim Cosmochim Acta 58:671–681

Russell AD, Hönisch B, Spero HJ, Lea DW (2004) Effects of seawater carbonate ion concentration and temperature on shell U, Mg, and Sr in cultured planktonic Foraminifera. Geochim Cosmochim Acta 68:4347–4361

Russell AD, Spero HJ (2000) Field examination of the oceanic carbonate ion effect on stable isotopes in planktonic Foraminifera. Paleoceanography 15:43–52

Sabine CL, Feely RA, Gruber N, Key RM, Lee K, Bullister JL, Wanninkhof R, Wong CS l, Wallace DWR, Tilbrook B (2004) The oceanic sink for anthropogenic CO_2. Science 305:367–371

Sadekov AY, Eggins SM, Klinkhammer GP, Rosenthal Y (2010) Effects of seafloor and laboratory dissolution on the Mg/Ca composition of *Globigerinoides sacculifer* and *Orbulina universa* tests—A laser ablation ICPMS microanalysis perspective. Earth Planet Sci Lett 292:312–324. doi:10.1016/j.epsl.2010.01.039

Sanyal A, Hemming NG, Broecker WS, Lea DW, Spero HJ, Hanson GN (1996) Oceanic pH control on the boron isotopic composition of Foraminifera: evidence from culture experiments. Paleoceanography 11:513–517

Sanyal A, Bijma J, Spero HJ, Lea DW (2001) Empirical relationship between pH and the boron isotopic composition of *Globigerinoides sacculifer*: Implications for the boron isotope paleo-pH proxy. Paleoceanography 16:515–519. doi:10.1029/2000PA000547

Sarmiento JL, Gruber N (2006) Ocean biogeochemical dynamics. Princeton University Press, Princeton, Oxford

Schiebel R (2002) Planktic foraminiferal sedimentation and the marine calcite budget. Glob Biogeochem Cycles. doi:10.1029/2001GB001459

Schiebel R, Barker S, Lendt R, Thomas H, Bollmann J (2007) Planktic foraminiferal dissolution in the twilight zone. Deep-Sea Res II 54:676–686

Schiebel R, Hemleben C (2005) Modern planktic Foraminifera. Paläontol Z 79:135–148

Schiffelbein P, Hills S (1984) Direct assessment of stable isotope variability in planktonic Foraminifera populations. Palaeogeogr Palaeoclimatol Palaeoecol 48:197–213

Schmid TW, Bernasconi SM (2010) An automated method for "clumped-isotope" measurements on small carbonate samples. Rapid Commun Mass Spectrom 24:1955–1963. doi:10.1002/rcm.4598

Shackleton NJ (1974) Attainment of isotopic equilibrium between ocean water and the benthonic Foraminifera genus *Uvigerina*: Isotopic changes in the ocean during the last glacial. Cent Nat Rech Sci Colloq Int 219:203–209

Shackleton NJ, Opdyke ND (1973) Oxygen isotope and palaeomagnetic stratigraphy of Equatorial Pacific core V28-238: oxygen isotope temperatures and ice volumes on a 10^5 year and 10^6 year scale. Quat Res 3:39–55

Simstich J, Sarnthein M, Erlenkeuser H (2003) Paired $\delta^{18}O$ signals of *Neogloboquadrina pachyderma* (s) and *Turborotalita quinqueloba* show thermal stratification structure in Nordic Seas. Mar Micropaleontol 48:107–125

Spero HJ (1992) Do planktic Foraminifera accurately record shifts in the carbon isotopic composition of seawater CO_2? Mar Micropaleontol 19:275–285

Spero HJ, Bijma J, Lea DW, Bemis BE (1997) Effect of seawater carbonate concentration on foraminiferal carbon and oxygen isotopes. Nature 390:497–500

Spero HJ, DeNiro MJ (1987) The influence of symbiont photosynthesis on the $\delta^{18}O$ and $\delta^{13}C$ values of planktonic foraminiferal shell calcite. Symbiosis 4:213–228

Spero HJ, Eggins SM, Russell AD, Vetter L, Kilburn MR, Hönisch B (2015) Timing and mechanism for intratest Mg/Ca variability in a living planktic foraminifer. Earth Planet Sci Lett 409:32–42. doi:10.1016/j.epsl.2014.10.030

Spero HJ, Lea DW (1996) Experimental determination of stable isotope variability in *Globigerina bulloides*: implications for paleoceanographic reconstructions. Mar Micropaleontol 28:231–246

Spivack AJ, You CF, Smith HJ (1993) foraminiferal boron isotope ratios as a proxy for surface ocean pH over the past 21 Myr. Nature 363:149–151

Stein M, Almogi-Labin A, Goldstein SL, Hemleben C, Starinsky A (2007) Late quaternary changes in desert dust inputs to the Red Sea and Gulf of Aden from $^{87}Sr/^{86}Sr$ ratios in deep-sea cores. Earth Planet Sci Lett 261:104–119. doi:10.1016/j.epsl.2007.06.008

Tachikawa K, Toyofuku T, Basile-Doelsch I, Delhaye T (2013) Microscale neodymium distribution in sedimentary planktonic foraminiferal tests and associated mineral phases. Geochim Cosmochim Acta 100:11–23. doi:10.1016/j.gca.2012.10.010

Ter Kuile BH, Erez J (1991) Carbon budgets for two species of benthonic symbiont-bearing Foraminifera. Biol Bull 180:489–495

Tett P, Droop MR, Heaney SI (1985) The Redfield ratio and phytoplankton growth rate. J Mar Biol Assoc UK 65:487–504

Tripati AK, Eagle RA, Thiagarajan N, Gagnon AC, Bauch H, Halloran PR, Eiler JM (2010) ^{13}C-^{18}O isotope signatures and "clumped isotope" thermometry in Foraminifera and coccoliths. Geochim Cosmochim Acta 74:5697–5717

Uchikawa J, Zeebe RE (2010) Examining possible effects of seawater pH decline on foraminiferal stable isotopes during the Paleocene-Eocene Thermal Maximum. Paleoceanography. doi:10.1029/2009PA001864

Uhle ME, Macko SA, Spero HJ, Engel MH, Lea DW (1997) Sources of carbon and nitrogen in modern planktonic Foraminifera: the role of algal symbionts as determined by bulk and compound specific stable isotopic analyses. Org Geochem 27:103–113

Uhle ME, Macko SA, Spero HJ, Lea DW, Ruddiman WF, Engel MH (1999) The fate of nitrogen in the *Orbulina universa* Foraminifera: Symbiont system determined by nitrogen isotope analyses of shell-bound organic matter. Limnol Oceanogr 44:1968–1977

Urey HC (1947) The thermodynamic properties of isotopic substances. J Chem Soc Resumed 562. doi:10.1039/jr9470000562

Vance D, Scrivner AE, Beney P, Staubwasser M, Henderson GM, Slowey N (2004) The use of Foraminifera as a record of the past neodymium isotope composition of seawater. Paleoceanography. doi:10.1029/2003PA000957

Van Raden UJ, Groeneveld J, Raitzsch M, Kucera M (2011) Mg/Ca in the planktonic Foraminifera *Globorotalia inflata* and *Globigerinoides bulloides* from Western Mediterranean plankton tow and core top samples. Mar Micropaleontol 78:101–112. doi:10.1016/j.marmicro.2010.11.002

Voelker AHL (2002) Global distribution of centennial-scale records for Marine Isotope Stage (MIS) 3: A database. Quat Sci Rev 21:1185–1212

Voelker AHL, Grootes PM, Nadeau MJ, Sarnthein M (2000) Radiocarbon levels in the Iceland Sea from 25–53 kyr and their link to the earth's magnetic field intensity. Radiocarbon 42:437–452

Von Langen PJ, Pak DK, Spero HJ, Lea DW (2005) Effects of temperature on Mg/Ca in neogloboquadrinid shells determined by live culturing. Geochem Geophys Geosystems. doi:10.1029/2005GC000989

Wacker U, Fiebig J, Tödter J, Schöne BR, Bahr A, Friedrich O, Tütken T, Gischler E, Joachimski MM (2014) Empirical calibration of the clumped isotope paleothermometer using calcites of various origins. Geochim Cosmochim Acta 141:127–144. doi:10.1016/j.gca.2014.06.004

Weldeab S, Lea DW, Schneider RR, Andersen N (2007) 155,000 years of West African monsoon and ocean thermal evolution. Science 316:1303–1307

Weldeab S, Schneider RR, Kölling M (2006) Deglacial sea surface temperature and salinity increase in the western tropical Atlantic in synchrony with high latitude climate instabilities. Earth Planet Sci Lett 241:699–706

Wolf-Gladrow DA, Bijma J, Zeebe RE (1999a) Model simulation of the carbonate chemistry in the microenvironment of symbiont bearing Foraminifera. Mar Chem 64:181–198

Wolf-Gladrow DA, Riebesell U, Burkhardt S, Bijma J (1999b) Direct effects of CO_2 concentration on growth and isotopic composition of marine plankton. Tellus Ser B-Chem Phys Meteorol 51:461–476

Zachos JC, Röhl U, Schellenberg SA, Sluijs A, Hodell DA, Kelly DC, Thomas E, Nicolo M, Raffi I, Lourens LJ, McCarren H, Kroon D (2005) Rapid acidification of the ocean during the Paleocene-Eocene thermal maximum. Science 308:1611–1615. doi:10.1126/science.1109004

Zahn R, Keir R (1994) Tracer-nutrient correlations in the upper ocean: Observational and box model constraints on the use of benthic foraminiferal $\delta^{13}C$ and Cd/Ca as paleo-proxies for the intermediate-depth ocean. In: Zahn R, Pedersen TF, Kaminski M, Labeyrie L (eds) Carbon cycling in the Glacial Ocean: constraints on the Ocean's role in global change. NATO ASI Series I, vol 17. Springer, Berlin

Zeebe RE (2012) History of seawater carbonate chemistry, atmospheric CO_2, and ocean acidification. Annu Rev Earth Planet Sci 40:141–165

Zeebe RE, Bijma J, Wolf-Gladrow DA (1999) A diffusion-reaction model of carbon isotope fractionation in Foraminifera. Mar Chem 64:199–227

Zeebe RE, Sanyal A (2002) Comparison of two potential strategies of planktonic Foraminifera for house building: Mg^{2+} or H^+ removal? Geochim Cosmochim Acta 66:1159–1169

Zeebe RE, Wolf-Gladrow D (2001) CO_2 in seawater: equilibrium, kinetics, isotopes. Elsevier, Amsterdam

Ziveri P, Thoms S, Probert I, Geisen M, Langer G (2012) A universal carbonate ion effect on stable oxygen isotope ratios in unicellular planktonic calcifying organisms. Biogeosciences 9:1025–1032. doi:10.5194/bg-9-1025-2012

Methods

10

Analyses of planktic foraminifers are targeted towards three main goals. (1) The understanding of biological prerequisites and ecological demands of modern species facilitated by collection of planktic foraminifers at sea, and culturing of specimens in laboratory experiments. Those samples and associated data provide means for the analyses of the temporal and regional distribution, including depth habitat, availability of food, temperature, salinity, a variety of chemical parameters (i.e., stable isotopes, element ratios, pH, and other parameters of the marine carbonate system), and the availability of light (i.e. quality and intensity) to the symbiotic algae of foraminifers. (2) Proxy calibration and application of foraminifers in biostratigraphy, paleoecology, paleoceanography, and paleoclimate reconstruction (see Fischer and Wefer 1999, and reference therein). In addition to analyzing population dynamics and faunal assemblages of planktic foraminifers, (3) technology for biogeochemical analyses of foraminifers has been developed since the early 1950s, and is still being improved and extended to define and calibrate new proxies on the foraminifer shell, as well as on population dynamics (e.g., transfer functions). In this chapter, those methods are presented and discussed, which are most applied in sampling and analyses of planktic foraminifers at sea and in the laboratory. Given the rapid development of analytical methods, we provide merely an introduction to the various methods applied today, meant as first step to find the latest information published in specialist journals.

10.1 Sampling

While sampling planktic foraminifers from the water column, and processing of samples in the laboratory, any alteration of the individuals, tests, and assemblages is to be avoided (in theory), or kept at a minimum. When storing and conserving samples obtained from the water column, care should be taken to avoid fragmentation and dissolution of the planktic foraminifer tests, and to keep the pH of processing and storing liquids ≥ 8.2 at all times. In particular, processed fresh water from shipboard tanks, and deionized waters from laboratory devices may be delivered at low pH. pH should hence be monitored to avoid irreparable damage to samples.

Assemblages can not be entirely sampled with plankton nets, and specimens smaller than the net gauze will not be quantitatively included in the samples. When sampling with sediment traps, trapping efficiency, which usually deviates from 100 %, impedes complete samples. In turn, dissolution of tests or precipitation of any substance from the sampling solution on top of the foraminifer test can be avoided through correct and careful treatment of samples. Whereas some sampling artefacts may be overlooked and become clear only during data analyses, inadequate sampling of live individuals for culturing experiments emerges at once through inactivity or death of individuals. Proper sampling and processing hence constitutes the basis of any scientific work, good results, and fun at work.

© Springer-Verlag Berlin Heidelberg 2017
R. Schiebel and C. Hemleben, *Planktic Foraminifers in the Modern Ocean*,
DOI 10.1007/978-3-662-50297-6_10

10.1.1 Manual Collection of Live Specimens by SCUBA Divers

Planktic foraminifer specimens for culturing are ideally being sampled by hand at 'blue water' locations. Wide-mouthed glass jars are used as sampling containers to guarantee for minimal disturbance of specimens, and to avoid damage of cytoplasm and test, and particularly of the fragile spines (Hemleben et al. 1989; Huber et al. 1996). Spinose species are relatively easy to detect with the naked eye due to their large diameter, and could be sampled from oligotrophic (blue) surface waters using standard SCUBA equipment and techniques. Foraminifer specimens are relatively easy to detect at a distance of 50–80 cm in sunlit waters, and against a dark background such as the hull of a ship. Opening the lid of the jar close to the specimen to be collected will suck the foraminifer into the jar. Glass jars of about 125 mL are large enough for later culturing of specimens in their original ambient seawater for the first days. During the transport to the culture laboratory, the jars should be kept at constant temperature in an insulated chest. Back at the laboratory, the specimens could be kept in the original jars, or may be transferred to other culture vessels. Ambient seawater should be collected together with the foraminifers to serve as replacement water, and treated in the same way (e.g., same temperature) as the culture vessels containing live specimens.

10.1.2 Assemblage Sampling

For assemblage analyses and biogeochemical analyses, planktic foraminifers are preferably sampled with plankton nets. Some of the seminal early studies of Bé and co-workers were carried out using nets with rather large mesh-sizes between 200 and 366 µm (e.g., Bé 1960; Tolderlund and Bé 1971; Bé and Hutson 1977). Those nets are comparatively inexpensive, robust, and allow quick hauls, but do not sufficiently capture small-sized species, as well as pre-adult specimens of most modern planktic foraminifer species. Attempts have been made to use small mesh-sizes between

30 and 80 µm (e.g., Schott 1935; Phleger 1945), and which need to be hauled very slowly (up to $0.3~\text{ms}^{-1}$) to avoid tearing of the gauze. In addition, fine-meshed nets easily get clogged with particulate matter in mesotrophic and eutrophic waters, and back-pressure of the water eventually impedes quantitative sampling. 100-µm nets have been proven good compromise between employability onboard research vessels, and applicability to faunistic studies, although some of the small-sized species (e.g., tenuitellids) might still be largely underrepresented in the samples, in comparison to the original planktic foraminifer populations. The volume of sampled seawater is quantified with flow meters (see text box on Flow meters).

Flow meters: Flow meters, analogous and digital, installed at the outside *and* inside of the plankton net and CPR, provide independent data on the volume of seawater sampled (e.g., Motoda et al. 1957). Flow meter data are particularly important to calibrate new sampling devices, and to measure volumes of sampled seawater under varying sampling conditions, such as different hauling speeds. Resulting calibration curves may later be applied to correct for sampling errors, which are most certain to occur when working at sea under sometimes unforeseen and difficult conditions. In addition, winches of vessels, which are poorly equipped for scientific sampling (like some 'ships of opportunity'), might not be manufactured for precise adjustment of hauling speeds. Resulting deviations in sampled seawater volumes may later be corrected by using flow meter data. Even winches of research vessels may turn out to be less adjustable than expected.

10.1.3 Single Nets

The smallest, lightest, and cheapest option for sampling planktic foraminifers from the upper water column is the Apstein net, optionally being hauled by hand and employable even from small zodiacs. Apstein nets of typically 25 cm diameter and 50 or 100 cm length are suited to sample well preserved specimens for culturing and geochemical analyses, but are not suited for quantitative sampling of the planktic foraminifer fauna. Larger ring-nets (e.g., 'bongo nets') may be employed for quantitative sampling, and require a vessel with an adequate winch for scientific sampling.

10.1.4   Multiple Opening-Closing Nets

Multiple opening-closing nets (MCNs) have been employed for vertical and horizontal tows of up to nine sampling intervals, depending on the design of the device (Bé et al. 1959; Bé 1962; Wiebe et al. 1976). Apertures of the multi-nets usually range between 0.125 and 1 m² (see, e.g., www.hydrobios.de). Multi-nets with 0.25 m² (50 cm × 50 cm) opening and five net bags have proven most suitable for sampling of planktic foraminifers, and to be deployed from ships of different size and equipped with different types of winches (Fig. 10.1). MCNs are large and heavy enough to be employed with long and (optionally, for manual release) conductive wire down to a maximum water depth of 3000 m (recommendation of the manufacturer), and may be hauled at a speed of 0.5 ms^{-1} for quantitative sampling of the water column when using 100-µm gauze.

Fig. 10.1 Multinet equipped with five 100-µm nets (type Hydrobios midi, 50 × 50 cm opening) returning from vertical haul on the French research vessel 'Marion Dufresne' in the southern Indian Ocean. The sampling cups (*red*) are placed in a rack below the plankton-nets. An addition weight below the cup-rack keeps system straight. *Photo* H. Howa, Angers University, France, with permission

New makes of nets can be employed off-line, and conductive wire is hence not imperative. Hauling speeds are adjusted to the size of net-gauze used with the MCN, and need to be adapted to weather conditions. Rough sea and rolling ship requires low hauling speeds (e.g., 0.3 ms^{-1}) to prevent tearing of the net gauze at sharp increases of back-pressure of the sampled water, which may be caused by rapid movement of the ship. Samples obtained under different from 'normal' conditions may not be directly comparable with samples obtained with other hauling speeds. If no Apstein net or MCN is available, any other type of net could be used for sampling such as, for example, a larger MOCNESS (please be aware of idiosyncratic terminology of net types).

10.1.5 Continuous Plankton Recorder (CPR)

A Hardy-Plankton Recorder (also Longhurst-Hardy Plankton-Recorder, LHPR; or Continuous Plankton Recorder, CPR) is towed through the surface water column behind a sailing ship at different water depths, and produces under-way samples for quantitative analyses (e.g., Hardy 1935; Longhurst et al. 1966; Reid et al. 2003; Sir Alister Hardy Foundation for Ocean Science, http://www.sahfos.ac.uk/). To avoid tearing of the sampling gauze (called silk) by back-pressure of the sampled waters, rather coarse mesh-sizes (≥ 200 μm) have usually been used. Pre-adult and small-sized planktic foraminifers may hence not be quantitatively sampled, and a large part of the fauna might is missed.

10.1.6 Seawater Pumps

An elegant method of sampling planktic foraminifers from surface waters is the use of shipboard seawater-pumps, which can be employed during sailing, and hence consume no additional (i.e. costly) on-site ship-time. For example, deck wash pumps and fire pumps may be employed, provided that the sampled water is neither contaminated nor compressed while being pumped, which could cause damage to fragile specimens. Air bubbles produced along the hull of the sailing ship may affect the water intake, which needs to be taken into account to correctly assess the volume of the sampled seawater. The pumped seawater may be sampled with an Apstein net (cf. Ottens 1992).

10.1.7 Sampling the Test Flux

Continuous sampling of settling planktic foraminifer tests is carried out with moored or drifting sediment traps of varying design (cf. Buesseler et al. 2007). Automated sediment traps of conical shape have been used since the 1970s, usually derived from traps of the WHOI PARFLUX design (see Honjo et al. 1980 for the PARFLUX Mark II trap). Moored sediment traps typically have a 1-m^2 opening (size of the sampling area), and are equipped with 24 sampling cups. A baffle grid covers the opening to keep large swimmers and large 'particles' off the inside of the trap to avoid any damage of the sampled matter (e.g., Wiebe et al. 1976; Honjo and Manganini 1993; Lampitt et al. 2008, and references therein). Depending on region and water depth, current strengths, and lateral transport of tests (Siegel and Deuser 1997; von Gyldenfeldt et al. 2000) (see Chap. 8), data from sediment traps need to be corrected for possible under-estimation (or over-estimation) of the 'real flux' (i.e., sampling efficiency) using thorium (^{230}Th, ^{234}Th) or lead (^{210}Pb) isotope based methods (e.g., Scholten et al. 2001; Lampitt et al. 2008; Schmidt et al. 2009; Kuhnt et al. 2013).

Sediment trap samples may be affected by alteration (e.g., dissolution) of particles within the sampling cups. To prevent degradation of the trapped matter, formaldehyde (3–4 %), sodium azide (50 g NaN$_3$ L^{-1}), or mercuric (II)-chlorid (3.3 g HgCl L^{-1}) may be added to poison the sampling vessels (e.g., Fischer and Wefer 1991; Koppelmann et al. 2000; O'Neill et al. 2005, and references therein Buesseler et al. 2007). A buffer (e.g., sodium borate) should be used to keep $pH \geq 8.2$, to prevent dissolution of calcareous particles including foraminifer tests. Sampling

Table 10.1 Planktic foraminifer analyses from sediment trap deployments from all of the major ocean basins

#	Lat.[N] Long.[E]	Location	Time of deployment	Sampling intervals (days)	Trap depth (m)	Test size (µm)	Data	References
1	21°9'–20°41'	NE Atlantic	4/1990-11/1991	10–21.5	730	n.s.	CaCO$_3$, δ^{18}O	[1]
	20°45'–19°45'	Off Cape	3/1988-3/1989	27	2195		G. ruber	
	21°9'–20°41'	Blanc	4/1990-11/1991	10–21.5	~3500		(200–300 µm)	
1	20°45.6'–18°41.9'	Off Cape	7/2005-9/2006	21.5	1277	~420	Mg/Ca, G. inflata	[2]
		Blanc					G. ruber	
2	28°42.5'–13°9.3'	Canary	1/1997-9/1997	~14	800	<1000	Test flux	[3]
	29°11'–15°27'	Islands	12/1996-9/1997		700			[4]
					900			
					3700			
	29°45.7'–17°57.3'		1/1997-10/1997		4200			
3	33°–20°	Madeira	2/2002-4/2003	5–61	2000	>63	Test flux, species	[5]
		Basin			3000			
	33°–22°		4/2003-4/2004	11–31	3000			
3		NE Atlantic	4/1989-4/1990	14	1000	150–	CaCO$_3$ flux	[6]
					2000	250		
4	47°–20°				1000			
					2000			
					3749			
4	47°50'–19°39'	NE Atlantic	4-9/1992	5–8, 28	~1000	>20	Test and	[7]
			9/1996	14	2000	>100	CaCO$_3$ flux	
			6/1992-5/1993	28	2030	>20		
			5/1992-5/1993	8, 14–28	~3500	>20		
5	54°–21°		6-7/1992	28	2200	>20		

(continued)

Table 10.1 (continued)

#	Lat.[N] Long.[E]	Location	Time of deployment	Sampling intervals (days)	Trap depth (m)	Test size (μm)	Data	References
6	44°33'–2°45'	Bay of Biscay	6/2006–7/2007	5–12	800	>150	Test flux, species	[8]
			6/2006–6/2008	5–12	1700			
7	~45°–6°	Bay of Biscay	7–19/4/2009	0.125	200	>100	Test flux, species	[9]
8	36°1'–4°16'	Alboran Sea	6-1997–5/1998	3–11	1004	>150	Test flux	[10]
	36°14'–4°28'	(Med. Sea)	3/1998–5/1998	3–10	958		Species	[11]
9	43.02° 5.18°	Gulf of Lions	10/1993–1/2006	14, 30	500	>150	Test flux, species	[12]
	42.41° 3.54°	(Med. Sea)						
10	–32°5'–64°15'	Off Bermuda	4/1978–5/1979	30–75	3200	>125	Test flux, species	[13]
			4/1978–5/1984	60 ± 7	3200	>125	$\delta^{18}O$, $\delta^{13}C$	[14]
					3200	>37		[15]
11	10°30'–65°31'	Cariaco Basin	5/2005–9/2008	14	150	>125	$[CO_3^{2-}]$	[16]
					230			
					410			
					800			
					1200			
11	10°30'–65°31'	Cariaco Basin	1/1997–12/1999	14	230	150–355	$\delta^{18}O$, $\delta^{13}C$	[17]
11	10°30'–64°40'	Cariaco Basin	5/2005–3/2008	14	150	>125	$\delta^{18}O$, $\delta^{13}C$	[18]
					230			
12	27°30'–90°18'	Gulf of Mexico	1/2008–5/2009	4–14	700	>150	Test flux, species	[19]
			9/2009–12/2010				Test $CaCO_3$ flux	

(continued)

Table 10.1 (continued)

#	Lat.[N] Long.[E]	Location	Time of deployment	Sampling intervals (days)	Trap depth (m)	Test size (μm)	Data	References
13	13°31'–54°0'	W Atlantic	12/1977-2/1978	98	389	>100	Test flux	[20]
		Station E			988		CaCO$_3$, species	
					3755			
					5068			
14	15°21'-151°28'	Central Pacific	9-10/1978	61	978	>100		
		Station P1			2778			
					4280			
					5582			
15	50°–145°	Subpolar	9/1982-10/1983	15–16	1000	>125	Test and mass flux	[21]
	(Station PAPA)	N Pacific		(11)	3858			
15	50°–145°	N Pacific	9/1982-10/1985	14	3800	63–1000	Test flux, CaCO$_3$	[22]
		Station P						
16	42.09°–125.77°	off Oregon	9/1987-9/1988	33–87	1000	>125	Test flux	[23]
	42.19°–127.58°			27–85			Species	
	41.54°–132.02°			30–94				
17	39.5°–128°	NE Pacific	3-8/1981	11, 79	1235	>150–1000	Test flux	[24]
			8/1979-6/1980	318	4050	1000		
18	34°14'–120°2'	St.Barbara B.	8/1993-9/1996	14	590	>125	Test flux	[25]
19	33°33'–118°20'	San Pedro	1-7/1988	7	500	180-	Test flux	[26]
		Basin				600	δ^{18}O, δ^{13}C	[27]
20	27°53'–111°40'	Guaymas B.	2/1991-10/1997	14	485	>125	δ^{18}O	[28]

(continued)

Table 10.1 (continued)

#	Lat.[N] Long.[E]	Location	Time of deployment	Sampling intervals (days)	Trap depth (m)	Test size (µm)	Data	References
21	5°21'-81°53'	Panama Basin	12/1979-12/1980	60	890	>125	Test flux, CaCO$_3$	[29]
					2590		fauna	
					3560			
22	~-30°-73°11'	off Chile	7/1993-6/1994	8–9	2173	n.s.	Test flux, CaCO$_3$	[30]
			11/1991-4/1992	8–9	3497	n.s.	δ^{18}O, δ^{13}C,	[31]
			7/1993-1/1994	8–9	3520	n.s.	Species	[32]
22	~-30°-73°11'	off Chile	7/1993-6/1998	6–13	2303–2578	>150	Test and CaCO$_3$ flux, species, δ^{18}O	[33]
23	-60°55'-57°6'	Drake Passage	12/1980-1/1981	52	965	n.s.	N. pachyderma,	[34]
					2540		δ^{18}O, δ^{13}C	
			25/1/1981		surface			
	-62°22'-57°59.9'	Bransfield S.	11/1985-5/1986	9, 18	687	<1000	Total flux, species,	[35]
24	-62°27'-34°46'	Weddell Sea	1/1985-3/1986	11–22	863	n.s.	δ^{18}O, δ^{13}C	[36]
						<1000		[35]
25	-64°55'-2°30'	Maud Rise	1/1987-11/1987	16	4456	<1000	Test flux, species,	[35]
	-64°53.3'-2° 33.2'		1/1988-2/1989	16, 32	360		δ^{18}O, δ^{13}C	
	-64°55.5'-2° 35.5'		3/1989-2/1990	18	352			
26	-46° 56°5'	Crozet	12/2004-12/2005	2–28	3195	>63	Test CaCO$_3$, flux,	[37]
	-49°3' 51°30.59'		1/2005-1/2006	2–28	3160		Species	
	-44°29.9' 49° 59.9'		12/2004-12/2005	11–28	2000			

(continued)

Table 10.1 (continued)

#	Lat.[N] Long.[E]	Location	Time of deployment	Sampling intervals (days)	Trap depth (m)	Test size (µm)	Data	References
27	46°45.6' 142°4.2'	S off Tasmania	9/1997-1/1999	4.25-15.5, (24.5)	1060	>150	Test flux	[38]
					3850		$\delta^{18}O$, $\delta^{13}C$	[39]
	51° 141°44.3'		9/1997-1/1999		3080			[40]
	53°44.8' 141° 45.5'		9/1997-1/1999		830			[41]
					1580			
28	-44°37' 178°37'	E off New Zealand	6/1996-5/1997	~16	300	>150	Test and CaCO$_3$ flux	[42]
					1000		$\delta^{18}O$, species	[40]
	-42°42' 178°38'		9/1996-5/1997	7-16	300			
					1000			
28	-50°S 171°	Campbell Plateau, SE off New Zealand	5/1998-7/1999	9.5, 10	415	>150	Test flux and weight	[43]
	-52°S 174°				442		Species	
					362			
29	46°7.2' 175°1.9'	NW Pacific	6/1993-4/1994	13-31	1412	>125	Test flux	[44]
30	37°24.2' 174° 56.7'		6/1993-4/1994	9-16	1482		Species	[45]
31	30°0.1' 174°59.7'		6/1993-5/1994	13-31	3873			
32	39°59.8' 165°0.1'	NW Pacific	12/1997-12/1998	17.38,	2986	>125	Test flux, species	[46]
33	43°58.1' 155°3.1'		12/1997-12/1998	15.04	2957			
34	50°0.6' 165°1.5'		12/1997-12/1998		3260			
			12/1997-11/2001		3260		Test flux, $\delta^{18}O$, $\delta^{13}C$	[47]

(continued)

Table 10.1 (continued)

#	Lat.[N] Long.[E]	Location	Time of deployment	Sampling intervals (days)	Trap depth (m)	Test size (µm)	Data	References
35	~25° 137°	NW Pacific	12/1997-8/1999	13, 18	917	>125	Test flux, species	[48]
			8/1998-8/1999		4336		$CaCO_3$ flux	[49]
					1388			
					4758			
36	~39° 147°		11/1997-8/1998		1371			
					4787			
			8/1998-8/1999		1586			
					4787			
36	~36° 147°		8/1998-8/1999	15, 20, 30	1108			
				(16)	5081			
36	~36° 154°		8/1999-8/2000		1191			
					5034			
37	27°23' 126°44'	Ryukyu Island Arc and	10/1994-8/1995	17	1000	>125	Test flux, species	[50]
	25°4' 127°35'				3000			
37	~28° to 31°N	East China Sea	3/1993-2/1994	1–16	1–1070			[51]
	~123° to 127°N							
38	~22.5° 120°	SW off Taiwan	10-11/2009	3	233	>150	Test flux and weight	[52]
			3-4/2010		250		$\delta^{18}O$, $\delta^{13}C$	
			7-8/2010		816		Species	
39	8°17.5' 108°2'	S off Java	11/2000-11/2002	16-18	2200	>63	Test flux, Mg/Ca,	[53]
	8°16.1' 108°8.5'	S off Java	11/2002-7/2003	(22, 28)	2460		$\delta^{18}O$, $\delta^{13}C$	
40	17°27' 89°37'	Bay of	2/1994-2/1995	14-38	684	>250	Test flux	[54]
	15°32' 89°13'	Bengal			731			

(continued)

Table 10.1 (continued)

#	Lat.[N] Long.[E]	Location	Time of deployment	Sampling intervals (days)	Trap depth (m)	Test size (µm)	Data	References
41	14°25' 64°35'	Arabian Sea	3-4 and 9-10/1996	26	2986	>20	Test $CaCO_3$ flux	[55]
42	10°45.4' 51°56.6'	Somalia Basin	6/1992-2/1993	7–14	1265	>100	*N. pachyderma*	[56]
	10°43.1'53°34.4'				1032	>125	Species, test size	[57]
							Test flux, $\delta^{18}O$	[58]
43	16.8° 40.8°	Mozambique Channel	11/2003-3/2006	21	2250	>150	Test flux and weight	[59]
							$\delta^{18}O$, $\delta^{13}C$, Mg/Ca	[60]
44	27° 3°51'	Walvis Ridge	8/2000-2/2001	8	2700	>150	Test flux, life cycle, species	[61]
								[62]
	27° 3°51'	Walvis Ridge	2/2000-2/2001	8	2700	>150	$\delta^{18}O$, species, test flux, habitat depth	[63]

Some of the traps are discontinuous. Consecutive numbering corresponds to numbers in Fig. 10.2. References (Refs.) in square parentheses are given in the table

References. [1] Fischer et al. (1996), [2] Haarmann et al. (2011), [3] Abrantes et al. (2002), [4] Wilke et al. (2009), [5] Storz et al. (2009), [6] Honjo and Manganini (1993), [7] Schiebel (2002), [8] Kuhnt et al. (2013), [9] Siccha et al. (2012), [10] Bárcena et al. (2004), [11] Hernández-Almeida et al. (2011), [12] Rigual-Hernández et al. (2012), [13] Deuser et al. (1981), [14] Deuser (1986), [15] Deuser and Ross (1989), [16] Marshall et al. (2013), [17] Tedesco et al. (2007), [18] Wejnert et al. (2013), [19] Poore et al. (2013), [20] Thunell and Honjo (1981), [21] Reynolds and Thunell (1985), [22] Thunell and Honjo (1987), [23] Ortiz and Mix (1992), [24] Fischer et al. (1983), [25] Kincaid et al. (2000), [26] Sautter and Thunell (1991), [27] Thunell and Sautter (1992), [28] Wejnert et al. (2010), [29] Thunell and Reynolds (1984), [30] Marchant (1995), [31] Marchant et al. (1998), [32] Hebbeln et al. (2000), [33] Marchant et al. (2004), [34] Wefer et al. (1982), [35] Donner and Wefer (1994), [36] Fischer et al. (1988), [37] Salter et al. (2014), [38] King and Howard (2003), [39] King and Howard (2004), [40] King and Howard (2005), [41] Moy et al. (2009), [42] King and Howard (2001), [43] Northcote and Neil (2005), [44] Eguchi et al. (1999), [45] Eguchi et al. (2003), [46] Kuroyanagi et al. (2002), [47] Kuroyanagi et al. (2011), [48] Mohiuddin et al. (2002), [49] Mohiuddin et al. (2004), [50] Xu et al. (2005), [51] Yamasaki and Oda (2003), [52] Lin (2014), [53] Mohtadi et al. (2009), [54] Stoll et al. (2007), [55] Koppelmann et al. (2000), [56] Ivanova et al. (1999), [57] Peeters et al. (1999), [58] Conan and Brummer (2000), [59] Fallet et al. (2010), [60] Fallet et al. (2011), [61] Lončarić et al. (2005), [62] Lončarić et al. (2006), [63] Lončarić et al. (2007)

vessels should ideally be filled with filtered in situ seawater obtained from deployment depths ahead of deployment, and salt (1 g NaCl L^{-1}) may be added to produce a dense solution, and to prevent leakage and loss of the sampled matter (e.g., O'Neill et al. 2005). After the recovery of the trap, samples should be stored cool, ideally at 4 °C.

First long-term records of the planktic foraminifer test flux of up to almost seven years from off Bermuda have led to an understanding of the seasonal and interannual population dynamics of planktic foraminifers (Deuser et al. 1981; Deuser 1986; Deuser and Ross 1989). Those early projects have stimulated sediment trap studies in all major ocean basins, from the equatorial to polar ocean, and across a wide range of trophic condition from oligotrophic to eutrophic waters (Table 10.1, Fig. 10.2). The Deuser-traps off Bermuda were deployed 35 times at 3200 m water depth between April 1978 and May 1984, i.e. at average sampling intervals of 60 days (Deuser 1986; Deuser and Ross 1989). The Ocean Flux Program (OFP) in the Sargasso Sea off Bermuda has been run for more than 35 years (see also Bermuda Atlantic Time-Series Study, BATS).

The long-term deployment run by R. Thunell's in the Guaymas Basin was operated at fortnightly sampling intervals (Wejnert et al. 2010; see also McConnell and Thunell 2005). The longest time-series of planktic foraminifer test flux over 12 years, from October 1993 to January 2006, were sampled with sediment traps in the Gulf of Lion, in the northwestern Mediterranean Sea (Rigual-Hernández et al. 2012). However, some of the time-series are discontinuous due to malfunction of the sampling gear, and problems while deploying or recovering the traps. In addition, 'swimmers' or any other 'matter' may block the sampling containers, or affect the samples in any other way (e.g., pH changes caused by degradation of organic matter), and may hence impede quantitative analyses.

Sediment traps with very short sampling intervals of 3 h, drifting at 200 m water depth within the same water body, were employed to sample the short-term flux of planktic foraminifers in the southern Bay of Biscay (Siccha et al. 2012). Those samples have revealed small-scale variability of hours and at a local range (patchiness) of planktic foraminifer tests flux, in contrast to large-scale variability

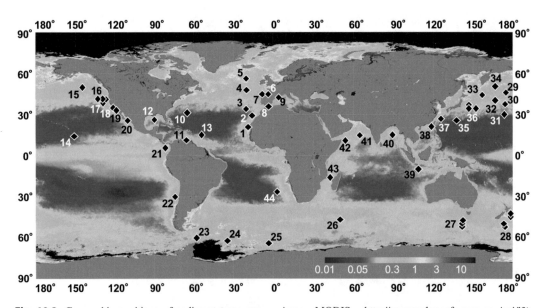

Fig. 10.2 Geographic positions of sediment traps analyzed for planktic foraminifers, and annual global aquatic chlorophyll a concentration (mg m^{-3}, 2013, from Aqua MODIS, http://oceancolor.gsfc.nasa.gov/cgi/l3). Numbering of the trap locations corresponds to Table 10.1

(seasonal to interannual, and regional) investigated by most other approaches (Table 10.1).

Data from sediment trap studies (see also compilations in Schiebel 2002; Žarić et al. 2005, 2006) add information to the systematic understanding of the temporal and regional population dynamics and biogeochemistry of planktic foraminifers (Table 10.1, Fig. 10.2). Although the global coverage of samples includes most of the range of environmental conditions (T, S, productivity) prevailing in the low to high latitude ocean basins (see Fig. 10.2), most remote regions like the central South Pacific have not yet been included in any long-term sampling program due to logistic limitations.

10.2 Processing of Samples

Net-collected samples should be fixed immediately after sampling in a 4 % formaldehyde solution or in alcohol (Ganssen 1981; Hemleben et al. 1989; Schiebel et al. 1995), i.e. addition of one part of concentrated (38 %) formaldehyde to ten parts of seawater sample, or two parts of alcohol to one part of seawater. To prevent dissolution of the shell calcite, wet samples need to be buffered at pH 8.2 using hexamethyltetramine (also called hexamine). Sodium-bicarbonate buffered-formaldehyde solution may be used for sample fixation in case no biogeochemical or morphometric analyses will be carried out, since crusts could precipitate from the bicarbonate-seawater solution, and alter the weight and chemical composition of tests. In case formaldehyde is not available, methyl alcohol or ethyl alcohol could be used for sample fixation. Rose Bengal should not be used to stain the samples (in contrast to processing benthic foraminifers, e.g., Lutze and Altenbach 1991), because the natural color of the planktic foraminifer cytoplasm would be lost, and with it some useful information on the pigmentation of test and cytoplasm. In case samples will not be analyzed for planktic foraminifers (and other calcareous plankton) immediately, buffering needs to be repeated after two month, six month, and from then on once per year, to make up for pH changes caused by degrading organic matter in the sample

solution. Samples should be stored at low temperature (ideally at $\sim 4\ °C$). All steps of preparation and observation should be noted, including changes in storing conditions (e.g., changes in pH, and temperature).

10.2.1 Fixation for Transmission Electron Microscopy (TEM)

Sample fixation for Transmission Electron Microscopy for imaging of the fine structure of the foraminifer cytoplasm is ideally carried out immediately after sampling with a protocol developed by Anderson and Bé (1978). The fixative is minimally disruptive of a wide range of cellular structures, as well as symbiotic algal cells embedded in the cytoplasm. For optimum preservation of the most delicate structures like microtubules, calcium and other interfering substances are excluded from the fixative (Hemleben et al. 1989).

Fixing the cytoplasm of live specimens while stabilizing the molecular structure is achieved with 2 % glutaraldehyde solution in 0.1 M cacodylate buffer (pH 8.2), with 1 % OsO_4 prepared in the same buffer. For optimum preservation of microtubules, the individual should be suspended in a minimum volume of seawater. The fixative should be prepared in a solution at salinity equivalent to the sampled seawater, to exclude as much calcium as possible during fixation.

Following to fixation in OsO_4, the foraminifer shell is removed through decalcification to facilitate subsequent sectioning. In particular, thick shells of mature specimens would disrupt sectioning. To maintain the delicate organic layers during decalcification, specimen are embedded in a 0.8 % agar sol at 40 °C, preferably within a shallow watch glass. The specimen can be isolated within a small (2-mm sized) cube of cold agar, using a line razor blade.

The shell is removed by treating the fixed organisms with 0.1 N HCl or 1 % EDTA (ethylenediamine tetraacetic acid) sufficiently long to remove the shell calcite. The progress of shell dissolution may be monitored with a polarizing light microscope. Alternative to decalcification, the shell can be dissolved after embedding: The plastic is cut off to expose some surface of shell. The remaining plastic block is immersed in 0.1 N HCl until all calcite is etched away. Subsequently, the block is cleaned with absolute alcohol or acetone to remove any water, re-infiltrated with epoxy, and polymerized.

Dehydration of fixed specimens prior to embedding in a graded series of acetone baths is preferable to alcohol dehydration as precipitation of residual OsO_4 in the specimen is less likely. Dehydrated specimens are embedded in a plastic polymer appropriate in hardness and quality required by the kind of Glas- or Diatome Diamond Knife used to prepare thin sections for TEM analyses.

10.2.2   Analysis of Wet Samples

In the laboratory, wet samples are decanted into a high-rimmed glass dish with a flat bottom, and a diameter sufficient for 'gravity sorting' (e.g., 9-cm Pyrex dish). Heavy particles including foraminifer tests will accumulate in the center of the dish when being carefully rotated. The tests can then be pipetted from the dish under an incident-light microscope of sufficient working distance, using a glass (Pasteur) pipette fitted with a rubber bulb. Tests should be transferred into an evaporation dish made of glass or porcelain. A black microscope table facilitates recognition of the usually whitish tests. Specimens are cleaned from particulate matter using as little water as possible. A minimum of 300 specimens should be enumerated for statistically sufficiently interpretable data (van der Plas and Tobi 1965; Patterson and Fishbein 1989, see Sect. 10.13). Remains of cytoplasm and internal structures of foraminifer tests are particularly well visible in wet samples.

10.2.3   Analysis of Dry Samples

Analysis of dry samples may have advantages over wet analysis, and provides similar results. Before being dried, as much water as possible should be pipetted of the sample. The sample

should then carefully be dried over night at room temperature (~ 20 °C) or in an oven at a maximum temperature of 50 °C. Dry samples are best transferred into 'Franke cells' or 'Plummer cells' with a black background, analyzed, and stored in a dry and clean place for many years. When analyzing very small tests, cardboard cells may be preferred over plastic cells to avoid electrostatic phenomena like any unwanted displacement of tests. A paintbrush may be used for manipulation of the foraminifer tests under the microscope. The finest and most pointed paintbrush should be selected, and which still needs to bear two filaments at the tip to allow for capillary action. Alternatively, a preparation needle can be used. The paintbrush may be used wet (clean tap water will do), the needle with care.

Equal aliquots of large dry samples are produced out with a micro-splitter, also called Otto-micro-splitter. An ideal split contains just above 300 specimens to be classified and counted to produce statistically significant assemblage data (see Sect. 10.13). Assemblage data on entire samples are produced by multiplication of count numbers and split-sizes. Faunal analysis of large samples is alleviated by size-fractionation (sieving) before splitting into aliquots. Sieve sizes of 63, 100, 125, 150, 200, 250, 315, 355, 400 μm, followed by 100-μm increments facilitate comparison of data with other studies. Often applied minimum size classes in assemblage analyses are 100 μm and 150 μm. In addition, size-fractionated samples facilitate balanced analysis of all size classes. In particular, increasingly large and rare tests are sufficiently considered when applying the size-classes given above.

Sieve-size analyses of planktic foraminifer tests usually start at a minimum size of 100 μm for practical reasons. Most plankton nets are equipped with 100-μm gauze, and tests smaller than 100 μm are not quantitatively sampled. Specimens <100 μm in test-size are difficult to classify using an incident light microscope, since those samples include many difficult-to-identify juvenile individuals of large-sized species. Most assemblage studies of planktic foraminifers therefore use the size fractions >100 μm. Tests <100 μm are usually either treated as uniform size class, or analyzed by means of automated image analyses (see Sect. 10.10).

10.2.4 Wet Oxidation of Organic-Rich Samples

In case samples are too rich in organic matter (e.g., algae or zooplankton) to allow efficient picking of planktic foraminifers for faunal analyses or analyses of test chemistry, oxidation of the organic matter may be advised. For dry oxidation, an oxygen-plasma low-temperature asher (LTA) may be employed. Wet oxidation can also be carried out using standard chemical solutions, i.e. hot (70 °C) 18 % H_2O_2, and 0.024 M NaOH at pH >8, and without any additional technology except of stainless steel sieves, standard laboratory glassware, and a fume hood (Fallet et al. 2009). In both methods, excess seawater should be removed, and the sample should briefly (to avoid calcite dissolution) be washed with deionized (e.g., MilliQ®) water) water to prevent precipitation of salt crystals on the foraminifer tests. While dry oxidation takes about 8 h for LTA alone, the entire process of wet oxidation takes only $\sim$3 h (Fallet et al. 2009). Both dry and wet oxidation have been shown to not significantly alter weight, stable isotope ratios, and element ratios (e.g., Mg/Ca, Sr/Ca, and Ba/Ca) of the shell calcite of *G. ruber*, *G. sacculifer* (*trilobus* morphotype), *N. dutertrei*, and *G. bulloides* (Fallet et al. 2009).

10.3 Methods in Molecular Genetics

10.3.1 DNA Isolation

The eukaryotic genome is composed of double-stranded desoxyribonucleic acid (DNA) localized in the nucleus and in mitochondria (and chloroplasts in plant cells). The DNA itself consists of the four desoxynucleotides adenine (A), thymine (T), guanosine (G), and cytosine (C). A and T, and G and C, respectively, are bound by hydrogen bonds. For nucleotide sequencing, i.e. for determining the sequence of these nucleotides of a given DNA region, the respective part of interest of the genome is amplified by polymerase chain reaction (PCR) using specific primers. These primers are short desoxynucleotide sequences reconstructed from a known sequence. The isolation of DNA is the initial step, and a necessary prerequisite for nucleotide sequencing. For single-cell foraminifers with a calcareous shell, different methods for DNA extraction have been developed. Merlé et al. (1994) used proteinase K and phenol-chloroform to digest the cells and extract the DNA. Pawlowski et al. (1994) modified the extraction procedure and used a sodium deoxicholate buffer (DOC), a method, which was subsequently applied for most of the foraminifer molecular studies. However, the calcareous shells of planktic foraminifers dissolve in this buffer, preventing any further taxonomic or morphometric classification of the specimens after DNA extraction. Therefore, DOC was later replaced by some workers by a guanidinium thyocyanat buffer (De Vargas et al. 2002), which does not destroy the shells of foraminifers. To date, new methods have been developed making possible the isolation of DNA even after preparation of the cell images (e.g., Seears and Wade 2014).

10.3.2 Selection of Primers and PCR

Before DNA (nucleotide) sequencing, a specific gene region has to be amplified by PCR. For this purpose, flanking primers are designed according to the known conserved regions of the selected gene (e.g., the partial ribosomal SSU (or 18S) RNA gene; see Darling et al. 1997). PCR is carried out with the purified total DNA following standard procedures with subsequent denaturing, annealing and replication steps using a specific DNA polymerase (*Taq* polymerase) isolated from the bacterium *Thermophilus aquaticus*, which is heat-stable and replicates DNA at high temperature. The amplified PCR products are subsequently purified by agarose gel electrophoresis, the respective bands are cut out of the gel, purified, and then either sequenced directly or cloned before sequencing.

10.3.3 Cloning and Nucleotide Sequencing

In order to gain a high-quality sequence read of the PCR amplification products from DNA of a single cell, these are often cloned before sequencing (Grimm et al. 2007; Aurahs et al. 2009b). The PCR products are purified using the QIAquick PCR purification and gel extraction kits (Qiagen) and cloned. For cloning, a PCR product is ligated into a plasmid vector (e.g., pUC18), and transformed into competent *Escherichia coli* cells (*E. coli* DH5α, bacteria strain). Genetic variability within single foraminifer individuals is determined by sequencing several clones. Nucleotide sequencing is carried out in both directions, for example, with an ABI 377 automatic sequencer (Perkin Elmer) using the standard vector primers M13uni and M13rev. Newly assembled sequences are uploaded to Genbank (http://www.ncbi.nlm.nih.gov/nuccore/), and the accession numbers specify the sequences. The nucleotide sequences obtained are then evaluated by computer analyses. For the different computer programs used in these studies see below.

PCR products can be also sequenced directly. While being considerably faster (from DNA isolation to sequence), the procedure gives rise to replication errors. Therefore, several readings are necessary to obtain the reliable sequence. Direct sequencing can be used if a sequence type is already known from other studies, and large numbers of individuals need to be genotyped.

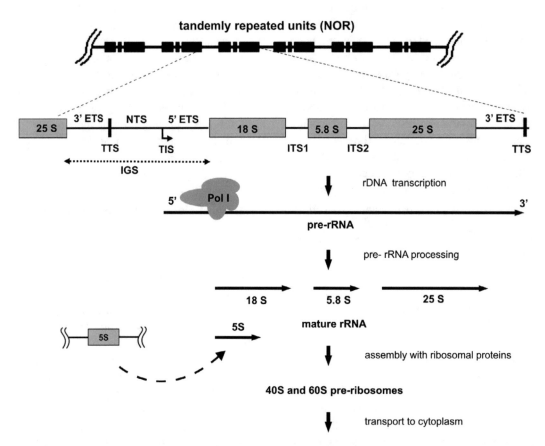

Fig. 10.3 Location and general structural organization of the eukaryotic nuclear encoded ribosomal RNA genes (rDNA) within the nucleolus of the cell nucleus, transcription and processing into the mature rRNA molecules. 18S rDNA corresponds to SSU (small subunit) rDNA, 25//28S rDNA corresponds to the LSU (large subunit) rDNA; ITS, internal transcribed spacer; ETS, external transcribed spacer; TIS, transcription initiation site; TTS, transcription termination site; IGS, intergenic spacer; Pol I, RNA polymerase I; pre-rRNA, rRNA precursor. Modified after Volkov et al. (2007)

10.3.4   Molecular Marker

Ribosomal DNA The nuclear encoded genes (rDNA) are mostly used as molecular marker for the phylogenetic and molecular genetic analyses of Foraminifera. Foraminifer ribosomal RNA genes generally exhibit a similar structure to those from other eukaryotic organisms (Fig. 10.3), although the internal structure of the respective gene regions is highly divergent. The foraminifer SSU (18S) rDNA sequence contains specific variable regions (see below; Hancock and Dover 1988), which can be used for differentiation of species or even different types of "cryptic" species (e.g., Pawlowski and Holzmann 2002; Darling and Wade 2008; Aurahs et al. 2009b; Ujiié and Lipps 2009). Therefore, the SSU rDNA became the standard genetic marker for the characterization of species and different genotypes, and for phylogenetic approaches in planktic Foraminifera. In the meantime, a SSU (18S) rDNA data bank has been established, i.e. PFR^2, Planktonic Foraminifera Ribosomal Reference database (Morard et al. 2015).

The SSU (or 18S rRNA) gene of foraminifers, and particularly of planktic foraminifers, is unique among eukaryotes due to the occurrence of characteristic variable regions 37/e1′, 41/e1′, 45/e1′ and 46/e1′ (Fig. 10.4). The variable region 37/e1′ corresponds to a universal variable region of the prokaryote structure model (De Vargas et al. 1997; Neefs et al. 1990). The other three length-variable regions of the SSU rDNA are also known from the SSU rDNA of other eukaryotes. However, the degree of variability in these gene regions varies greatly between the different groups of foraminifers, and only a few species of planktic foraminifers (e.g. the non-spinose

Globorotaliidae) can be aligned in these regions to benthic foraminifers. For the spinose planktic foraminifers this is only possible within conserved regions of the gene (e.g., De Vargas et al. 1997). Therefore, manual alignments of SSU rDNA of planktic foraminifers were modified based on the SSU rRNA universal secondary structure model (e.g., Van de Peer et al. 1996; Wuyts et al. 2002), in order to include only homologous nucleotide positions in the phylogenetic reconstructions (Darling et al. 2006; De Vargas et al. 1997; Pawlowski et al. 1997; Aurahs et al. 2009b).

In a new approach, the automatical multiple alignment of the sequences gave rather reliable results (Aurahs et al. 2009a). This method has the advantage that the corresponding sequenced gene region, containing both rather conserved and more variable sequences (see Fig. 10.4), can be directly aligned and used for phylogenetic analyses.

Population genetic studies and differentiation of cryptic species can be further defined by using also the even more variable internal transcribed spacers, ITS I and ITS 2, including the conserved 5.8S rDNA, of the rRNA gene (see Fig. 10.3) (Ujiié et al. 2010).

RFLP, Restriction Fragment Length Polymorphism This method allows rapid analysis of a large number of DNA samples from related species or populations. It is faster and cheaper than cloning and sequencing of the respective gene regions. For this purpose, the purified, PCR amplified SSU rDNA products are digested with the respective restriction enzymes at specific short nucleotide sequences, resulting in a number of DNA fragments of different sizes that show genotype specific patterns after agarose gel-electrophoresis (e.g., De Vargas et al. 2001;

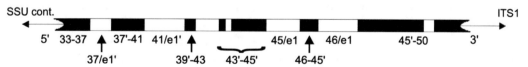

Fig. 10.4 Schematic representation of the 3′ SSU rDNA fragment used for the genetic identification of planktic (and benthic) foraminifers. Black areas represent the relatively conserved regions, white regions correspond to the more variable parts of the fragment. The numbering refers to a hypothetical secondary structure model for the 3′ SSU rDNA according to Wuyts et al. (2002), labeled after the SSU rRNA helices they are encoding for. From Aurahs (2010), modified after Grimm et al. (2007)

Morard et al. 2009). However, it requires previous knowledge of the respective sequences, and minor variations are not detected in the SSU rDNA between closely related genetic types of planktic foraminifers.

Protein-Coding Genes Gene coding for specific proteins have as yet seldom been sequenced and applied to phylogenetic studies of foraminifers. Actin genes, which are rather conserved throughout the eukaryotic kingdom offer a possibility, but have the disadvantage that several paralogs normally exist in the genome, and respective homologs may be compared between different foraminifers (Flakowski 2005).

10.3.5   Next Generation Sequencing (NGS)

With the enormous improvement in nucleotide sequencing methods, rapid increase in knowledge about the genetic constitution and genome evolution of foraminifers is expected in the near future (Pawlowski et al. 2014). Next Generation Sequencing will allow sequencing and comparison of whole genomes. This will facilitate broader information about phylogenetic relationships among different foraminifer species, and verify the occurrence of cryptic species (for a review see Metzker 2010).

10.3.6   Computer Evaluation of the Nucleotide Sequences used for Phylogenetic Studies

ABGD: "Automatic Barcode Gap Discovery" allows calculation of genetic distances within and among genetic types delimitated according to each possible species-level threshold (Puillandre et al. 2012). ABGD is an automatic procedure that sorts sequences into putative species based on a barcode gap, i.e., the gap in genetic distances distribution between intraspecific and interspecific diversity. The barcode gap is observed whenever the divergence among organisms belonging to the same species is

smaller than divergence among organisms from different species (André et al. 2014).

GMYC: "General Mixed Yule Coalescent" uses phylogenetic trees to identify transitions from coalescent to speciation branching patterns (Pons et al. 2006). The GMYC approach identifies boundaries between evolutionary units on the basis of shifts in branching rates. Branching within species is the result of coalescent processes, whereas branching between species reflects the timing of speciation events. These methods provide alternative delimitations, and offer the opportunity to analyze sequences that lack former assignation of their genetic type. Finally, these alternative delimitations are confronted in an attempt to connect SSU rDNA sequences to identified genuine species (André et al. 2014).

ML: "Maximum Likelihood", a statistical probability method, is used to estimate the phylogenetic trees for a set of species. The probabilities of DNA base substitutions are modeled by continuous-time Markov chains (Felsenstein 1981, 2004). PhyML trees used for patristic distance calculation, BEAST ultrametric trees and patristic distance matrices (André et al. 2014).

MP: This method in phylogenetics, "Maximum Parsimony" estimates the parameters of a statistical model. It provides estimations for the model's parameters. As an optimal criterion under which the phylogenetic tree has minimized, the total number of character-state changes is to be preferred. The shortest possible tree that explains the data is considered best.

MrBayes: "Bayesian inference (BI)" is a program for Bayesian phylogenetic analysis. The program uses Markov Chain Monte Carlo (MCMC) techniques to sample from the posterior probability distribution (Huelsenbeck and Ronquist 2001; Ronquist and Huelsenbeck 2003).

NJ: "Neighbor Joining" is a bottom-up cluster method for producing unrooted phylogenetic trees based on DNA sequence data. The algorithm requires knowledge of the distance between each pair of taxa (e.g., species or sequences) to form the tree (Saitou and Nei 1987). In contrast, UPGMA (Unweighted Pair

Group Method with Arithmetic mean) produces rooted trees.

PAUP: "Phylogenetic Analysis Using Parsimony" (Swofford 2001).

RAxML: "Randomized Axelerated Maximum Likelihood" is a method used for phylogenetic studies based on large data sets (Stamatakis 2014).

"SplitsTree" is a program for inferring phylogenetic (split) networks (Huson and Bryant 2006).

10.4   Culturing in the Laboratory

Culturing of any biota that serves as proxy in Earth system-science is an indispensable prerequisite for calibration of any proxy (e.g., size, weight, stable isotopes, and element ratios). Planktic foraminifers have been cultured in the laboratory for various analytical purposes since the early 1970s (Bé et al. 1977; Hemleben et al. 1989). Analyses of, for example, radiocarbon (^{14}C) and trace elements (see, e.g., Lea 1999 for a review), boron isotopes (Sanyal et al. 1996), Cd/Ca ratios (Ripperger et al. 2008), and clumped isotopes (Tripati et al. 2010, 2014; Eiler 2011; Wacker et al. 2014) from planktic foraminifer tests requires large sample volumes (several milligrams of $CaCO_3$, i.e. several hundreds to thousands of tests), which could be provided if foraminifer tests could be grown under controlled conditions (e.g., temperature, light) in the laboratory over multiple generations. New designs for culturing of marine micro-biota would allow investigation of planktic foraminifer growth under constant chemical and physical conditions such as, for example, pH, and $[CO_2]$. The rather inexpensive chemostat set-up developed for culturing of benthic foraminifers by Hintz et al. (2004) would potentially allow parallel culturing of large numbers of specimens of any species in time-series experiments (see also Hemleben et al. 1989). A chemostat set-up adopted from the one developed by Hintz et al. (2004) was used for culturing planktic foraminifers (*Globigerina bulloides*, *G. sacculifer*, *G. siphonifera*, *T. quinqueloba*, *N. dutertrei*, *N. incompta*, and *G. inflata*) at JAMSTEC, Yokosuka, Japan (Fig. 10.5).

Culture-protocols are discussed in detail by Hemleben et al. (1989). Culturing of planktic foraminifers has been developed as standard method by H. Spero at the Wrigley Institute for Environmental Science on Santa Catalina Island, California, USA (see, e.g., Spero 1992). The Wrigley Institute for Environmental Science on Santa Catalina Island is situated close to waters where sampling of a variety of abundant live planktic foraminifer species by SCUBA diving is possible. Successful culture experiments are facilitated at laboratories sited close to deep marine waters for sampling of planktic foraminifers, and with infrastructure for culturing experiments. Planktic foraminifers have unfortunately not yet been successfully cultured over an entire generation. Although offspring of *O. universa*, *G. bulloides*, *G. truncatulinoides*, and *G. glutinata* have been kept in laboratory culture (e.g., Hemleben et al. 1987; Spero 1992; Spero and Lea 1996; Bijma et al. 1998; K. Kimoto, personal communication, 2007), a second generation has not yet reproduced in culture.

10.4.1   Preparation of Specimens for Culture Experiments

Undamaged specimens should be transferred to the laboratory immediately, i.e. within a couple of hours after sampling. Specimens need to be identified and described using an inverted microscope or incident light microscope. Specimens should ideally be photographed, and transferred to clean culture vessels with the least possible delay. Culture vessels should have a flat bottom to allow for observation with an inverted microscope. Lids of culture vessels may be sealed with Parafilm® to impede gas exchange between culture and atmosphere (Allen et al. 2012). Standard digital cameras are suited for documentation of, for example, chamber formation, changes in cytoplasm color, preservation of spines, gametogenesis, and general behavior under laboratory conditions.

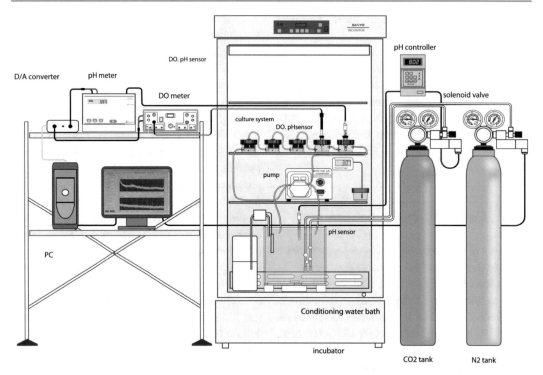

Fig. 10.5 Design of chemostat culture system developed from the design of Hintz et al. (2004). Environmentally controlled (T, S) 20-L seawater reservoir (*light blue*, large) monitored by *p*H and CO_2 electrodes installed at the top of reservoir. Culture vessels (*light blue*, small) are serially connected. Reservoir and culture vessels are placed in an incubator. Water circulates from the reservoir to culture vessels via Taigon tubes by a peristaltic pump. Taigon tubes protect the culturing water from gas exchange with the outside environment. Culture vessels are closed by screw-lids. The volume of water flow is variable (e.g., 5 mL min^{-1}) to provide equal water quality to all compartments of the culturing system at all times. Circulating water enters culture vessels at the bottom and leaves vessel at the top. To reduce contamination of the system by waste products and particles of all sorts, the inlet of culture vessels is covered with 8-µm gauze. To prevent the incubated foraminifer specimens from escaping culture vessels, the water outlet is closed with a 40-µm mesh. The bottom of vessels is covered with a porous polystyrene cell, which allows constant and balanced water circulation. A 12:12 h dark-light cycle is applied. Light levels within the culture vessel range between 70 and 140 µEinstein m^{-2} s^{-1} (cf. Bemis et al. 1998; Spero and Parker 1985). Culture experiments have been conducted at air-conditioned laboratories at JAMSTEC (Natsushima, Japan). From T. Toyofuku, JAMSTEC, Natsushima, Japan, 2014, with permission

Species should ideally be cultured in their natural ambient seawater, but which needs to be filtered to remove large particles and other plankton organisms (see, e.g., Spero and Williams 1988, 0.45 µm; Allen et al. 2012, 0.8 µm filter). A mix of natural and artificial seawater (1:1) was used to perform a low-DIC experiment (Allen et al. 2012). Filters should be wide enough to not remove fine particles, which could potentially serve as food source for some of the cultured planktic foraminifer species. Water of the culture vessels should preferably be replaced by freshly filtered seawater during days when no food is provided to the foraminifers to not disturb feeding (Spero and Lea 1993).

10.4.2   Feeding in Laboratory Culture

Quality and quantity of food is essential for successful culturing of planktic foraminifer individuals in the laboratory. In addition, chemical and physical parameters need to be carefully chosen and monitored while culturing. Although planktic foraminifers can survive for some time

Table 10.2 Indications of vitality of spinose and non-spinose species, and *H. pelagica* in culture (from Hemleben et al. 1989)

	Healthy	Poorly nourished	Unhealthy
Spinose	Spine length 3× max. test diameter	Reduced spine length	Short spines or no spines
	Network of rhizopodia at or between spines	Same as in healthy individuals	Rhizopodia generally shorter than test diameter
	All chambers filled with cytoplasm	Last formed chamber partially filled or empty	Several chambers only partially filled or empty
	Floating in culture vessel	Same as in healthy	Resting at bottom of culture vessel
H. pelagica	Bubble capsule surrounding test and doubling total diameter	Same as in healthy individuals	Bubble capsule irregular in shape and few bubbles only
	Reddish cytoplasm	Pale reddish to white cytoplasm	Same as in poorly nourished
Non-spinose	Many long rhizopodia may extend at several times test diameter	Same as in healthy individuals	Few short rhizopodia
	All chambers filled with cytoplasm	Final chamber only partially filled with cytoplasm or empty	Several chambers only partially filled or empty
	Attached to and actively moving on bottom of culture vessel	Same as in healthy individuals	Not moving on bottom of culture vessel

without being fed, optimum growth and maturation only occurs if food is appropriately provided (Hemleben et al. 1989). The optimum diet of most species is unfortunately unknown, but both algal and animal prey is consumed by most surface dwelling species (Bé et al. 1977; Anderson et al. 1979; Spindler et al. 1984; Hemleben et al. 1989). Subsurface dwelling species (e.g., *G. truncatulinoides*, *G. scitula*, *G. hirsuta*) possibly prefer rather degraded organic matter (Itou and Noriki 2002; Schiebel et al. 2002) at an unknown concentration and quality, which might be the reason for largely unsuccessful culture attempts of any planktic foraminifer species so far.

Most spinose planktic foraminifer species are omnivorous and tend to favor animal prey over algal prey. Adult *G. ruber* and *G. sacculifer* are fed live *Artemia* nauplii, for example, every 48 h (Hemleben et al. 1989; Spero 1992; Allen et al. 2012). The *Artemia* nauplii should not be older than one day. Other foraminifer species may be offered food of different kind or at different frequency (Spindler et al. 1984). Juvenile foraminifer specimens may be fed with small pieces of *Artemia* nauplii (Hemleben et al. 1989). Nauplii food is transferred to the culture dish with a Pasteur pipette and placed near the foraminifer rhizopods, where the food might be accepted within several hours. The feeding process should be monitored, and food might need to be offered several times before being accepted by the foraminifer. Unconsumed food remains need to be removed from the culture dish after feeding (Hemleben et al. 1989).

Non-spinose foraminifer species prefer algal over animal prey, and cultured algae (e.g., *Dunaliella* or *Chlorella*) may be offered to the foraminifers (see Hemleben et al. 1989). In general, appropriate food should be provided at optimal time-intervals to the different species of foraminifers in culture, to keep specimens active and at good health, and enhance the possibility of chamber formation and reproduction (Table 10.2).

10.4.3 Illumination of Symbiont-Bearing Species in Culture

One of the parameters particularly important for culturing of symbiont-bearing planktic foraminifers is an appropriate quality and quantity of light (Jørgensen et al. 1985; Spero and Williams 1988). Illumination may be chosen according to the goal of experiment, and may vary between diurnal 12-h light and 12-h dark cycles, and more or less rapid changes in illumination (e.g., Caron et al. 1982; Jørgensen et al. 1985; Hemleben et al. 1987; Hönisch et al. 2011; Allen et al. 2012). Sufficient light intensity is provided by cool fluorescent light bulbs, and should be monitored with a light meter (e.g., Allen et al. 2012). Compensation light levels where foraminifer respiration exceeds symbiont photosynthesis start at 26–30 µEinstein m^{-2} s^{-1} (Spero and Lea 1993). Maximum symbiont activity occurs at 350–400 µEinstein m^{-2} s^{-1}, and does not significantly increase at higher light levels (e.g., Jørgensen et al. 1985; Spero and Parker 1985; Spero and Lea 1993). Natural illumination at 5–10 m water depth at Barbados during midday in April ranges at 400–500 µEinstein m^{-2} s^{-1} (Caron et al. 1982). In addition to light intensity, the quality of light affects the endosymbiotic activity of planktic foraminifers (Jørgensen et al. 1985), and light sources should be chosen accordingly. Maximum symbiont activity of dinoflagellates in *G. sacculifer* occurs at wavelength of about 450 and 690 nm (Jørgensen et al. 1985).

10.5 Microsensor Analysis

Microsensor analysis of planktic foraminifers was applied to measure photosynthetic rates of symbionts in cultured *G. sacculifer* as early as 1982 by Jørgensen et al. (1985). Oxygen and *p*H were measured with microelectrodes, and manipulated with a micromanipulator at ±5 µm precision. Measurements were carried out under a dissecting microscope, between the spines at the immediate surface of the test of *G. sacculifer*. A similar approach was followed to measure respiration rates of *O. universa* (Fig. 10.6) and *G. sacculifer* (Rink et al. 1998; Lombard et al. 2009a).

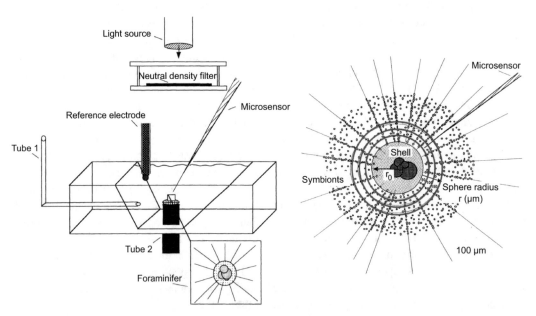

Fig. 10.6 Left panel: Schematic drawing of the measuring chamber (10 mL volume) with a single foraminifer placed on a nylon mesh. Microsensors were positioned with a micromanipulator. The incident light was adjusted by neutral density filters. Right panel: *Red circles* at 50-µm distances indicate microsensor positioning for the photosynthesis measurements (r_o is the radius of a spherical adult *O. universa*). Schematic drawing, after Rink et al. (1998)

10.6   Micro X-Ray Imaging and Computer Tomography (CT)

Micro-CT is a non-destructive method, which provides morphometric information on internal chamber volume, test calcite mass, and dissolution of tests walls (Johnstone et al. 2010, 2014; Görög et al. 2012). Resolution of X-Ray micro-CT ranges between 0.5 and 7 µm depending on the employed scanning system and voltage (Speijer et al. 2008; Johnstone et al. 2010). Experiments in the early 1950s had already shown that microradiography provides sufficient resolution to visualize internal structures of foraminifer tests (cf. Schmidt 1952; Schmidt et al. 2013). Most importantly, X-ray microscopy was employed to visualize the internal test architecture, and the early ontogenetic development of 23 modern planktic foraminifer species (Bé et al. 1969), and several Upper Cretaceous species (Huber 1987). X-Ray diffraction (XRD) was developed to analyze calcite crystallinity of the foraminifer shell as measure of $[CO_3^{2-}]$ (Bassinot et al. 2004).

10.7   Analyses of the Chemical Composition of Tests

10.7.1   Analyses of Stable Isotopes

The ratio of stable carbon and oxygen isotopes, as well as stable isotopes of a suite of other elements of planktic foraminifer shell calcite are major proxies in paleoceanography (e.g., Rohling and Cooke 1999; Fischer and Wefer 1999; Henderson 2002; Katz et al. 2010, and references therein). Pioneering works in the development of mass spectrometry (isotope chemistry) in paleoceanography, and analyses of stable isotopes in foraminifer calcite were started by Epstein et al. (1951, 1953) by developing a paleotemperature equation based on the carbonate of molluscs. Those equations were subsequently refined and applied to foraminifers by Emiliani (1954) when isotope chemistry became an important tool in paleoceanography. Those approaches were then accomplished by N. Shackleton from the 1960s onward (e.g., Shackleton 1968; Shackleton and Opdyke 1973). In addition to paleoceanographic data, stable isotopes add information on the paleo-ecology of planktic foraminifers (e.g., Mulitza et al. 1997; Rohling et al. 2004). For example, from interpretation of the temperature effect on the stable isotope composition of Paleogene planktic foraminifers, Shackleton et al. (1985) could show that depths preferences in the habitat of spinose (globigerinid) and non-spinose (globorotalid) species were opposite from the modern distribution pattern, and Paleogene globorotalids preferred a shallower habitat than globigerinids on average.

Various types of Inductively Coupled Plasma Mass Spectrometers (ICP-MSs) are employed to measure stable isotope ratios from different calcite volumes, and at different reproducibility. Depending on the foraminifer test size and calcite mass, as well as specifications of the employed mass spectrometer, about 3–25 (at least 10 µg $CaCO_3$) specimens are needed for a $\delta^{18}O$ and $\delta^{13}C$ analysis (e.g., Niebler et al. 1999; Rohling et al. 2004). To reduce deviation of results caused by 'vital effects', analysis of stable isotopes should be carried out on mono-specific samples, and from as narrow test size classes of adult specimens (>200 µm) as possible. Standardized analysis of tests of adult individuals reduces the possibility of metabolic effects on the isotope ratio, which can vary significantly between individuals of different ontogenetic stages (Niebler et al. 1999).

Much less volume of calcite is needed in LASER-Ablation Inductively Coupled Plasma–Mass Spectrometry (LA-ICP-MS) and Secondary Ion Mass Spectrometry (SIMS). Between 10 and 100 ng of test calcite are ablated in a helium atmosphere with LASER pulses over some seconds, and measured with an ICP-MS. Test walls ablated by LASER ideally measure some 20–40 µm in diameter, and 0.2–10 µm in depth. Therefore, single chambers of planktic foraminifer tests can be analyzed using LA-ICP-MS (Eggins et al. 2003; Reichart et al. 2003) (Fig. 10.7). Horizontal and vertical (i.e. depth) resolution of (Nano-) SIMS analysis

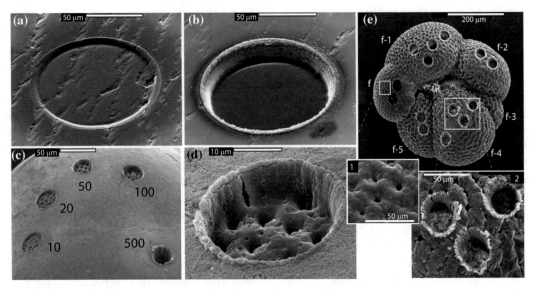

Fig. 10.7 SEM images of LASER ablation pits formed in (**a**, **b**) gem-quality Iceland spar using 10 and 100 LASER pulses, and (**c**) a fossil *P. obliquiloculata* test by 10, 20, 50, 100 and 500 LASER pulses using a LASER fluence of 5 J/cm^2. (**d**) Detail of the 50 pulses pit shown in panel C, which is approximately 7.5 μm deep. (**e**) Test of *N. dutertrei* in which 14 separate composition profiles have been analyzed by LA-ICP-MS and up to four replicates on each chamber. Inset e1 shows detail of the reticulate surface texture present on the final chamber. *Inset* e2 shows detail of 30 μm diameter pits in chamber f-4 and the surrounding blocky calcite textured test surface. Labels f, f-1, f-2, etc. indicate the chamber calcification order, counting back from the final chamber (**f**). Note scale bars. From Eggins et al. (2003)

ranges at 6–10 μm and ∼1 μm, respectively (Kunioka et al. 2006; Vetter et al. 2014).

All chemical analyses of foraminifer tests, in particular the high-resolution analyses of small sample volumes, do critically depend on the preparation of samples, i.e. on the cleaning steps to expose the original calcite to be analyzed (e.g., Boyle and Keigwin 1985; Barker et al. 2003; Eggins et al. 2003; Vetter et al. 2013). A flow-through method for cleaning (dissolving) foraminifer tests was developed by Haley and Klinkhammer (2002). The method employs chromatographic equipment, and is assumed to produce reproducible results. In addition, the method provides information on the contaminant phases.

10.7.2   Analyses of Element Ratios

Element ratios of planktic foraminifer test calcite are measured with mass spectrometers such as, for example, Inductively Coupled Plasma-Optical Emission Spectrometers (ICP-OESs; e.g., Friedrich et al. 2012), or Multi Collector ICP-MS (e.g., Fietzke et al. 2004). In general, cleaning methods are similar to those in stable isotope analyses (see above), but might need to account for a much wider range of contaminants depending on the element ratio to be analyzed. High-resolution MC-ICP-MS is employed for analyses of trace elements with very low concentrations (see, e.g., Paris et al. 2014; Ripperger et al. 2008). Ratios of rare elements may also be analyzed applying Thermal Ionisation Mass Spectrometry (TIMS, or isotope dilution TIMS, ID-TIMS) like cadmium-to-calcium ratios (Rickaby et al. 2000). TIMS is widely applied to obtain U-Th ages (e.g., Bard et al. 1993).

Electron microprobe or ion microprobe analysis using an Electron Probe Micro Analyzer (EPMA, or EMPA) allows high-resolution mapping of element ratios of foraminifer test walls (e.g., Duckworth 1977). EPMA is a non-destructive method, widely applied to measure Mg/Ca ratios (Sadekov et al. 2005;

Toyofuku and Kitazato 2005; Kozdon et al. 2011). For EPMA analysis, specimens are embedded in epoxy resin on glass slides, polished to produce a cross section of the test wall, and coated with carbon. The size of each spot-measurement is ~ 2 µm. Standard deviation (2σ) is 1.2 % for Mg, and 1.6 % for Ca (Toyofuku and Kitazato 2005). EPMA and LA-ICP-MS data from the same samples are comparable by applying a constant calibration factor (Eggins et al. 2004; Sadekov et al. 2005; Fehrenbacher et al. 2015). A similar resolution of ~ 2 µm is achieved with Particle-Induced X-ray Emission (PIXE) in multi-element analysis of planktic foraminifer tests (Gehlen et al. 2004).

Secondary Ion Mass Spectrometry (SIMS, and NanoSIMS) allows measurement of metal-to-calcium (Me/Ca) and stable isotope ratios of planktic foraminifer tests at ~ 1–10 µm resolution, and from small sample volumes <2 µg (Bice et al. 2005; Kunioka et al. 2006; Vetter et al. 2014). Tests need to be cleaned, mounted on slides using ethyl cyanoacrylate instant adhesive and low viscosity epoxy resin, and polished to expose the test wall to be analyzed (Bice et al. 2005). An even surface is produced by repeated application of the adhesive and polishing. Samples need to be cleaned between each step by sonication. Standard deviation of replicate Mg/Ca measurements is <1 %. SIMS are in good agreement with ICP-MS data produced from the same samples (Bice et al. 2005; Vetter et al. 2014). Accuracy of $\delta^{18}O$ data from Ion Microprobe analyses is affected by preparation and geometry, as well as instrumental characteristics, and need to be corrected before being compared to ICP-MS data (Kozdon et al. 2009, 2011).

Cleaning protocols: Cleaning of planktic foraminifer tests for analyses of trace metal ratios is essential to generate accurate and reproducible results (e.g., Boyle and Keigwin 1985). To properly clean the tests from the outside and inside, they are gently broken open between two glass slides (e.g., Barker et al. 2003; Sexton et al. 2006). Ultrasonication may be applied with care (for some seconds) to not disintegrate test fragments. Oxide coatings are to be removed in particular for Cd/Ca analyses (e.g., Boyle and Keigwin 1985; Ripperger and Rehkämper 2007). For analyses of Mg/Ca and Sr/Ca ratios, most importantly silicate contamination needs to be removed, as well as clay, Mn-oxides, and Fe-oxides by reductive treatment (Barker et al. 2003). Organic matter is removed by oxidation (Barker et al. 2003). For analyses of live planktic foraminifers from plankton-tow samples the oxidative step using hydrogen peroxide may be repeatedly applied to entirely remove cytoplasm from within the tests. In turn, the reductive and oxidative steps may be omitted because they may remove significant portions of calcite from shell surfaces (Vetter et al. 2013). The reducing reagent alone may causes partial dissolution of carbonate resulting in up to 15 % reduced Mg/Ca values on average compared to studies without reductive step (see in-depth discussions by Barker et al. 2003; Sexton et al. 2006; Bian and Martin 2010). To remove any re-adsorbed contaminants, a final weak acid 'polish' may be performed (e.g., Friedrich et al. 2012).

Calibration for temperature calculation from Mg/Ca ratio: Species-specific calibrations are applied to calculate ambient seawater temperature from the Mg/Ca ratio planktic foraminifers. Calibrations are available for *G. bulloides* and *G. ruber* from Elderfield and Ganssen (2000) and Anand et al. (2003), respectively. For other species, the multi-species calibration of Anand et al. (2003) may be applied (Friedrich et al. 2012). Those calibrations indicate a temperature sensitivity for Mg/Ca of ~ 10 % for a 1 °C change in temperature for almost all planktic foraminifer species (e.g., Anand et al. 2003; Elderfield and Ganssen 2000; Lea et al. 1999).

10.7.3   Radiocarbon Analyses

Radiocarbon (^{14}C) is measured from planktic foraminifer tests for absolute dating of late Quaternary sediments (e.g., Bard 1988; Voelker et al. 1998; Barker et al. 2007). About 800–1000 tests of medium sized (~ 250 μm in test diameter) planktic foraminifers equal 10 mg of calcite needed for one ^{14}C Accelerator Mass Spectrometry (AMS) measurement (e.g., Voelker et al. 2000). Although surface dwelling planktic foraminifers produce their test calcite in relative vicinity to the atmospheric ^{14}C pool, large deviation of their ^{14}C AMS signal from calendar ages have been detected (e.g., Reimer et al. 2013). Those deviations result from reservoir effects, i.e. the age of ambient water body in which test calcite is precipitated. Consequently, radiocarbon should preferably be analyzed from mono-specific samples, since different planktic foraminifer species may calcify their tests at different water depths, different seasons, and different ecologic conditions, i.e. in waters of different age. Depending on ocean basin and region, the most abundant species may be selected for ^{14}C AMS analysis, still taking its ecology into consideration. For example, *G. bulloides* are most frequent in high-productive waters like upwelling regions, i.e. waters with relatively old ^{14}C AMS ages, and high reservoir ages. In contrast, *G. ruber* is more productive in waters marginal to upwelling cells and more stratified surface waters (e.g., Schiebel et al. 2004), and would hence represent waters of lower reservoir age. In case a sufficient amount of mono-specific tests is not available from a sample, tests from species with similar ecologies could be combined for ^{14}C AMS dating. In addition to species-specific, as well as regional and seasonal differences, reservoir ages change over time (e.g., Bard 1988; Reimer et al. 2013). To account for all of the different effects, which affect the ^{14}C AMS age of planktic foraminifer calcite (Barker et al. 2007; Mekik 2014, and references therein), and which cause deviation from calendar age, raw radiocarbon data need to be calibrated (Reimer et al. 2013, and references therein).

10.8   Biomass Analysis

A non-destructive method for biomass analysis of individual foraminifers was developed and calibrated by Movellan et al. (2012). The method employs nano-spectrophotometry and a standard bicinchoninic method for protein quantification (Smith et al. 1985), assuming that foraminifer protein-biomass equals carbon-biomass (Zubkov et al. 1999; Movellan 2013). Following protein measurement, tests are dried and stored for fuurther analyses.

Foraminifer individuals are isolated immediately after sampling. Each individual is transferred into a bath of micro-filtered seawater, and gently cleaned with a brush to remove particles. Specimens are then immersed in deionized water for less than a second to remove remaining seawater. Each foraminifer is individually stored in an Eppendorf cup and immediately analyzed for biomass, or stored frozen at −80 °C to prevent degradation of organic matter, and facilitate later protein-biomass quantification.

For biomass analysis, 20 μl of micro-filtered tap water is added to each Eppendorf cup including fresh or unfrozen foraminifers for 30 min. Immersion of foraminifers in micro-filtered tap water causes an osmotic shock, and quantitatively exposes the foraminifer cytoplasm to the working reagent (400 μL), which is then added to each Eppendorf cup (Movellan et al. 2012). Efficiency and yield of the osmotic shock method for cytoplasm exposure was tested on specimens of *Globorotalia hirsuta*, *Globorotalia scitula*, and *Globigerinella siphonifera*. The three species were chosen for their differences in test architectures, i.e. globular chambers with wide apertures (*G. siphonifera*), compressed chambers with intermediate-sized apertures (*G. hirsuta*), and compressed chambers with small apertures (*G. scitula*).

Protein-biomass analyses with the bicinchoninic acid (BCA) method employ a mix of copper solution (4 % (w/v) $CuSO_4$ $5H_2O$ solution; Sigma-Aldrich) and BCA (Sigma) solution (Smith et al. 1985; Zubkov and Sleigh 1999; Mojtahid et al. 2011). In contact with proteins the Cu^{2+} ions of the copper solution are reduced to Cu^+. The Cu^+

ions react with the BCA, and a purple color is produced. The intensity of the color increases proportionally with the protein concentration. Protein standard solution consists of bovine serum albumin (BSA) of known concentration. Each sample and standard solution is measured in triplicate (Movellan et al. 2012). Foraminifer samples and protein standard solutions are prepared simultaneously, to make sure that the incubation time and temperature are identical. The reaction and resulting coloration of the sample solution depends on incubation time and temperature. An optimum color spectrum is obtained at an incubation time of 24 h at room temperature (20 ± 2 °C).

After incubation, each sample is centrifuged for 3 s at 5000 rpm, and the absorbance of the 562-nm wavelength is measured with a nano-spectrophotometer on 2 µL of sample or standard solution (NanoDrop 2000®, Thermo Scientific). The absorbance of the working reagent is affected both by color and brightness resulting from the concentration of proteins. Each absorbance value is measured three times, and standard curves are constructed using polynomial regression.

10.9   Determination of Test Calcite Mass

Calcite mass of planktic foraminifer tests is a measure both of production and dissolution of shell, and hence provides information on environmental conditions of ambient seawater of live individuals, and settling tests (e.g., Barker and Elderfield 2002; de Moel et al. 2009; Moy et al. 2009). Among the parameters affecting production and remineralisation of test calcite are, in final consequence, carbonate chemistry ($[CO_3^{2-}]$, and other carbon species) and pH, which are affected by light and symbiont activity (i.e. $[CO_2]$) in the symbiont-bearing foraminifer species. Therefore, different methods were developed to determine planktic foraminifer calcite mass.

The most obvious method appears to be simple weighing of clean, empty, and well-preserved (i.e. unbroken) tests of similar ontogenetic stage. To produce comparable results, size-normalized test weights are determined (Lohmann 1995; Broecker and Clark 2001a, b; Beer et al. 2010a, and references therein). Batches of tests from narrow size intervals (e.g., 200–250 µm) may be produced by sieving. To compensate for any variability in size and mass of tests from the same sieve-size interval, a sufficient number of tests (e.g., 10–50 tests) may be combined for weighing (Broecker and Clark 2001b). Alternatively, tests may be analyzed for their discrete size and weight (Broecker and Clark 2001a, b). Both methods are inexpensive and fast, and produce interpretable results.

A microbalance (e.g., Mettler Toledo XP2U, readability of 0.1 µg) may be employed to weigh individual foraminifer tests, or batches of tests (Moy et al. 2009; Movellan et al. 2012). Weighing should be carried out after a minimum of 12 h of acclimatisation in an air-conditioned weighing-room at constant temperature and humidity. Repeated weighing (three times) of individual foraminifer tests (>100 µm) is advised to enhance precision of data (Schiebel and Movellan 2012).

Unfortunately, fossil tests are often filled with sediment, and impossible to clean without causing damage to the original shell. Therefore, methods independent of test size and weight were developed to determine shell calcite mass. Crystallinity of test calcite as measure of dissolution is analyzed using X-ray diffraction (Bassinot et al. 2004). The method provides quantitative results for past $[CO_3^{2-}]_3$, given that conditions of production and sedimentation are analogous to modern conditions (Bassinot et al. 2004). Measurement of shell-thickness of equivalent cross-sections (i.e. of the same species, and same chamber) with a Scanning Electron Microscope (SEM, see below) provides information on calcite mass (de Moel et al. 2009), but would possibly not be suited for analyses of large sample volumes, since rather time-consuming and costly.

Shell calcite mass determination: A variety of different methods have been developed for the determination of the planktic foraminifer test calcite mass as proxy of shell production and dissolution. (1) Weighing seems to be the most obvious method, but it is limited by the precision of weighing balances within the range of 0.1 µg at the best, and the weight of small tests (<100 µm) below 0.6–1.2 µg even for well preserved modern specimens (Schiebel and Hemleben 2000; Barker and Elderfield 2002; Schiebel et al. 2007). Weighing, hence, would not be suitable to detect differences between individual small tests, which are calcified or dissolved to a different degree. In addition, any kind of contamination within, or on the surface of, the test, and any sediment infill, would not be detected by weighing. The same would possibly be true for any titration method. (2) Analyses of the crystallinity of the planktic foraminifer test calcite, inferred from X-ray diffraction, requires crushing of a large number of tests; i.e., for example about 80 *G. ruber* of the 250–315 µm size fraction (Bassinot et al. 2004). Analysis of crystallinity by particle-induced X-ray emission (PIXE) requires only single tests, but is nonetheless a destructive and laborious method (Gehlen et al. 2004). Therefore, application of the method is limited by the availability of tests, as well as manpower. (3) X-ray computed tomography (CT) provides images from the outside and inside of the tests at a resolution of 7 µm. Taking only ~50 min per specimen for CT scanning, the method is still not suited for analyses of entire assemblages (Johnstone et al. 2010). (4) SEM analyses are suited to visualize encrustation and dissolution of the primary shell calcite at high detail, but this method requires expensive technology, and possibly cannot be quantitatively applied to assemblages, because it is too costly. (5) A combination

of some of the above given methods may be suited to resolve the test-calcite-mass problem to a satisfactory degree.

10.10   Automated Microscopy

Microfossils have played a key role in palaeoceanographic reconstructions, largely as proxies of changing water mass properties traceable by their faunal and stable isotopic compositions and their trace-element chemistry. Although major effects on the population structures and evolutionary developments of associated assemblages are expected, little work has been done so far, largely because of the time-consuming morphometric and taxonomic data collection. This problem has been overcome by automated acquisition and processing of data (Schmidt et al. 2003; Bollmann et al. 2004, and references therein; Schmidt et al. 2004a, b, c; Beer et al. 2010a, b).

Automated particle analysis in palaeoceanography and micropalaeontology is carried out with a fully automated incident light microscope system (Bollmann et al. 2004). Images are acquired and particles are analyzed with analySIS FIVE (SIS/Olympus©) software supported by a custom made software add-in. Samples are prepared on up to six glass trays, and are automatically moved under a Leica© Z16APO monocular microscope with a plan-apochromatic objective using a motorized xy-stage and Lstep-PCI controller manufactured by Märzhäuser© (Germany). Manual positioning of the xy-stage with a joystick for analyses of particular objects is also facilitated via analySIS. Images are captured with a 12-megapixel CC12 colour camera (SIS©). Constant illumination of samples is provided by a Leica© CLS100X light source and a Leica© ring-light (Clayton et al. 2009). Resolution of the system ranges from 1.44 × 1.44 µm to 24.5 × 24.5 µm per pixel. Depending on average particle size, between ~2000 and ~10,000 particles per sample tray can be analyzed

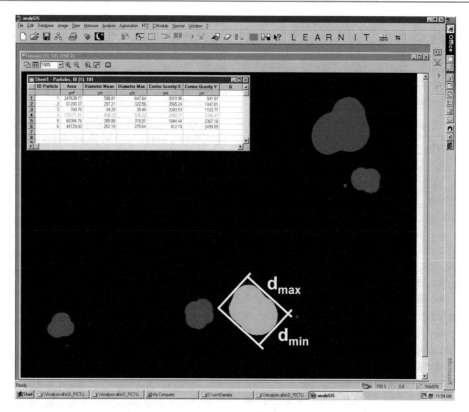

Fig. 10.8 Screen-shot of automated image analysis system (analySIS FIVE, SIS/Olympus©). Planktic foraminifer tests are sorted for size (color scheme) in the example shown here. Minimum test diameter (d_{min}) rectangular to maximum test diameter (d_{max}) acquired automatically. An additional 111 morphometric measures and color parameters can be automatically acquired by the system from image series obtained from up to six strew mounted samples at high efficiency

(Schmidt 2002). Acquisition of images and morphometric analyses of the images takes between 15 min and 1.5 h per sample, respectively. Each particle can be analyzed for up to 65 morphometric parameters, 29 color and gray-scale parameters, and additional 20 parameters to be user-defined (analysis FIVE, SIS/Olympus©) (Fig. 10.8). The data are automatically saved, for example, as an Excel spreadsheet.

Minimum test diameter: The minimum test diameter is applied as a measure of test size, which is easy to acquire and robust. The minimum test diameter is therefore acquired and applied in morphometric analyses of test assemblages. Minimum test diameter is the longest (!) distance measured rectangular to the line of maximum diameter of the test, whereas maximum diameter of the test is the longest distance of the two-dimensional silhouette-area of the entire test (Fig. 10.8). Minimum diameter is a more robust size-measure of the test than maximum diameter, and is therefore used in most morphometric analyses discussing test-size. Minimum diameter does well display test size, being highly correlated to (two-dimensional) silhouette area, i.e., the way tests are viewed from above through a binocular microscope, and which is a good representative of test volume (Beer et al. 2010a). In addition, minimum test diameter is comparable to sieve-size, since particles including foraminifer tests pass through the mesh of a sieve with their smallest diameter.

10.11   Electron Microscopy

Scanning Electron Microscopes (SEMs) and Transmission Electron Microscopes (TEMs) of various makes are used to analyze fine structures of test and cytoplasm, respectively. Classical SEM and Environmental SEM (ESEM) are employed for high-resolution imaging of hard surfaces, i.e. tests. Tests are analyzed in near vacuum conditions, and hence need to be dry. Classical SEM allows high-quality imaging at high resolution of up to about 1 nm. Objects need to be coated with graphite, gold, platinum, or other conductive materials, though, and may not further be used for chemical analyses. In turn, coating of objects in ESEM is not necessary, and objects stay unchanged during scanning. ESEM is a non-destructive imaging method, which may be employed if objects are to be further used, for example, for stable isotope or element analyses. ESEM can even be employed on wet objects, because vacuum conditions are not applied. In turn, resolution of high-quality images in ESEM is much lower than in classical SEM, and limited to objects >1 μm.

Transmission Electron Microscopy (TEM) is applied for visualization of cytoplasmic fine structures, at a resolution of several nanometers. The valid visualization of delicate and labile cytoplasmic components requires fixation of the live matter in as natural a state as is possible. Following fixation of the cytoplasm (see above), the shell is removed for subsequent sectioning. Dehydrated specimens are then embedded in a plastic polymer appropriate in hardness and quality required by the kind of Diatome Diamond Knife used for sectioning, and the degree of stability needed during examination with the TEM.

10.12   Modeling

Numerical Modeling of planktic foraminifers follows different avenues to better understand physiology and population dynamics, and finally the biology and ecological needs of modern species and assemblages, and the effect of planktic foraminifers on the marine carbon turn-over (biogeochemical modeling). Another approach including sensitivity analyses (Žarić et al. 2005), and modeling ('prediction') of the species richness and diversity, relative abundance of species, and test flux, uses empirical input data from sediment traps and surface sediments (Žarić et al. 2006). Modeling of the global distribution and seasonal bias of surface dwelling species in fossil assemblages using a dynamic ecosystems approach is targeted at a better understanding of planktic foraminifers in paleoceanographic records (Fraile et al. 2008, 2009a, b).

Ecophysiological modeling has been empirically based, utilizing input data from laboratory observations and natural distributions of live individuals, and aims at a more complete qualitative and quantitative use of planktic foraminifer as proxy in paleoecology (Lombard et al. 2009b, 2011; Roy et al. 2015).

Modeling in planktic foraminifer research had started much earlier, though. A 'computer method' to calculate planktic foraminifer test architecture and shell growth from simple spheres was designed in the late 1980s (Ott et al. 1992; Signes et al. 1993; Łabaj et al. 2003; Tyszka and Topa 2005). The model includes assumptions on allometric shell growth, protoplasmic growth, and ontogeny of planktic foraminifers, and was designed with a biogeochemical perspective, i.e. to explain the carbon budget of planktic foraminifer shell calcite and biomass (Signes et al. 1993). A following empirical model of planktic foraminifer carbonate flux in the central Red Sea includes biological and ecological information, such as reproduction rate and length of the reproductive period at the species level (*G. sacculifer*). Final goal of the approach was to enumerate calcite flux pulses, and to quantify annual calcite budgets (Bijma et al. 1994).

10.13   Census Data for Assemblage Analysis

Assemblages of live individuals or empty tests are analyzed for population dynamics by counting a certain number of individuals. Those analyses are preferably conveyed at the species level, or at a higher systematic level (i.e. morpho-types) if possible. In case of standard counts carried out with an incident light microscope 80× to 120× magnification, planktic foraminifer assemblages are analyzed for morpho-types or morpho-species. The number of individuals to be counted depends on the number of morpho-species in a sample, their relative abundance, and the level of statistical significance and confidence to be achieved. It is generally suggested to count at least 300 specimens per (whole) sample, i.e. all test-size fractions of an entire sample or a representative split of a sample (Patterson and Fishbein 1989).

For example, in case 300 specimens are counted from a sample, and the relative abundance of any species is found to be 15 %, the corresponding value of 2σ is 4 %. The relative abundance of the species hence ranges at 15 ± 4 %, i.e. between 11 and 19 %, at a 95 % confidence (van der Plas and Tobi 1965; Patterson and Fishbein 1989). The relative significance of data increases with increasing relative abundance of a species (>15 %), and decreases with decreasing relative abundance (<15 %). Statistically interpretable data are limited to about 4 % when counting 300 specimens, and to about 2 % when counting 500 specimens (Fig. 10.9). For reasonable interpretation of the distribution of rare species, large numbers of specimens need to be classified and counted, a task, which is rather time consuming. Automated methods in microscopy and image analysis have been developed to speed up and facilitate otherwise time consuming analyses (see chapter on Automated Microscopy above).

Species diversity is one of the basic measures of assemblages, which can be deduced from count data. The simplest measure of diversity is 'species richness', i.e. the number of species in a sample. 'Species richness' does neither account

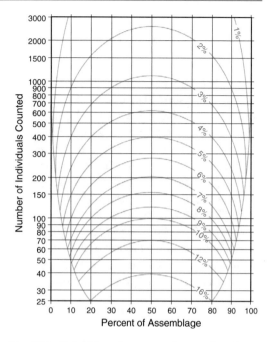

Fig. 10.9 Reliability of test counting results. Curves with percentages give 2σ values. At 300 tests counted, lowest interpretable numbers range at 4 %, i.e. 4 ± 2 % at 50 % rel. σ). Fields in the lower right and left corner are not valid. Redrawn after van der Plas and Tobi (1965)

for the size of sample, i.e. number of specimens counted in total, nor for the relative frequency of species within the sample. A more complete description of species diversity is provided by indices such as the Fisher α index, and the Shannon-Wiener index.

The Fisher α index is used to assess species diversity in a sample, and to estimate species diversity of large samples from numbers obtained from smaller sub-samples (Fisher et al. 1943; Murray 2006). The same is done in ecology by the 'rarefaction' method to assess species richness ('rarefaction curves'). Rarefaction curves are produced by continuously plotting the numbers of specimens of each classified species while counting.

The Shannon-Wiener index (H') is easy to calculate, and in combination with the 'evenness' (E) provides a rather complete description of diversity (Shannon 1948; Shannon and Weaver 1963; Hayek and Buzas 1997).

$$H' = -\sum_{i=1}^{n} pi \ln pi \quad (i = 1 \text{ to } n) \qquad (10.1)$$

with p_i being the proportion (numbers ≤ 1, i.e. per cent divided by 100) of the ith species in a sample, and ln being the natural logarithm. H' hence combines information on the number of species present in a sample, and the relative abundance of species. Similar H' may be produced by different combinations of species distributions. For a complete and unequivocal description of species diversity in a sample, 'evenness' ($0 \geq E_H \leq 1$) or 'equitability' provides a measure of the balance of the distribution of species in a given sample, with S being the total number of species present in a sample.

$$E_H = H'/H_{\max} = H'/\ln S \qquad (10.2)$$

10.13.1 Statistical Analysis of Assemblage Data

Simple linear regression of least squares is the basic statistical method applied for the comparison of data. Resulting correlation coefficient, standard error, standard deviation, and probability (p-values) are used for a statistical description of the distribution of data. Student's t-test and F-test are applied for the comparison of two populations of data. In case more than two populations data are to be compared, an Analysis of Variance (ANOVA) may be carried out.

Multivariate analyses may be employed to group information from data, and to visualize information of large and complex data sets. Cluster analyses produce dendrograms, in which data are grouped in clusters. Differences between clusters (i.e. groups) are expressed as distances. Multidimensional correlation of data is called factor analysis. Factor analysis produces groups of data called factors. Factors and clusters may be produced by different methods, and by the use of different algorithms. Software packages allow easy application of multivariate methods, and production of multivariate data. In turn,

interpretation of the resulting data might be more difficult than the production of results, and it is strongly advised to seek the help of an expert for reasonable interpretation of data.

10.13.2 Analyses of Test Size Data

Ontogenetic development of planktic foraminifer tests occurs at intervals by adding new chambers to the test. Test size of individuals, and size distribution of assemblages may hence be analyzed either from sieve-size classes or discrete size data (Peeters et al. 1999; Schiebel and Hemleben 2000; Schmidt 2002; Beer et al. 2010b). To account for smaller test-size increments when adding smaller chambers at earlier ontogenetic stages, and larger increments later in ontogeny, sieve-size intervals should increase with foraminifer test size (see above). Discrete size measurement such as, for example, from image analyses, provides more detailed data than sieve-size analyses. However, sieve-size effects may be averaged out when large numbers of specimens are analyzed. In addition, any methodological affects caused by sieving, and physical damage of tests, are largely avoided in image analyses.

Size-distributions of planktic foraminifer test assemblages are inherently incomplete to some degree for test-sizes close to the sampling mesh-size (e.g., ≥ 100 μm). Small specimens near the sampling mesh-size may be missed, and very small specimens just below the sampling size may be included. The latter are easily identified during later analytical steps, and may be excluded from further analyses. Missing of the former may be detected by cohort analysis: The number of individuals should increase with decreasing size (Fig. 10.10), or decrease to a reproducible degree (Peeters et al. 1999). If this is not the case, and the smallest sampled size-class contains fewer individuals than the second smallest size class, a methodological (sampling) error may be the reason (Schmidt 2002). An introduction to the theoretical background of natural, i.e. biological and ecological effects on body size is given by Schmidt et al. (2006).

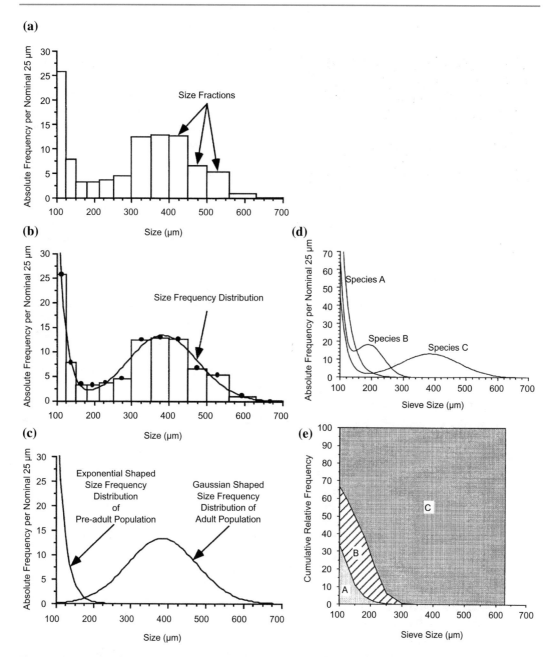

Fig. 10.10 Interpretation of size frequency distributions of planktic foraminifer species and assemblages from plankton net samples, sediment traps, and surface sediments. **a** Frequency normalization per size fraction may be applied if widths of size fractions are not equidistant. **b** Curve fitting to obtain a size frequency distribution in large assemblages. The size of any foraminifer species within a size fraction may finally be represented by a single value, i.e. the mean of all size fractions. **c** Exponential and normal distribution may explain 'hidden' cohorts in count data. The sum of both cohorts yields the size frequency distribution of the whole assemblage. **d** Cohorts of three species (A, B, C) with different size-frequency distributions, caused by (**e**) differences in relative abundance and test size within a hypothetical sample. After Peeters et al. (1999)

When analyzing test-size data at the species level, it should be accounted for mortality rates and reproduction rates of pre-adult and adult cohorts, respectively (Schiebel et al. 1995; Peeters et al. 1999). Assemblage size-analyses may include size-effects caused by both cohorts of the same species, and size differences between species (Fig. 10.10). Bimodel or polymodal size-distribution of the same species within a sample may also indicate mixing of populations, and expatriation/immigration of individuals by currents. The same is true for size-sorted assemblages, which may be cut off at either side, and hence lack either small (juvenile) or large (adult) individuals.

10.13.3 Transfer Functions

Transfer functions are a suite of statistical methods used in paleoceanography to reconstruct past environmental conditions from the distribution of microfossils (e.g., Imbrie and Kipp 1971; Hutson 1977; Sachs et al. 1977; Vincent and Berger 1981; Fischer and Wefer 1999; Guiot and de Vernal 2007, and references therein). Planktic foraminifers are employed in transfer calculations due to their wide distribution, and relatively well-known paleo-biogeography in relation to modern distribution patterns and environmental parameters. Transfer functions have classically been used to reconstruct Sea Surface Temperature (SST). Today, transfer calculations are also employed to reconstruct any other (paleo-) environmental parameter, which is sufficiently resolved in the paleo-record (i.e. down-core) and modern assemblages (i.e. regional coverage of surface sediment samples), provided sufficient sensitivity at the species to assemblage level.

The most simple equation to calculate average temperature (T_{est}) from planktic foraminifer assemblage data, i.e. the ratios of species (p_i) and their optimum temperature conditions (t_i) is given by Berger (1969) as

$$T_{est} = \sum p_i t_i / \sum p_i (i = 1 \text{ to } n)$$

Imbrie and Kipp (1971) developed a conceptual ecological model of species abundance in relation to environmental parameters. The transfer function of Imbrie and Kipp (1971) includes coefficients, which account for various environmental and biological effects other than temperature, which affect the differential distribution of planktic foraminifer species:

$$P_{est} = k_0 + k_1 A + k_2 B + \cdots + k_n X \quad (10.4)$$

with P being the environmental parameter to be reconstructed, k being the empirically derived regression coefficients, and A to X being the ratios of species (of statistically significant abundance!) from census counts. The Imbrie and Kipp (1971) model applies the results of factor analysis (multivariate statistical regression) of planktic forminifer census counts, and synthetic variables characteristic of five assemblage groups (i.e. tropical, subtropical, polar, sub-polar, and 'gyre margin'), to obtain more refined temperature reconstruction (see, e.g., CLIMAP 1976). Apart from the quality of down-core census counts, the quality of transfer calculations crucially depends on the accuracy of the modern dataset compiled from surface sediment samples (e.g., Hilbrecht 1996; Pflaumann et al. 1996; Kucera et al. 2005).

New transfer methods have been developed from the classical method of Imbrie and Kipp (1971), including the Modern Analogue Techniques (MAT, Hutson 1980; and SIMMAX, Pflaumann et al. 1996), Artificial Neural Networks (ANN, e.g., Malmgren and Nordlund 1996), and the Revised Analogue Method (RAM, Waelbroeck et al. 1998) (for a review see Guiot and de Vernal 2007). All of those methods are based on modern analogue data from surface sediments. Unfortunately, the geographical and temporal coverage of data on live planktic foraminifers is too incomplete to be applied as modern analogue in transfer calculations. In addition, assemblage data from surface sediments do better represent down-core assemblages both of which having experienced alteration during sedimentation. In turn, ecological data directly derived from live planktic foraminifers (e.g.,

Lombard et al. 2011) would possibly improve the accuracy of transfer calculations.

Transfer calculations using planktic foraminifers are largely limited to the Quaternary. Modern analogues could possibly not be applied to time-intervals much older than Quaternary, since ecological demands of species, and the composition of species assemblages have evolved over geological periods of time (cf. De Vargas and Pawlowski 1998). In addition, transfer calculations are limited to the regional scale, or the scale of oceans basins at maximum, depending on the coverage of the surface (analogue) data (cf. Pflaumann et al. 1996). The regional distribution of planktic foraminifer species, i.e. morphotypes, and more importantly genotypes (e.g., Darling and Wade 2008, see Chaps. 2 and 7) with varying ecological demands, further limits the regional applicability of transfer calculations. Transfer functions are hence inherently based on simplification, since it is impossible to account for the entire complexity of abiotic and biotic parameters. However, transfer calculations may still produce non-analogue situations at the regional scale, resulting from the degree of (falsely) assumed analogy and model calibration, and so far unidentified changes of environmental and biological prerequisites over time (e.g., Guiot and de Vernal 2007).

Transfer calculations on planktic foraminifers have been applied to the Quaternary North and South Atlantic Ocean, and Indian Ocean with great success, facilitated by the good preservation of planktic foraminifer tests (e.g., Vincent and Berger 1981; Dittert et al. 1999, and references therein). In general, transfer calculations based on planktic foraminifers have been among the most valuable tools in paleoceanography over the past 40 years, and have greatly advanced our understanding of the changing oceans and climates during the Quaternary. In addition to temperature reconstruction, other parameters like primary productivity have been reconstructed with transfer functions (Ivanova et al. 2003). Like any other tool in paleoceanography, transfer functions are ideally applied in a multi-proxy approach, i.e. in combination with data on, for example, stable and radioactive isotopes, and

element ratios (e.g., Fischer and Wefer 1999, and references therein; see Chap. 9).

10.14 Applications

Planktic foraminifers are widely used proxies in many fields of academic and commercial applications such as, for example, paleoceanography, biostratigraphy, and hydrocarbon exploration. Planktic foraminifer tests are ubiquitously used in paleoceanograhy, and have been reported 'intelligent design for paleocenography' (Jonathan Erez, Hebrew University of Jerusalem, oral communication), and 'paleo-argo floats' (Andy Ridgwell, UC Riverside, oral communication). The application of planktic foraminifers goes beyond the use in biostratigraphy and paleoceanography, facilitated by technological and new methodological approaches. New approaches employ planktic foraminfers for the monitoring of ecological impacts of wastewater disposal by, for example, hydrocarbon industries. In addition, test production of planktic foraminifers may provide a measure of ocean acidification, and anthropogenic impact other than CO_2 emissions. Since planktic foraminifer production is affected by, and does affect, the global carbon cycle, planktic foraminifers may indicate and mitigate environmental change on various temporal and spatial scales.

Considering biological, biogeochemical, ecological, and sedimentological processes, planktic foraminifers provide powerful tools to reconstruct ancient marine systems and climatic conditions (e.g., Vincent and Berger 1981; Shackleton 1987; Sarnthein et al. 2003; Kucera et al. 2005; Kucera 2007). In addition to the obvious use of planktic foraminifer in paleoceanographic and paleoecological analyses, planktic foraminifers provide useful proxy data in all kinds of studies of the marine carbonate system, over centennial to orbital (Milankovitch) time-scales (e.g., Rohling et al. 2012). Contributing a significant amount to the marine planktic biomass at the lower heterotrophic level (Buitenhuis et al. 2013), planktic foraminifers are actively contributing to the marine carbon turnover, and are not only 'passive' recorders of the hydrology of ambient seawater. Considering this, planktic foraminifers may be taken into account as active mediators of the past CO_2 budget, marine carbon turnover, and for their specie-specific effects on the regional biogeochemistry and ecology. More specifically, planktic foraminifers counteract the CO_2 drawdown of the non-calcareous plankton in (iron) fertilized Southern Ocean waters (Salter et al. 2014), and register decreasing pH of ambient seawater (Ocean Acidification, OA) at the same time (de Moel et al. 2009; Moy et al. 2009).

10.14.1   First Example: Ocean Acidification (OA)

Ocean Acidification (OA) caused by increasing atmospheric and surface water CO_2 concentration potentially affects production and dissolution of planktic foraminifer tests. Calcification of modern planktic foraminifer tests has reduced by $\sim 30~\%$ compared tests from below the surface mixed sediment layer in the Arabian Sea (de Moel et al. 2009, *G. ruber*) and pre-industrial sediments south of Tasmania (Moy et al. 2009, *G. bulloides*), the latter of which were sampled from Southern Ocean waters being major sink of modern atmospheric CO_2 (Khatiwala et al. 2009). A similar negative feedback of planktic foraminifer test weight has been shown for glacial-interglacial CO_2 changes using *G. bulloides* from the temperate eastern North Atlantic, but which was affected by an additional change in calcification temperature (Barker and Elderfield 2002). However, all three studies (Barker and Elderfield 2002; de Moel et al. 2009; Moy et al. 2009) were carried out at sites of different surface marine pCO_2 and atmospheric CO_2 uptake of modern surface ocean waters (Khatiwala et al. 2009), and hence being source of CO_2 (Arabian Sea) or sink of atmospheric CO_2 (North Atlantic and Southern Ocean) on an annual average (Takahashi et al. 2002).

The size-normalized test weight of the symbiont bearing *G. ruber* from Arabian Sea waters shows only very slight positive relation to CO_3^{2-} concentration between 170 and 280 $\mu mol~kg^{-1}$ (Beer et al. 2010a). In contrast, calcification of symbiont-barren *G. bulloides* from the same water is strongly related to $[CO_3^{2-}]$ and $[CO_2]$ to the opposite direction as *G. ruber* (Beer et al. 2010a). The same CO_2-related loss in test weight and calcite production of $\sim 30~\%$ of the two species *G. ruber* and *G. bulloides*, although reported from different water masses, is hence not easy to explain. An alternative and much easier explanation of decreasing test calcite mass from the pre-industrial to modern ocean would be the dissolution of tests during sedimentation (e.g., Berger and Piper 1972; Lohmann 1995; Broecker and Clark 2001a). Dissolution of tests at decreasing $[CO_3^{2-}]$ and Ω, and increasing pH in the subsurface water column (Schiebel et al. 2007) and in surface sediments would result in weight-loss and shell-thinning of all species only depending on their dissolution susceptibility (see Dittert et al. 1999, and references therein). In addition, dissolution at deeper water bodies would be much less regional and much less affected by seasonal changes and hence more balanced than changes in calcite production in surface waters.

The effect of OA and decreasing pH on the calcite production of planktic foraminifers, and between different symbiont-barren and bearing species is not yet well understood. In case planktic foraminifers would be able to adjust to increasing CO_2 in the same way as coccolithophores by selecting for those species (clones), which are capable to sustain (or enhance) calcification (Lohbeck et al. 2012), OA might not affect planktic foraminifer calcite production at the global scale. Future planktic foraminifer calcite production might hence be even more dominated by symbiont bearing species capable of compensating for CO_2 increase (Köhler-Rink and Kühl 2005), and shift towards subtropical and tropical waters of high year-round radiation sustaining symbiont activity.

10.14.2 Second Example: Sapropel Formation

Formation of Mediterranean sapropels during anoxic events has been reconstructed in detail from planktic foraminifer population dynamics, and stable isotope analysis of major planktic foraminifer species in combination with other structural (e.g., alkenones, TEX$_{86}$) and chemical (e.g., Ti/Al ratio) proxies of temperature and terrestrial input (e.g., Weldeab et al. 2002; Rohling et al. 2004; Hayes et al. 2005; Castañeda et al. 2010; Hennekam et al. 2014; Mojtahid et al. 2015). Diachronous shifts of stable isotope values across the Eemian Sapropel S5, and the presence/absence of different planktic foraminifer species (Fig. 10.11) are assessed to reconstruct changes in seasonality (*Globigerinoides ruber* white and *Globigerinoides sacculifer* relative to *Neogloboquadrina incompta*), stratification of surface to subsurface water masses (*G. ruber* white relative to *G. scitula*), surface water salinity and riverine runoff (δ^{18}O of *G. ruber* white relative to *O. universa*), and trophic state of water masses (δ^{13}C of *O. universa* and *G. sacculifer*).

The multi-species planktic foraminifer study of Rohling et al. (2004) confirms significantly increased freshwater input, enhanced biological productivity, shoaling of the pycnocline, and stagnation of subsurface circulation during sapropel formation, relative to non-sapropel

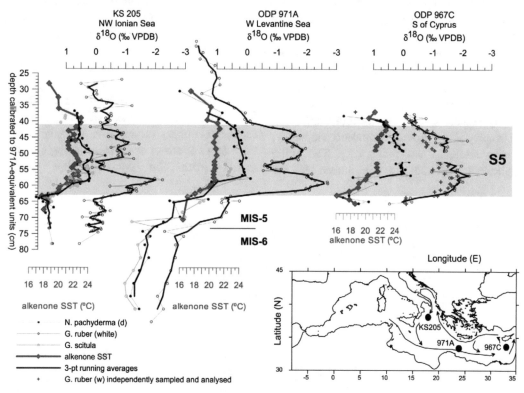

Fig. 10.11 Stable oxygen isotope and alkenone SST records of Sapropel S5 from three sites in the eastern Mediterranean. δ^{18}O of the three planktic foraminifer species *N. pachyderma*, *G. ruber*, and *G. scitula* show differential reactions to changing environmental conditions, resulting from different dwelling depth and autecological prerequisites. The subsurface-dwelling *G. scitula* (*green symbols*) disappears during S5, possibly cause by increasing oxygen deficiency in the subsurface water column. δ^{18}O values of surface-dwelling *G. ruber* (symbiont-bearing) and *N. incompta* (i.e. *N. pachyderma* d, symbiont-barren) indicate different synecological and autecological reactions, which may display differences in seasonality, ambient water temperature, salinity, and trophic conditions. The scale of SST records is adjusted to 1 °C corresponding to 0.23 ‰ on the δ^{18}O scales. From Rohling et al. (2004), and references therein

conditions in the eastern Mediterranean (cf. Rossignol-Strick et al. 1982). A similar scenario is assumed from planktic foraminifer assemblage counts and morphometric data during formation of the Holocene Sapropel S1 (Mojtahid et al. 2015). Significantly increased test sizes of both types of *G. ruber* white sensu stricto and sensu lato (see Chap. 2 Classification) during sapropel conditions indicate increased Nile River freshwater runoff, in combination with Ti/Al ratios (Hennekam et al. 2014). It is assumed that freshening of surface waters off the Nile River delta caused impaired ecological conditions, and delayed reproduction of planktic foraminifers, which led to prolonged maturity and growth of large individuals (Fig. 10.12). Finally, planktic foraminifer based proxies are applied in combination with additional chemical and structural proxies such as Sr and Nd isotope ratios, $U_{37}^{k'}$ and TEX_{86} records to achieve maximum information, and facilitate comprehensive syntheses of the paleo-environment and paleoclimate.

The two examples on Ocean Acidification and Sapropel Formation presented above in brief merely indicate to which extent foraminifers can be employed as proxies in paleoceanography, climate research, and stratigraphy. The entire application spectrum is not limited to the chemical elements and isotopes discussed above, but includes a wide range of chemical elements and isotopes (e.g., Henderson 2002), and beyond the limits of current knowledge and feasibility. Options multiply when applying the range of methods (chemical and physical) and proxies to the different foraminifer species including morphotypes and genotypes. Moreover, certain proxies are applicable as multi-purpose tools. For example, stable oxygen isotopes yield information on the environment (e.g., temperature, salinity, and ice volume) and stratigraphy at the same time (Fig. 10.11). When adding data on population dynamics (e.g., species' abundance) and the morphometry of individual tests and entire assemblages (e.g., calcite mass, test size, and porosity; Fig. 10.12), information again multiplies. The ultrastructure and composition of the organic tissues (e.g., N isotopes; Ren et al. 2009, 2012a) of foraminifers has not yet been analyzed to its full extent, and will add another new scope to the understanding of foraminifers and their applicability. Modern technology such as LA-ICP-MS and NanoSIMS provides detailed fine-scale data, for example, on diurnal changes in calcification under varying environmental conditions (e.g., Spero et al. 2015). Finally, complementary data from non-foraminifer proxies

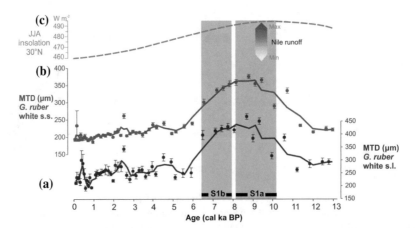

Fig. 10.12 Changes in minimum test diameter (MTD) of a *G. ruber* white sensu lato (s.l.) and *G. ruber* white sensu stricto (s.s.), in relation summer (June, July, August, JJA) insolation at 30°N. Summer insolation at 30°N affects Indian monsoons, precipitation at the sources of the White Nile and Blue Nile, and runoff of Nile waters into the eastern Mediterranean Sea. S1a and S1b indicate time-periods of early and late Sapropel S1 formation, respectively. After Mojtahid et al. (2015)

comprise important information (e.g., Fischer and Wefer 1999), and synergetic effects foster a better systematic understanding and quantification of processes and budgets of the changing ocean. Along with the rapid technological development, new questions and challenges will arise, and remedy may be provided. Ultimate goal of the community effort in (paleo-) environment and climate research are implementation in programs for a more sustainable management of the ocean and climate, and to preserve a habitable planet.

References

Abrantes F, Meggers H, Nave S, Bollmann J, Palma S, Sprengel C, Henderiks J, Spies A, Salgueiro E, Moita T, Neuer S (2002) Fluxes of micro-organisms along a productivity gradient in the Canary Islands region (29°N): Implications for paleoreconstructions. Deep-Sea Res II 49:3599–3629. doi:10.1016/S0967-0645(02)00100-5

Allen KA, Hönisch B, Eggins SM, Rosenthal Y (2012) Environmental controls on B/Ca in calcite tests of the tropical planktic foraminifer species *Globigerinoides ruber* and *Globigerinoides sacculifer*. Earth Planet Sci Lett 351–352:270–280. doi:10.1016/j.epsl.2012.07.004

Anand P, Elderfield H, Conte MH (2003) Calibration of Mg/Ca thermometry in planktonic Foraminifera from a sediment trap time series. Paleoceanography 18:1050. doi:10.1029/2002PA000846

Anderson OR, Bé AWH (1978) Recent advances in foraminiferal fine structure research. In: Hedley RH, Adams CG (eds) Foraminifera 1. Academic Press, London

Anderson OR, Spindler M, Bé AWH, Hemleben C (1979) Trophic activity of planktonic Foraminifera. J Mar Biol Assoc U K 59:791–799. doi:10.1017/S002531540004577X

André, Quillévéré F, Morard R, Ujiié Y, Escarguel G, de Vargas C, de Garidel-Thoron T, Douady CJ (2014) SSU rDNA divergence in planktonic Foraminifera: Molecular taxonomy and biogeographic implications. PLoS ONE. doi:10.1371/journal.pone.0104641

Aurahs R (2010) Genetic diversity and cryptic speciation in Planktonic Foraminiferal Morphotaxa. PhD Thesis, Universität Tübingen

Aurahs R, Göker M, Grimm GW, Hemleben V, Hemleben C, Schiebel R, Kučera M (2009a) Using the multiple analysis approach to reconstruct phylogenetic relationships among planktonic Foraminifera from highly divergent and length-polymorphic SSU rDNA sequences. Bioinforma Biol Insights 3:155–177

Aurahs R, Grimm GW, Hemleben V, Hemleben C, Kucera M (2009b) Geographical distribution of cryptic genetic types in the planktonic foraminifer *Globigerinoides ruber*. Mol Ecol 18:1692–1706. doi:10.1111/j.1365-294X.2009.04136.x

Bárcena MA, Flores JA, Sierro FJ, Pérez-Folgado M, Fabres J, Calafat A, Canals M (2004) Planktonic response to main oceanographic changes in the Alboran Sea (Western Mediterranean) as documented in sediment traps and surface sediments. Mar Micropaleontol 53:423–445

Bard E (1988) Correction of accelerator mass spectrometry ^{14}C ages measured in planktonic Foraminifera: paleoceanographic implications. Paleoceanography 3:635–645

Bard E, Arnold M, Fairbanks RG, Hamelin B (1993) ^{230}Th, ^{234}U and ^{14}C ages obtained by mass spectrometry on corals. Radiocarbon 35:191–199

Barker S, Elderfield H (2002) Foraminiferal calcification response to glacial-interglacial changes in atmospheric CO_2. Science 297:833–836. doi:10.1126/science.1072815

Barker S, Greaves M, Elderfield H (2003) A study of cleaning procedures used for foraminiferal Mg/Ca paleothermometry. Geochem Geophys Geosystems 4 (9). doi:10.1029/2003GC000559

Barker S, Broecker W, Clark E, Hajdas I (2007) Radiocarbon age offsets of Foraminifera resulting from differential dissolution and fragmentation within the sedimentary bioturbated zone. Paleoceanography. doi:10.1029/2006PA001354

Bassinot FC, Mélières F, Gehlen M, Levi C, Labeyrie L (2004) Crystallinity of Foraminifera shells: a proxy to reconstruct past bottom water CO_3^{2-}-changes? Geochem Geophys Geosystems 5:Q08D10. doi:10.1029/2003GC000668

Bé AWH (1960) Ecology of recent planktonic Foraminifera: part 2: bathymetric and seasonal distributions in the Sargasso Sea off Bermuda. Micropaleontology 6:373–392. doi:10.2307/1484218

Bé AWH (1962) Quantitative multiple opening-and-closing plankton samplers. Deep-Sea Res 9:144–151. doi:10.1016/0011-7471(62)90007-4

Bé AWH, Hutson WH (1977) Ecology of planktonic Foraminifera and biogeographic patterns of life and fossil assemblages in the Indian Ocean. Micropaleontology 23:369–414. doi:10.2307/1485406

Bé AWH, Ewing M, Linton LW (1959) A quantitative multiple opening-and-closing plankton sampler for vertical towing. J Cons 25:36–46. doi:10.1093/icesjms/25.1.36

Bé AWH, Jongebloed WL, McIntyre A (1969) X-ray microscopy of recent planktonic Foraminifera. J Paleontol 43:1384–1396

Bé AWH, Hemleben C, Anderson OR, Spindler M, Hacunda J, Tuntivate-Choy S (1977) Laboratory and field observations of living planktonic Foraminifera. Micropaleontology 23:155–179

Beer CJ, Schiebel R, Wilson PA (2010a) Testing planktic foraminiferal shell weight as a surface water $[CO_3^{2-}]$

proxy using plankton net samples. Geology 38:103–106. doi:10.1130/G30150.1

Beer CJ, Schiebel R, Wilson PA (2010b) Technical Note: On methodologies for determining the size-normalised weight of planktic Foraminifera. Biogeosciences 7:2193–2198. doi:10.5194/bg-7-2193-2010

Bemis BE, Spero HJ, Bijma J, Lea DW (1998) Reevaluation of the oxygen isotopic composition of planktonic Foraminifera: experimental results and revised paleotemperature equations. Paleoceanography 13:150–160

Berger WH (1969) Ecologic patterns of living planktonic Foraminifera. Deep-Sea Res 16:1–24. doi:10.1016/0011-7471(69)90047-3

Berger WH, Piper DJW (1972) Planktonic Foraminifera: differential settling, dissolution, and redeposition. Limnol Oceanogr 17:275–287. doi:10.4319/lo.1972.17.2.0275

Bian N, Martin PA (2010) Investigating the fidelity of Mg/Ca and other elemental data from reductively cleaned planktonic Foraminifera. Paleoceanography. doi:10.1029/2009PA001796

Bice KL, Layne GD, Dahl K (2005) Application of secondary ion mass spectrometry to the determination of Mg/Ca in rare, delicate, or altered planktonic Foraminifera: examples from the Holocene, Paleogene, and Cretaceous. Geochem Geophys Geosystems. doi:10.1029/2005GC000974

Bijma J, Hemleben C, Wellnitz K (1994) Lunar-influenced carbonate flux of the planktic foraminifer Globigerinoides sacculifer (Brady) from the central Red Sea. Deep-Sea Res I 41:511–530. doi:10.1016/0967-0637(94)90093-0

Bijma J, Hemleben C, Huber BT, Erlenkeuser H, Kroon D (1998) Experimental determination of the ontogenetic stable isotope variability in two morphotypes of Globigerinella siphonifera (d'Orbigny). Mar Micropaleontol 35:141–160

Bollmann J, Quinn PS, Vela M, Brabec B, Brechner S, Cortés MY, Hilbrecht H, Schmidt DN, Schiebel R, Thierstein HR (2004) Automated particle analysis: calcareous microfossils. In: Francus P (ed) Image analysis, sediments and paleoenvironments. Kluwer Academic Publishers, Dordrecht, pp 229–252

Boyle EA, Keigwin LD (1985) Comparison of Atlantic and Pacific paleochemical records for the last 215,000 years: changes in deep ocean circulation and chemical inventories. Earth Planet Sci Lett 76:135–150

Broecker WS, Clark E (2001a) An evaluation of Lohmann's Foraminifera weight dissolution index. Paleoceanography 16:531–534. doi:10.1029/2000PA000600

Broecker WS, Clark E (2001b) Reevaluation of the CaCO$_3$ size index paleocarbonate ion proxy. Paleoceanography 16:669–671

Buesseler KO, Lamborg CH, Boyd PW, Lam PJ, Trull TW, Bidigare RR, Bishop JKB, Casciotti KL, Dehairs F, Elskens M, Honda M, Karl DM, Siegel DA, Silver MW, Steinberg DK, Valdes J,

Mooy BV, Wilson S (2007) Revisiting carbon flux through the ocean's twilight zone. Science 316:567–570. doi:10.1126/science.1137959

Buitenhuis ET, Vogt M, Moriarty R, Bednaršek N, Doney SC, Leblanc K, Le Quéré C, Luo YW, O'Brien C, O'Brien T, Peloquin J, Schiebel R, Swan C (2013) MAREDAT: towards a world atlas of MARine Ecosystem DATa. Earth Syst Sci Data 5:227–239. doi:10.5194/essd-5-227-2013

Caron DA, Bé AWH, Anderson OR (1982) Effects of variations in light intensity on life processes of the planktonic foraminifer Globigerinoides sacculifer in laboratory culture. J Mar Biol Assoc UK 62:435–451

Castañeda IS, Schefuß E, Pätzold J, Sinninghe Damsté JS, Weldeab S, Schouten S (2010) Millennial-scale sea surface temperature changes in the eastern Mediterranean (Nile River Delta region) over the last 27,000 years. Paleoceanography. doi:10.1029/2009PA001740

Clayton CRI, Abbireddy COR, Schiebel R (2009) A method of estimating the form of coarse particulates. Géotechnique 59:493–501

Conan SMH, Brummer GJA (2000) Fluxes of planktic Foraminifera in response to monsoonal upwelling on the Somalia Basin margin. Deep-Sea Res II 47:2207–2227

Darling KF, Wade CM (2008) The genetic diversity of planktic Foraminifera and the global distribution of ribosomal RNA genotypes. Mar Micropaleontol 67:216–238

Darling KF, Wade CM, Kroon D, Brown AJL (1997) Planktic foraminiferal molecular evolution and their polyphyletic origins from benthic taxa. Mar Micropaleontol 30:251–266

Darling KF, Kucera M, Kroon D, Wade CM (2006) A resolution for the coiling direction paradox in Neogloboquadrina pachyderma. Paleoceanography. doi:10.1029/2005PA001189

De Moel H, Ganssen GM, Peeters FJC, Jung SJA, Kroon D, Brummer GJA, Zeebe RE (2009) Planktic foraminiferal shell thinning in the Arabian Sea due to anthropogenic ocean acidification? Biogeosciences 6:1917–1925. doi:10.5194/bg-6-1917-2009

Deuser WG (1986) Seasonal and interannual variations in deep-water particle fluxes in the Sargasso Sea and their relation to surface hydrography. Deep-Sea Res I 33:225–246

Deuser WG, Ross EH (1989) Seasonally abundant planktonic Foraminifera of the Sargasso Sea; succession, deep-water fluxes, isotopic compositions, and paleoceanographic implications. J Foraminifer Res 19:268–293

Deuser WG, Ross EH, Hemleben C, Spindler M (1981) Seasonal changes in species composition, numbers, mass, size, and isotopic composition of planktonic Foraminifera settling into the deep Sargasso Sea. Palaeogeogr Palaeoclimatol Palaeoecol 33:103–127

De Vargas C, Pawlowski J (1998) Molecular versus taxonomic rates of evolution in planktonic Foraminifera. Mol Phylogenet Evol 9:463–469

De Vargas C, Zaninetti L, Hilbrecht H, Pawlowski J (1997) Phylogeny and rates of molecular evolution of planktonic Foraminifera: SSU rDNA sequences compared to the fossil record. J Mol Evol 45:285–294

De Vargas C, Renaud S, Hilbrecht H, Pawlowski J (2001) Pleistocene adaptive radiation in *Globorotalia truncatulinoides*: genetic, morphologic, and environmental evidence. Paleobiology 27:104–125

De Vargas C, Bonzon M, Rees NW, Pawlowski J, Zaninetti L (2002) A molecular approach to biodiversity and biogeography in the planktonic foraminifer *Globigerinella siphonifera* (d'Orbigny). Mar Micropaleontol 45:101–116

Dittert N, Baumann KH, Bickert T, Henrich R, Huber R, Kinkel H, Meggers H (1999) Carbonate dissolution in the deep-sea: methods, quantification and paleoceanographic application. In: Fischer G, Wefer G (eds) Use of proxies in paleoceanography. Springer, Berlin, pp 255–284

Donner B, Wefer G (1994) Flux and stable isotope composition of *Neogloboquadrina pachyderma* and other planktonic foraminifers in the Southern Ocean (Atlantic sector). Deep-Sea Res I 41:1733–1743

Duckworth DL (1977) Magnesium concentration in the tests of the planktonic foraminifer *Globorotalia truncatulinoides*. J Foraminifer Res 7:304–312

Eggins S, De Dekker P, Marshall J (2003) Mg/Ca variation in planktonic Foraminifera tests: implications for reconstructing palaeo-seawater temperature and habitat migration. Earth Planet Sci Lett 212:291–306

Eggins S, Sadekov A, De Deckker P (2004) Modulation and daily banding of Mg/Ca in *Orbulina universa* tests by symbiont photosynthesis and respiration: a complication for seawater thermometry? Earth Planet Sci Lett 225:411–419

Eguchi NO, Kawahata H, Taira A (1999) Seasonal response of planktonic Foraminifera to surface ocean condition: sediment trap results from the central North Pacific Ocean. J Oceanogr 55:681–691

Eguchi NO, Ujiié H, Kawahata H, Taira A (2003) Seasonal variations in planktonic Foraminifera at three sediment traps in the subarctic, transition and subtropical zones of the central North Pacific Ocean. Mar Micropaleontol 48:149–163

Eiler JM (2011) Paleoclimate reconstruction using carbonate clumped isotope thermometry. Quat Sci Rev 30:3575–3588. doi:10.1016/j.quascirev.2011.09.001

Elderfield H, Ganssen G (2000) Past temperature and δ^{18}O of surface ocean waters inferred from foraminiferal Mg/Ca ratios. Nature 405:442–445

Emiliani C (1954) Depth habitats of some species of pelagic Foraminifera as indicated by oxygen isotope ratios. Am J Sci 252:149–158

Epstein S, Buchsbaum R, Lowenstam H, Urey HC (1951) Carbonate-water isotopic temperature scale. Geol Soc Am Bull 62:417–426

Epstein S, Buchsbaum R, Lowenstam HA, Urey HC (1953) Revised carbonate-water isotopic temperature scale. Geol Soc Am Bull 64:1315–1326

Fallet U, Boer W, van Assen C, Greaves M, Brummer GJA (2009) A novel application of wet oxidation to retrieve carbonates from large organic-rich samples for ocean-climate research. Geochem Geophys Geosystems. doi:10.1029/2009GC002573

Fallet U, Brummer GJA, Zinke J, Vogels S, Ridderinkhof H (2010) Contrasting seasonal fluxes of planktonic Foraminifera and impacts on paleothermometry in the Mozambique Channel upstream of the Agulhas current. Paleoceanography. doi:10.1029/2010PA001942

Fallet U, Ullgren JE, Castañeda IS, van Aken HM, Schouten S, Ridderinkhof H, Brummer GJA (2011) Contrasting variability in foraminiferal and organic paleotemperature proxies in sedimenting particles of the Mozambique Channel (SW Indian Ocean). Geochim Cosmochim Acta 75:5834–5848. doi:10.1016/j.gca.2011.08.009

Fehrenbacher JS, Spero HJ, Russell AD, Vetter L, Eggins S (2015) Optimizing LA-ICP-MS analytical procedures for elemental depth profiling of Foraminifera shells. Chem Geol 407–408:2–9. doi:10.1016/j.chemgeo.2015.04.007

Felsenstein J (1981) Evolutionary trees from DNA sequences: a maximum likelihood approach. J Mol Evol 17:368–376. doi:10.1007/BF01734359

Felsenstein J (2004) Inferring phylogenies. Sinauer Associates Inc, Sunderland, MA

Fietzke J, Eisenhauer A, Gussone N, Bock B, Liebetrau V, Nägler TF, Spero HJ, Bijma J, Dullo C (2004) Direct measurement of ^{44}Ca/^{40}Ca ratios by MC–ICP–MS using the cool plasma technique. Chem Geol 206:11–20. doi:10.1016/j.chemgeo.2004.01.014

Fischer G, Wefer G (1991) Sampling, preparation and analysis of marine particulate matter. Mar Part Anal Charact 63:391–397

Fischer G, Wefer G (1999) Use of proxies in paleoceanography: examples from the South Atlantic. Springer, Berlin, Heidelberg

Fisher RA, Corbet AS, Williams CB (1943) The relation between the number of species and the number of individuals in a random sample of an animal population. J Anim Ecol 42–58

Fischer K, Dymond J, Moser C, Murray D, Matherne A (1983) Seasonal variation in particulate flux in an offshore area adjacent to coastal upwelling. In: Suess E, Thiede J (eds) Coastal upwelling its sediment record. Springer, New York, pp 209–224

Fischer G, Fütterer D, Gersonde R, Honjo S, Ostermann D, Wefer G (1988) Seasonal variability of particle flux in the Weddell Sea and its relation to ice cover. Nature 335:426–428

Fischer G, Donner B, Ratmeyer V, Davenport R, Wefer G (1996) Distinct year-to-year particle flux variations off Cape Blanc during 1988–1991: relation to δ^{18}O-deduced sea-surface temperatures and trade winds. J Mar Res 54:73–98

Flakowski J (2005) Actin phylogeny of Foraminifera. J Foraminifer Res 35:93–102. doi:10.2113/35.2.93

Fraile I, Schulz M, Mulitza S, Kucera M (2008) Predicting the global distribution of planktonic Foraminifera using a dynamic ecosystem model. Biogeosciences 5:891–911

Fraile I, Mulitza S, Schulz M (2009a) Modeling planktonic foraminiferal seasonality: implications for sea-surface temperature reconstructions. Mar Micropaleontol 72:1–9. doi:10.1016/j.marmicro.2009.01.003

Fraile I, Schulz M, Mulitza S, Merkel U, Prange M, Paul A (2009b) Modeling the seasonal distribution of planktonic Foraminifera during the last glacial maximum. Paleoceanography. doi:10.1029/2008PA001686

Friedrich O, Schiebel R, Wilson PA, Weldeab S, Beer CJ, Cooper MJ, Fiebig J (2012) Influence of test size, water depth, and ecology on Mg/Ca, Sr/Ca, $\delta^{18}O$ and $\delta^{13}C$ in nine modern species of planktic foraminifers. Earth Planet Sci Lett 319–320:133–145. doi:10.1016/j.epsl.2011.12.002

Ganssen G (1981) Isotopic analysis of Foraminifera shells: interference from chemical treatment. Palaeogeogr Palaeoclimatol Palaeoecol 33:271–276

Gehlen M, Bassinot F, Beck L, Khodja H (2004) Trace element cartography of Globigerinoides ruber shells using particle-induced X-ray emission. Geochem Geophys Geosystems. doi:10.1029/2004GC000822

Görög Á, Szinger B, Tóth E, Viszkok J (2012) Methodology of the micro-computer tomography on Foraminifera. Palaeontol Electron 15:15

Grimm GW, Stögerer K, Topaç Ertan K, Kitazato H, Kučera M, Hemleben V, Hemleben C (2007) Diversity of rDNA in chilostomella: molecular differentiation patterns and putative hermit types. Mar Micropaleontol 62:75–90. doi:10.1016/j.marmicro.2006.07.005

Guiot J, de Vernal A (2007) Transfer functions: methods for quantitative paleoceanography based on microfossils. In: Hillaire-Marcel C, de Vernal A (eds) Developments in marine geology. Elsevier, pp 523–563

Haarmann T, Hathorne EC, Mohtadi M, Groeneveld J, Kölling M, Bickert T (2011) Mg/Ca ratios of single planktonic foraminifer shells and the potential to reconstruct the thermal seasonality of the water column. Paleoceanography. doi:10.1029/2010PA002091

Haley BA, Klinkhammer GP (2002) Development of a flow-through system for cleaning and dissolving foraminiferal tests. Chem Geol 185:51–69

Hancock JM, Dover GA (1988) Molecular coevolution among cryptically simple expansion segments of eukaryotic 26S/28S rRNAs. Mol Biol Evol 5:377–391

Hardy AC (1935) The continuous plankton recorder: a new method of survey. Rapports et proces-verbaux des réunions. J Cons Perm Int Pour L'Exploration Mer 95:36–47

Hayek LAC, Buzas MA (1997) Surveying natural populations. Columbia University Press, New York

Hayes A, Kucera M, Kallel N, Sbaffi L, Rohling EJ (2005) Glacial Mediterranean sea surface temperatures based on planktonic foraminiferal assemblages. Quat Sci Rev 24:999–1016. doi:10.1016/j.quascirev.2004.02.018

Hebbeln D, Marchant M, Wefer G (2000) Seasonal variations of the particle flux in the Peru-Chile current at 30°S under "normal" and El Niño conditions. Deep-Sea Res II 47:2101–2128

Hemleben C, Spindler M, Breitinger I, Ott R (1987) Morphological and physiological responses of Globigerinoides sacculifer (Brady) under varying laboratory conditions. Mar Micropaleontol 12:305–324

Hemleben C, Spindler M, Anderson OR (1989) Modern planktonic Foraminifera. Springer, Berlin

Henderson GM (2002) New oceanic proxies for paleoclimate. Earth Planet Sci Lett 203:1–13

Hennekam R, Jilbert T, Schnetger B, de Lange GJ (2014) Solar forcing of Nile discharge and sapropel S1 formation in the early to middle Holocene eastern Mediterranean. Paleoceanography 29:343–356. doi:10.1002/2013PA002553

Hernández-Almeida I, Bárcena MA, Flores JA, Sierro FJ, Sanchez-Vidal A, Calafat A (2011) Microplankton response to environmental conditions in the Alboran Sea (Western Mediterranean): one year sediment trap record. Mar Micropaleontol 78:14–24

Hilbrecht H (1996) Extant planktic Foraminifera and the physical environment in the Atlantic and Indian Oceans: an atlas based on Climap and Levitus (1982). In: Mitteilungen aus dem Geologischen Institut der Eidgen. Technischen Hochschule und der Universität Zürich. Zürich, p 93

Hintz CJ, Chandler GT, Bernhard JM, McCorkle DC, Havach SM, Blanks JK, Shaw TJ (2004) A physicochemically constrained seawater culturing system for production of benthic Foraminifera. Limnol Oceanogr-Methods 2:160–170

Hönisch B, Allen KA, Russell AD, Eggins SM, Bijma J, Spero HJ, Lea DW, Yu J (2011) Planktic foraminifers as recorders of seawater Ba/Ca. Mar Micropaleontol 79:52–57

Honjo S, Manganini SJ (1993) Annual biogenic particle fluxes to the interior of the North Atlantic Ocean; studied at 34°N 21°W and 48°N 21°W. Deep-Sea Res II 40:587–607

Honjo S, Connell JF, Sachs PL (1980) Deep-ocean sediment trap: Design and function of PARFLUX Mark II. Deep-Sea Res I 27:745–753

Huber BT (1987) Ontogenetic morphometrics of some Upper Cretaceous Foraminifera from the southern high latitudes. Antarct J U S 22:15–17

Huber BT, Bijma J, Spero HJ (1996) Blue-water SCUBA collection of planktonic Foraminifera. Am Acad Underw Sci 127–132

Huelsenbeck JP, Ronquist F (2001) MRBAYES: Bayesian inference of phylogenetic trees. Bioinformatics 17:754–755. doi:10.1093/bioinformatics/17.8.754

Huson DH, Bryant D (2006) Application of phylogenetic networks in evolutionary studies. Mol Biol Evol 23:254–267. doi:10.1093/molbev/msj030

Hutson WH (1977) Transfer functions under no-analog conditions: experiments with Indian Ocean planktonic Foraminifera. Quat Res 8:355–367

Hutson WH (1980) The Agulhas Current during the Late Pleistocene: analysis of modern faunal analogs. Science 207:64–66

Imbrie J, Kipp NG (1971) A new micropaleontological method for quantitative paleoclimatology: application to a late Pleistocene Caribbean core. In: Turekian KK (ed) The late Cenozoic glacial ages. Yale University Press, New Haven, pp 71–181

Itou M, Noriki S (2002) Shell fluxes of solution-resistant planktonic foraminifers as a proxy for mixed-layer depth. Geophys Res Lett. doi:10.1029/2002GL014693

Ivanova E, Conan SMH, Peeters FJ, Troelstra SR (1999) Living Neogloboquadrina pachyderma sin and its distribution in the sediments from Oman and Somalia upwelling areas. Mar Micropaleontol 36:91–107

Ivanova E, Schiebel R, Singh AD, Schmiedl G, Niebler HS, Hemleben C (2003) Primary production in the Arabian Sea during the last 135 000 years. Palaeogeogr Palaeoclimatol Palaeoecol 197:61–82

Johnstone HJH, Schulz M, Barker S, Elderfield H (2010) Inside story: an X-ray computed tomography method for assessing dissolution in the tests of planktonic Foraminifera. Mar Micropaleontol 77:58–70. doi:10.1016/j.marmicro.2010.07.004

Johnstone HJH, Kiefer T, Elderfield H, Schulz M (2014) Calcite saturation, foraminiferal test mass, and Mg/Ca-based temperatures dissolution corrected using XDX-A 150 ka record from the western Indian Ocean. Geochem Geophys Geosystems 15:781–797. doi:10.1002/2013GC004994

Jørgensen BB, Erez J, Revsbech NP, Cohen Y (1985) Symbiotic photosynthesis in a planktonic foraminiferan, Globigerinoides sacculifer (Brady), studied with microelectrodes. Limnol Oceanogr 30:1253–1267

Katz ME, Cramer BS, Franzese A, Hönisch B, Miller KG, Rosenthal Y, Wright JD (2010) Traditional and emerging geochemical proxies in Foraminifera. J Foraminifer Res 40:165–192

Khatiwala S, Primeau F, Hall T (2009) Reconstruction of the history of anthropogenic CO_2 concentrations in the ocean. Nature 462:346–349. doi:10.1038/nature08526

Kincaid E, Thunell RC, Le J, Lange CB, Weinheimer AL, Reid FMH (2000) Planktonic foraminiferal fluxes in the Santa Barbara Basin: response to seasonal and interannual hydrographic changes. Deep-Sea Res II 47:1157–1176

King AL, Howard WR (2001) Seasonality of foraminiferal flux in sediment traps at Chatham Rise, SW Pacific: implications for paleotemperature estimates. Deep-Sea Res I 48:1687–1708

King AL, Howard WR (2003) Planktonic foraminiferal flux seasonality in Subantarctic sediment traps: a test for paleoclimate reconstructions. Paleoceanography. doi:10.1029/2002PA000839

King AL, Howard WR (2004) Planktonic foraminiferal $\delta^{13}C$ records from Southern Ocean sediment traps: new estimates of the oceanic Suess effect. Glob Biogeochem Cycles. doi:10.1029/2003GB002162

King AL, Howard WR (2005) $\delta^{18}O$ seasonality of planktonic Foraminifera from Southern Ocean sediment traps: latitudinal gradients and implications for paleoclimate reconstructions. Mar Micropaleontol 56:1–24

Köhler-Rink S, Kühl M (2005) The chemical microenvironment of the symbiotic planktonic foraminifer Orbulina universa. Mar Biol Res 1:68–78. doi:10.1080/17451000510019015

Koppelmann R, Schäfer P, Schiebel R (2000) Organic carbon losses measured by heterotrophic activity of mesozooplankton and $CaCO_3$ flux in the bathypelagic zone of the Arabian Sea. Deep-Sea Res II 47:169–187

Kozdon R, Ushikubo T, Kita NT, Spicuzza M, Valley JW (2009) Intratest oxygen isotope variability in the planktonic foraminifer N. pachyderma: Real vs. apparent vital effects by ion microprobe. Chem Geol 258:327–337. doi:10.1016/j.chemgeo.2008.10.032

Kozdon R, Kelly DC, Kita NT, Fournelle JH, Valley JW (2011) Planktonic foraminiferal oxygen isotope analysis by ion microprobe technique suggests warm tropical sea surface temperatures during the Early Paleogene. Paleoceanography. doi:10.1029/2010PA002056

Kucera M (2007) Chapter Six: Planktonic Foraminifera as tracers of past oceanic environments. In: Hillaire-Marcel C, de Vernal A (eds) Developments in Marine Geology. Elsevier, pp 213–262

Kucera M, Rosell-Melé A, Schneider R, Waelbroeck C, Weinelt M (2005) Multiproxy approach for the reconstruction of the glacial ocean surface (MARGO). Quat Sci Rev 24:813–819. doi:10.1016/j.quascirev.2004.07.017

Kuhnt T, Howa H, Schmidt S, Marié L, Schiebel R (2013) Flux dynamics of planktic foraminiferal tests in the south-eastern Bay of Biscay (northeast Atlantic margin). J Mar Syst 109–110:169–181. doi:10.1016/j.jmarsys.2011.11.026

Kunioka D, Shirai K, Takahata N, Sano Y, Toyofuku T, Ujiie Y (2006) Microdistribution of Mg/Ca, Sr/Ca, and Ba/Ca ratios in Pulleniatina obliquiloculata test by using a NanoSIMS: implication for the vital effect mechanism. Geochem Geophys Geosystems 7:Q12P20. doi:10.1029/2006GC001280

Kuroyanagi A, Kawahata H, Nishi H, Honda MC (2002) Seasonal changes in planktonic Foraminifera in the northwestern North Pacific Ocean: sediment trap experiments from subarctic and subtropical gyres. Deep-Sea Res II 49:5627–5645

Kuroyanagi A, Kawahata H, Nishi H (2011) Seasonal variation in the oxygen isotopic composition of different-sized planktonic foraminifer Neogloboquadrina pachyderma (sinistral) in the northwestern North Pacific and implications for reconstruction of the paleoenvironment. Paleoceanography. doi:10.1029/2011PA002153

Łabaj P, Topa P, Tyszka J, Alda W (2003) 2D and 3D numerical models of the growth of Foraminiferal shells. In: Sloot PMA, Abramson D, Bogdanov AV,

Dongarra JJ, Zomaya AY, Gorbachev YE (eds) Computational Science—ICCS 2003. Springer, Berlin, pp 669–678

Lampitt RS, Boorman B, Brown L, Lucas M, Salter I, Sanders R, Saw K, Seeyave S, Thomalla SJ, Turnewitsch R (2008) Particle export from the euphotic zone: estimates using a novel drifting sediment trap,[234]Th and new production. Deep-Sea Res I 55:1484–1502

Lea DW (1999) Trance elements in foraminiferal calcite. In: Sen Gupta B (ed) Modern Foraminifera. Kluwer Academic Publishers, Dordrecht, pp 259–277

Lea DW, Mashiotta TA, Spero HJ (1999) Controls on magnesium and strontium uptake in planktonic Foraminifera determined by live culturing. Geochim Cosmochim Acta 63:2369–2379

Lin HL (2014) The seasonal succession of modern planktonic Foraminifera: sediment traps observations from southwest Taiwan waters. Cont Shelf Res 84: 13–22

Lohbeck KT, Riebesell U, Reusch TBH (2012) Adaptive evolution of a key phytoplankton species to ocean acidification. Nat Geosci 5:346–351. doi:10.1038/ngeo1441

Lohmann GP (1995) A model for variation in the chemistry of planktonic Foraminifera due to secondary calcification and selective dissolution. Paleoceanography 10:445–457

Lombard F, Erez J, Michel E, Labeyrie L (2009a) Temperature effect on respiration and photosynthesis of the symbiont-bearing planktonic Foraminifera *Globigerinoides ruber*, *Orbulina universa*, and *Globigerinella siphonifera*. Limnol Oceanogr 54:210–218

Lombard F, Labeyrie L, Michel E, Spero HJ, Lea DW (2009b) Modelling the temperature dependent growth rates of planktic Foraminifera. Mar Micropaleontol 70:1–7. doi:10.1016/j.marmicro.2008.09.004

Lombard F, Labeyrie L, Michel E, Bopp L, Cortijo E, Retailleau S, Howa H, Jorissen F (2011) Modelling planktic foraminifer growth and distribution using an ecophysiological multi-species approach. Biogeosciences 8:853–873. doi:10.5194/bg-8-853-2011

Lončarić N, Brummer GJA, Kroon D (2005) Lunar cycles and seasonal variations in deposition fluxes of planktic foraminiferal shell carbonate to the deep South Atlantic (central Walvis Ridge). Deep-Sea Res I 52:1178–1188

Lončarić N, Peeters FJC, Kroon D, Brummer GJA (2006) Oxygen isotope ecology of recent planktic Foraminifera at the central Walvis Ridge (SE Atlantic). Paleoceanography. doi:10.1029/2005PA001207

Lončarić N, van Iperen J, Kroon D, Brummer GJA (2007) Seasonal export and sediment preservation of diatomaceous, foraminiferal and organic matter mass fluxes in a trophic gradient across the SE Atlantic. Prog Oceanogr 73:27–59

Longhurst AR, Reith AD, Bower RE, Seibert DLR (1966) A new system for the collection of multiple serial plankton samples. Deep Sea Res 13:213–222

Lutze GF, Altenbach A (1991) Technik und Signifikanz der Lebendfärbung benthischer Foraminiferen mit Bengalrot. Geol Jahrb Reihe A 128:165

Malmgren BA, Nordlund U (1996) Application of artificial neural networks to chemostratigraphy. Paleoceanography 11:505–512

Marchant M (1995) Die Sedimentation planktischer Foraminiferen im Auftriebsgebiet vor Chile heute und während der letzten ca. 15.000 Jahre. Berichte Fachbereich Geowiss Univ Brem 69:311–323

Marchant M, Hebbeln D, Wefer G (1998) Seasonal flux patterns of planktic Foraminifera in the Peru-Chile Current. Deep-Sea Res I 45:1161–1185

Marchant M, Hebbeln D, Giglio S, Coloma C, González HE (2004) Seasonal and interannual variability in the flux of planktic Foraminifera in the Humboldt Current System off central Chile (30°S). Deep-Sea Res II 51:2441–2455

Marshall BJ, Thunell RC, Henehan MJ, Astor Y, Wejnert KE (2013) Planktonic foraminiferal area density as a proxy for carbonate ion concentration: a calibration study using the Cariaco Basin ocean time series. Paleoceanography 28:363–376. doi:10.1002/palo. 20034

McConnell MC, Thunell RC (2005) Calibration of the planktonic foraminiferal Mg/Ca paleothermometer: sediment trap results from the Guaymas Basin, Gulf of California. Paleoceanography. doi:10.1029/2004PA001077

Mekik F (2014) Radiocarbon dating of planktonic foraminifer shells: a cautionary tale. Paleoceanography 29:13–29. doi:10.1002/2013PA002532

Merlé C, Moullade M, Lima O, Perasso R (1994) Essai de caractérisation phylogénétique de foraminifères planctoniques à partir de séquences partielles d'ADNr28S. Comptes Rendus Académie Sci Sér 2 Sci Terre Planètes 319:149–153

Metzker ML (2010) Sequencing technologies—the next generation. Nat Rev Genet 11:31–46. doi:10.1038/nrg2626

Mohiuddin MM, Nishimura A, Tanaka Y, Shimamoto A (2002) Regional and interannual productivity of biogenic components and planktonic foraminiferal fluxes in the northwestern Pacific Basin. Mar Micropaleontol 45:57–82

Mohiuddin MM, Nishimura A, Tanaka Y, Shimamoto A (2004) Seasonality of biogenic particle and planktonic Foraminifera fluxes: response to hydrographic variability in the Kuroshio Extension, northwestern Pacific Ocean. Deep-Sea Res I 51:1659–1683. doi:10.1016/j.dsr.2004.06.002

Mohtadi M, Steinke S, Groeneveld J, Fink HG, Rixen T, Hebbeln D, Donner B, Herunadi B (2009) Low-latitude control on seasonal and interannual changes in planktonic foraminiferal flux and shell geochemistry off south Java: a sediment trap study. Paleoceanography. doi:10.1029/2008PA001636

Mojtahid M, Zubkov MV, Hartmann M, Gooday AJ (2011) Grazing of intertidal benthic Foraminifera on bacteria: assessment using pulse-chase radiotracing.

J Exp Mar Biol Ecol 399:25–34. doi:10.1016/j.jembe. 2011.01.011

Mojtahid M, Manceau R, Schiebel R, Hennekam R, de Lange GJ (2015) Thirteen thousand years of south-eastern Mediterranean climate variability inferred from an integrative planktic foraminiferal-based approach: Holocene climate in the SE Mediterranean. Paleoceanography 30:402–422. doi:10.1002/2014PA002705

Morard R, Quillévéré F, Escarguel G, Ujiie Y, de Garidel-Thoron T, Norris RD, de Vargas C (2009) Morphological recognition of cryptic species in the planktonic foraminifer *Orbulina universa*. Mar Micropaleontol 71:148–165

Morard R, Darling KF, Mahé F, Audic S, Ujiié Y, Weiner AKM, André A, Seears HA, Wade CM, Quillévéré F, Douady CJ, Escarguel G, de Garidel-Thoron T, Siccha M, Kucera M, de Vargas C (2015) PFR2: a curated database of planktonic Foraminifera 18S ribosomal DNA as a resource for studies of plankton ecology, biogeography and evolution. Mol Ecol Resour. doi:10.1111/1755-0998.12410

Motoda S, Anraku M, Minoda T (1957) Experiments on the performance of plankton samples with nets. In: Bulletin Faculty Fisheries. Hokkaido University, pp 1–22

Movellan A (2013) La biomasse des foraminifères planctoniques actuels et son impact sur la pompe biologique de carbone. PhD Thesis, University of Angers

Movellan A, Schiebel R, Zubkov MV, Smyth A, Howa H (2012) Quantification of protein biomass of individual foraminifers using nano-spectrophotometry. Biogeosciences Discuss 9:6651–6681. doi:10.5194/bgd-9-6651-2012

Moy AD, Howard WR, Bray SG, Trull TW (2009) Reduced calcification in modern Southern Ocean planktonic Foraminifera. Nat Geosci 2:276–280. doi:10.1038/ngeo460

Mulitza S, Dürkoop A, Hale W, Wefer G, Niebler HS (1997) Planktonic Foraminifera as recorders of past surface-water stratification. Geology 25:335–338

Murray JW (2006) Ecology and applications of benthic Foraminifera. Cambridge University Press, Cambridge

Neefs JM, Peer Y, Hendriks L, de Wachter R (1990) Compilation of small ribosomal subunit RNA sequences. Nucleic Acids Res 18:2237–2317

Niebler HS, Hubberten HW, Gersonde R (1999) Oxygen isotope values of planktic Foraminifera: a tool for the reconstruction of surface water stratification. In: Fischer G, Wefer G (eds) Use of Proxies in Paleoceanography: examples from the South Atlantic. Springer, Berlin, Heidelberg, pp 165–189

Northcote LC, Neil HL (2005) Seasonal variations in foraminiferal flux in the Southern Ocean, Campbell Plateau, New Zealand. Mar Micropaleontol 56:122–137. doi:10.1016/j.marmicro.2005.05.001

O'Neill LP, Benitez-Nelson CR, Styles RM, Tappa E, Thunell RC (2005) Diagenetic effects on particulate phosphorus samples collected using formalin poisoned sediment traps. Limnol Oceanogr Methods 3:308–317

Ortiz JD, Mix AC (1992) The spatial distribution and seasonal succession of planktonic Foraminifera in the California Current off Oregon, September 1987–September 1988. In: Summerhays CP, Prell WS, Emeis KC (eds) Upwelling systems: evolution since the Early Miocene. pp 197–213

Ottens JJ (1992) Planktic Foraminifera as indicators of ocean environments in the northeast Atlantic. PhD Thesis, Free University, Amsterdam

Ott R, Bijma J, Hemleben C (1992) A computer method for estimating volumes and surface areas of complex structures consisting of overlapping spheres. Math Comput Model 16:83–98

Paris G, Fehrenbacher JS, Sessions AL, Spero HJ, Adkins JF (2014) Experimental determination of carbonate-associated sulfate $\delta^{34}S$ in planktonic Foraminifera shells. Geochem Geophys Geosystems 15:1452–1461. doi:10.1002/2014GC005295

Patterson RT, Fishbein E (1989) Re-examination of the statistical methods used to determine the number of point counts needed for micropaleontological quantitative research. J Paleontol 63:245–248

Pawlowski J, Holzmann M (2002) Molecular phylogeny of Foraminifera: a review. Eur J Protistol 38:1–10

Pawlowski J, Bolivar I, Fahrni J, Zaninetti L (1994) Taxonomic identification of Foraminifera using ribosomal DNA sequences. Micropaleontology 40:373. doi:10.2307/1485942

Pawlowski J, Bolivar I, Fahrni JF, de Vargas C, Gouy M, Zaninetti L (1997) Extreme differences in rates of molecular evolution of Foraminifera revealed by comparison of ribosomal DNA sequences and the fossil record. Mol Biol Evol 14:498–505

Pawlowski J, Lejzerowicz F, Esling P (2014) Next-generation environmental diversity surveys of Foraminifera: preparing the future. Biol Bull 227: 93–106

Peeters F, Ivanova E, Conan S, Brummer GJA, Ganssen G, Troelstra S, van Hinte J (1999) A size analysis of planktic Foraminifera from the Arabian Sea. Mar Micropaleontol 36:31–63

Pflaumann U, Duprat J, Pujol C, Labeyrie LD (1996) SIMMAX: a modern analog technique to deduce Atlantic sea surface temperatures from planktonic Foraminifera in deep-sea sediments. Paleoceanography 11:15–35

Phleger FB (1945) Vertical distribution of pelagic Foraminifera. Am J Sci 243:377–383

Pons J, Barraclough T, Gomez-Zurita J, Cardoso A, Duran D, Hazell S, Kamoun S, Sumlin W, Vogler A (2006) Sequence-based species delimitation for the DNA taxonomy of undescribed insects. Syst Biol 55:595–609. doi:10.1080/10635150600852011

Poore RZ, Tedesco KA, Spear JW (2013) Seasonal flux and assemblage composition of planktic foraminifers from a sediment-trap study in the Northern Gulf of Mexico. J Coast Res 63:6–19. doi:10.2112/SI63-002.1

Puillandre N, Lambert A, Brouillet S, Achaz G (2012) ABGD, automatic barcode gap discovery for primary species delimitation. Mol Ecol 21:1864–1877. doi:10.1111/j.1365-294X.2011.05239.x

Reichart GJ, Jorissen F, Anschutz P, Mason PRD (2003) Single foraminiferal test chemistry records the marine environment. Geology 31:355–358

Reid PC, Colebrook JM, Matthews JBL, Aiken J (2003) The continuous plankton recorder: concepts and history, from plankton indicator to undulating recorders. Prog Oceanogr 58:117–173

Reimer PJ, Bard E, Bayliss A, Beck JW, Blackwell PG, Ramsey CB, Buck CE, Cheng H, Edwards RL, Friedrich M (2013) IntCal13 and Marine13 radiocarbon age calibration curves 0–50,000 years cal BP. Radiocarbon 55:1869–1887

Ren H, Sigman DM, Meckler AN, Plessen B, Robinson RS, Rosenthal Y, Haug GH (2009) Foraminiferal isotope evidence of reduced nitrogen fixation in the ice age Atlantic Ocean. Science 323:244–248. doi:10.1126/science.1165787

Ren H, Sigman DM, Thunell RC, Prokopenko MG (2012a) Nitrogen isotopic composition of planktonic Foraminifera from the modern ocean and recent sediments. Limnol Oceanogr 57:1011–1024. doi:10.4319/lo.2012.57.4.1011

Reynolds L, Thunell RC (1985) Seasonal succession of planktonic Foraminifera in the subpolar North Pacific. J Foraminifer Res 15:282–301

Rickaby REM, Greaves MJ, Elderfield H (2000) Cd in planktonic and benthic foraminiferal shells determined by thermal ionisation mass spectrometry. Geochim Cosmochim Acta 64:1229–1236

Rigual-Hernández AS, Sierro FJ, Bárcena MA, Flores JA, Heussner S (2012) Seasonal and interannual changes of planktic foraminiferal fluxes in the Gulf of Lions (NW Mediterranean) and their implications for paleoceanographic studies: two 12-year sediment trap records. Deep-Sea Res I 66:26–40

Rink S, Kühl M, Bijma J, Spero HJ (1998) Microsensor studies of photosynthesis and respiration in the symbiotic foraminifer *Orbulina universa*. Mar Biol 131:583–595

Ripperger S, Rehkämper M (2007) A highly sensitive MC-ICPMS method for Cd/Ca analyses of foraminiferal tests. J Anal At Spectrom 22:1275–1283

Ripperger S, Schiebel R, Rehkämper M, Halliday AN (2008) Cd/Ca ratios of in situ collected planktonic foraminiferal tests. Paleoceanography. doi:10.1029/2007PA001524

Rohling EJ, Cooke S (1999) Stable oxygen and carbon isotopes in foraminiferal carbonate shells. In: Sen Gupta BS (ed) Modern Foraminifera. Kluwer Academic Publishers, Dordrecht, pp 239–258

Rohling EJ, Sprovieri M, Cane T, Casford JSL, Cooke S, Bouloubassi I, Emeis KC, Schiebel R, Rogerson M, Hayes A, Jorissen FJ, Kroon D (2004) Reconstructing past planktic foraminiferal habitats using stable isotope data: a case history for Mediterranean sapropel S5. Mar Micropaleontol 50:89–123. doi:10.1016/S0377-8398(03)00068-9

Rohling EJ, Sluijs A, Dijkstra HA, Köhler P, van de Wal RSW, von der Heydt AS, Beerling DJ, Berger A, Bijl PK, Crucifix M, DeConto R, Drijfhout SS, Fedorov A, Foster GL, Ganopolski A, Hansen J, Hönisch B, Hooghiemstra H, Huber M, Huybers P, Knutti R, Lea DW, Lourens LJ, Lunt D, Masson-Demotte V, Medina-Elizalde M, Otto-Bliesner B, Pagani M, Pälike H, Renssen H, Royer DL, Siddall M, Valdes P, Zachos JC, Zeebe RE (2012) Making sense of palaeoclimate sensitivity. Nature 491:683–691. doi:10.1038/nature11574

Ronquist F, Huelsenbeck JP (2003) MrBayes 3: Bayesian phylogenetic inference under mixed models. Bioinformatics 19:1572–1574. doi:10.1093/bioinformatics/btg180

Rossignol-Strick M, Nesteroff W, Olive P, Vergnaud-Grazzini C (1982) After the deluge: Mediterranean stagnation and sapropel formation. Nature 295:105–110. doi:10.1038/295105a0

Roy T, Lombard F, Bopp L, Gehlen M (2015) Projected impacts of climate change and ocean acidification on the global biogeography of planktonic Foraminifera. Biogeosciences 12:2873–2889. doi:10.5194/bg-12-2873-2015

Sachs HM, Webb T III, Clark DR (1977) Paleoecological transfer functions. Annu Rev Earth Planet Sci 5:159

Sadekov AY, Eggins SM, de Deckker P (2005) Characterization of Mg/Ca distributions in planktonic Foraminifera species by electron microprobe mapping. Geochem Geophys Geosystems. doi:10.1029/2005GC000973

Saitou N, Nei M (1987) The neighbor-joining method: a new method for reconstructing phylogenetic trees. Mol Biol Evol 4:406–425

Salter I, Schiebel R, Ziveri P, Movellan A, Lampitt R, Wolff GA (2014) Carbonate counter pump stimulated by natural iron fertilization in the Polar frontal zone. Nat Geosci 7:885–889. doi:10.1038/ngeo2285

Sanyal A, Hemming NG, Broecker WS, Lea DW, Spero HJ, Hanson GN (1996) Oceanic pH control on the boron isotopic composition of Foraminifera: evidence from culture experiments. Paleoceanography 11:513–517

Sarnthein M, Gersonde R, Niebler S, Pflaumann U, Spielhagen R, Thiede J, Wefer G, Weinelt M (2003) Overview of Glacial Atlantic Ocean Mapping (GLAMAP 2000). Paleoceanography. doi:10.1029/2002PA000769

Sautter LR, Thunell RC (1991) Seasonal variability in the $\delta^{18}O$ and $\delta^{13}C$ of planktonic Foraminifera from an upwelling environment: sediment trap results from the San Pedro Basin, Southern California Bight. Paleoceanography 6:307–334

Schiebel R (2002) Planktic foraminiferal sedimentation and the marine calcite budget. Glob Biogeochem Cycles. doi:10.1029/2001GB001459

Schiebel R, Hemleben C (2000) Interannual variability of planktic foraminiferal populations and test flux in the eastern North Atlantic Ocean (JGOFS). Deep-Sea Res II 47:1809–1852

Schiebel R, Movellan A (2012) First-order estimate of the planktic foraminifer biomass in the modern ocean. Earth Syst Sci Data 4:75–89. doi:10.5194/essd-4-75-2012

Schiebel R, Hiller B, Hemleben C (1995) Impacts of storms on Recent planktic foraminiferal test production and CaCO$_3$ flux in the North Atlantic at 47°N, 20°W (JGOFS). Mar Micropaleontol 26:115–129

Schiebel R, Schmuker B, Alves M, Hemleben C (2002) Tracking the recent and late Pleistocene Azores front by the distribution of planktic foraminifers. J Mar Syst 37:213–227

Schiebel R, Zeltner A, Treppke UF, Waniek JJ, Bollmann J, Rixen T, Hemleben C (2004) Distribution of diatoms, coccolithophores and planktic foraminifers along a trophic gradient during SW monsoon in the Arabian Sea. Mar Micropaleontol 51:345–371

Schiebel R, Barker S, Lendt R, Thomas H, Bollmann J (2007) Planktic foraminiferal dissolution in the twilight zone. Deep-Sea Res II 54:676–686

Schmidt RAM (1952) Microradiography of microfossils with X-ray diffraction equipment. Science 115:94–95

Schmidt DN (2002) Size variability in planktic foraminifers. PhD Thesis, ETH Zürich

Schmidt DN, Renaud S, Bollmann J (2003) Response of planktic foraminiferal size to late Quaternary climate change. Paleoceanography. doi:10.1029/2002PA000831

Schmidt DN, Renaud S, Bollmann J, Schiebel R, Thierstein HR (2004a) Size distribution of Holocene planktic foraminifer assemblages: biogeography, ecology and adaptation. Mar Micropaleontol 50:319–338

Schmidt DN, Thierstein HR, Bollmann J (2004b) The evolutionary history of size variation of planktic foraminiferal assemblages in the Cenozoic. Palaeogeogr Palaeoclimatol Palaeoecol 212:159–180

Schmidt DN, Thierstein HR, Bollmann J, Schiebel R (2004c) Abiotic forcing of plankton evolution in the Cenozoic. Science 303:207–210

Schmidt DN, Lazarus D, Young JR, Kucera M (2006) Biogeography and evolution of body size in marine plankton. Earth-Sci Rev 78:239–266

Schmidt S, Howa H, Mouret A, Lombard F, Anschutz P, Labeyrie L (2009) Particle fluxes and recent sediment accumulation on the Aquitanian margin of Bay of Biscay. Cont Shelf Res 29:1044–1052. doi:10.1016/j.csr.2008.11.018

Schmidt DN, Rayfield EJ, Cocking A, Marone F (2013) Linking evolution and development: synchrotron radiation X-ray tomographic microscopy of planktic foraminifers. Palaeontology 56:741–749

Scholten JC, Fietzke J, Vogler S, Rutgers van der Loeff MM, Mangini A, Koeve W, Waniek J, Stoffers P, Antia A, Kuss J (2001) Trapping efficiencies of sediment traps from the deep Eastern North Atlantic: the ^{230}Th calibration. Deep-Sea Res II 48:2383–2408. doi:10.1016/S0967-0645(00)00176-4

Schott W (1935) Die Foraminiferen des äquatorialen Teil des Atlantischen Ozeans: Deutsche Atlantische Expeditionen Meteor 1925–1927. Wiss Ergeb 43–134

Seears HA, Wade CM (2014) Extracting DNA from within intact foraminiferal shells. Mar Micropaleontol 109:46–53. doi:10.1016/j.marmicro.2014.04.001

Sexton PF, Wilson PA, Pearson PN (2006) Microstructural and geochemical perspectives on planktic foraminiferal preservation: "Glassy" versus "Frosty". Geochem Geophys Geosystems. doi:10.1029/2006GC001291

Shackleton NJ (1968) Depth of pelagic Foraminifera and isotopic changes in Pleistocene oceans. Nature 218:79–80

Shackleton NJ (1987) Oxygen isotopes, ice volume and sea level. Quat Sci Rev 6:183–190. doi:10.1016/0277-3791(87)90003-5

Shackleton NJ, Corfield RM, Hall MA (1985) Stable isotope data and the ontogeny of Paleocene planktonic Foraminifera. J Foraminifer Res 15:321–336

Shackleton NJ, Opdyke ND (1973) Oxygen isotope and palaeomagnetic stratigraphy of Equatorial Pacific core V28-238: oxygen isotope temperatures and ice volumes on a 10^5 year and 10^6 year scale. Quat Res 3:39–55

Shannon CE (1948) A mathematical theory of communication. Bell Syst Tech J 27(379–423):623–656

Shannon CE, Weaver W (1963) The mathematical theory of communication. University of Illinois Press, Urbana

Siccha M, Schiebel R, Schmidt S, Howa H (2012) Short-term and small-scale variability in planktic Foraminifera test flux in the Bay of Biscay. Deep-Sea Res I 64:146–156. doi:10.1016/j.dsr.2012.02.004

Siegel DA, Deuser WG (1997) Trajectories of sinking particles in the Sargasso Sea: modeling of statistical funnels above deep-ocean sediment traps. Deep-Sea Res I 44:1519–1541

Signes M, Bijma J, Hemleben C, Ott R (1993) A model for planktic foraminiferal shell growth. Paleobiology 19:71–91

Smith PK, Krohn RI, Hermanson GT, Mallia AK, Gartner FH, Provenzano MD, Fujimoto EK, Goeke NM, Olson BJ, Klenk DC (1985) Measurement of protein using bicinchoninic acid. Anal Biochem 150:76–85

Speijer RP, Van Loo D, Masschaele B, Vlassenbroeck J, Cnudde V, Jacobs P (2008) Quantifying foraminiferal growth with high-resolution X-ray computed tomography: new opportunities in foraminiferal ontogeny, phylogeny, and paleoceanographic applications. Geosphere 4:760. doi:10.1130/GES00176.1

Spero HJ (1992) Do planktic Foraminifera accurately record shifts in the carbon isotopic composition of seawater CO$_2$? Mar Micropaleontol 19:275–285

Spero HJ, Parker SL (1985) Photosynthesis in the symbiotic planktonic foraminifer *Orbulina universa*, and its potential contribution to oceanic primary productivity. J Foraminifer Res 15:273–281

Spero HJ, Williams DF (1988) Extracting environmental
information from planktonic foraminiferal δ^{13}C data.
Nature 335:717–719

Spero HJ, Lea DW (1993) Intraspecific stable isotope
variability in the planktic Foraminifera *Globigerinoides sacculifer*: results from laboratory experiments.
Mar Micropaleontol 22:221–234

Spero HJ, Lea DW (1996) Experimental determination of
stable isotope variability in *Globigerina bulloides*:
implications for paleoceanographic reconstructions.
Mar Micropaleontol 28:231–246

Spero HJ, Eggins SM, Russell AD, Vetter L, Kilburn MR,
Hönisch B (2015) Timing and mechanism for intratest
Mg/Ca variability in a living planktic foraminifer.
Earth Planet Sci Lett 409:32–42. doi:10.1016/j.epsl.
2014.10.030

Spindler M, Hemleben C, Salomons JB, Smit LP (1984)
Feeding behavior of some planktonic foraminifers in
laboratory cultures. J Foraminifer Res 14:237–249

Stamatakis A (2014) RAxML version 8: a tool for
phylogenetic analysis and post-analysis of large
phylogenies. Bioinformatics 30:1312–1313. doi:10.
1093/bioinformatics/btu033

Stoll HM, Arevalos A, Burke A, Ziveri P, Mortyn G,
Shimizu N, Unger D (2007) Seasonal cycles in
biogenic production and export in Northern Bay of
Bengal sediment traps. Deep-Sea Res II 54:558–580

Storz D, Schulz H, Waniek JJ, Schulz-Bull DE, Kučera M
(2009) Seasonal and interannual variability of the
planktic foraminiferal flux in the vicinity of the Azores
Current. Deep-Sea Res I 56:107–124

Swofford DL (2001) Paup*: phylogenetic analysis using
parsimony (and other methods) 4.0. B5. National
Illinois History Survey, Champaign, USA

Takahashi T, Sutherland SC, Sweeney C, Poisson A,
Metzl N, Tilbrook B, Bates N, Wanninkhof R, Feely
RA, Sabine C (2002) Global sea-air CO_2 flux based on
climatological surface ocean pCO_2, and seasonal
biological and temperature effects. Deep-Sea Res II
2:1601–1622

Tedesco KA, Thunell RC, Astor Y, Muller-Karger F
(2007) The oxygen isotope composition of planktonic
Foraminifera from the Cariaco Basin, Venezuela:
seasonal and interannual variations. Mar Micropaleontol 62:180–193

Thunell RC, Honjo S (1981) Planktonic foraminiferal flux
to the deep ocean: sediment trap results from the
tropical Atlantic and the central Pacific. Mar Geol
40:237–253

Thunell RC, Reynolds LA (1984) Sedimentation of planktonic Foraminifera: seasonal changes in species flux in
the Panama Basin. Micropaleontology 30:243–262

Thunell RC, Honjo S (1987) Seasonal and interannual
changes in planktonic foraminiferal production in the
North Pacific. Nature 328:335–337

Thunell R, Sautter LR (1992) Planktonic foraminiferal
faunal and stable isotopic indices of upwelling: a

sediment trap study in the San Pedro Basin, Southern
California Bight. In: Summerhays CP, Prell WL,
Emeis KC (eds) Upwelling systems: evolution since
the early miocene. Geological Society. Special Publications, London, pp 77–91

Tolderlund DS, Bé AWH (1971) Seasonal distribution of
planktonic Foraminifera in the western North Atlantic.
Micropaleontology 17:297–329

Toyofuku T, Kitazato H (2005) Micromapping of Mg/Ca
values in cultured specimens of the high-magnesium
benthic Foraminifera. Geochem Geophys Geosystems.
doi:10.1029/2005GC000961

Tripati AK, Eagle RA, Thiagarajan N, Gagnon AC,
Bauch H, Halloran PR, Eiler JM (2010) ^{13}C-^{18}O
isotope signatures and "clumped isotope" thermometry in Foraminifera and coccoliths. Geochim Cosmochim Acta 74:5697–5717

Tripati AK, Sahany S, Pittman D, Eagle RA, Neelin JD,
Mitchell JL, Beaufort L (2014) Modern and glacial
tropical snowlines controlled by sea surface temperature and atmospheric mixing. Nat Geosci 7:205–209.
doi:10.1038/ngeo2082

Tyszka J, Topa P (2005) A new approach to modeling of
foraminiferal shells. Paleobiology 31:522–537. doi:10.
1666/0094-8373(2005)031[0522:ANATMO]2.0.CO;2

Ujiié Y, Lipps JH (2009) Cryptic diversity in planktic
Foraminifera in the northwest Pacific Ocean.
J Foraminifer Res 39:145–154

Ujiié Y, de Garidel-Thoron T, Watanabe S, Wiebe P, de
Vargas C (2010) Coiling dimorphism within a genetic
type of the planktonic foraminifer *Globorotalia truncatulinoides*. Mar Micropaleontol 77:145–153

Van de Peer Y, Nicolaï S, de Rijk P, de Wachter R (1996)
Database on the structure of small ribosomal subunit
RNA. Nucleic Acids Res 24:86–91. doi:10.1093/nar/
24.1.86

Van der Plas L, Tobi AC (1965) A chart for judging the
reliability of point counting results. Am J Sci 263:87–90

Vetter L, Spero HJ, Russell AD, Fehrenbacher JS (2013)
LA-ICP-MS depth profiling perspective on cleaning
protocols for elemental analyses in planktic
foraminifers. Geochem Geophys Geosystems
14:2916–2931. doi:10.1002/ggge.20163

Vetter L, Kozdon R, Valley JW, Mora CI, Spero HJ
(2014) SIMS measurements of intrashell δ^{13}C in the
cultured planktic foraminifer *Orbulina universa*.
Geochim Cosmochim Acta 139:527–539. doi:10.
1016/j.gca.2014.04.049

Vincent E, Berger WH (1981) Planktonic Foraminifera
and their use in paleoceanography. Ocean Lithosphere
Sea 7:1025–1119

Voelker AHL, Sarnthein M, Grootes PM, Erlenkeuser H,
Laj C, Mazaud A, Nadeau MJ, Schleicher M (1998)
Correlation of marine ^{14}C ages from the Nordic Seas
with the GISP2 isotope record; implications for ^{14}C
calibration beyond 25 ka BP. Radiocarbon 40:517–
534

Voelker AHL, Grootes PM, Nadeau MJ, Sarnthein M (2000) Radiocarbon levels in the Iceland Sea from 25-53 kyr and their link to the earth's magnetic field intensity. Radiocarbon 42:437–452

Volkov RA, Komarova NY, Hemleben V (2007) Ribosomal DNA in plant hybrids: inheritance, rearrangement, expression. Syst Biodivers 5:261–276. doi:10.1017/S1477200007002447

Von Gyldenfeldt AB, Carstens J, Meincke J (2000) Estimation of the catchment area of a sediment trap by means of current meters and foraminiferal tests. Deep-Sea Res II 47:1701–1717

Wacker U, Fiebig J, Tödter J, Schöne BR, Bahr A, Friedrich O, Tütken T, Gischler E, Joachimski MM (2014) Empirical calibration of the clumped isotope paleothermometer using calcites of various origins. Geochim Cosmochim Acta 141:127–144. doi:10.1016/j.gca.2014.06.004

Waelbroeck C, Labeyrie L, Duplessy JC, Guiot J, Labracherie M, Leclaire H, Duprat J (1998) Improving past sea surface temperature estimates based on planktonic fossil faunas. Paleoceanography 13:272–283

Wefer G, Suess E, Balzer W, Liebezeit G, Müller PJ, Ungerer CA, Zenk W (1982) Fluxes of biogenic components from sediment trap deployment in circumpolar waters of the Drake passage. Nature 299:145–147

Wejnert KE, Pride CJ, Thunell RC (2010) The oxygen isotope composition of planktonic Foraminifera from the Guaymas Basin, Gulf of California: seasonal, annual, and interspecies variability. Mar Micropaleontol 74:29–37

Wejnert KE, Thunell RC, Astor Y (2013) Comparison of species-specific oxygen isotope paleotemperature equations: sensitivity analysis using planktonic Foraminifera from the Cariaco Basin, Venezuela. Mar Micropaleontol 101:76–88. doi:10.1016/j.marmicro.2013.03.001

Weldeab S, Emeis KC, Hemleben C, Vennemann TW, Schulz H (2002) Sr and Nd isotope composition of Late Pleistocene sapropels and nonsapropelic sediments from the Eastern Mediterranean Sea: implications for detrital influx and climatic conditions in the source areas. Geochim Cosmochim Acta 66:3585–3598. doi:10.1016/S0016-7037(02)00954-7

Wiebe PH, Boyd SH, Winget CL (1976) Particulate matter sinking to the deep-sea floor at 2000 m in the Tongue of the Ocean, Bahamas, with a description of a new sedimentation trap. J Mar Res 34:341–354

Wilke I, Meggers H, Bickert T (2009) Depth habitats and seasonal distributions of recent planktic foraminifers in the Canary Islands region (29°N) based on oxygen isotopes. Deep-Sea Res Part Oceanogr Res Pap 56:89–106. doi:10.1016/j.dsr.2008.08.001

Wuyts J, van de Peer Y, Winkelmans T, de Wachter R (2002) The European database on small subunit ribosomal RNA. Nucleic Acids Res 30:183–185. doi:10.1093/nar/30.1.183

Xu X, Yamasaki M, Oda M, Honda MC (2005) Comparison of seasonal flux variations of planktonic Foraminifera in sediment traps on both sides of the Ryukyu Islands, Japan. Mar Micropaleontol 58:45–55. doi:10.1016/j.marmicro.2005.09.002

Yamasaki M, Oda M (2003) Sedimentation of planktonic Foraminifera in the East China Sea: evidence from a sediment trap experiment. Mar Micropaleontol 49:3–20. doi:10.1016/S0377-8398(03)00024-0

Žarić S, Donner B, Fischer G, Mulitza S, Wefer G (2005) Sensitivity of planktic Foraminifera to sea surface temperature and export production as derived from sediment trap data. Mar Micropaleontol 55:75–105

Žarić S, Schulz M, Mulitza S (2006) Global prediction of planktic foraminiferal fluxes from hydrographic and productivity data. Biogeosciences 3:187–207

Zubkov MV, Sleigh MA (1999) Growth of amoebae and flagellates on bacteria deposited on filters. Microb Ecol 37:107–115. doi:10.1007/s002489900135

Zubkov MV, Fuchs BM, Eilers H, Burkill PH, Amann R (1999) Determination of total protein content of bacterial cells by SYPRO staining and flow cytometry. Appl Environ Microbiol 65:3251–3257

Glossary

Allometry The study of body shape and size.

Amino acid Organic compound for the production of protein polymers composed of carbon, nitrogen, oxygen, hydrogen, and sometimes sulfur.

Amphipod Crustacean with laterally compressed bodies.

Annulate lamella (*pl.* lamellae) Concentric or parallel arrays of membranous flattened saccules interrupted by pores produced prior to gametogenesis and providing the double membrane envelope of the gamete nuclei.

Aperture Opening in last formed chamber.

Aposymbiosis Obligatory symbiotic organisms living independently, usually involving negative effects for both host and symbiont.

***Artemia* nauplius (*pl.* nauplii)** The early developmental stages of the brine shrimp belonging to the genus *Artemia*.

Asexual reproduction Reproduction by division of the parent organism without formation of gametes, producing offspring almost identical to the parent.

Autecology The relationship of an individual organism to its abiotic and biotic environment.

Benthic Living on the substratum or attached to the surface of fixed or floating matter.

Bilamellar wall Primary wall composed of a primary organic membrane separating two calcite lamellae. The outer layer is deposited concurrently over the exterior of previously built chambers.

Bubble capsule Cytoplasmic alveoli surrounding the test of *Hastigerina pelagica*, resembling a mass of soap bubbles.

Calanoid copepod Abundant small marine crustacean with elongated ovoid forebody, which is clearly distinct from the abdomen. Only one thoracic segment is fused with the head.

Calcite crust Secretion of calcite over the exterior of the foraminifer test late in ontogeny.

Carnivore An organism that preys upon animals.

CCD, Calcite Compensation Depth Level in the water column where the deposition of calcite equals the dissolution of calcite.

Chlorophyll Green pigment in plants (including algae), which absorbs light, and serves as a photocatalyst in photosynthesis.

Chrysomonad Flagellate belonging to the Order Chrysomonadina.

Chrysophycophyte Eclectic group of algae characterized by yellow-green pigmentation.

Coccoid alga (*pl.* algae) Altered state compared to free-living algae, often characterized by loss of changes in the thecal wall (when present). Symbiotic algae in planktic foraminifers.

© Springer-Verlag Berlin Heidelberg 2017
R. Schiebel and C. Hemleben, *Planktic Foraminifers in the Modern Ocean*,
DOI 10.1007/978-3-662-50297-6

Coccolithophorid Photosynthetic marine gold-brown alga secreting a sphere of calcite platelets (liths) of species-specific morphology (from mid Triassic to Recent).

Commensalism Association between two organisms where neither member is harmed, and only one of the associates may benefit.

Copepod Small marine crustacean divided into head, thorax bearing appendices, and abdomen without appendices. The head smoothly merges into the thorax without distinct segmentation.

Cope's rule Following Cope's rule, species increase in body size over evolutionary time.

Cristae Inward folds or sacs attached to the inner membrane of the mitochondrion.

Cyanophyte Bacterium with blue-green pigmentation composed of chlorophylls, carotenoids, c-phycocyanin, and c-phycoerythrin.

Cyclopoid copepod Marine crustaceans resembling calanoid copepods, but usually producing a more ovoid body. The first and sometimes the second thoracic segment is fused with the head.

Cytoplasm Living substance of the cell including the nucleus and other cell organelles.

Cytoskeletal structures Intracytoplasmic supportive structures including microtubules, and in some cases microfilaments.

Deep Chlorophyll Maximum Maximum in chlorophyll concentration, usually at the base of the surface mixed layer of the ocean or in a lake.

Deuteroconch Second chamber formed following the proloculus.

Diatom Alga producing siliceous shells called frustrules composed of two halves resembling a Camembert (cheese from Normandy, France) box (from Jurassic to Recent).

Dinoflagellate Alga bearing a flagellated theca (wall) composed of closely intercalated segments (thecal plates), and non-flagellated

cysts formed of dinosporin, the latter being well fossilized from the upper Triassic.

Dioecious Reproduction from gametes of two different parents.

DNA Desoxyribonucleic acid, Nitrogen-containing sugar-phosphate compound forming the chromatin polymers in the chromosomes.

Endoplasmatic reticulum (ER) Intracellular canal-like, membranous network penetrating deeply into the cytoplasm, and possibly providing cisternae for the transport of substances throughout the cell.

Enzyme Biological catalyst composed of protein, which controls the rate and direction of metabolic processes.

Euhedral Crystal or crystallite with well-developed crystal boundaries.

Euphausid Marine planktic crustacean ranging from 5 to 30 cm in length. Most species possess luminous organs.

Euphotic Zone Sunlit surface waters supporting photosynthesis.

Euryhaline Organism tolerating a wide range of salinities.

Eurythermal Organism tolerating a wide range of temperatures.

Eutrophic Environment with high nutrient concentration and high biological production.

Fibrillar body Vacuolar-bound fibrillar mass within the cytoplasm of planktic foraminifers, with species-specific fine structural features. May serve as flotation device.

Filopodium (*pl.* filopodia) Long, thin rhizopodium without internal stiffened rod. Micro-tubule bundles forming axonemes.

Galactose A six-carbon sugar.

Gamete Haploid reproductive cell, which fuses with another gamete to yield a zygote and initiate the next diploid generation.

Gametogenesis Production of gametes during reproduction.

Gametogenic calcification Calcification prior to gametogenesis (facultative) producing an additional, more or less continuous layer of calcite on top of the test wall.

Genetic code Inherited information contained within DNA in a cell or the base sequence (nucleotide sequence) of a gene, which contains the information to direct the synthesis of a specific protein.

Genotype Specific genetic composition of an organism.

Glucose A six-carbon sugar.

Glutaraldehyde A pentane dialdehyde $(CH_2(CH_2CHO)_2)$ used as a fixative of organic matter for Transmission Electron Microscopy (TEM).

Glycerol A three carbon tri-hydroxy alcohol, serving in esterification of fatly acids to form storage lipids.

Golgi apparatus Golgi complex, Golgi body, Intracellular organelle forming a horseshoe-shaped or fan-shaped stack of cisternae, which produce secretory vesicles near the periphery. Origin of some lysosomes and secretory vesicles distributed throughout the cell.

Halocline Horizon with strong vertical salinity gradient over a certain water depth.

Harpacticoid copepod Marine crustacean less than 1 mm in size, often benthic and some pelagic, with long slender bodies, and lacking distinctly visible body segmentation.

Herbivore An organism that preys upon plants.

Heteropod Pelagic gastropod (snail) with a translucent body, wing-like appendages, and some with an aragonite shell.

HNLC High-Nutrient Low-Chlorophyll, oceanic region, where primary production is limited in micronutrient (often iron, Fe) concentration, or dominated by grazers. Vast HNLC regions are the Southern Ocean and the equatorial Pacific.

Holoplankton Planktic organism, which floats in the open water column during the entire ontogenetic development.

Holothurian Echinoderms with elongate cucumber shaped body (sea cucumber).

Hydrolysate Chemically degraded organic compound reduced to the individual monomeric component.

Inner Organic Lining (IOL) Thick, dense layer between the inner test surface and the internal cytoplasm of planktic foraminifers.

Isotope Atom with the same number of protons, and differing numbers of neutrons.

K-selection Species with a survival strategy based on carrying capacity, i.e. low growth rate and long life expectancy (see also r-selection).

Lipid Oily substance forming food reserves within the cytoplasm, and a significant component of cellular membranes.

Lunar periodicity The synodic cycle of the moon phases with a period of 29.5 days.

Lysocline The level in the water column where commencing dissolution of calcite (calcite lysocline) or aragonite (aragonite lysocline) reaches 10 % (or 20 %, depending on definition).

Lysosome Digestive vacuole possessing hydrolytic enzymes, which perform digestion. The primary lysosome originates from the Golgi complex or from the endoplasmatic reticulum.

Mesoscale Expression in oceanography to classify hydrologic structures such as eddies with an average diameter of about 100 to 200 km.

Mesotrophic Environment with medium levels of nutrients and intermediate biological production.

Microfilament A fine intracellular protein filament (perhaps composed of actin, a contractile protein) approximately 6 nm in diameter.

Microtubule A slender intracellular tubule about 30 nm in diameter composed of protein and forming a cytoskeletal framework within the cell.

Miliolid foraminifer Benthic foraminifer with test of high magnesium calcite and needle-like ultrastructure.

Mitochondrion (*pl.* mitochondria) Subcellular organelle enveloped by a double membrane enclosing enzyme systems. Mediating aerobic metabolism including glucose and fat metabolism, resulting in the production of high-energy compounds (e.g., ATP) utilized to drive cellular processes.

Monoecious Reproduction from gametes of the same parent.

Morphogenesis The origin of form in a living system, and the pattern of development of an organism during ontogeny.

Morphology Form and structure of an organism.

Mucocyst Cytoplasmic organelle containing an ejectable mucoid mass.

Neanic Growth stage between juvenile and adult stage.

Nucleus Cell organelle containing much of the genetic information (chromosomes), and is the center for coordination of cellular activity.

Nutrient Substance used to obtain energy or to sustain metabolic activity.

Nutrition Mode of gaining energy, source of nutrients and food and their transformation to sustain life activities.

Oligotrophic Environment with low levels of nutrients and limited biological production

Omnivore An organism that preys upon plants, animals, and other food sources.

Ontogeny Process of growth and development of an organism from inception of growth to maturity.

Organelle Structure within a cell that serves a particular function (e.g., mitochondrion, lysosome, Golgi body).

Osmium tetroxide (OsO_4) Oxide of the heavy metal osmium, which is used in buffered aqueous solution as a fixative and stain in electron microscopy.

Ostracod Small crustacean with mostly calcareous carapace formed of two valves, from Ordovician to Recent.

Paleoecology The reconstructed ancient environment.

Parapatric Occurring next to each other. Close neighbors (see also sympatric).

Parasitism Association between two organisms with one of them profiting to the expense of the other.

Perialgal vacuole Vacuole enclosing an algal symbiont surrounded by a cellular membrane, which may serve as a specialized barrier for appropriate isolation of the alga from the host cytoplasm.

Peroxisome Intracytoplasmic organelle surrounded by a single membrane, and containing a lightly granular matrix sometimes with a membranous or crystalline inclusion.

Phenotype Physical characteristics of an organism resulting from the combined effects of generic and environmental factors during development of the organism. In contrast to genotype, and genetic characteristics of an organism.

Photosynthesis Production of carbon-containing compounds from inorganic carbon using the energy of light.

Phylogeny The evolutionary development of taxa.

Planispiral Test coiling in one plane producing a bilateral symmetry.

Plankton Organism with a floating life habit, and limited and undirected motility, hence being moved by currents.

Plasma membrane Outer membrane surrounding a cell and regulating the exchange of material between cell and environment.

Polychaete Annelid worm (related to the earthworm). A group of organisms, which includes most of the marine segmented worms.

Population All individuals of a species in an area.

Population dynamics Changes within populations (individuals or groups of individuals) related to environmental change.

Pore Small canal penetrating the shell wall and closed by an organic membrane sometimes containing micropores.

Prasinophyte Alga belonging to the class Prasinophyceae characterized by tiny scales covering the flagella, and bearing a nucleus, which protrudes into the pyrenoid in some species.

Primary lysosome Vesicle containing digestive enzymes, which are destined to catalyze the hydrolytic decomposition of food in secondary lysosomes produced by fusion of the primary lysosome with a food vacuole.

Primary Organic Membrane (POM) Organic layer between the two first-formed calcite layers of rotaliid test walls in continuation with the pore plate. Nucleation site of calcite deposition in a bilamellar wall.

Proloculus First-formed chamber in foraminifers.

Protista Group of organisms encompassing the classical 'protozoa' possessing a true nucleus enclosed within a membranous envelope.

Pseudopodium (*pl.* pseudopodia) Specialized cytoplasmic projections, which serve locomotion, feeding, and other physiological functions.

Pteropods Obsolete term used for a polyphyletic group of pelagic gastropods combining the systematic orders Thecosomata and Gymnosomata, in which the foot is modified for swimming, and an aragonite shell may be present or absent (from Miocene to Recent).

Pustule Small protuberance of the test wall with internal layers following the contour of the wall, serving as attachment site for the rhizopodia.

Pycnocline Horizon with strong vertical density gradient over a certain water depth.

Pyrenoid Subcellular organelle associated with the plastid, and a site of starch accumulation during photosynthesis.

Radiolaria Marine protozoa secreting siliceous skeletons with axopodia radiating from a central cell body surrounded by a porous organic wall (central capsule).

Rhizopodium (*pl.* rhizopodia) Fine pseudopodia with a branching or reticulate pattern.

Rhodophyceae Red-pigmented algae.

Ribosome Small cellular organelle about 20 nm in size, composed of a small and a large subunit, and aiding in protein synthesis.

Rough endoplasmatic reticulum Site of protein synthesis. Ribosomes are attached to the cytoplasmic surface of the endoplasmatic reticulum, causing a granular appearance.

r-selection Species with a survival strategy based on high growth rate and production of a large number of offspring (see also K-selection).

Salp (*pl.* salps) Translucent planktic tunicate up to 10 cm in size.

Schizogamy Production of gametes by nuclear proliferation and cytoplasmic fission.

SCUBA Self-contained underwater breathing apparatus.

Secondary aperture Additional aperture on spiral side of the test.

Sediment trap Device to collect particles from the water column.

Sergestid (*pl.* sergestids) Group of marine polychaete worms.

Sexual reproduction Reproduction by gamete formation and their fusion (syngamy) to form a zygote, the earliest stage of a new individual.

Smooth endoplasmatic reticulum A site of lipid synthesis. Endoplasmatic reticulum without ribosomes.

Spine Thin calcite projection anchored within the test wall like a fence pole, leaving characteristic holes after spine shedding during gametogenesis. Evolved in the early Cenozoic.

Spinose (*adj.*) Presence of surface spines on the test of planktic foraminifers.

Streptospiral Irregular coiling.

Subhedral Crystals or crystallites with partially developed boundaries.

Suture Depression between two adjacent chambers.

Symbiosis Structural and physiological association between two organisms and mutual benefit. Symbiont and host are usually of different taxonomic status. In planktic foraminifers, symbionts are exclusively algal cells.

Sympatric Occurring together in the same area (see also parapatric).

Synecology Interactions of species with one another and their environment affecting their abundance and reproductive continuity.

Syngamy Fusion of gametes to produce a zygote.

Terebellid Group of marine polychaete worms.

Thermocline Horizon with strong vertical temperature gradient over a certain water depth

Thylakoid Internal membrane in plastids containing photosynthetic pigments.

Tintinnid Marine ciliate typically forming a conical or trumpet-shaped lorica, either formed by secreted organic substance, or by small agglutinated particles gathered from the environment.

Trochospiral Coiling in a spiral resembling a conical snail shell, producing an asymmetric test with different umbilical and spiral sides.

Trophic activity Mode of feeding activity including the quantity, kind, and range of prey consumed, and the physiological mechanisms for prey apprehension, ingestion, and digestion.

Tunicate Sac-like animal, either benthic or pelagic, belonging to the chordate subphylum.

Tychopelagic Adopted planktic (pelagic) life-style of generally benthic organisms, which are transported into the open water column, for example, by currents.

Vesicle Small membrane-bound secretory body about 0.1 μm in size, with the appearance of a small vacuole.

Zooxantella (*pl.* **Zooxantellae**) Alga associated with a host, usually in symbiosis, and possessing a yellow-green pigment. For comparison, Zoochlorellae produce a green pigment.

Zygote Diploid cell produced by the fusion of gametes, capable of developing into a mature organism.

Index

© Springer-Verlag Berlin Heidelberg 2017
R. Schiebel and C. Hemleben, *Planktic Foraminifers in the Modern Ocean*,
DOI 10.1007/978-3-662-50297-6

Printed in the United States
By Bookmasters